Trivariate Local Lagrange Interpolation and Macro Elements of Arbitrary Smoothness

Michael A. Matt

Trivariate Local Lagrange Interpolation and Macro Elements of Arbitrary Smoothness

Foreword by Prof. Dr. Ming-Jun Lai

Springer Spektrum RESEARCH

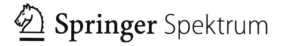

Michael A. Matt
Mannheim, Germany

Dissertation Universität Mannheim, 2011

ISBN 978-3-8348-2383-0 ISBN 978-3-8348-2384-7 (eBook)
DOI 10.1007/978-3-8348-2384-7

The Deutsche Nationalbibliothek lists this publication in the Deutsche Nationalbibliografie; detailed bibliographic data are available in the Internet at http://dnb.d-nb.de.

Springer Spektrum

Cover design: KünkelLopka GmbH, Heidelberg

Printed on acid-free paper

Springer Spektrum is a brand of Springer DE. Springer DE is part of Springer Science+Business Media.
www.springer-spektrum.de

Foreword

Multivariate splines are multivariate piecewise polynomial functions defined over a triangulation (in $\mathbb{R}^2$) or tetrahedral partition (in $\mathbb{R}^3$) or a simplicial partition (in $\mathbb{R}^d, d \geq 3$) with a certain smoothness. These functions are very flexible and can be used to approximate any known or unknown complicated functions. Usually one can divide any domain of interest into triangle/tetrahedron/simplex pieces according to a given function to be approximated and then using piecewise polynomials to approximate the given function. They are also computer compatible in the sense of computation, as only additions and multiplications are needed to evaluate these functions on a computer. Thus, they are a most efficient and effective tool for numerically approximating known or unknown functions. In addition, multivariate spline functions are very important tools in the area of applied mathematics as they are widely used in approximation theory, computer aided geometric design, scattered data fitting/interpolation, numerical solution of partial differential equations such as flow simulation and image processing including image denoising.

These functions have been studied in the last fifty years. In particular, univariate splines have been studied thoroughly. One understands univariate splines very well in theory and computation. One typical monograph on univariate spline functions is the well-known book written by Carl de Boor. It is called "A Practical Guide to Splines" published first in 1978 by Springer Verlag. Another popular book is Larry Schumaker's monograph "Spline Functions: Basics and Applications" which is now in second edition published by Cambridge University Press in 2008. Theory and computation of bivariate splines are also understood very well. For approximation properties of bivariate splines, there is a monograph "Spline Functions on Triangulation" authored by Ming-Jun Lai and Larry Schumaker, published by Cambridge University Press in 2007. For using bivariate splines for scattered data fitting/interpolation and numerical solution of partial differential equations, one can find the paper by G. Awanou, Ming-Jun Lai and P. Wenston in 2006 useful. In their paper, they explained how to compute bivariate splines without constructing explicit basis functions and use them for data fitting and numerical solutions of

PDEs including 2D Navier-Stokes equations and many flow simulations and image denoising simulations. These numerical results can be found at www.math.uga.edu/~mjlai.

Approximation properties of spline functions defined on spherical triangulations can also be found in the monograph by M.-J. Lai and L. L. Schumaker mentioned above. A numerical implementation of spherical spline functions to reconstruct geopotential around the Earth can be found in the paper by V. Baramidze, M.-J. Lai, C. K. Shum, and P. Wenston in 2009. In addition, some basic properties of trivariate spline functions can also be found in the monograph by Lai and Schumaker. For example, Lai and Schumaker describe in their monograph how to construct C^1 and C^2 macro-element functions over the Alfeld split of tetrahedral partitions and the Worsey-Farin split of tetrahedral partitions. However, how to construct locally supported basis functions such as macro-elements for smoothness $r > 2$ was not discussed in the monograph. In addition, how to find interpolating spline functions using trivariate spline functions of lower degree without using any variational formulation is not known.

In his thesis, these two questions are carefully examined and fully answered. That is, Michael Matt considers trivariate macro-elements and trivariate local Lagrange interpolation methods. The macro-elements of arbitrary smoothness, based on the Alfeld and the Worsey-Farin split of tetrahedra, are described very carefully and are illustrated with many examples. Michael Matt has spent a great deal of time and patience to write down in detail how to determine for C^r macro-elements for $r = 3,4,5,6$ and etc.. One has to point out that macro-elements are locally supported functions and they form a superspline subspace which has much smaller dimension than the whole spline space defined on the same tetrahedral partition. As this superspline subspace possesses the same full approximation power as the whole spline space, they are extremely important as these are enough to use for approximate known and unknown functions with a much smaller dimension. Thus, they are most efficient. The results in this thesis are an important extension of the literature on trivariate macro-elements known so far. Michael Matt also describes two methods for local Lagrange interpolation with trivariate splines. Both methods, for once continuously differentiable cubic splines based on a type-4 partition and for splines of degree nine on arbitrary tetrahedral partitions, are examined very detailed. Especially the method based on arbitrary tetrahedral partitions is a crucial contribution to this area of researchsince it is the first method for trivariate Lagrange interpolation for two times continuously differentiable splines.

To summarize, the results in this dissertation extend the construction theory of trivariate splines of C^r macro-elements for any $r \geq 1$. The construction schemes for finding interpolatory splines over arbitrary tetrahedral partition presented in this thesis are highly recommended for any application practitioner.

Prof. Dr. Ming-Jun Lai
Department of Mathematics
University of Georgia
Athens, GA, U.S.A.

Acknowledgments

I would like to thank Prof. Dr. Günther Nürnberger and Prof. Dr. Ming-Jun Lai for their support in writing this dissertation. I want to thank them for interesting discussions and suggestions that helped to improve this work. I also would like to thank my parents, my sister, and all other persons who encouraged and supported me while writing this dissertation and accompanied me in this time.

Michael Andreas Matt

Abstract

In this work, we construct two trivariate local Lagrange interpolation methods which yield optimal approximation order and C^r macro-elements based on the Alfeld and the Worsey-Farin split of a tetrahedral partition. The first interpolation method is based on cubic C^1 splines over type-4 cube partitions, for which numerical tests are given. The other one is the first trivariate Lagrange interpolation method using C^2 splines. It is based on arbitrary tetrahedral partitions using splines of degree nine. In order to obtain this method, several new results on C^2 splines over partial Worsey-Farin splits are required. We construct trivariate macro-elements based on the Alfeld, where each tetrahedron is divided into four subtetrahedra, and the Worsey-Farin split, where each tetrahedron is divided into twelve subtetrahedra, of a tetrahedral partition. In order to obtain the macro-elements based on the Worsey-Farin split we construct minimal determining sets for C^r macro-elements over the Clough-Tocher split of a triangle, which are more variable than those in the literature.

Zusammenfassung

In dieser Arbeit konstruieren wir zwei Methoden zur lokalen Lagrange-Interpolation mit trivariaten Splines, welche optimale Approximationsordnung besitzen, sowie C^r Makro-Elemente, welche auf der Alfeld- und der Worsey-Farin-Unterteilung einer Tetraederpartition basieren. Eine Interpolationsmethode basiert auf kubischen C^1 Splines auf Typ-4 Würfelpartitionen. Für diese werden auch numerische Tests angegeben. Die nächste Methode ist die erste zur lokalen Lagrange-Interpolation mit trivariaten C^2 Splines. Sie basiert auf beliebigen Tetraederpartitionen und verwendet Splines vom Grad neun. Um diese Methode zu erhalten, werden einige neue Resultate zu C^2 Splines über partiellen Worsey-Farin-Unterteilungen benötigt. Wir konstruieren trivariate Makro-Elemente beliebiger Glattheit, die auf der Alfeld-Unterteilung, bei der jeder Tetraeder in vier Subtetraeder unterteilt ist, und der Worsey-Farin-Unterteilung, bei welcher jeder Tetraeder in zwölf Subtetraeder unterteilt ist, einer Tetraederpartition beruhen. Um die Makro-Elemente, die auf der Worsey-Farin-Unterteilung basieren, zu erzeugen, konstruieren wir minimal bestimmende Mengen für C^r Makro-Elemente über der Clough-Tocher-Unterteilung eines Dreiecks, welche, im Vergleich zu den bereits in der Literatur bekannten, variabler sind.

Contents

1 Introduction

Multivariate splines play an important role in several areas of applied mathematics. Due to their efficient computability and their approximation properties, they are widely used for the construction and reconstruction of surfaces and volumes, the interpolation and approximation of scattered data, and many other fields in computer aided geometric design and numerical analysis, such as the solution of partial differential equations.

In this thesis we consider the space of multivariate splines of degree d defined on a tessellation Δ of a polyhedral domain $\Omega \subset \mathbb{R}^n$ into n-simplices, which is given by

$$\mathcal{S}_d^r(\Delta) := \{s \in C^r(\Omega) : s|_T \in \mathcal{P}_d^n \text{ for all } T \in \Delta\},$$

where $C^r(\Omega)$ is the space of functions of C^r smoothness and $\mathcal{P}_d^n$ is the space of multivariate polynomials of degree d. We are mainly interested in the case $n = 3$, and to some extend $n = 2$, and there we mostly consider certain subspaces, the spaces of supersplines, which fulfill supersmoothness conditions at vertices and edges of the tetrahedra or triangles, and in some cases also additional individual smoothness conditions.

A common approach for the construction and reconstruction of volumes is interpolation. For a tessellation Δ of a domain $\Omega \subset \mathbb{R}^n$, a set $\mathcal{L} := \{\kappa_1, \ldots, \kappa_m\}$ is called a Lagrange interpolation set for an m-dimensional spline space $\mathcal{S}_d^r(\Delta)$, provided that for each function $f \in C(\Omega)$ there exists a unique spline $s_f \in \mathcal{S}_d^r(\Delta)$, such that

$$s_f(\kappa_i) = f(\kappa_i), \quad i = 1, \ldots, m,$$

holds. If also derivatives of a sufficiently smooth function f are interpolated, the corresponding set $\mathcal{H}$ is called a Hermite interpolation set.

An important property of interpolation sets is locality. An interpolation set is called local, provided that the value of an interpolant s_f at a point $\xi \in \Omega$ only depends on data values in a finite environment of ξ. Thus, the interpolant can be determined by solving several smaller linear system of equations at a time, which is very useful for the implementation of interpolation methods. A further desirable property is the stability of an interpolation method. Roughly speaking, a method is called stable if a small modification of a data value only leads to a small change of the corresponding interpolant. These two properties are important in order to achieve optimal approximation order. The approximation order of a spline space $\mathcal{S}_d^r(\Delta)$ is the biggest natural number k, such that

$$dist(f,\mathcal{S}_d^r(\Delta)) := \inf\{\|f-s\| : s \in \mathcal{S}_d^r(\Delta)\} \leq K|\Delta|^k$$

holds for a constant $K > 0$ that only depends on f, d, and the smallest angles of Δ, where $|\Delta|$ is the mesh size of Δ. Then, for an interpolation method, the approximation error $|f - s_f|$ is considered, where s_f is the interpolant of f constructed by the method. Thus, the approximation order gives the rate of convergence of the error for refinements of the tessellation Δ. It was shown by Ciarlet and Raviart [23] that $d+1$ is an upper bound for the approximation order, which implies that $k = d+1$ is the optimal approximation order. Following de Boor and Jia [36], the optimal approximation order cannot be obtained by every interpolation method.

The terminology "spline function" was first used in Schoenberg [91, 92], though earlier papers were concerned with splines without actually using this name (see [86, 88]). Univariate splines were extensively studied and many results on approximation and interpolation, as well as on the dimension of univariate spline spaces, are known (see [32, 69, 93], and references therein).

In recent years a lot of research has been done on bivariate splines (see [2–8, 8, 14, 15, 20, 22, 25–29, 31, 34, 38, 39, 44–53, 56–61, 70, 71, 73–80, 82, 85, 87, 89, 90, 94, 100, 101]).

In contrast, much less is known about trivariate splines, especially for spline spaces with a low degree of polynomials compared to the degree of smoothness (see [12, 16, 17, 21, 41, 43, 62, 105]). Due to the complex structure of these spline spaces many problems, such as the construction of local interpolation sets, are still open. One approach are macro-element methods, which are mostly based on the refinement of the tetrahedra of a given partition into smaller subtetrahedra (see [1, 9–11, 13, 18, 55, 63, 64, 96–99, 103, 104]). Another approach is based on regular tetrahedral partitions (see [40, 42, 72, 81, 83, 95]).

In this thesis, we consider local Lagrange interpolation methods for C^1 and C^2 splines and C^r macro-elements based on the Alfeld and the Worsey-Farin split of tetrahedra.

Local Lagrange interpolation methods can be used to interpolate scattered data and to construct and reconstruct volumes. We consider local Lagrange interpolation with C^1 cubic splines on type-4 cube partitions. There are already a few articles on trivariate local Lagrange interpolation on various tetrahedral partitions (see [41–43, 81, 95]). Though the method by Matt and Nürnberger [67] presented here is the first one that was actually implemented. Hence, it is the first time that it can also be shown numerically that the spline space corresponding to the trivariate Lagrange interpolation method yields optimal approximation order. We also construct local and stable Lagrange interpolation sets for C^2 splines on arbitrary tetrahedral partitions. It can be seen from above that there exists some literature on bivariate Lagrange interpolation methods based on C^2 and also some literature on local Lagrange interpolation with trivariate C^1 splines. Here, we construct the first trivariate Lagrange interpolation method for C^2 splines. We also show that the interpolation set is local and stable and that the corresponding spline space yields optimal approximation order.

We also examine C^r macro-element methods based on the Alfeld and the Worsey-Farin refinement of a tetrahedron. Macro-elements can be used for example to numerically solve partial differential equations, to simulate properties of materials or economic models. In the bivariate case there exists a vast literature on macro-elements, also for those of type C^r (see [6, 7, 59, 61, 106]). In the trivariate setting there exists some literature on C^1 and C^2 macro-elements (see [1, 9–11, 13, 18, 55, 64, 96–99, 103, 104]), though only polynomial C^r macro-elements are known so far (see [63]). Their degree of polynomials is $d = 8r + 1$, which is quite high. By splitting the tetrahedra with the Alfeld and the Worsey-Farin split, we obtain C^r macro-elements with a significantly lower degree of polynomials. We also show that for the corresponding Hermite interpolation methods the considered superspline spaces yield optimal approximation order.

This thesis is divided into the following chapters: In chapter 2, we consider the fundamentals of multivariate spline theory. First, we investigate tessellations of a polyhedral domain $\Omega \subset \mathbb{R}^n$ into n-simplices, especially tetrahedral partitions, as well as some special tessellations for $n = 2$ and $n = 3$. Subsequently, we introduce multivariate splines and supersplines, and their basis, the multivariate polynomials. Then, we examine the Bernstein-Bézier techniques for multivariate splines. Using barycentric coordinates, the multivariate Bernstein polynomials, and thus the B-form of polynomials, can be defined. Following, we consider the de Casteljau algorithm for efficient evaluation of polynomials in the B-form, which can also be used for subdivision. Next, we describe smoothness conditions for polynomials in the B-form on adjacent n-simplices, especially for $n = 2$ and $n = 3$, which were introduced by Farin [39] and de Boor [33]. Subsequently, we consider the concept of minimal determining sets, which can be used to characterize spline spaces. Minimal determining sets also play an important role, since their cardinality is equal to the dimension of the corresponding spline space. Next, we define the Lagrange and Hermite interpolation and some properties of interpolation sets, such as locality and stability. Finally, we introduce nodal minimal determining sets and define the approximation order of spline spaces.

In chapter 3, we consider minimal determining sets for C^1 and C^2 splines on partial Worsey-Farin splits of a tetrahedron, which were defined in chapter 2. First, we review the results of Hecklin, Nürnberger, Schumaker, and Zeilfelder [42] for C^1 splines. Subsequently, we construct several minimal determining sets for different C^2 superspline spaces. We also state some lemmata that show how a spline defined on a tetrahedral partition $\Delta \setminus T$ can be extended to a spline defined on Δ, where the tetrahedron T is refined with a partial Worsey-Farin split. To this end, additional smoothness conditions are used. The minimal determining sets and the lemmata considered in this chapter are then applied in chapters 4 and 5.

In chapter 4, we present the work on local Lagrange interpolation with trivariate C^1 splines on type-4 cube partitions by Matt and Nürnberger [67]. At first the type-4 cube partition is defined. Therefore, the cubes of a cube partition $\diamond$ are divided into five classes. Afterwards, each cube is split into five tetrahedra, according to the classification of the cubes. Then, following this classification and the location of the tetrahedra in the cubes, we chose the interpolation points and refine some of the tetrahedra with a partial Worsey-Farin split. Subsequently, it is shown that the set of interpolation points chosen is a Lagrange interpolation set for the space of C^1 splines of degree three on the final tetrahedral partition. The final partition is obtained by refining some of the tetrahedra even further, though by the time a spline is determined on these tetrahedra this is not needed and thus omitted at the moment, in order to keep the computation of the spline less complex. We also give a nodal minimal determining set for the interpolation method. Following that, we show that the interpolation method yields optimal approximation order. Finally, numerical tests and visualizations of the Marschner-Lobb test function (see [65]) are given.

In chapter 5, we construct a local Lagrange interpolation method for C^2 splines of degree nine on arbitrary tetrahedral partitions. The construction is based on a decompositions of a tetrahedral partition Δ into classes of tetrahedra. This decomposition induces an order of the tetrahedra in Δ, according to their number of common vertices, edges, and faces. In contrast to the decomposition used in [43] to create a local Lagrange interpolation method for C^1 splines, the number of common vertices, edges, and faces has to be considered simultaneously in order to construct a local Lagrange interpolation set for C^2 splines. Next, as with the Lagrange interpolation method considered in chapter 4, some of the tetrahedra of the partition Δ have to be refined with partial Worsey-Farin splits according to the number of common edges with the previous tetrahedra in the order imposed by the decomposition. Then, we construct a superspline space based on the refined partition, which is endowed with several additional smoothness conditions corresponding to the lemmata in chapter 3. Subsequently, we construct a local and stable Lagrange interpolation set. It is notable that the interpolation set is 11-local which is very low, compared to the number of 24 classes needed to create the method and especially when compared to the C^1 method for local Lagrange interpolation on arbitrary tetrahedral partitions by Hecklin, Nürnberger, Schumaker, and Zeilfelder [43] which is 10-local. We also give a nodal minimal determining set for the superspline space considered in this chapter. Finally, we examine the approximation order of the spline space considered in this chapter and show that it is optimal.

In chapter 6, we consider the minimal determining sets for bivariate C^r macro-elements based on the Clough-Tocher split of a triangle constructed by Matt [66]. First, we consider conditions on the minimal degree of polynomials and the minimal degrees of supersmoothnesses in order to construct C^r macro-elements based on non-split triangles and triangles refined with a Clough-Tocher or a Powell-Sabin split, respectively. These are used throughout this chapter and in chapters 7 and 8, in order to derive the minimal conditions for macro-elements based on the Alfeld and the Worsey-Farin split of a tetrahedron. Then, minimal determining sets for C^r splines with various degrees of polynomials and supersmoothnesses based on the Clough-Tocher split of a triangle are examined. In case the minimal degrees are applied, the macro-elements reduce to those constructed in [59]. These minimal determining sets are needed for the construction of trivariate C^r macro-elements over the Worsey-Farin split of a tetrahedron in chapter 8. Subsequently, we illustrate the minimal determining sets of the C^r

macro-elements with several examples for $r = 0,\dots,9$. These are also used in the examples for minimal determining sets for macro-elements based on the Worsey-Farin split of a tetrahedron in chapter 8.

In chapter 7, we consider the trivariate C^r macro-elements based on the Alfeld split of a tetrahedron by Lai and Matt [54]. It is firstly shown which restrictions for the degree of polynomials, as well as the supersmoothness conditions, have to be fulfilled in order to construct macro-elements over the Alfeld split of a tetrahedron. These can be derived from the restrictions for bivariate macro-elements considered in chapter 6. Thus, it can be seen that the degree of polynomials and supersmoothnesses is the lowest possible to construct such macro-elements. Subsequently, a corresponding superspline space and minimal determining sets for the macro-elements are examined, first on one tetrahedron divided with the Alfeld split and then on a refined tetrahedral partition. Since these minimal determining sets are quite complex, we illustrate them for C^r macro-elements on a single tetrahedron for $r = 1,\dots,6$. It can be seen that for $r = 1,2$ the macro-elements reduce to those in [62] in section 18.3 and 18.7. In the following, we consider nodal minimal determining sets for the macro-elements. First, nodal minimal determining sets for the macro-element defined on one Alfeld split tetrahedron are analyzed, and then for macro-elements over a refined tetrahedral partition. Finally, we examine a Hermite interpolation set for C^r splines over the Alfeld split of tetrahedra and show that it yields optimal approximation order.

In chapter 8, we examine the C^r macro-elements based on the Worsey-Farin split of tetrahedra by Matt [66]. We first consider the conditions for the minimal degree of polynomials and the minimal degrees of supersmoothnesses needed to construct C^r macro-elements based on the Worsey-Farin split. Again, these can be derived from the corresponding conditions for bivariate macro-elements considered in chapter 6. Following, we present a superspline space which can be used to define C^r macro-elements based on the Worsey-Farin split of a tetrahedron and a corresponding minimal determining set. We also give a minimal determining set for a superspline space defined on a tetrahedral partition, where each tetrahedron is refined with a Worsey-Farin split. Subsequently, we illustrate minimal determining sets for C^r macro-elements based on the Worsey-Farin split of one tetrahedron for $r = 1,\dots,6$. For $r = 1$ and $r = 2$ the macro-elements reduce to those considered by Lai and Schumaker [62] in sections 18.4 and 18.8, respectively. Next, we examine nodal minimal determining sets for the C^r macro-elements, first for macro-elements based on a single Worsey-

Farin split tetrahedron and then on a whole tetrahedral partitions that has been refined with the Worsey-Farin split. We conclude this chapter with a Hermite interpolation set for C^r splines over the Worsey-Farin split of tetrahedra and consider the approximation order, which is optimal.

The next chapter contains the references. It is followed by appendix A, where we state several bivariate lemmata concerned with minimal determining sets for splines based on triangles refined with the Clough-Tocher split. These lemmata are needed in chapter 3 in order to proof the theorems on minimal determining sets for splines on tetrahedra refined with partial Worsey-Farin splits. For a better understanding we illustrate the minimal determining sets constructed here.

2 Preliminaries

In this chapter, we consider some results and techniques of multivariate spline theory. Since we are mainly interested in trivariate splines in this work, and as an aid to some extent also bivariate splines, we state most of the results and definitions shown in this chapter combined in the multivariate setting. However, to ease the understanding, we also show some of theses results for the trivariate and bivariate case. In section 2.1, we define tessellations of a domain in $\mathbb{R}^n$. Furthermore, we show some refinement schemes of triangles and tetrahedral partitions and introduce some notation and the Euler relations for tetrahedra. In section 2.2, we examine multivariate polynomials, which form a basis for the subsequently considered multivariate splines and supersplines. In the next section, we describe the Bernstein-Bézier techniques for multivariate splines. These are based on the barycentric coordinates, which are needed to define Bernstein polynomials and the resulting B-form of polynomials that is used throughout this dissertation. Subsequently the de Casteljau algorithm, which can be used to efficiently evaluate polynomials in the B-form, as well as smoothness conditions between two polynomials are considered. Finally, the concept of minimal determining sets is introduced. In the last section, we define the problem of interpolation and give another characterization of spline spaces, nodal minimal determining sets. In the end we define the approximation order of spline spaces.

2.1 Tessellations of $\mathbb{R}^n$

In this section we define tessellations of a polyhedral subset of $\mathbb{R}^n$ into n-simplices. Especially the cases $n = 2,3$ are of interest here. Then, we define certain special refinements of triangles and tetrahedral partitions. Moreover, some notation concerning the relations of tetrahedra are established, as well as the trivariate Euler relations.

2.1.1 Tessellations of $\mathbb{R}^n$ into n-simplices

First, since the methods considered in this dissertation are mostly in the trivariate setting, tetrahedral partitions are denominated. Thereafter, tessellations of $\mathbb{R}^n$, $n \in \mathbb{N}$, into $n-$simplices are defined, which are also needed later on in this chapter.

Definition 2.1:
Let Ω be a polyhedral subset of $\mathbb{R}^3$. If Ω is divided into tetrahedra T_i, $i = 1,\dots,N$, such that the intersection of two different tetrahedra is either empty, a common vertex, a common edge, or a common face, then $\Delta := \{T_1,\dots,T_N\}$ is called a **tetrahedral partition** of Ω.

A tetrahedron $T := \langle v_1, v_2, v_3, v_4 \rangle$ is called non-degenerated provided that it has nonzero volume.

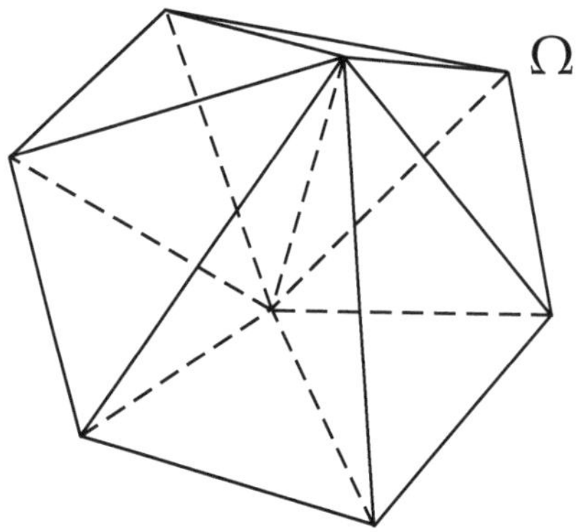

Figure 2.1: Tetrahedral partition of a polyhedral domain Ω.

Definition 2.2:
Let Ω be a polyhedral subset of $\mathbb{R}^n$, $n \in \mathbb{N}$. If Ω is divided into n-simplices T_i, $i = 1,\dots,N$, such that the intersection of two different n-simplices is either empty or a common $k-$simplex, for $k = 0,\dots,n-1$, then $\Delta := \{T_1,\dots,T_N\}$ is called a **tessellation** of Ω into **n-simplices**.

An n-simplex $T := \langle v_1,\dots,v_{n+1} \rangle$ is called non-degenerated if the points $v_i \in \mathbb{R}^n$, $i = 1,\dots,n+1$, are linear independent.

In the bivariate case a tessellation of Ω into 2-simplices is called a **triangulation**.

2.1.2 Special n-simplices

In this subsection, special refinements of n-simplices are considered. These are, the Clough-Tocher and the Powell-Sabin split of a triangle, as well the Alfeld and Worsey-Farin split of a tetrahedron and a tetrahedral partition. Moreover, partial Worsey-Farin splits of a tetrahedron are considered.

2.1.2.1 Refinements of triangles

Definition 2.3 (Clough-Tocher split (cf. [24])):
Let $F := \langle v_1, v_2, v_3 \rangle$ be a triangle in $\mathbb{R}^2$, and let v_F be a point strictly inside F. Then the Clough-Tocher split of F is obtained by connecting v_F to the three vertices of F. The resulting refinement is denoted by F_{CT}, which consists of the three subtriangles $F_i := \langle v_i, v_{i+1}, v_F \rangle$, $i = 1,2,3$, where $v_4 := v_1$.

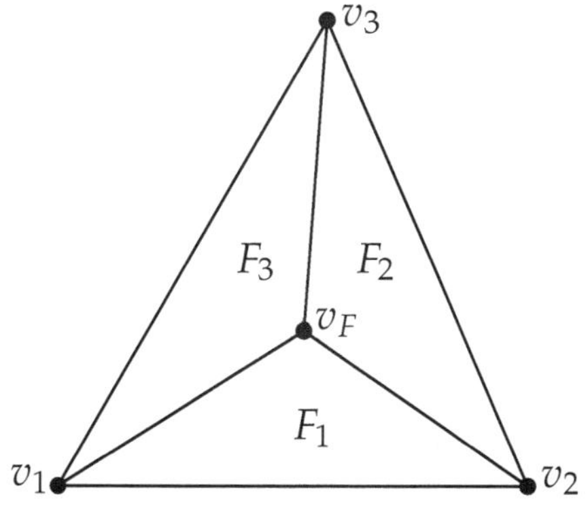

Figure 2.2: Clough-Tocher split of a triangle $F := \langle v_1, v_2, v_3 \rangle$ at the split point v_F in the interior of F.

Definition 2.4 (Powell-Sabin split (cf. [85]):
Let $F := \langle v_1, v_2, v_3 \rangle$ be a triangle in $\mathbb{R}^2$, v_F be a point strictly inside F, and let $v_{e,i}$, $i = 1,2,3$, be points strictly in the interior of the edges $e_i := \langle v_i, v_{i+1} \rangle$, $i = 1,2,3$, respectively, where $v_4 := v_1$. Then the Powell-Sabin split of F is obtained by connecting v_F to the three vertices of F and to the three points $v_{e,i}$, $i = 1,2,3$. The resulting refinement is denoted

by F_{PS} and consists of the six subtriangles $F_i := \langle v_i, v_{e,i}, v_F \rangle$, $i = 1,2,3$, and $\widetilde{F}_i := \langle v_{e,i}, v_{i+1}, v_F \rangle$, $i = 1,2,3$, where $v_4 := v_1$.

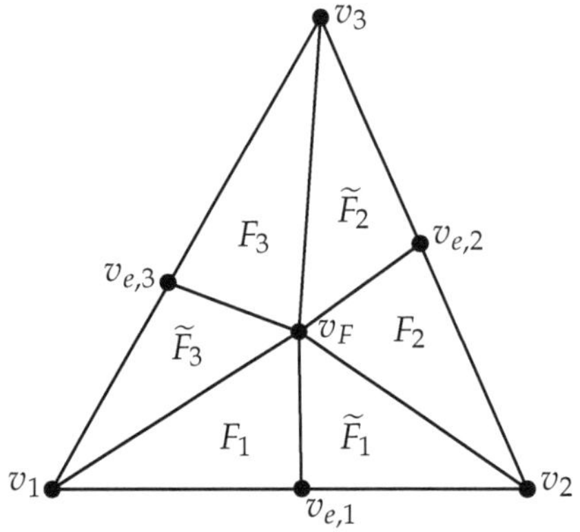

Figure 2.3: Powell-Sabin split of a triangle $F := \langle v_1, v_2, v_3 \rangle$ at the split point v_F in the interior of F, and the points $v_{e,i}$, $i = 1,2,3$, in the interior of the edges of F.

Remark 2.5:
In order to construct the Clough-Tocher split of a triangle $F := \langle v_1, v_2, v_3 \rangle$, usually the barycenter $v_F := \frac{v_1+v_2+v_3}{3}$ of F is chosen as split point.

To construct the Powell-Sabin split of a triangle F, the incenter of F is chosen as the split point v_F in the interior of F. As split points in the interior of the edges of F, the centers of the edges are chosen. In case F shares an edge with another triangle $\widetilde{F}$, then the split point in the common edge is chosen as the intersection of the line connecting the incenters v_F of F and $v_{\widetilde{F}}$ of $\widetilde{F}$ with this edge. To chose the incenters of the triangles as split points ensures, that the line connecting two interior split points of neighboring triangles intersects the common edge. This property is needed in order to obtain C^r, $r > 0$ smoothness, which is explained later on in this chapter, across this edge.

2.1.2.2 Refinements of tetrahedral partitions

Definition 2.6 (Alfeld split (cf. [1])):
Let $T := \langle v_1, v_2, v_3, v_4 \rangle$ be a tetrahedron in $\mathbb{R}^3$, and let v_T be the barycenter of T. Then the Alfeld split of T is constructed by connecting v_T to the four vertices of T. The obtained refinement is denoted by T_A, which consists of the four subtetrahedra $T_i := \langle v_i, v_{i+1}, v_{i+2}, v_T \rangle$, $i = 1, \ldots, 4$, where $v_5 := v_1$ and $v_6 := v_2$. For a tetrahedral partition Δ, the partition obtained by applying the Alfeld split to each tetrahedron in Δ is denoted by Δ_A.

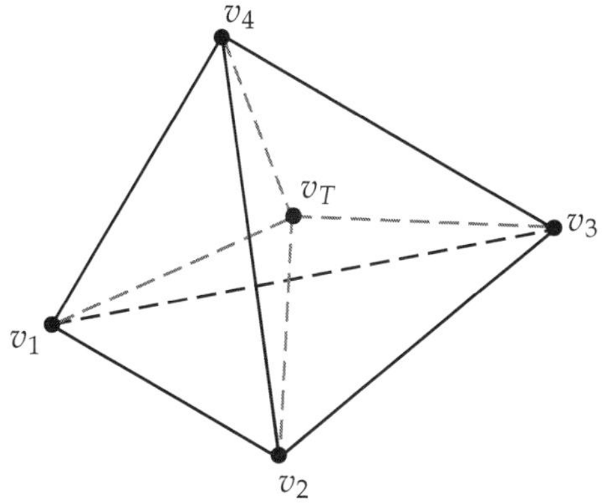

Figure 2.4: Alfeld split of a tetrahedron $T := \langle v_1, v_2, v_3, v_4 \rangle$ at the split point v_T in the interior of T.

Definition 2.7 (Worsey-Farin split (cf. [103])):
Let Δ be a tetrahedral partition and for each tetrahedron T in Δ let v_T be the incenter of T. For each interior face F of Δ, let v_F be the intersection of F and the straight line connecting the two incenters of the tetrahedra sharing F. In case F is on the boundary of Δ, the point v_F is chosen as the barycenter of F. Then the Worsey-Farin split of a tetrahedron $T := \langle v_1, v_2, v_3, v_4 \rangle$ is constructed by connecting v_T to the four vertices of T and the four points $v_{F,i}$, $i = 1, \ldots, 4$, in the interior of the faces $F_i := \langle v_i, v_{i+1}, v_{i+2} \rangle$, where $v_5 := v_1$ and $v_6 := v_2$, and by connecting the points $v_{F,i}$, $i = 1, \ldots, 4$, to the three vertices of the corresponding face F_i, respectively. The obtained refinement is denoted by T_{WF}, which consists of the

twelve subtetrahedra $T_{i,j} := \langle u_{i,j}, u_{i,j+1}, v_{F,i}, v_T \rangle$, $i = 1,\dots,4$, $j = 1,2,3$, where $u_{i,j} := v_{i+j-1}$, with $u_{i,4} := u_{i,1}$ and $u_{i,j} := v_{i+j-5}$ for $i + j > 5$ and $j \neq 4$. We define Δ_{WF} to be the refined tetrahedral partition obtained by dividing each tetrahedron of Δ with the Worsey-Farin split.

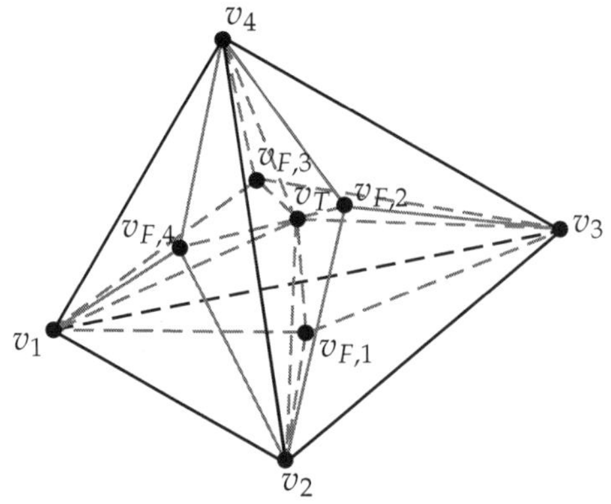

Figure 2.5: Worsey-Farin split of a tetrahedron $T := \langle v_1, v_2, v_3, v_4 \rangle$ at the split point v_T in the interior of T and the split points $v_{F,i}$, $i = 1,\dots,4$, in the interior of the faces of T.

Note that by choosing the incenters of the tetrahedra as interior split points, it is ensured that the line connecting two of these split points from neighboring tetrahedra intersects the common face F (cf. Lemma 16.24 in [62]).

Definition 2.8 (Partial Worsey-Farin splits (cf. [42])):
Let T be a tetrahedron in $\mathbb{R}^3$, and let v_T be the incenter of T. Given an integer $0 \leq m \leq 4$, let $F_1,\dots,F_m$ be distinct faces of T, and for each $i = 1,\dots,m$, let $v_{F,i}$ be a point in the interior of F_i. Then the m-th order partial Worsey-Farin split T^m_{WF} of T is defined as the refinement obtained by the following steps:

1. connect v_T to each of the four vertices of T;
2. connect v_T to the points $v_{F,i}$ for $i = 1,\dots,m$;
3. connect $v_{F,i}$ to the three vertices of F_i for $i = 1,\dots,m$.

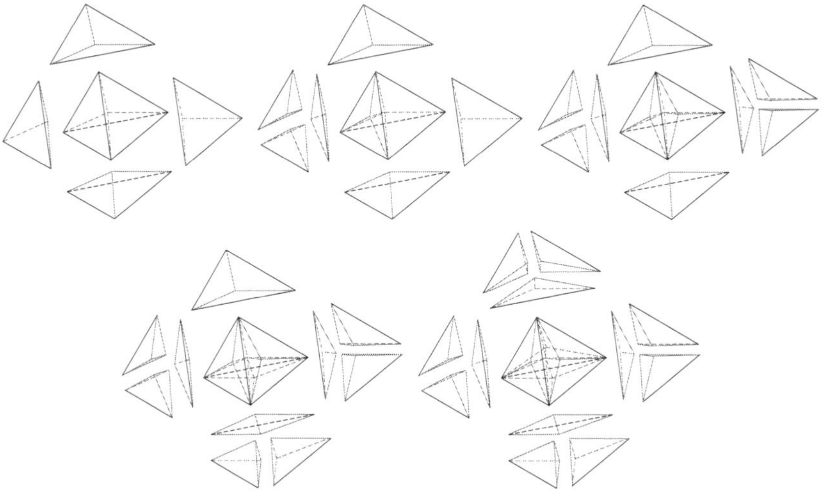

Figure 2.6: Partial Worsey-Farin splits of order m for $m = 0,\dots,4$, of a tetrahedron.

Note that in case T shares a face F with another tetrahedron, that is refined with a partial Worsey-Farin split, the point v_F is chosen as the intersection of the straight line connecting the incenters of the two tetrahedra and the face F. Else, the point v_F is chosen as the barycenter of F. Thus, the split points are chosen in the same way as for the Worsey-Farin split of a tetrahedron (cf. Definition 2.7).

It can easily be seen that the m-th order partial Worsey-Farin split of a tetrahedron results in $4 + 2m$ subtetrahedra. For $m = 0$ the partial Worsey-Farin split reduces to the Alfeld split (cf. Definition 2.6), although here the incenter is chosen as split point and not the barycenter. For the case $m = 4$ the partial Worsey-Farin split results in the Worsey-Farin split (cf. Definition 2.7).

2.1.3 Notation and Euler relations

In this subsection some notation are established and the Euler relations are stated for tetrahedral partitions Δ.

Notation 2.9:
For a tetrahedral partition Δ we define the following sets:

$\mathcal{V}_I, \mathcal{V}_B, \mathcal{V}$	:	Set of inner, boundary, and all vertices of Δ
$\mathcal{E}_I, \mathcal{E}_B, \mathcal{E}$	:	Set of inner, boundary, and all edges of Δ
$\mathcal{F}_I, \mathcal{F}_B, \mathcal{F}$	:	Set of inner, boundary, and all faces of Δ
N	:	Cardinality of the set of tetrahedra in Δ

Following Leonhard Euler, the following properties hold:

$$\#\mathcal{V}_B = 2N - \#\mathcal{F}_I + 2$$
$$N = \#\mathcal{V}_I - \#\mathcal{E}_I + \#\mathcal{F}_I + 1$$

In order to characterize the tetrahedra of a partition Δ, the following terms and definitions are used.

Two tetrahedra $T, \widetilde{T} \in \Delta$ **touch** each other, if they share a common vertex, or a common edge. They are called **neighbors**, if they share a common face.

Two triangular faces $F, \widetilde{F} \in \mathcal{F}$ are called **neighbors**, if there is a tetrahedron $T \in \Delta$ where $F, \widetilde{F} \in T$. Two faces F and $\widetilde{F}$ of two tetrahedra T and $\widetilde{T}$ are called **degenerated** if they lie in a common plane.

The next definition deals with certain subsets of a tetrahedral partition Δ. It is equivalent for tessellations Δ of a domain $\Omega \subseteq \mathbb{R}^n$.

Definition 2.10:
Let $T \in \Delta$ and

$$\text{star}^0(T) := T.$$

Then, for $n \geq 1$, $\text{star}^n(T)$ is defined inductively as

$$\text{star}^n(T) := \bigcup \{\widetilde{T} \in \Delta : \widetilde{T} \cap \text{star}^{n-1}(T) \neq \emptyset\}.$$

2.2 Splines

In this section first multivariate polynomials are defined. These form a basis for the space of multivariate splines, which are introduced afterwards. Furthermore, special subspaces of the spline space, the spaces of multivariate supersplines, are defined.

2.2.1 Multivariate polynomials

Definition 2.11:
Let $d \in \mathbb{N}_0$ and $(x_1, \ldots, x_n) \in \mathbb{R}^n$. Then

$$\mathcal{P}_d^n = \text{span}\{x_1^{i_1} \cdot \ldots \cdot x_n^{i_n} : i_1, \ldots, i_n \geq 0, i_1 + \ldots + i_n \leq d\}$$

is the space of polynomials in n dimensions of total degree $\leq d$, where $\dim(\mathcal{P}_d^n) = \binom{d+n}{n}$.

Thus, each multivariate polynomial $p \in \mathcal{P}_d^n$ can be written in the form

$$p(x_1, \ldots, x_n) = \sum_{\substack{i_1+\ldots+i_n \leq d \\ i_1,\ldots,i_n \geq 0}} a_{i_1 \ldots i_n} x_1^{i_1} \cdot \ldots \cdot x_n^{i_n},$$

where the $a_{i_1 \ldots i_n} \in \mathbb{R}$ are linear factors. This representation of multivariate polynomials is also called the monomial form.

2.2.2 Multivariate splines

In this subsection multivariate splines are defined. Therefore, first the space of smooth functions has to be denominated.

Definition 2.12:
Let Ω be a domain in $\mathbb{R}^n$, $n \in \mathbb{N}$, and let $r \in \mathbb{N}_0$. Then

$$C^r(\Omega) := \{f : \Omega \longrightarrow \mathbb{R} : f \text{ is } r\text{-times continuous differentiable}\}$$

is the space of functions of C^r smoothness, or also C^r continuity. For $C^0(\Omega)$, we also write $C(\Omega)$.

Now, the space of multivariate splines can be defined.

Definition 2.13:
Let $r, d \in \mathbb{N}_0$, with $0 \leq r < d$, and let $\Omega \subset \mathbb{R}^n, n \in \mathbb{N}$, be a polyhedral domain and Δ a tessellation of Ω into $n-$simplices. Then

$$\mathcal{S}_d^r(\Delta) = \{s \in C^r(\Omega) : s_{|T} \in \mathcal{P}_d^n \; \forall \, T \in \Delta\}$$

is called the **space of multivariate splines** of C^r smoothness of degree d.

Functions contained in $\mathcal{S}_d^r(\Delta)$ are piecewise polynomials of degree d which have C^r continuity at each intersection of two simplices in Δ.

2.2.3 Multivariate supersplines

For certain problems, as the partial Worsey-Farin splits in chapter 3, therewith also the Lagrange interpolation methods in chapter 4 and 5, or the macro-elements presented in chapter 7 and 8, additional smoothness conditions are needed, i.e. at the vertices and edges of a tetrahedral partition. The subspace of splines that fulfill these additional conditions is called the space of multivariate superspline. Thus, for a given tessellation Δ of a domain Ω into $n-$simplices, $n \in \mathbb{N}$, a superspline subspace of $\mathcal{S}_d^r(\Delta)$ is to be of $C^{\rho_{k,i}}$ smoothness at the $k-$simplices $v_{k,i}$ in Δ, for $k = 0,\dots,n-1$, with $r \leq \rho_{n-1,i} \leq \dots \leq \rho_{0,i} < d$.

Definition 2.14:
Let r,d be in $\mathbb{N}_0$, with $0 \leq r < d$, Ω a polyhedral domain in $\mathbb{R}^n$, Δ a tessellation of Ω into $n-$simplices, $n \in \mathbb{N}$, N_k the cardinality of the set of $k-$simplices in Δ, for $k = 0,\dots,n-1$, and let $\{v_{k,1},\dots,v_{k,N_k}\}$ be the set of $k-$simplices in Δ. Let $\rho_k := (\rho_{k,1},\dots,\rho_{k,N_k})$, $\rho_{k,i} \in \mathbb{N}$, for $i = 1,\dots,N_k$ and $k = 0,\dots,n-1$, where the $\rho_{k,i}$ are to fulfill $r \leq \min_{i=1,\dots,N_{n-1}} \rho_{n-1,i} \leq \max_{i=1,\dots,N_{n-1}} \rho_{n-1,i} \leq \dots \leq \min_{i=1,\dots,N_0} \rho_{0,i} \leq \max_{i=1,\dots,N_0} \rho_{0,i} < d$. Then the space

$$\mathcal{S}_d^{r,\rho_0,\dots\rho_{n-1}}(\Delta) = \{s \in \mathcal{S}_d^r(\Delta) : s \in C^{\rho_{k,i}}(v_{k,i}),\ i = 1,\dots,N_k,\ k = 0,\dots,n-2\}$$

is called a **superspline subspace** of $\mathcal{S}_d^r(\Delta)$.

Following Definition 2.14, for $\Omega \subset \mathbb{R}^2$ and a corresponding triangulation Δ a spline in $\mathcal{S}_d^{r,\rho}$ consists of bivariate polynomials of degree d that join with C^r smoothness at the edges of Δ and C^ρ smoothness at the vertices of Δ. Accordingly, for $\Omega \subset \mathbb{R}^3$ and a tetrahedral partition Δ, splines in $\mathcal{S}_d^{r,\rho_1,\rho_2}$ are piecewise polynomials of degree d that join with C^r smoothness at the triangular faces of Δ, C^{ρ_2} smoothness at the edges of Δ, and C^{ρ_1} smoothness at the vertices of Δ.

2.3 Bernstein-Bézier techniques

In this section, we consider the well known Bernstein-Bézier techniques, which are used to examine multivariate splines. The basis for these techniques can be found in [14, 33, 39, 62], among others. We first introduce barycentric coordinates, that form a basis for the Bernstein-Bézier techniques. Subsequently, we introduce the Bernstein polynomials and the

B-form of a polynomial, which will be used throughout this work. Then, the de Casteljau algorithm for the evaluation of polynomials in the B-form, and also the subdivision of polynomials, is considered (cf. [37]). In the next subsection, smoothness conditions between adjacent n-simplices are investigated. Finally, the concept of minimal determining sets is introduced, which can be used to characterize spline spaces and to consider their dimension.

2.3.1 Barycentric coordinates

In this subsection, we consider barycentric coordinates. These are very efficient for the representation of points in $\mathbb{R}^n$ relative to an n-simplex T, since they are affine invariant, in contrast to the usual Cartesian coordinates. Note that here and in the course of this work, we will assume that T is a non-degenerated n-simplex.

Definition 2.15:
Let $T := \langle v_1, \ldots, v_{n+1} \rangle$ be an n-simplex in $\mathbb{R}^n$ with vertices $v_1, \ldots, v_{n+1}$. Then there exist unique **barycentric coordinates** $\phi_1, \ldots, \phi_{n+1} \in \mathcal{P}_1^n$, that satisfy

$$\sum_{i=1}^{n+1} \phi_i = 1,$$

as well as the interpolation property

$$\phi_i(v_j) = \delta_{i,j} = \begin{cases} 1, & \text{for } i = j, \\ 0, & \text{for } i \neq j, \end{cases} \qquad i,j = 1, \ldots, n+1.$$

They can be explicitly computed as

$$\phi_i(v) = \frac{\left| \begin{pmatrix} 1 & \ldots & 1 & \ldots & 1 \\ v_1 & \ldots & v & \ldots & v_{n+1} \end{pmatrix} \right|}{\left| \begin{pmatrix} 1 & \ldots & 1 & \ldots & 1 \\ v_1 & \ldots & v_i & \ldots & v_{n+1} \end{pmatrix} \right|}, \qquad \text{for } i = 1, \ldots, n+1, \text{ and for all } v \in \mathbb{R}^n.$$

Moreover, for each point $v \in \mathbb{R}^n$

$$v = \phi_1(v)v_1 + \ldots + \phi_{n+1}(v)v_{n+1}$$

holds.

In the bivariate setting the barycentric coordinates ϕ_1,ϕ_2, and ϕ_3 relative to a triangle $F := \langle v_1,v_2,v_3\rangle$ can be visualized as planes in $\mathbb{R}^3$ (see Figure 2.7).

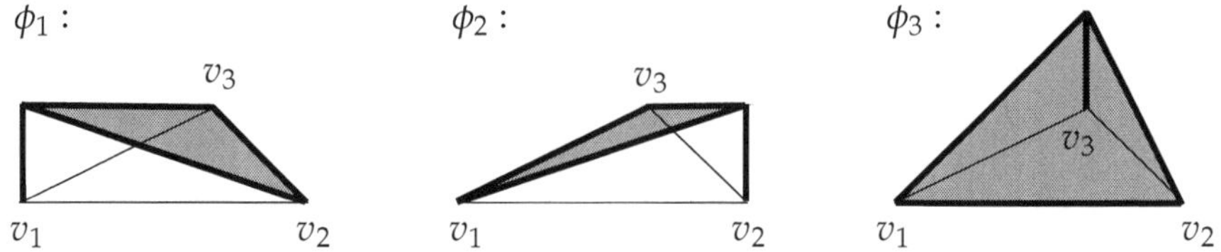

Figure 2.7: Barycentric coordinates relative to a triangle $F := \langle v_1,v_2,v_3\rangle$.

2.3.2 Bernstein polynomials and the B-form

In this subsection, we define the Bernstein polynomials, that form a basis of the space $\mathcal{P}_d^n$. Moreover, we define the equally spaced domain points.

Definition 2.16:
Let $T := \langle v_1,\ldots,v_{n+1}\rangle$ be an n-simplex in $\mathbb{R}^n$ with barycentric coordinates $\phi_1,\ldots,\phi_{n+1}$, and let $d \in \mathbb{N}_0$. Then the polynomials $B^{d,T}_{i_1,\ldots,i_{n+1}} \in \mathcal{P}_d^n$, with

$$B^{d,T}_{i_1,\ldots,i_{n+1}} := \frac{d!}{i_1! \cdot \ldots \cdot i_{n+1}!}\phi_1^{i_1} \cdot \ldots \cdot \phi_{n+1}^{i_{n+1}},$$

$$i_1,\ldots,i_{n+1} \geq 0,\ i_1 + \ldots + i_{n+1} = d,$$

are called **Bernstein polynomials** of degree d relative to T.

The Bernstein polynomials have several interesting properties. They form a partition of unity,

$$\sum_{\substack{i_1+\ldots+i_{n+1}=d \\ i_1,\ldots,i_{n+1}\geq 0}} B^{d,T}_{i_1,\ldots,i_{n+1}} \equiv 1,$$

and they satisfy the recurrence formula

$$B^{d,T}_{i_1,\ldots,i_{n+1}} = \phi_1 B^{d-1,T}_{i_1-1,\ldots,i_{n+1}} + \ldots + \phi_{n+1} B^{d-1,T}_{i_1-1,\ldots,i_{n+1}-1}. \tag{2.1}$$

Moreover, they form a basis for $\mathcal{P}_d^n$:

Theorem 2.17:
Let $T := \langle v_1, \ldots, v_{n+1} \rangle$ be an n-simplex in $\mathbb{R}^n$. Then the set

$$\{B^{d,T}_{i_1,\ldots,i_{n+1}}\}_{i_1+\ldots+i_{n+1}=d}$$

of Bernstein polynomials, is a basis for the space $\mathcal{P}_d^n$ of n-dimensional polynomials of degree d.

Following Theorem 2.17, there exists a unique representation of each polynomial $p \in \mathcal{P}_d^n$ in terms of Bernstein polynomials relative to an n-simplex T.

Definition 2.18:
Let $T := \langle v_1, \ldots, v_{n+1} \rangle \in \mathbb{R}^n$, $d \in \mathbb{N}_0$, and $B^{d,T}_{i_1,\ldots,i_{n+1}}$, $i_1 + \ldots + i_{n+1} = d$, the Bernstein polynomials of degree d relative to T. Then every polynomial $p \in \mathcal{P}_d^n$ can be uniquely written as

$$p = \sum_{\substack{i_1+\ldots+i_{n+1}=d \\ i_1,\ldots,i_{n+1}\geq 0}} c^T_{i_1,\ldots,i_{n+1}} B^{d,T}_{i_1,\ldots,i_{n+1}}.$$

This representation is called **Bernstein-Bézier-form (B-form)** and the coefficients $c^T_{i_1,\ldots,i_{n+1}}$ are called **Bernstein-Bézier-coefficients (B-coefficients)**.

This leads to the following definition of equally spaced points $\xi^T_{i_1,\ldots,i_{n+1}}$ in an n-simplex T that are associated with the B-coefficients $c^T_{i_1,\ldots,i_{n+1}}$ of a polynomial p defined on T:

Definition 2.19:
Let $d \in \mathbb{N}_0$ and $T := \langle v_1, \ldots, v_{n+1} \rangle$ an n-simplex in $\mathbb{R}^n$. Then we define

$$\mathcal{D}_{T,d} := \{\xi^T_{i_1,\ldots,i_{n+1}} := \frac{i_1 v_1 + \ldots + i_{n+1} v_{n+1}}{d}, \ i_1 + \ldots + i_{n+1} = d\}$$

to be the set of **domain points** relative to T.

For $m \geq 0$ and $1 \leq n$, we say that the domain point $\xi^T_{i_1,\ldots,i_{n+1}}$ has a distance of $i_1 + \ldots + i_k = m$ from the (n-k)-simplex $\langle v_{k+1}, \ldots, v_{n+1} \rangle$, and the domain points with $i_1 + \ldots + i_k \leq m$ are within a distance of m from $\langle v_{k+1}, \ldots, v_{n+1} \rangle$. In certain cases we need m to be negative. So for $m = -1$ the set of domain points with or within a distance of m from an (n-k)-simplex is empty. The definitions are analogue for all other combinations of vertices of T.

For a tessellation Δ, we define

$$\mathcal{D}_{\Delta,d} := \bigcup_{T \in \Delta} \mathcal{D}_{T,d}.$$

Then, for a domain point $\xi := \xi^T_{i_1,\dots,i_{n+1}}$ let c_ξ be the B-coefficient $c^T_{i_1,\dots,i_{n+1}}$.

In the following we define certain sets of domain points for the bivariate and trivariate setting. For a clearer distinction between the indices of the domain points, we will use i,j,k in the bivariate and i,j,k,l in the trivariate setting, instead of i_1,i_2,i_3 and i_1,i_2,i_3,i_4.

Thus, for a triangle $F := \langle v_1, v_2, v_3 \rangle$, let

$$R^F_m(v_1) := \{\xi^T_{i,j,k} : \ i = d - m\}$$

be the ring of radius m around v_1. Moreover, let

$$D^F_m(v_1) := \{\xi^T_{i,j,k} : \ i \geq d - m\}$$

be the disk of radius m around v_1. The definitions are analogue for the other vertices of F. Then, for a vertex v of a triangulation Δ, we have

$$R_m(v) := \bigcup_{\{F \in \Delta: \ v \in F\}} R^F_m(v)$$

and

$$D_m(v) := \bigcup_{\{F \in \Delta: \ v \in F\}} D^F_m(v).$$

In the trivariate setting, for a tetrahedron $T := \langle v_1, v_2, v_3, v_4 \rangle$, let

$$R^T_m(v_1) := \{\xi^T_{i,j,k,l} : \ i = d - m\}$$

be the shell of radius m around v_1, and let

$$D^T_m(v_1) := \{\xi^T_{i,j,k,l} : \ i \geq d - m\}$$

be the ball of radius m around v_1. Furthermore, let

$$E^T_m(\langle v_1, v_2 \rangle) := \{\xi^T_{i,j,k,l} : \ k + l \leq m\}$$

be the tube of radius m around the edge $\langle v_1, v_2 \rangle$. In addition, let

$$F_m^T(\langle v_1, v_2, v_3 \rangle) := \{\xi_{i,j,k,l}^T : l = m\}$$

be the set of domain points with a distance of m from the face $\langle v_1, v_2, v_3 \rangle$ of T. The definitions are similar for the other vertices, edges, and faces of T. Then, for a vertex v, an edge e, and a face F of a tetrahedral partition Δ, we have

$$R_m(v) := \bigcup_{\{T \in \Delta:\ v \in T\}} R_m^T(v),$$

$$D_m(v) := \bigcup_{\{T \in \Delta:\ v \in T\}} D_m^T(v),$$

$$E_m(e) := \bigcup_{\{T \in \Delta:\ e \in T\}} E_m^T(e),$$

and

$$F_m(F) := \bigcup_{\{T \in \Delta:\ F \in T\}} F_m^T(F).$$

2.3.3 De Casteljau algorithm

In this subsection, we present an efficient algorithm to evaluate a polynomial given in the B-form (cf. [37]). Moreover, we show an algorithm for subdividing polynomials defined on a single n-simplex.

In the following we assume that expressions with negative subscripts are equal to zero.

Theorem 2.20:
Let $p \in \mathcal{P}_d^n$ be a polynomial defined on an n-simplex T with B-coefficients

$$c_{i_1,\ldots,i_{n+1}}^{T(0)} := c_{i_1,\ldots,i_{n+1}}^T, \quad i_1 + \ldots + i_{n+1} = d.$$

Let $\phi_1, \ldots, \phi_{n+1}$ be the barycentric coordinates of T, $v \in \mathbb{R}^n$, and let

$$c_{i_1,\ldots,i_{n+1}}^{T(j)} := \phi_1(v) c_{i_1+1,\ldots,i_{n+1}}^{T(j-1)} + \ldots + \phi_{n+1}(v) c_{i_1,\ldots,i_{n+1}+1}^{T(j-1)}, \quad i_1 + \ldots + i_{n+1} = d - j. \tag{2.2}$$

Then

$$p(v) = \sum_{i_1+\ldots+i_{n+1}=d-j} c^{T(j)}_{i_1,\ldots,i_{n+1}} B^{d-j,T}_{i_1,\ldots,i_{n+1}}(v), \tag{2.3}$$

for all $0 \leq j \leq d$. Especially,

$$p(v) = c^{T(d)}_{0,\ldots,0}.$$

Proof:
We prove (2.3) by induction on j. The equation holds for $j = 0$, since it reduces to the normal B-form of p in this case. Now, assume that (2.3) holds for $j-1$. Following the recurrence formula (2.1), we have

$$\begin{aligned} p(v) &= \sum_{i_1+\ldots+i_{n+1}=d-j+1} c^{T(j-1)}_{i_1,\ldots,i_{n+1}} B^{d-j+1,T}_{i_1,\ldots,i_{n+1}}(v) \\ &= \sum_{i_1+\ldots+i_{n+1}=d-j+1} c^{T(j-1)}_{i_1,\ldots,i_{n+1}} \Big(\phi_1(v) B^{d-j,T}_{i_1-1,\ldots,i_{n+1}}(v) + \ldots + \\ &\qquad + \phi_{n+1}(v) B^{d-j,T}_{i_1,\ldots,i_{n+1}-1}(v) \Big). \end{aligned}$$

This sum can be divided into $n+1$ sums of the form

$$\begin{aligned} &\sum_{\substack{i_1+\ldots+i_{n+1}=d-j+1 \\ i_k \geq 1}} c^{T(j-1)}_{i_1,\ldots,i_{n+1}} \phi_k(v) B^{d-j,T}_{i_1,\ldots,i_k-1,\ldots,i_{n+1}}(v) \\ &= \sum_{i_1+\ldots+i_{n+1}=d-j} c^{T(j-1)}_{i_1,\ldots,i_k+1,\ldots,i_{n+1}} \phi_k(v) B^{d-j,T}_{i_1,\ldots,i_{n+1}}(v), \end{aligned}$$

for $k = 1,\ldots,n+1$. Then, combining these sums we obtain

$$\begin{aligned} p(v) = \sum_{i_1+\ldots+i_{n+1}=d-j} &\left(\phi_1(v) c^{T(j-1)}_{i_1+1,\ldots,i_{n+1}} + \ldots + \phi_{n+1}(v) c^{T(j-1)}_{i_1,\ldots,i_{n+1}+1} \right) \\ &\cdot B^{d-j,T}_{i_1,\ldots,i_{n+1}}(v). \end{aligned}$$

Applying (2.2), we get (2.3). Then, for $j = d$, we get

$$p(v) = c^{T(d)}_{0,\ldots,0} B^{0,T}_{0,\ldots,0}(v).$$

Thus, since $B^{0,T}_{0,\ldots,0}(v) = 1$, $p(v) = c^{T(d)}_{0,\ldots,0}$. □

This leads to the well known de Casteljau algorithm for evaluating polynomials $p \in \mathcal{P}_d^n$ in the B-form, which are defined on an n-simplex T:

Algorithm 2.21:
For $j = 1,\dots,d$,
for all $i_1 + \ldots + i_{n+1} = d - j$,

$$c_{i_1,\dots,i_{n+1}}^{T(j)} := \phi_1(v)c_{i_1+1,\dots,i_{n+1}}^{T(j-1)} + \ldots + \phi_{n+1}(v)c_{i_1,\dots,i_{n+1}+1}^{T(j-1)}.$$

The de Casteljau algorithm can also be used to subdivide a polynomial $p \in \mathcal{P}_d^n$ defined on an n-simplex $T := \langle v_1,\dots,v_{n+1}\rangle$ to polynomial pieces defined on the n-simplices $T_j := \langle u, v_{j+1},\dots,v_{j+n}\rangle$, $j = 1,\dots,n+1$, where u is point in the interior of T and $v_{n+1+k} := v_k$ for all k.

Theorem 2.22:
Let $p \in \mathcal{P}_d^n$ be a polynomial in the B-form defined on T with B-coefficients $\{c_{i_1,\dots,i_{n+1}}^{T(0)}\}_{i_1+\ldots+i_{n+1}=d}$. Then for all $v \in T$,

$$p(v) = \begin{cases} \sum\limits_{i_1+\ldots+i_{n+1}=d} c_{0,i_2,\dots,i_{n+1}}^{T_1(i_1)} B_{i_1,\dots,i_{n+1}}^{d,T_1}(v), & v \in T_1, \\ \vdots & \\ \sum\limits_{i_1+\ldots+i_{n+1}=d} c_{i_1,\dots,i_n,0}^{T_{n+1}(i_{n+1})} B_{i_1,\dots,i_{n+1}}^{d,T_{n+1}}(v), & v \in T_{n+1}, \end{cases}$$

where the coefficients $c_{i_1,\dots,i_{n+1}}^{T_j(k)}$ are generated by applying k steps of the de Casteljau algorithm based on the barycentric coordinates of u relative to T.

Proof:
Let $v \in T_1$ and $\widetilde{\phi}_j$, $j = 1,\dots,n+1$ the barycentric coordinates relative to T_1. Then v can be written as

$$v = \widetilde{\phi}_1(v)u + \widetilde{\phi}_2(v)v_2 + \ldots + \widetilde{\phi}_{n+1}(v)v_{n+1}.$$

Let ϕ_j, $j = 1,\dots,n+1$, be the barycentric coordinates relative to T. Then, since

$$u = \phi_1(u)v_1 + \ldots + \phi_{n+1}(u)v_{n+1},$$

we get

$$\begin{aligned} v &= \widetilde{\phi}_1(v)\left(\phi_1(u)v_1 + \ldots + \phi_{n+1}(u)v_{n+1}\right) + \widetilde{\phi}_2(v)v_2 + \ldots + \widetilde{\phi}_{n+1}(v)v_{n+1} \\ &= \widetilde{\phi}_1(v)\phi_1(u)v_1 + \left(\widetilde{\phi}_1(v)\phi_2(u) + \widetilde{\phi}_2(v)\right)v_2 \\ &\quad + \ldots + \left(\widetilde{\phi}_1(v)\phi_{n+1}(u) + \widetilde{\phi}_{n+1}(v)\right)v_{n+1} \end{aligned}$$

Thus, for the Bernstein polynomials relative to T we get

$$\begin{aligned} B^{d,T}_{l_1,\ldots,l_{n+1}}(v) &= \frac{d!}{l_1! \cdot \ldots \cdot l_{n+1}!}\left(\widetilde{\phi}_1(v)\phi_1(u)\right)^{l_1} \cdot \left(\widetilde{\phi}_1(v)\phi_2(u) + \widetilde{\phi}_2(v)\right)^{l_2} \\ &\qquad \cdot \ldots \cdot \left(\widetilde{\phi}_1(v)\phi_{n+1}(u) + \widetilde{\phi}_{n+1}(v)\right)^{l_{n+1}} \\ &= \sum_{\nu_2}^{l_2} \cdots \sum_{\nu_{n+1}}^{l_{n+1}} B^{d,T_1}_{l_1+\nu_2+\ldots+\nu_{n+1}, l_2-\nu_2,\ldots,l_{n+1}-\nu_{n+1}}(v) \\ &\qquad \cdot B^{l_1+\nu_2+\ldots,\nu_{n+1},T}_{l_1,\nu_2,\ldots,\nu_{n+1}}(u), \end{aligned}$$

for all $l_1 + \ldots + l_{n+1} = d$. Then, substituting this in the B-form

$$p(v) = \sum_{l_1+\ldots+l_{n+1}=d} c^T_{l_1,\ldots,l_{n+1}} B^{d,T}_{l_1,\ldots,l_{n+1}}(v)$$

of p, we get

$$\begin{aligned} p(v) = \sum_{l_1+\ldots+l_{n+1}=d} c^T_{l_1,\ldots,l_{n+1}} \sum_{\nu_2}^{l_2} \cdots \sum_{\nu_{n+1}}^{l_{n+1}} & B^{d,T_1}_{l_1+\nu_2+\ldots+\nu_{n+1}, l_2-\nu_2,\ldots,l_{n+1}-\nu_{n+1}}(v) \\ & \cdot B^{l_1+\nu_2+\ldots,\nu_{n+1},T}_{l_1,\nu_2,\ldots,\nu_{n+1}}(u). \end{aligned}$$

Now, for $l_j = i_j + \nu_j,\ j = 2,\ldots,n+1$, and $l_1 + \nu_2 + \ldots, \nu_{n+1} = i_1$, the B-coefficient of $B^{d,T_1}_{i_1,\ldots,i_{n+1}}$ is

$$\sum_{l_1+\nu_2+\ldots+\nu_{n+1}=i_1} c^T_{l_1,i_2+\nu_2,\ldots,i_{n+1}+\nu_{n+1}} B^{i_1,T_1}_{l_1,\nu_2,\ldots,\nu_{n+1}}(u),$$

which is equal to $c^{T(i_1)}_{0,i_2,\ldots,i_{n+1}}$ by (2.2). The proof is similar if v is in another n-simplex in $\{T_2,\ldots,T_{n+1}\}$. □

2.3.4 Smoothness conditions

In this subsection, we consider conditions for smooth joints of two polynomials p and $\widetilde{p}$ defined on n-simplices T and $\widetilde{T}$, respectively, that share at least a common 0-simplex. Moreover, we present some lemmata which make use of the smoothness conditions in order to compute B-coefficients of two polynomials defined on adjacent triangles and tetrahedra. These lemmata are then illustrated.

The following C^r smoothness conditions between two polynomials p and $\widetilde{p}$ were established by Farin [39] and de Boor [33]:

Theorem 2.23:
Let $T := \langle v_1, \dots, v_{n+1} \rangle$ and $\widetilde{T} := \langle \widetilde{v}_1, \dots, \widetilde{v}_{n+1} \rangle$ be two n-simplices in $\mathbb{R}^n$, and let $v_1 := \widetilde{v}_1$. Moreover, let

$$p(v) := \sum_{i_1+\ldots+i_{n+1}=d} c^T_{i_1,\ldots,i_{n+1}} B^{d,T}_{i_1,\ldots,i_{n+1}}(v)$$

and

$$\widetilde{p}(v) := \sum_{i_1+\ldots+i_{n+1}=d} c^{\widetilde{T}}_{i_1,\ldots,i_{n+1}} B^{d,\widetilde{T}}_{i_1,\ldots,i_{n+1}}(v)$$

be two polynomials defined on T and $\widetilde{T}$, respectively, where $B^{d,T}_{i_1,\ldots,i_{n+1}}$ and $B^{d,\widetilde{T}}_{i_1,\ldots,i_{n+1}}$ are the corresponding Bernstein polynomials.

Then the two polynomials p and $\widetilde{p}$ join with C^r smoothness if and only if

$$c^{\widetilde{T}}_{i_1,\ldots,i_{n+1}} = \sum_{\substack{\nu_1^{i_2}+\ldots+\nu_{n+1}^{i_2}=i_2 \\ \vdots \\ \nu_1^{i_{n+1}}+\ldots+\nu_{n+1}^{i_{n+1}}=i_{n+1}}} \Bigg(c^T_{i_1+\nu_1^{i_2}+\ldots+\nu_1^{i_{n+1}},\nu_2^{i_2}+\ldots+\nu_2^{i_{n+1}},\ldots,\nu_{n+1}^{i_2}+\ldots+\nu_{n+1}^{i_{n+1}}} \tag{2.4}$$

$$\cdot B^{d,T}_{\nu_1^{i_2},\ldots,\nu_{n+1}^{i_2}}(\widetilde{v}_2) \cdot \ldots \cdot B^{d,T}_{\nu_1^{i_{n+1}},\ldots,\nu_{n+1}^{i_{n+1}}}(\widetilde{v}_{n+1}) \Bigg),$$

for $i_2 + \ldots + i_{n+1} \leq r$ and $i_1 + \ldots + i_{n+1} = d$.

In case T and $\widetilde{T}$ have more common points $v_j = \widetilde{v}_j$, $j = 2, \ldots, k$, then the terms $\nu_1^{i_2} + \ldots + \nu_{n+1}^{i_2} = i_2, \ldots, \nu_1^{i_k} + \ldots + \nu_{n+1}^{i_k} = i_k$ in the sum in (2.4) can be omitted, since then

$$B^{d,T}_{\substack{i_j \\ v_1^j,\ldots,v_{n+1}^{i_j}}}(\widetilde{v}_j) = B^{d,T}_{\substack{i_j \\ v_1^j,\ldots,v_{n+1}^{i_j}}}(v_j) = \begin{cases} 0, & \text{for } i_j \neq d, \\ 1, & \text{for } i_j = d. \end{cases}$$

We are especially interested in 2- and 3-simplices that share certain numbers of vertices.

Corollary 2.24:
Let $F := \langle v_1, v_2, v_3 \rangle$ and $\widetilde{F} := \langle v_1, v_2, \widetilde{v}_3 \rangle$ be two triangles in $\mathbb{R}^2$ and let

$$p(v) := \sum_{i+j+k=d} c^T_{i,j,k} B^{d,F}_{i,j,k}(v)$$

and

$$\widetilde{p}(v) := \sum_{i+j+k=d} c^{\widetilde{T}}_{i,j,k} B^{d,\widetilde{F}}_{i,j,k}(v)$$

be two polynomials defined on F and $\widetilde{F}$, respectively.

Then the two polynomials p and $\widetilde{p}$ join with C^r smoothness across the common edge $\langle v_1, v_2 \rangle$ if and only if

$$c^{\widetilde{F}}_{i,j,k} = \sum_{\nu_1^k+\nu_2^k+\nu_3^k=k} c^T_{i+\nu_1^k, j+\nu_2^k, \nu_3^k} B^{d,T}_{\nu_1^k,\nu_2^k,\nu_3^k}(\widetilde{v}_3),$$

for $k \leq r$ and $i + j = d - k$.

In Figure 2.8, we illustrate C^1 and C^2 smoothness conditions between two polynomials of degree four defined on triangles $F := \langle v_1, v_2, v_3 \rangle$ and $\widetilde{F} := \langle v_1, v_2, \widetilde{v}_3 \rangle$ sharing the common edge $\langle v_1, v_2 \rangle$. The gray quadrilaterals containing some of the domain points in $\mathcal{D}_{F \cup \widetilde{F},4}$ indicate that the corresponding B-coefficients are connected by smoothness conditions. Thus, by a C^1 smoothness condition, exactly four B-coefficients correlate (see Figure 2.8 (left)). In Figure 2.8 (right) a C^2 smoothness condition is shown.

In the appendix in chapter A, we also need certain additional smoothness conditions. These are conditions for individual B-coefficients.

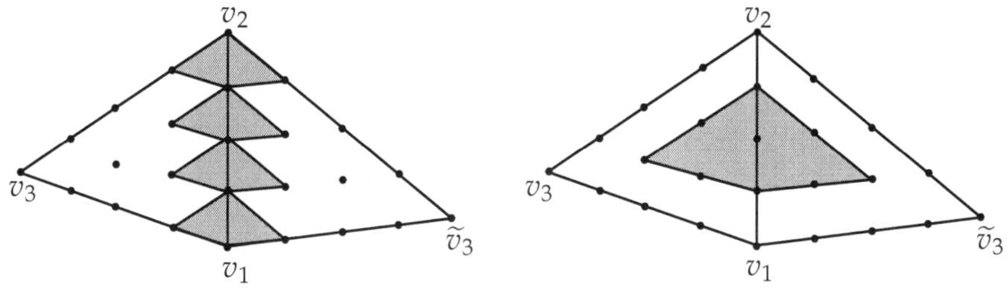

Figure 2.8: C^1 and C^2 smoothness conditions between two polynomials of degree four defined on triangles $F := \langle v_1, v_2, v_3 \rangle$ and $\widetilde{F} := \langle v_1, v_2, \widetilde{v}_3 \rangle$, respectively.

Definition 2.25:
Let $F := \langle v_1, v_2, v_3 \rangle$ and $\widetilde{F} := \langle v_1, v_2, \widetilde{v}_3 \rangle$ be two triangles in $\mathbb{R}^2$ and let

$$p(v) := \sum_{i+j+k=d} c^T_{i,j,k} B^{d,F}_{i,j,k}(v)$$

and

$$\widetilde{p}(v) := \sum_{i+j+k=d} c^{\widetilde{T}}_{i+j+k} B^{d,\widetilde{F}}_{i+j+k}(v)$$

be two polynomials defined on F and $\widetilde{F}$, respectively. Moreover, let

$$g(v) := \begin{cases} p(v), & \text{for } v \in F, \\ \widetilde{p}(v), & \text{for } v \in \widetilde{F} \setminus F. \end{cases}$$

Then, let

$$\tau^{F,\widetilde{F}}_{i,j,k} g := c^{\widetilde{F}}_{i,j,k} = \sum_{\nu_1^k+\nu_2^k+\nu_3^k=k} c^T_{i+\nu_1^k, j+\nu_2^k, \nu_3^k} B^{d,T}_{\nu_1^k,\nu_2^k,\nu_3^k}(\widetilde{v}_3),$$

for $i,j,k \geq 0$ and $i+j+k=d$.

Corollary 2.26:
Let $T := \langle v_1, v_2, v_3, v_4 \rangle$ and $\widetilde{T} := \langle \widetilde{v}_1, \widetilde{v}_2, \widetilde{v}_3, \widetilde{v}_4 \rangle$ be two tetrahedra in $\mathbb{R}^3$ and let

$$p(v) := \sum_{i+j+k+l=d} c^T_{i,j,k,l} B^{d,T}_{i,j,k,l}(v)$$

and

$$\widetilde{p}(v) := \sum_{i+j+k+l=d} c^{\widetilde{T}}_{i,j,k,l} B^{d,\widetilde{T}}_{i,j,k,l}(v)$$

be two polynomials defined on T and $\widetilde{T}$, respectively.

Then for $v_1 = \widetilde{v}_1$ the two polynomials p and $\widetilde{p}$ join with C^r smoothness at the common vertex v_1 if and only if

$$c^{\widetilde{T}}_{i,j,k,l} = \sum_{\substack{\nu^j_1+\nu^j_2+\nu^j_3+\nu^j_4=j \\ \nu^k_1+\nu^k_2+\nu^k_3+\nu^k_4=k \\ \nu^l_1+\nu^l_2+\nu^l_3+\nu^l_4=l}} c^T_{i+\nu^j_1+\nu^k_1+\nu^l_1,\nu^j_2+\nu^k_2+\nu^l_2,\nu^j_3+\nu^k_3+\nu^l_3,\nu^j_4+\nu^k_4+\nu^l_4} \cdot B^{d,T}_{\nu^j_1,\nu^j_2,\nu^j_3,\nu^j_4}(\widetilde{v}_2) B^{d,T}_{\nu^k_1,\nu^k_2,\nu^k_3,\nu^k_4}(\widetilde{v}_3) B^{d,T}_{\nu^l_1,\nu^l_2,\nu^l_3,\nu^l_4}(\widetilde{v}_4),$$

for $j+k+l \le r$ and $i = d-j-k-l$.

For $v_1 = \widetilde{v}_1$ and $v_2 = \widetilde{v}_2$ the two polynomials p and $\widetilde{p}$ join with C^r smoothness across the common edge $\langle v_1, v_2 \rangle$ if and only if

$$c^{\widetilde{T}}_{i,j,k,l} = \sum_{\substack{\nu^k_1+\nu^k_2+\nu^k_3+\nu^k_4=k \\ \nu^l_1+\nu^l_2+\nu^l_3+\nu^l_4=l}} c^T_{i+\nu^k_1+\nu^l_1,j+\nu^k_2+\nu^l_2,\nu^k_3+\nu^l_3,\nu^k_4+\nu^l_4} B^{d,T}_{\nu^k_1,\nu^k_2,\nu^k_3,\nu^k_4}(\widetilde{v}_3) B^{d,T}_{\nu^l_1,\nu^l_2,\nu^l_3,\nu^l_4}(\widetilde{v}_4),$$

for $k+l \le r$ and $i+j = d-k-l$.

For $v_1 = \widetilde{v}_1$, $v_2 = \widetilde{v}_2$, and $v_3 = \widetilde{v}_3$ the two polynomials p and $\widetilde{p}$ join with C^r smoothness across the common face $\langle v_1, v_2, v_3 \rangle$ if and only if

$$c^{\widetilde{T}}_{i,j,k,l} = \sum_{\nu^l_1+\nu^l_2+\nu^l_3+\nu^l_4=l} c^T_{i+\nu^l_1,j+\nu^l_2,k+\nu^l_3,\nu^l_4} B^{d,T}_{\nu^l_1,\nu^l_2,\nu^l_3,\nu^l_4}(\widetilde{v}_4),$$

for $l \le r$ and $i+j+k = d-l$.

In Figure 2.9, we illustrate some C^1 and C^2 smoothness conditions between two polynomials defined on tetrahedra the $T := \langle v_1, v_2, v_3, v_4 \rangle$ and $\widetilde{T} := \langle v_1, v_2, v_3, \widetilde{v}_4 \rangle$ sharing the common face $\langle v_1, v_2, v_3 \rangle$. The gray polyhedrons containing some of the domain points in T and $\widetilde{T}$ denote that the corresponding B-coefficients are connected by smoothness conditions. For the sake of clarity we omit the remaining domain points, whose B-coefficients are not involved in the smoothness conditions shown here. It can be seen that exactly five B-coefficients are correlated by a C^1 smoothness condition (see Figure 2.9 (left)). In Figure 2.9 (right) a C^2 smoothness condition is depicted.

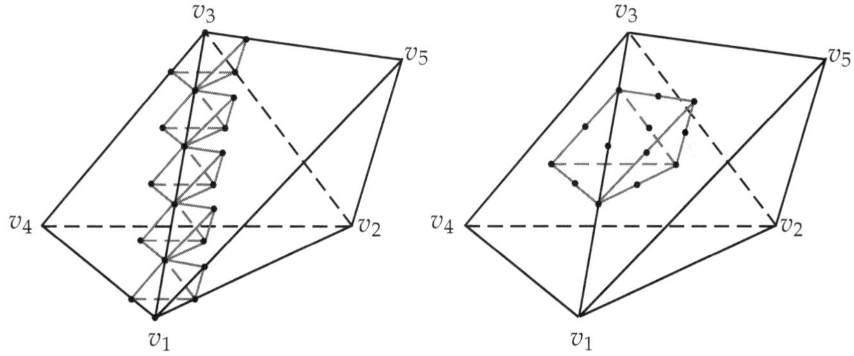

Figure 2.9: C^1 and C^2 smoothness conditions between two polynomials defined on tetrahedra $T := \langle v_1, v_2, v_3, v_4 \rangle$ and $\widetilde{T} := \langle v_1, v_2, v_3, \widetilde{v}_4 \rangle$, respectively.

In chapter 3 and 5 we also need certain trivariate additional smoothness conditions for individual B-coefficients of polynomials.

Definition 2.27:
Let $T := \langle v_1, v_2, v_3, v_4 \rangle$ and $\widetilde{T} := \langle v_1, \widetilde{v}_2, \widetilde{v}_3, \widetilde{v}_4 \rangle$ be two tetrahedra in $\mathbb{R}^3$ and let

$$p(v) := \sum_{i+j+k+l=d} c^T_{i,j,k,l} B^{d,T}_{i,j,k,l}(v)$$

and

$$\widetilde{p}(v) := \sum_{i+j+k+l=d} c^{\widetilde{T}}_{i,j,k,l} B^{d,\widetilde{T}}_{i,j,k,l}(v)$$

be two polynomials defined on T and $\widetilde{T}$, respectively. Moreover, let

$$g(v) := \begin{cases} p(v), & \text{for } v \in T, \\ \widetilde{p}(v), & \text{for } v \in \widetilde{T} \setminus T. \end{cases}$$

Then, let

$$\tau_{i,j,k,l}^{T,\widetilde{T}} g := c_{i,j,k,l}^{\widetilde{T}} = \sum_{\substack{\nu_1^j+\nu_2^j+\nu_3^j+\nu_4^j=j \\ \nu_1^k+\nu_2^k+\nu_3^k+\nu_4^k=k \\ \nu_1^l+\nu_2^l+\nu_3^l+\nu_4^l=l}} c^T_{i+\nu_1^j+\nu_1^k+\nu_1^l,\nu_2^j+\nu_2^k+\nu_2^l,\nu_3^j+\nu_3^k+\nu_3^l,\nu_4^j+\nu_4^k+\nu_4^l}$$
$$\cdot B^{d,T}_{\nu_1^j,\nu_2^j,\nu_3^j,\nu_4^j}(\widetilde{v}_2) B^{d,T}_{\nu_1^k,\nu_2^k,\nu_3^k,\nu_4^k}(\widetilde{v}_3) B^{d,T}_{\nu_1^l,\nu_2^l,\nu_3^l,\nu_4^l}(\widetilde{v}_4),$$

for $i,j,k,l \geq 0$ and $i+j+k+l=d$.

The smoothness conditions considered above can also be used in cases where some of the B-coefficients of p and $\widetilde{p}$ are known and others are still undetermined.

The following lemma is a generalization of Lemma 2.1 in [6] and makes use of bivariate smoothness conditions, in order to compute B-coefficients of two bivariate polynomials p and $\widetilde{p}$. The generalization is based on the usage of the two parameters l and $\widetilde{l}$ for the specification of the unknown B-coefficients, instead of just the parameter l. Since the proof of Lemma 2.28 works analogue to the one of Lemma 2.1 in [6], it is omitted here.

Lemma 2.28:
Let $F := \langle v_1, v_2, v_3 \rangle, \widetilde{F} := \langle v_1, v_2, v_4 \rangle$, and p and $\widetilde{p}$ two bivariate polynomials of degree d defined on F and $\widetilde{F}$, respectively. Suppose that all B-coefficients $c^F_{i,j,k}$ of p and $c^{\widetilde{F}}_{i,j,k}$ of $\widetilde{p}$ are known except for

$$c_\nu := c^F_{m-\nu,d-m,\nu}, \quad \nu = l+1,\ldots,q, \tag{2.5}$$
$$\widetilde{c}_{\widetilde{\nu}} := c^{\widetilde{F}}_{m-\widetilde{\nu},d-m,\widetilde{\nu}}, \quad \widetilde{\nu} = \widetilde{l}+1,\ldots,\widetilde{q}, \tag{2.6}$$

for some $l,\widetilde{l},q,\widetilde{q}$ with $0 \leq q,\widetilde{q}$, $-1 \leq l \leq q$, $-1 \leq \widetilde{l} \leq \widetilde{q}$, and $q+\widetilde{q}-\max\{l,\widetilde{l}\} \leq m \leq d$. Then these B-coefficients are uniquely and stably determined by the smoothness conditions

$$\begin{aligned} c^{\widetilde{F}}_{m-n,d-m,n} &= \sum_{i+j+k=n} c_{i+m-n,j+d-m,i} B^{n,F}_{i,j,k}, \\ \min\{l,\widetilde{l}\}+1 &\le n \le q+\widetilde{q}-\max\{l,\widetilde{l}\}. \end{aligned} \tag{2.7}$$

Next, we give some examples for Lemma 2.28.

Example 2.29:
Let $d=6$, $m=3$, $l=1$, $\widetilde{l}=0$, and $q=\widetilde{q}=2$. Then all B-coefficients of p and $\widetilde{p}$ are already known, except for $c^{F}_{1,3,2}, c^{\widetilde{F}}_{1,3,2}$, and $c^{\widetilde{F}}_{2,3,1}$, which can be uniquely and stably determined from C^1, C^2, and C^3 smoothness conditions. The already determined B-coefficients are associated with the domain points indicated by ■ and ○, and the undetermined ones correspond to domain points marked with ✚ in Fig. 2.10. The C^1, C^2, and C^3 smoothness conditions connect all B-coefficients corresponding to domain points marked with ✚ and ■ in Figure 2.10. Then, following Lemma 2.28, the B-coefficients associated with the domain points indicated by ✚ in Figure 2.10 can be uniquely and stably determined by the C^1, C^2, and C^3 smoothness conditions.

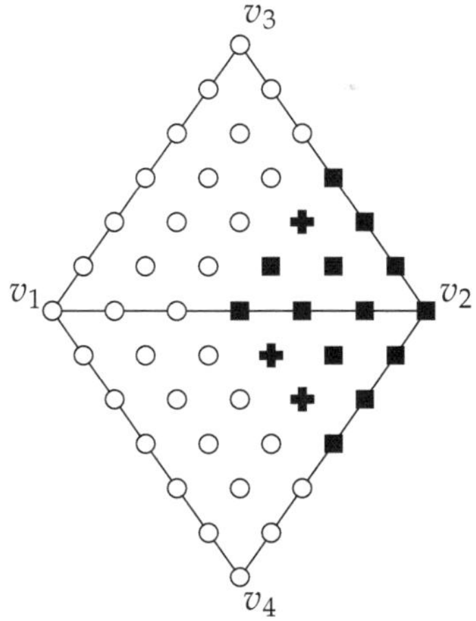

Figure 2.10: Domain points of the polynomials $p, \widetilde{p} \in \mathcal{P}^2_6$ defined on the triangles $F := \langle v_1, v_2, v_3 \rangle$ and $\widetilde{F} := \langle v_1, v_2, v_4 \rangle$, respectively.

Example 2.30:
Let $d=5$, $m=3$, $l=\widetilde{l}=-1, q=0$ and $\widetilde{q}=2$. Then all B-coefficients of p and $\widetilde{p}$ are already known, except for $c^F_{3,2,0}, c^{\widetilde{F}}_{3,2,0}, c^{\widetilde{F}}_{2,2,1}$, and $c^{\widetilde{F}}_{1,2,2}$, which can be uniquely and stably determined from C^0, C^1, C^2, and C^3 smoothness conditions. The already determined B-coefficients are associated with the domain points indicated by ■ and ○, and the undetermined ones correspond to domain points marked with ✚ in Fig. 2.11. From the C^0 smoothness conditions we get that $c^F_{3,2,0}$ and $c^{\widetilde{F}}_{3,2,0}$ are equal. Now, the C^1, C^2, and C^3 smoothness conditions connect all B-coefficients corresponding to domain points marked with ✚ and ■ in Figure 2.11. Then, following Lemma 2.28, the B-coefficients associated with the domain points indicated by ✚ in Figure 2.11 can be uniquely and stably determined by the C^1, C^2, and C^3 smoothness conditions.

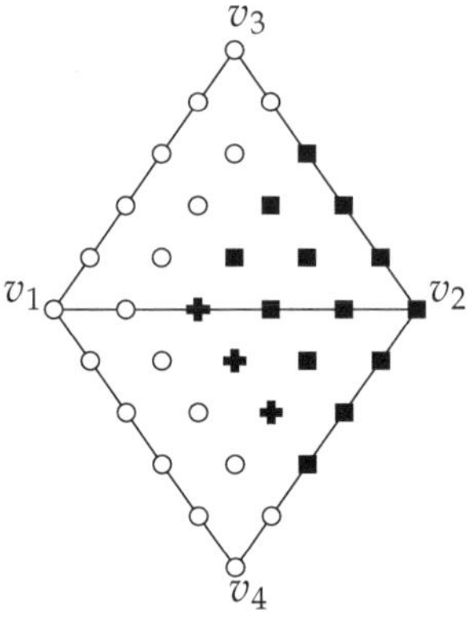

Figure 2.11: Domain points of the polynomials $p, \widetilde{p} \in \mathcal{P}^2_5$ defined on the triangles $F := \langle v_1, v_2, v_3 \rangle$ and $\widetilde{F} := \langle v_1, v_2, v_4 \rangle$, respectively.

Example 2.31:
Let $d=4$, $m=2$, $l=\widetilde{l}=0$, and $q=\widetilde{q}=1$. Then all B-coefficients of p and $\widetilde{p}$ are already known, except for $c^F_{1,2,1}$ and $c^{\widetilde{F}}_{1,2,1}$, which can be uniquely and stably determined from C^1 and C^2 smoothness conditions. The already determined B-coefficients are associated with the domain points indicated by ■ and ○, and the undetermined ones correspond to domain points marked with ✚ in Fig. 2.12. Now, the C^1 and C^2 smoothness conditions connect

all B-coefficients corresponding to domain points marked with ✚ and ■ in Figure 2.12. Then, following Lemma 2.28, the B-coefficients associated with the domain points indicated by ✚ in Figure 2.12 can be uniquely and stably determined by the C^1 and C^2 smoothness conditions.

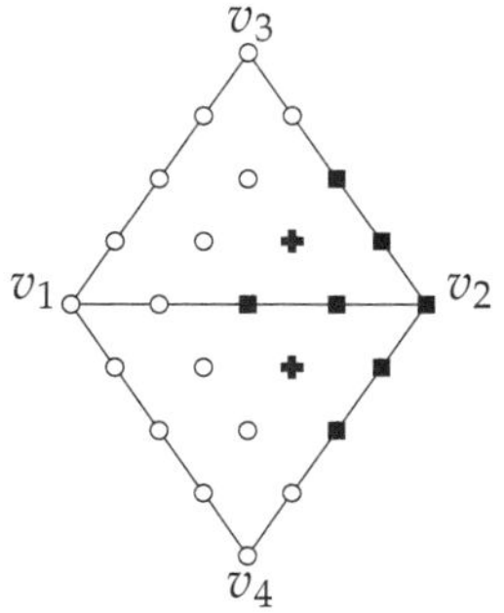

Figure 2.12: Domain points of the polynomials $p, \widetilde{p} \in \mathcal{P}_4^2$ defined on the triangles $F := \langle v_1, v_2, v_3 \rangle$ and $\widetilde{F} := \langle v_1, v_2, v_4 \rangle$, respectively.

Example 2.32:
Let $d = 4$, $m = 3$, $l = \widetilde{l} = -1$, and $q = \widetilde{q} = 1$. Then all B-coefficients of p and $\widetilde{p}$ are already known, except for $c^F_{3,1,0}, c^F_{2,1,1}, c^{\widetilde{F}}_{3,1,0}$ and $c^{\widetilde{F}}_{2,1,1}$, which can be uniquely and stably determined from C^0, C^1, C^2, and C^3 smoothness conditions. The already determined B-coefficients are associated with the domain points indicated by ■ and ○, and the undetermined ones correspond to domain points marked with ✚ in Fig. 2.13. By the C^0 smoothness conditions the B-coefficients $c^F_{3,1,0}$ and $c^{\widetilde{F}}_{3,1,0}$ have to be equal. Then, the C^1, C^2, and C^3 smoothness conditions connect all B-coefficients corresponding to domain points marked with ✚ and ■ in Figure 2.13. Thus, following Lemma 2.28, the B-coefficients associated with the domain points indicated by ✚ in Figure 2.13 can be uniquely and stably determined by the C^1 and C^2 smoothness conditions. Thus, following Lemma 2.28, the B-coefficients associated with the domain points indicated by ✚ in Figure 2.13 can be uniquely and stably determined by the C^1 and C^2 smoothness conditions.

Another example, for $d = 10$, $m = 8$, $l = \widetilde{l} = -1, q = 2$, and $\widetilde{q} = 3$ can be found in [6].

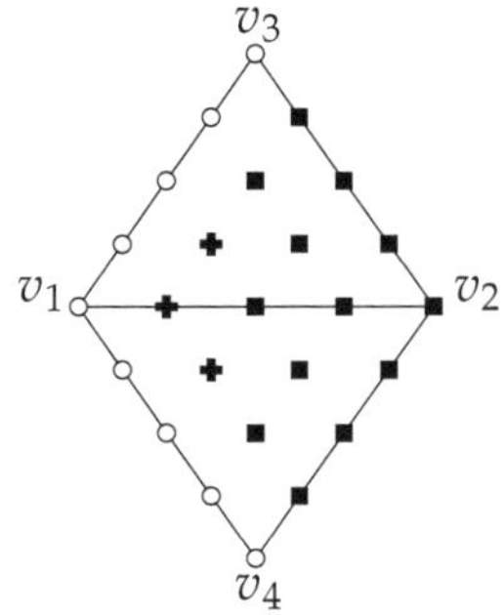

Figure 2.13: Domain points of the polynomials $p, \widetilde{p} \in \mathcal{P}_4^2$ defined on the triangles $F := \langle v_1, v_2, v_3 \rangle$ and $\widetilde{F} := \langle v_1, v_2, v_4 \rangle$, respectively.

The following lemma is a generalization of Lemma 1 in [54] and [66]. It is concerned with the computation of B-coefficients of two trivariate polynomials p and $\widetilde{p}$, where some B-coefficients of both polynomials are already known. The generalization is based on the usage of the two parameters m and $\widetilde{m}$ for the specification of the unknown B-coefficients, instead of just $m + l$.

Lemma 2.33:
Let $T := \langle v_1, v_2, v_3, v_4 \rangle$, $\widetilde{T} := \langle v_1, v_2, v_3, v_5 \rangle$, and p and $\widetilde{p}$ two polynomials of degree d. Suppose that all B-coefficients of p and $\widetilde{p}$ are already known except for the B-coefficients

$$c_i := c^T_{m+\widetilde{m}-i+1,d-m-\widetilde{m}-n-1,n,i}, \; i = j, \ldots, m + j, \tag{2.8}$$

$$\widetilde{c}_i := c^{\widetilde{T}}_{m+\widetilde{m}-i+1,d-m-\widetilde{m}-n-1,n,i}, \; i = j, \ldots, \widetilde{m} + j, \tag{2.9}$$

for some $m, \widetilde{m} \geq 0$, $0 \leq n \leq d - m - \widetilde{m} - j - 1$ and $j = 0,1$. Then these B-coefficients are uniquely and stably determined by the smoothness conditions

$$c^{\widetilde{T}}_{m+\widetilde{m}-i+1,d-m-\widetilde{m}-n-1,n,i} = \sum_{\substack{\alpha+\beta+\gamma+\delta=i \\ \alpha,\beta,\gamma,\delta \geq 0}} c^T_{m+\widetilde{m}-i+1+\alpha,d-m-\widetilde{m}-n-1+\beta,n+\gamma,\delta} B^{i,T}_{\alpha,\beta,\gamma,\delta}(v_5), \tag{2.10}$$

$i = j, \ldots, m + \widetilde{m} + j + 1$.

Proof:
To make the proof easier to understand, we only show the statement for $j = 0$. The proof is analog for the case $j = 1$.

Let $c := (c_0, \ldots, c_m, \widetilde{c}_0, \ldots, \widetilde{c}_{\widetilde{m}})^t$. Then the equations in (2.10) can be written in the form

$$Mc = b,$$

with

$$M := \begin{pmatrix} A & -I \\ B & O \end{pmatrix},$$

where I is the $(\widetilde{m}+1) \times (\widetilde{m}+1)$ identity matrix and O is the $(m+1) \times (\widetilde{m}+1)$ zero matrix. The $(m+1) \times (m+1)$ matrix A and the $(\widetilde{m}+1) \times (m+1)$ matrix B are defined by

$$A_{ij} := \binom{i-1}{j-1} \Phi_1^{i-j}(v_5) \Phi_4^{j-1}(v_5), \quad i,j = 1, \ldots, m+1,$$

and

$$B_{ij} := \binom{\widetilde{m}+i}{j-1} \Phi_1^{\widetilde{m}+1+i-j}(v_5) \Phi_4^{j-1}(v_5), \quad \begin{matrix} i = 1, \ldots, \widetilde{m}+1, \\ j = 1, \ldots, m+1, \end{matrix}$$

where Φ_ν, $\nu = 1, \ldots, 4$, are the barycentric coordinates associated with T. The vector b is given by

$$b_i = \begin{cases} -a_i & i = 1, \ldots, \widetilde{m}+1, \\ c^{\widetilde{T}}_{m+\widetilde{m}-i+2,d-m-\widetilde{m}-n-1,n,i-1} - a_i & i = \widetilde{m}+2, \ldots, m+\widetilde{m}+2, \end{cases}$$

with

$$a_i := \sum_{\substack{\alpha+\beta+\gamma+\delta=(i-1) \\ \alpha,\beta,\gamma,\delta \geq 0}}' c^T_{m+\widetilde{m}-i+2+\alpha,d-m-\widetilde{m}-n-1+\beta,n+\gamma,\delta} B^{i-1,T}_{\alpha,\beta,\gamma,\delta}(v_5).$$

The prime on the sum means that the sum is taken over all α, β, γ and δ such that $c^T_{m+\widetilde{m}-i+2+\alpha,d-m-\widetilde{m}-n-1+\beta,n+\gamma,\delta}$ is not one of the B-coefficients from (2.8).

Because of the block structure of M, it suffices to consider B to show that M is nonsingular. Therefore, let $\widetilde{B}$ be the matrix obtained by factoring $\frac{\Phi_4(v_5)^{j-1}}{(j-1)!}$ from the j-th column of B. Considering the functions $\{x^{\widetilde{m}+i}\}_{i=1}^{\widetilde{m}+1}$ and the linear functionals $\{\epsilon_{\Phi_4(v_5)}D^{j-1}\}_{j=1}^{m+1}$, it can be seen that the matrix $\widetilde{B}$ is the corresponding Gram matrix.

Suppose $\det(\widetilde{B}) = 0$. Then there would exist a nontrivial polynomial $g := \sum_{i=1}^{m+1} \alpha_i x^{\widetilde{m}+i}$ satisfying $D^{j-1}g(\Phi_4(v_5)) = 0$, $j = 1,\ldots,m+1$. Thus, g would be a nontrivial polynomial of degree $m+\widetilde{m}+1$ which vanishes $\widetilde{m}+1$ times at 0 and $m+1$ times at $\Phi_4(v_5)$. Since this is impossible, $\det(\widetilde{B}) \neq 0$, and thus the matrix M is nonsingular. □

In the following we give three examples for Lemma 2.33, where the first two are taken from [54]:

Example 2.34:
Let $d = 5$, $m = \widetilde{m} = 1$, and $j = 0$. Then the B-coefficients associated with the domain points marked with ✚ and ✖ in Fig. 2.14 can be uniquely and stably determined from C^0, C^1, C^2, and C^3 smoothness conditions. The B-coefficients of p and $\widetilde{p}$ associated with the domain points indicated by ■, ▲, and ○ in Fig. 2.14 are already known. Let us now consider the B-coefficients of p and $\widetilde{p}$ corresponding to the domain points marked with ✚. In each figure, the common edge of the two shown triangles is in fact just the common face $F := \langle v_1, v_2, v_3\rangle$ of the two tetrahedra T and $\widetilde{T}$, which becomes an edge when restricted to $R_i(v_1)$, $i = 2,3,4,5$. There are two undetermined B-coefficients associated with the domain point marked with ✚ in Fig. 2.14 (right). From the C^0 smoothness condition at F, we know that these two B-coefficients must be equal. Now, we consider the remaining trivariate smoothness conditions at F associated with this B-coefficient. The C^1 smoothness condition connects all B-coefficients corresponding to domain points in Fig. 2.14 (mid right and right) marked with ✚ and ▲. The C^2 smoothness condition connects all B-coefficients corresponding to domain points in Fig. 2.14 (mid left, mid right, and right) marked with ✚ and ▲. The C^3 smoothness condition connects all B-coefficients corresponding to domain points in Fig. 2.14 marked with ✚ and ▲. Thus, we are left with three unknown B-coefficients, whose domain points are marked with ✚ in Fig. 2.14 (mid right and right), and three smoothness conditions. By Lemma 2.33, these undetermined B-coefficients can be uniquely and stably determined by the C^1, C^2, and C^3 smoothness conditions. In the same way,

the B-coefficients associated with the domain points marked with ✖ can be uniquely determined.

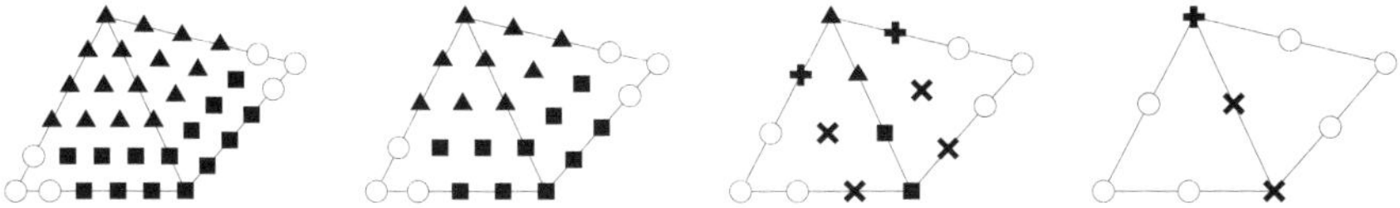

Figure 2.14: Domain points in the layers $R_5(v_1)$ (left), $R_4(v_1)$ (mid left), $R_3(v_1)$ (mid right), and $R_2(v_1)$ (right).

Example 2.35:
Let $d = 5$, $m = \widetilde{m} = 1$, and $j = 1$. Then the B-coefficients associated with the domain points marked with ✚ and ✖ in Fig. 2.15 can be uniquely and stably determined from C^1, C^2, C^3, and C^4 smoothness conditions. The B-coefficients of p and $\widetilde{p}$ associated with the domain points indicated by ■, ▲, and ○ in Fig. 2.15 are already determined. Let us now consider the B-coefficients of p and $\widetilde{p}$ corresponding to the domain points marked with ✚. In each figure, the common edge of the two shown triangles is in fact just the common face $F := \langle v_1, v_2, v_3 \rangle$ of the two tetrahedra T and $\widetilde{T}$, which becomes an edge when restricted to $R_i(v_1)$, $i = 1,2,3,4,5$. Thus, we have trivariate smoothness conditions here. There are two undetermined B-coefficients associated with the domain points marked with ✚ in Fig. 2.15 (mid right). The C^1 smoothness condition connects these B-coefficients with the B-coefficient associated with the domain point indicated by ▲ in Fig. 2.15 (right). The C^2 smoothness condition connects all B-coefficients corresponding to domain points in Fig. 2.15 (middle, mid right, and right) marked with ✚ and ▲. The C^3 smoothness condition connects all B-coefficients corresponding to domain points in Fig. 2.15 (mid left, middle, mid right, and right) marked with ✚ and ▲. The C^4 smoothness condition connects all B-coefficients corresponding to domain points in Fig. 2.15 marked with ✚ and ▲. Thus, we are left with four unknown B-coefficients, whose domain points are marked with ✚ in Fig. 2.15 (middle and mid right), and four smoothness conditions. By Lemma 2.33, these undetermined B-coefficients can be uniquely and stably determined by the C^1, C^2, C^3, and C^4 smoothness conditions. In the same way, the B-coefficients associated with the domain points marked with ✖ can

be uniquely determined.

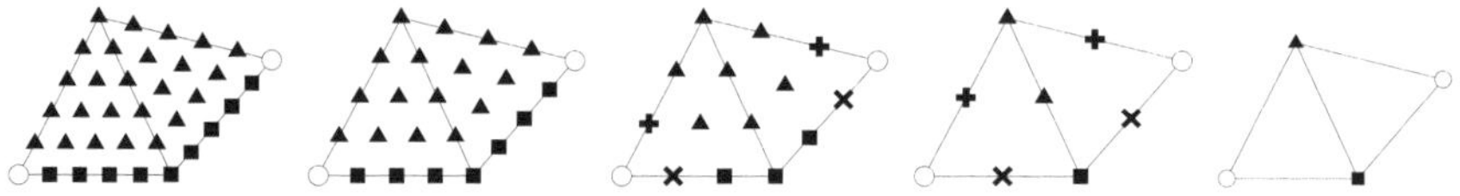

Figure 2.15: Domain points in the layers $R_5(v_1)$ (left), $R_4(v_1)$ (mid left), $R_3(v_1)$ (middle), $R_2(v_1)$ (mid right), and $R_1(v_1)$ (right).

Example 2.36:
Let $d = 5$, $m = 1$, $\widetilde{m} = 2$, and $j = 0$. Then the B-coefficients associated with the domain points marked with ✚ and ✖ in Fig. 2.16 can be uniquely and stably determined from C^0, C^1, C^2, C^3, and C^4 smoothness conditions. The B-coefficients of p and $\widetilde{p}$ associated with the domain points indicated by ■, ▲, and ○ in Fig. 2.16 are already known. Let us now consider the B-coefficients of p and $\widetilde{p}$ corresponding to the domain points marked with ✚. In each figure, the common edge of the two shown triangles is in fact just the common face $F := \langle v_1, v_2, v_3 \rangle$ of the two tetrahedra T and $\widetilde{T}$, which becomes an edge when restricted to $R_i(v_1)$, $i = 1, \ldots, 5$. There are two undetermined B-coefficients associated with the domain point marked with ✚ in Fig. 2.16 (right). From the C^0 smoothness condition at F, we know that these two B-coefficients must be equal. Next, we examine the remaining trivariate smoothness conditions at F associated with this B-coefficient. The C^1 smoothness condition connects all B-coefficients corresponding to domain points in Fig. 2.16 (mid right and right) marked with ✚ and ▲. The C^2 smoothness condition connects all B-coefficients corresponding to domain points in Fig. 2.16 (middle, mid right, and right) marked with ✚ and ▲. The C^3 smoothness condition connects all B-coefficients corresponding to domain points in Fig. 2.16 (mid left, middle, mid right, and right) marked with ✚ and ▲. The C^4 smoothness condition connects all B-coefficients corresponding to domain points in Fig. 2.16 marked with ✚ and ▲. Thus, we are left with four unknown B-coefficients, whose domain points are marked with ✚ in Fig. 2.16 (middle, mid right, and right), and four smoothness conditions. By Lemma 2.33, these undetermined B-coefficients can be uniquely and stably determined by the C^1, C^2, C^3, and C^4 smoothness conditions. In the same way, the B-coefficients associated with the domain points marked with ✖ can be uniquely determined.

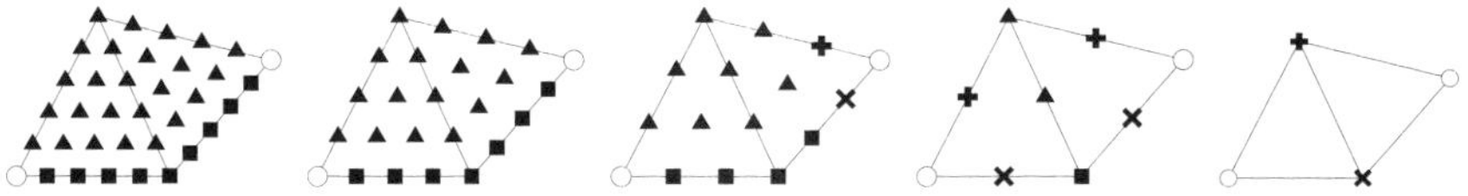

Figure 2.16: Domain points in the layers $R_5(v_1)$ (left), $R_4(v_1)$ (mid left), $R_3(v_1)$ (middle), $R_2(v_1)$ (mid right), and $R_1(v_1)$ (right).

2.3.5 Minimal determining sets

In this subsection we define determining and minimal determining sets of spline spaces. We also consider some properties of these sets.

Following Theorem 2.17 and Definition 2.19, each polynomial $p \in \mathcal{P}_d^n$ defined on an n-simplex T can be written in the B-form with B-coefficients $c^T_{i_1,\ldots,i_{n+1}}$. Since each spline is in fact just a polynomial when restricted to an n-simplex T, for a tessellation Δ of $\Omega \subset \mathbb{R}^n$ into n-simplices, each spline $s \in \mathcal{S}_d^0\Delta$ is uniquely determined by its set of B-coefficients $\{c_\xi\}_{\xi \in \mathcal{D}_{\Delta,d}}$.

Definition 2.37:
Let Δ be a tessellation of $\Omega \subset \mathbb{R}^n$ into n-simplices and let $\mathcal{D}_{\Delta,d}$ be the set of domain points relative to Δ. Moreover, let $\mathcal{S}_d^r(\Delta)$ be the space of C^r splines of degree d defined on Δ. Then, for all $\xi \in \mathcal{D}_{\Delta,d}$ let

$$\begin{aligned} \lambda_\xi : \mathcal{S}_d^r(\Delta) &\longrightarrow \mathbb{R} \\ s &\longrightarrow \lambda_\xi(s) = c_\xi. \end{aligned}$$

Then, the set $\mathcal{M} \subseteq \mathcal{D}_{\Delta,d}$ is called a **determining set** for $\mathcal{S}_d^r(\Delta)$ if

$$\lambda_\xi(s) = 0 \ \forall \xi \in \mathcal{M} \quad \Longrightarrow \quad \lambda_\xi(s) = 0 \ \forall \xi \in \mathcal{D}_{\Delta,d}$$

holds.

The set $\mathcal{M}$ is called a **minimal determining set** if $\mathcal{M}$ has the smallest cardinality of all possible determining sets of $\mathcal{S}_d^r(\Delta)$.

It is obvious that the set of domain points $\mathcal{D}_{\Delta,d}$ is always a determining set for a spline space $\mathcal{S}_d^r(\Delta)$. For any spline space $\mathcal{S}_d^r(\Delta)$ with $r > 0$, there are smaller determining sets than $\mathcal{D}_{\Delta,d}$. In general, there is more than one minimal determining set for a spline space $\mathcal{S}_d^r(\Delta)$.

Theorem 2.38:
For a spline space $\mathcal{S}_d^r(\Delta)$ and a determining set $\mathcal{M}$

$$\dim \mathcal{S}_d^r(\Delta) \leq \#\mathcal{M}$$

holds. If $\mathcal{M}$ is a minimal determining set for $\mathcal{S}_d^r(\Delta)$, then

$$\dim \mathcal{S}_d^r(\Delta) = \#\mathcal{M}.$$

Definition 2.39:
A minimal determining set $\mathcal{M}$ for a spline space $\mathcal{S}_d^r(\Delta)$ is called **local**, if for each n-simplex $T \in \Delta$ and for each domain point $\xi_{i_1,\ldots,i_{n+1}}^T \in T$ there exists a natural number m and a set $\Gamma_{\xi_{i_1,\ldots,i_{n+1}}^T} \subseteq (\mathcal{M} \cap \operatorname{star}^m(T))$, such that the B-coefficients $c_{i_1,\ldots,i_{n+1}}^T$ of a spline s in $\mathcal{S}_d^r(\Delta)$ only depend on the B-coefficients

$$\{c_\xi : \xi \in \Gamma_{\xi_{i_1,\ldots,i_{n+1}}^T}\}.$$

Then the minimal determining set $\mathcal{M}$ is also called m-**local**.

The minimal determining set is called **stable**, if there exists a constant C depending only on the smallest solid and face angles in Δ, such that for all $\xi \in \mathcal{D}_{\Delta,d}$

$$|c_\xi| \leq C \max_{\tilde{\xi} \in \Gamma_\xi} |c_{\tilde{\xi}}| \tag{2.11}$$

holds.

2.4 Interpolation with multivariate splines

In this section Lagrange and Hermite interpolation are considered, as well as their locality and stability. Moreover, some lemmata concerned with the solution of some interpolation problems are given. We also investigate the concept of nodal minimal determining sets and their locality and stability, as well as the approximation order of multivariate splines. The basis for these considerations can be found in the book on approximation theory by Powell [84], in the book on approximation theory by Nürnberger [69], and in the book on bivariate and trivariate splines by Lai and Schumaker [62].

2.4.1 Interpolation

Definition 2.40:
Let $\mathcal{L} := \{\kappa_1, \ldots, \kappa_n\}$ be a set of distinct points in a polyhedral domain Ω in $\mathbb{R}^3$, and let G be an n-dimensional subspace of $C(\Omega)$. Then $\mathcal{L}$ is called a **Lagrange interpolation set** for G, if for each function $f \in C(\Omega)$, a unique interpolant g_f exists, such that

$$g_f(\kappa_i) = f(\kappa_i), \quad i = 1, \ldots, n,$$

holds.

In case also partial derivatives of a sufficiently smooth function f are interpolated, a set $\mathcal{H}$ consisting of n such interpolation conditions, such that a unique interpolant $\widetilde{g}_f$ exists, is called **Hermite interpolation set**.

Definition 2.41:
An interpolation set P for a spline space $\mathcal{S}_d^r(\Delta)$ is called **local**, if for each tetrahedron $T \in \Delta$ and for each domain point $\xi_{i,j,k,l}^T \in T$ there exists a natural number m and a set $\Gamma_{\xi_{i,j,k,l}^T} \subseteq (P \cap \text{star}^m(T))$, such that the B-coefficients $c_{i,j,k,l}^T$ of a spline s in $\mathcal{S}_d^r(\Delta)$ only depend on the values

$$\{s(\kappa) : \kappa \in \Gamma_{\xi_{i,j,k,l}^T}\}.$$

The interpolation set is then also called m-**local**.

The interpolation set is called **stable**, if there exists a constant C depending only on the smallest angles in Δ, such that for all $\xi \in \mathcal{D}_{\Delta,d}$

$$|c_\xi| \leq C \max_{\kappa \in \Gamma_{\xi_{i,j,k,l}^T}} |s(\kappa)| \tag{2.12}$$

holds.

In the next lemma, we determine some B-coefficients corresponding to domain points in the middle of a triangle F or tetrahedron T by certain derivatives at a point in the interior of F or T, respectively. It is a generalization of Lemma 2 in [54] (see also Lemma 2 in [66]).

Lemma 2.42:
Let p be a polynomial of degree $d \geq |m| + 1$, $m := (m_1, \ldots, m_n)$, $m_i \geq -1$, $i = 1, \ldots, n$, and $|m| := m_1 + \ldots + m_n$, defined on a triangle F for $n = 3$, or

a tetrahedron T for $n = 4$, respectively. Suppose we already know the B-coefficients of p corresponding to the domain points within a distance of $m_1,\dots,m_n$ from the edges of F, or the faces of T, respectively. Let v_F be any point in the interior of F, or v_T any point in the interior of T. Then p is uniquely and stably determined by the values of

$$\{D^\alpha p(v_F)\}_{|\alpha|\le d-(|m|+n)} \tag{2.13}$$

with $\alpha = (\alpha_1,\dots,\alpha_{n-1})$.

Proof:
The cardinality of the set of undetermined B-coefficients of p is $\binom{d-|m|-1}{n-1}$, which is equal to the number of derivatives in (2.13). Thus, to show that the derivatives in (2.13) uniquely determine p, it suffices to show that $p \equiv 0$ if we set the B-coefficients of p corresponding to the domain points within a distance of $m_1,\dots,m_n$ from the edges of F, or the faces of T, respectively, and the values of the derivatives to zero.

Since the B-coefficients of p corresponding to the domain points within a distance of $m_1,\dots,m_n$ from the edges of F, or the faces of T, respectively, are zero, the derivatives normal to these edges, or faces, are also zero up to order $m_1,\dots,m_n$, respectively. Thus, the polynomial p can be written as $p = l_1^{m_1+1}\cdot\ldots\cdot l_n^{m_n+1}q$ where $q \in \mathcal{P}^{n-1}_{d-(|m|+n)}$ and the l_i, $i = 1,\dots,n$, are nontrivial linear polynomials which vanish on one edge of F, or one face of T, respectively. Now, $D^\alpha p(v_F) = 0$, or $D^\alpha p(v_T) = 0$, respectively, for $|\alpha| \le d-(|m|+n)$ implies $D^\alpha q(v_F) = 0$, or $D^\alpha q(v_T) = 0$, for $|\alpha| \le d-(|m|+n)$, which implies $q \equiv 0$. Therefore, we have $p \equiv 0$. □

We also need to solve certain bivariate interpolation problems where the undetermined B-coefficients are directly computed by data values. Concerning the solution of the corresponding linear system Larry L. Schumaker made the following conjecture (see [10] and [62]):

Conjecture 2.43:
Given d and a triangle $F := \langle v_1,v_2,v_3\rangle$, let Γ be an arbitrary subset of $\mathcal{D}_{F,d}$. Then the matrix

$$M := \left(B^{d,F}_\xi(\kappa)\right)_{\xi,\kappa\in\mathcal{D}_{F,d}} \tag{2.14}$$

is nonsingular. Thus, for any real numbers $\{z_\kappa\}_{\kappa\in\Gamma}$, there is a unique $p := \sum_{\xi\in\Gamma} c_\xi B^{d,F}_\xi$ such that $p(\kappa) = z_\kappa$ for all $\kappa \in \Gamma$.

It has already been shown that the conjecture holds for certain configurations of Γ:

Lemma 2.44 ([62], Lemma 2.25):
Let $\Gamma := \mathcal{D}_{F,d} \setminus \{\xi^F_{i,j,k} : i \geq m_1, j \geq m_2, k \geq m_3\}$ for some $m_1, m_2, m_3 \geq 0$ with $m := m_1 + m_2 + m_3 < d$. Then the matrix (2.14) is nonsingular.

We also need the following lemma, that is concerned with the determination of some B-coefficients of a cubic spline defined on a Clough-Tocher split triangle. It will be used in chapter 4.

Lemma 2.45 ([42]):
Suppose that we are given all of the B-coefficients of $s \in S^1_3(F_{CT})$ except for $c^{F_1}_{3,0,0}, c^{F_1}_{2,1,0}, c^{F_1}_{2,0,1}, c^{F_1}_{1,1,1}$. Then for any given real number z, there exists a unique choice of these B-coefficients so that $s(v_F) = z$.

2.4.2 Nodal minimal determining sets

In this subsection we describe nodal determining and nodal minimal determining sets of spline spaces. Moreover, we consider some attributes of these sets.

When considering splines, especially macro-elements, it is sometimes very useful to parametrize these in terms of nodal degrees of freedom. Thus, for a derivative D^α of degree $|\alpha|$, let

$$\lambda := \epsilon_u \sum_{|\alpha| \leq m} a^\alpha D^\alpha,$$

where ϵ_u is the point evaluation at u.

Definition 2.46:
Let Δ be a tessellation of $\Omega \subset \mathbb{R}^n$ into n-simplices and let $\mathcal{S}^r_d(\Delta)$ be the space of C^r splines of degree d defined on Δ. Moreover, let $\mathcal{N} := \{\lambda_i\}_{i=1}^k$, for some $k \geq 1$. Then $\mathcal{N}$ is called a **nodal determining set** for $\mathcal{S}^r_d(\Delta)$ provided that

$$\lambda s = 0 \; \forall \lambda \in \mathcal{N} \quad \Longrightarrow \quad s \equiv 0$$

holds. The set $\mathcal{N}$ is called a **nodal minimal determining set** if there exists no nodal determining set with a smaller cardinality.

Analogue to Theorem 2.38, for nodal determining sets the following Theorem holds:

Theorem 2.47:
For a spline space $\mathcal{S}_d^r(\Delta)$ and a nodal determining set $\mathcal{N}$

$$\dim \mathcal{S}_d^r(\Delta) \leq \#\mathcal{N}$$

holds. If $\mathcal{N}$ is a nodal minimal determining set for $\mathcal{S}_d^r(\Delta)$, then

$$\dim \mathcal{S}_d^r(\Delta) = \#\mathcal{N}.$$

Definition 2.48:
A nodal minimal determining set $\mathcal{N}$ for a spline space $\mathcal{S}_d^r(\Delta)$ is called **local**, provided that for each n-simplex $T \in \Delta$ and for each domain point $\xi^T_{i_1,\dots,i_{n+1}} \in T$ there exists a natural number m and a set $\Gamma_{\xi^T_{i_1,\dots,i_{n+1}}} \subseteq (\mathcal{N} \cap \text{star}^m(T))$, such that the B-coefficients $c^T_{i_1,\dots,i_{n+1}}$ of a spline s in $\mathcal{S}_d^r(\Delta)$ only depend on the nodal data in $\Gamma_{\xi^T_{i_1,\dots,i_{n+1}}}$. Then the nodal minimal determining set $\mathcal{N}$ is also called m-**local**.

The nodal minimal determining set is called **stable**, if there exists a constant C depending only on the smallest angles in Δ, such that for all $\xi \in \mathcal{D}_{\Delta,d}$

$$|c_\xi| \leq C \sum_{j=0}^{l} |T|^j |s|_{j,\text{star}^m(T)} \tag{2.15}$$

holds, where l is the order of the highest derivative involved in the nodal data in $\mathcal{N}$ and $|s|_{j,\text{star}^m(T)} := \max_{|\alpha|=j} \|D^\alpha f\|_{\text{star}^m(T)}$.

2.4.3 Approximation order of spline spaces

In this subsection we consider the approximation order of multivariate spline spaces.

When interpolating a sufficiently smooth function f at corresponding interpolation points, it is desirable that the interpolant s_f approximates the whole function f. Therefore, we consider the error of the interpolation $\|f - s_f\|_\Omega := \max_{z\in\Omega} |(f - s_f)(z)|$ in the infinity norm.

Definition 2.49:
Let Ω be a polyhedral domain in $\mathbb{R}^n$ and let Δ a tessellation of Ω into n-simplices. Then the space $\mathcal{S}_d^r$ has **approximation order** $k \in \mathbb{N}$ provided that

for each function $f \in C^{d+1}(\Omega)$ there exists a constant $K > 0$, only depending on f, d, and the smallest angles of Δ, such that

$$dist(f, \mathcal{S}_d^r(\Delta)) := \inf\{\|f - s\|_\Omega : s \in \mathcal{S}_d^r(\Delta)\} \leq K|\Delta|^k$$

holds, where $|\Delta|$ is the mesh size of Δ.

Following Ciarlet and Raviart [23], $k \leq d + 1$ holds, and thus the approximation order is **optimal** for $k = d + 1$. It was shown by de Boor and Jia [36] that not every interpolation method yields optimal approximation order. Important factors for a spline space to have optimal approximation order are the degree of polynomials, the degree of smoothness, the geometry of the tessellation, and the locality of the interpolation method (cf. [30, 31, 34, 35, 57]).

3 Minimal determining sets for splines on partial Worsey-Farin splits

In this chapter, we consider minimal determining sets for splines on partial Worsey-Farin splits of a tetrahedron. These are needed to construct the Lagrange interpolation methods described in chapter 4 and 5. In section 3.1 the minimal determining sets for C^1 splines on partial Worsey-Farin splits are considered. In the following section 3.2 the minimal determining sets for C^2 splines are constructed.

3.1 Minimal determining sets for cubic C^1 splines on partial Worsey-Farin splits

In this section, minimal determining sets for splines in $\mathcal{S}_3^1(T_{WF}^m)$ are considered. These are used in chapter 4, to construct a local Lagrange interpolation set on a tetrahedral partition, where some of the tetrahedra are refined with partial Worsey-Farin splits. They were first considered by Hecklin et al. [42] and were also used in [43, 72].

In order to proof the result on the minimal determining sets for the C^1 splines on partial Worsey-Farin splits shown in this section, the following Lemma and Corollary are needed.

Lemma 3.1 ([42]):
For all $0 \leq m \leq 4$, $\mathcal{S}_2^1(T_{WF}^m) \equiv \mathcal{P}_2^2$.

Corollary 3.2 ([42]):
Suppose $s \in \mathcal{S}_3^1(T_{WF}^m)$. Then $s \in C^2(v_T)$, i.e., all of the polynomial pieces on subtetrahedra sharing the vertex v_T have common derivatives at v_T up to order 2.

Now the result on the minimal determining sets for splines in $\mathcal{S}_3^1(T_{WF}^m)$ can be stated.

Theorem 3.3 (cf. [42]):
Fix $0 \le m \le 4$, let $T := \langle v_1, v_2, v_3, v_4 \rangle$ be a tetrahedron and T_{WF}^m the corresponding m-th order partial Worsey-Farin split. Let $\mathcal{M}_m^1$ be the union of the following sets of domain points in $\mathcal{D}_{T_{WF}^m,3}$:

1. for each $i = 1,\dots,4$, $D_1^{T_i}(v_i)$ for some tetrahedron $T_i \in T_{WF}^m$ containing the vertex v_i
2. for each face F of T that is not split, the point $\xi_{1,1,1}^F$
3. for each face F of T that has been subjected to a Clough-Tocher split, the points $\{\xi_{1,1,1}^{F_i}\}_{i=1}^3$, where F_1, F_2, F_3 are the subtriangles of F

Then $\mathcal{M}_m^1$ is a stable minimal determining set for $\mathcal{S}_3^1(T_{WF}^m)$.

Proof:
In order to show that $\mathcal{M}_m^1$ is a stable minimal determining set for the spline space $\mathcal{S}_3^1(T_{WF}^m)$, the B-coefficients c_ξ of a spline $s \in \mathcal{S}_3^1(T_{WF}^m)$ are set to arbitrary values for each domain point $\xi \in \mathcal{M}_m^1$ and it is shown that all the remaining B-coefficients of s are uniquely and stably determined.

From the B-coefficients of s associated with the domain points in $D_1^{T_i}(v_i)$, $i = 1,\dots,4$, all remaining B-coefficients of s corresponding to domain points in $D_1(v_i)$, $i = 1,\dots,4$, can be uniquely and stably determined using the C^1 smoothness conditions at the vertices of T.

Then, if F is a face of T that is not refined with a Clough-Tocher split, all B-coefficients of $s|_F$ are already uniquely and stably determined, since the domain point $\xi_{1,1,1}^F$ is contained in the minimal determining set $\mathcal{M}_m^1$.

In case F is a face of T that is refined with the Clough-Tocher split, setting the B-coefficients of s corresponding to the domain points $\{\xi_{1,1,1}^{F_i}\}_{i=1}^3$ uniquely determines the remaining B-coefficients associated with the domain points of the minimal determining set of the classical Clough-Tocher macro-element (see [6, 24, 59, 62]). Thus, also following Theorem 6.6 from chapter 6, the remaining B-coefficients of $s|_F$ can be uniquely and stably determined using smoothness conditions.

So far all B-coefficients of s associated with the domain points on the faces of T, those in $R_3(v_T)$, are uniquely and stably determined.

Then, using the C^1 smoothness conditions across the interior faces of T_{WF}^m, the B-coefficients of s corresponding to the domain points on the edges of the tetrahedron formed by the domain points in $D_2(v_T)$ can be uniquely and stably determined. Then, following Lemma 3.1 and Corollary 3.2, it can be seen that the remaining undetermined B-coefficients of s can be uniquely and stably determined using smoothness conditions. □

3.2 Minimal determining sets for C^2 splines of degree nine on partial Worsey-Farin splits

In this section, we consider minimal determining sets for C^2 splines of degree nine on partial Worsey-Farin splits of a tetrahedron, which are applied in chapter 5, in order to extend a spline to some refined tetrahedra. To this end, we state several lemmata which can be used to extend a superspline s defined on a tetrahedral partition $\Delta \setminus T$ to a superspline $\widetilde{s}$ defined on Δ, where the tetrahedron T is refined with a partial Worsey-Farin split. Therefore, we have to require some additional smoothness conditions, such that $\widetilde{s}|_T$ is uniquely and stably determined by these conditions and the normal smoothness and supersmoothness conditions across the vertices, edges, faces and some interpolation points on vertices of T. Since the only cases that occur in chapter 5 are m-th order partial Worsey-Farin splits for $m = 1, \ldots, 4$, the case $m = 0$ is not examined here.

3.2.1 Minimal determining sets for C^2 supersplines based on first order partial Worsey-Farin splits

In this subsection minimal determining sets for C^2 splines on first order partial Worsey-Farin splits of a tetrahedron $T := \langle v_1, v_2, v_3, v_4 \rangle$ and corresponding lemmata are considered. Therefore, let T_{WF}^1 be the first order partial Worsey-Farin split of T at the split point v_T in the interior of T, and let the face $\langle v_1, v_2, v_3 \rangle$ be refined with a Clough-Tocher split at the split point v_F. The corresponding subtetrahedra are denoted by

$$\begin{aligned} T_{1,1} &:= \langle v_1, v_2, v_F, v_T \rangle, & T_{1,2} &:= \langle v_2, v_3, v_F, v_T \rangle, \\ T_{1,3} &:= \langle v_3, v_1, v_F, v_T \rangle, & T_2 &:= \langle v_2, v_3, v_4, v_T \rangle, \\ T_3 &:= \langle v_3, v_1, v_4, v_T \rangle, & T_4 &:= \langle v_1, v_2, v_4, v_T \rangle. \end{aligned} \tag{3.1}$$

Theorem 3.4:
Let $\mathcal{M}_1^2$ be the union of the following sets of domain points in $\mathcal{D}_{T_{WF}^1,9}$:

(M$1^2_{T^1_{WF}}$) $D_4^{T_{v_i}}(v_i)$, $i=1,\dots,4$, with $v_i \in T_{v_i} \in T^1_{WF}$

(M$2^2_{T^1_{WF}}$) $\bigcup_{i=1}^{3} E_3^{T_{\langle v_i,v_{i+1}\rangle}}(\langle v_i,v_{i+1}\rangle) \setminus (D_4^{T_{\langle v_i,v_{i+1}\rangle}}(v_i) \cup D_4^{T_{\langle v_i,v_{i+1}\rangle}}(v_{i+1}))$,
with $\langle v_i,v_{i+1}\rangle \in T_{\langle v_i,v_{i+1}\rangle}$, and $v_4 := v_1$

(M$3^2_{T^1_{WF}}$) $\{\xi^{T_{1,1}}_{3,0,4,2}, \xi^{T_{1,2}}_{3,0,4,2}, \xi^{T_{1,3}}_{3,0,4,2}\}$

(M$4^2_{T^1_{WF}}$) $\{\xi^{T_2}_{4-i,1+i,4,0}, \xi^{T_3}_{4-i,1+i,4,0}, \xi^{T_4}_{4-i,1+i,4,0}\}_{i=0}^{3}$

(M$5^2_{T^1_{WF}}$) $\{\xi^{T_2}_{4-i,0,3+i,2}, \xi^{T_3}_{4-i,0,3+i,2}, \xi^{T_4}_{4-i,0,3+i,2}\}_{i=0}^{1}$

(M$6^2_{T^1_{WF}}$) $\{\xi^{T_j}_{3,2,3,1}, \xi^{T_j}_{2,3,3,1}, \xi^{T_j}_{2,2,4,1}\}_{j=2}^{4}$

(M$7^2_{T^1_{WF}}$) $\{\xi^{T_j}_{2,2,3,2}, \xi^{T_j}_{3,2,2,2}, \xi^{T_j}_{2,3,2,2}\}_{j=2}^{4}$

(M$8^2_{T^1_{WF}}$) $\{\xi^{T_{1,i}}_{4,0,2,3}, \xi^{T_{1,i}}_{3,0,3,3}, \xi^{T_{1,i}}_{3,1,2,3}, \xi^{T_{1,i}}_{3,0,2,4}, \xi^{T_{1,i}}_{2,2,1,4}, \xi^{T_{1,i}}_{2,0,2,5}\}_{i=1}^{3} \cup \{\xi^{T_{1,1}}_{2,0,4,3}, \xi^{T_{1,1}}_{0,0,0,9}\}$

Then $\mathcal{M}_1^2$ is a stable minimal determining set for

$$\begin{aligned}\widetilde{\mathcal{S}}_9^2(T^1_{WF}) := \{s \in C^2(T^1_{WF}) : s|_{\widetilde{T}} \in \mathcal{P}_9^3,\ \forall \widetilde{T} \in T^1_{WF},\\ s \in C^4(v_i),\ i=1,\dots,4,\\ s \in C^3(e),\ \text{for all edges } e \in T,\\ s \in C^7(\langle v_F, v_T\rangle),\\ s \in C^7(v_T),\\ \tau^{T_{1,i},T_{1,i+1}}_{3,5,1,0} s = 0,\ i=1,2,3,\ \text{with } T_{1,4} := T_{1,1},\\ \tau^{T_{1,i},T_{1,i+1}}_{3,5,0,1} s = 0,\ i=1,2,3,\ \text{with } T_{1,4} := T_{1,1}\},\end{aligned}$$

and the dimension of $\widetilde{\mathcal{S}}_9^2(T^1_{WF})$ is equal to 259.

Proof:
First, we show that $\mathcal{M}_1^2$ is a stable minimal determining set for $\widetilde{\mathcal{S}}_9^2(T^1_{WF})$. Therefore, we set the B-coefficients c_ξ of a spline $s \in \widetilde{\mathcal{S}}_9^2(T^1_{WF})$ to arbitrary values for each domain point $\xi \in \mathcal{M}_1^2$. Then we show that all other B-coefficients of s are uniquely and stably determined.

First, the B-coefficients of s associated with the domain points in the balls $D_4(v_i)$, $i=1,\dots,4$, are considered. Using the C^4 supersmoothness conditions at the vertices of T, we can uniquely and stably determine all these

B-coefficients from those corresponding to the domain points in $(\mathrm{M1}^2_{T^1_{WF}})$.

The remaining undetermined B-coefficients of s associated with domain points in the tubes $E_3(\langle v_1,v_2\rangle), E_3(\langle v_2,v_3\rangle)$, and $E_3(\langle v_3,v_1\rangle)$ can be uniquely and stably determined from the B-coefficients corresponding to the domain points in $(\mathrm{M2}^2_{T^1_{WF}})$ using the C^3 supersmoothness conditions at the corresponding three edges of T.

Next, we consider the B-coefficients associated with the domain points in the Clough-Tocher split face $F := \langle v_1,v_2,v_3\rangle$. The three subtriangles forming this face are degenerated faces of the three subtetrahedra $T_{1,1}, T_{1,2}$, and $T_{1,3}$, since they lie in one common plane. Thus, we can regard the B-coefficients of $s|_F$ as those of a bivariate spline $g_0 \in \widetilde{\mathcal{S}}^2_9(F_{CT})$ defined in (A.20). Considering g_0, we can see that the B-coefficients corresponding to the domain points within a distance of four from the vertices of F and corresponding to those within a distance of three from the edges of F are already uniquely and stably determined. Thus, following Lemma A.13 and Remark A.16, all B-coefficients of g_0 are uniquely and stably determined, and therefore also all B-coefficients of $s|_F$.

Subsequently, we examine the B-coefficients of s associated with the domain points in the shells $R_8^{T_{1,1}}(v_T), R_8^{T_{1,2}}(v_T)$, and $R_8^{T_{1,3}}(v_T)$, which form a triangle F^1. Following the same arguments as above, we can consider these B-coefficients as those of a bivariate spline g_1 in the space $\widetilde{\mathcal{S}}^2_8(F^1_{CT})$ defined in (A.10). Investigating this spline, we see that the B-coefficients corresponding to the domain points within a distance of three from the vertices and within a distance of two from the edges of F^1 are already uniquely and stably determined. Then, following Lemma A.8 and Remark A.16, all other B-coefficients of g_1 are also uniquely and stably determined. Thus, we have also uniquely and stably determined all B-coefficients of $s|_{F^1}$.

Now, we consider the B-coefficients of s associated with the domain points in the shells $R_7^{T_{1,1}}(v_T), R_7^{T_{1,2}}(v_T)$, and $R_7^{T_{1,3}}(v_T)$. These domain points form the triangle F^2. Again, we can regard the B-coefficients of $s|_{F^2}$ as those of a bivariate spline $g_2 \in \widetilde{\mathcal{S}}^2_7(F^2_{CT})$ defined in (A.6). Investigating this bivariate spline, it can be seen that the B-coefficients corresponding to the domain points within a distance of two from the vertices and within a distance of one from the edges of F^2 are already uniquely and stably determined. Then, together with the B-coefficients associated with the domain points in $(\mathrm{M3}^2_{T^1_{WF}})$, all B-coefficients corresponding to the domain points in the minimal determining set of Lemma A.6 are uniquely and stably determined. Thus, all B-coefficients of g_2, and consequently also all B-coefficients of $s|_{F^2}$, are uniquely and stably determined.

Next, we consider the other three faces of T. So far, all B-coefficients corresponding to the domain points on these faces, except those associated with the domain points in (M4$^2_{T^1_{WF}}$), are uniquely and stably determined. Since these domain points are contained in the set $\mathcal{M}^2_1$, the spline s is uniquely and stably determined on all faces of T.

Subsequently, we consider the B-coefficients of s corresponding to the domain points within a distance of two from the three edges $\langle v_i, v_4\rangle$, $i = 1,2,3$. Together with the B-coefficients of s associated with the domain points in the set (M5$^2_{T^1_{WF}}$), we have already uniquely and stably determined all B-coefficients of s corresponding to the domain points on the interior faces of T^1_{WF} within a distance of two from these edges. Moreover, we have already uniquely and stably determined the B-coefficients of s restricted to the faces of T. Thus, the only undetermined B-coefficients of s associated with the domain points in $E_2(\langle v_i, v_4\rangle)$, $i = 1,2,3$, correspond to domain points in the shell $R_8(v_T)$ with a distance of one from the interiors faces of T^1_{WF} containing the edges $\langle v_i, v_4\rangle$, $i = 1,2,3$. Then, using the C^1 and C^2 smoothness conditions across these faces and following Lemma 2.33 for the case $j = 1$, we can uniquely and stably determine the last undetermined B-coefficients of $E_2(\langle v_i, v_4\rangle)$, $i = 1,2,3$.

Next, we consider the B-coefficients of s corresponding to the domain points with a distance of one from the three non-split faces of T. Let $\widetilde{F}^1_i$ be the triangle formed by the domain points in the shell $R^{T_i}_8(v_T)$, $i = 2,3,4$, respectively. We have already uniquely and stably determined the B-coefficients of $s|_{\widetilde{F}^1_i}$ corresponding to the domain points within a distance of one from the two edges of $\widetilde{F}^1_i$, that are formed by the interior faces of T^1_{WF} that contain the edges $\langle v_j, v_4\rangle$, $j = 1,2,3$ of T, for all $i = 2,3,4$. Moreover, we have also already uniquely and stably determined the B-coefficients of $s|_{\widetilde{F}^1_i}$ corresponding to the domain points within a distance of two from the third edge of $\widetilde{F}^1_i$, $i = 2,3,4$. Thus, for each triangle $\widetilde{F}^1_i$, $i = 2,3,4$, the only undetermined B-coefficients of s restricted to $\widetilde{F}^1_i$ are those corresponding to the domain points contained in (M6$^2_{T^1_{WF}}$).

Now, we examine the B-coefficients of s corresponding to the domain points within a distance of three from the edges $\langle v_i, v_4\rangle$, $i = 1,2,3$. Here, the only undetermined B-coefficients correspond to domain points in the shells $R_7(v_T)$ and $R_6(v_T)$ with a distance of one and on the interior faces of T^1_{WF} containing the edges $\langle v_i, v_4 \rangle$, $i = 1,2,3$, respectively. Thus, we can use the C^3 supersmoothness conditions at these edges in order to uniquely and stably determine the corresponding B-coefficients, following Lemma 2.33 for the case $j = 0$. Then, the only undetermined B-coefficients of s restricted to the shell $R_7(v_T)$ correspond to the domain points in (M$7^2_{T^1_{WF}}$). Thus, the B-coefficients of s associated with the domain points in $R_7(v_T)$ are uniquely and stably determined.

Next, we consider the B-coefficients of $s|_{F^3}$, where F^3 is the triangle formed by the domain points in the three shells $R_6^{T_{1,1}}(v_T)$, $R_6^{T_{1,2}}(v_T)$, and $R_6^{T_{1,3}}(v_T)$. Note that due to the partial Worsey-Farin split of T, the triangle F^3 has been refined with a Clough-Tocher split. Following the same arguments as above and due to the C^7 supersmoothness at the edge $\langle v_F, v_T \rangle$, we can regard the B-coefficients of $s|_{F^3}$ as the B-coefficients of a bivariate polynomial g_3 defined on F^3. Considering g_3 we see that we have already uniquely and stably determined the B-coefficients corresponding to the domain points within a distance of one from the vertices of F^3 and those corresponding to the domain points on the edges of F^3. Now, together with the B-coefficients associated with the domain points $\{\xi^{T_{1,i}}_{4,0,2,3}, \xi^{T_{1,i}}_{3,0,3,3}, \xi^{T_{1,i}}_{3,1,2,3}\}_{i=1}^{3}$ and the domain point $\xi^{T_{1,1}}_{2,0,4,3}$ in (M$8^2_{T^1_{WF}}$), we have uniquely and stably determined all B-coefficients corresponding to the domain points in the minimal determining set of Lemma A.5 of a bivariate polynomial in $\mathcal{P}^2_6$ defined on F^3_{CT}. Thus, following Lemma A.5, the remaining undetermined B-coefficients of g_3 can be uniquely and stably determined.

Then, using the C^2 smoothness conditions at the interior faces $\langle v_1, v_2, v_T \rangle$, $\langle v_2, v_3, v_T \rangle$, and $\langle v_3, v_1, v_T \rangle$ of T^1_{WF}, we can uniquely and stably compute the remaining undetermined B-coefficients of s corresponding to the domain points in the tubes $E_4(\langle v_1, v_2 \rangle)$, $E_4(\langle v_2, v_3 \rangle)$, and $E_4(\langle v_3, v_1 \rangle)$, following Lemma 2.33 for $j = 0$.

Next, we consider the B-coefficients of s restricted to the triangle F^4, which is formed by the domain points in the shells $R_5^{T_{1,1}}(v_T)$, $R_5^{T_{1,2}}(v_T)$, and $R_5^{T_{1,3}}(v_T)$. Note that the triangle F^4 is also subjected to the Clough-Tocher split. Again, due to the C^7 supersmoothness at the edge $\langle v_F, v_T \rangle$, we can consider the B-coefficients of $s|_{F^4}$ as those of a bivariate polynomial g_4 defined on F^4. Investigating g_4, we can see that the B-coefficients corresponding to the domain points on the edges of F^4 are already uniquely and stably determined. Then, using the smoothness conditions at the vertices of F^4 we can uniquely and stably determine the B-coefficients of g_4 corresponding to the domain points within a distance of one from the vertices of F^4. Now, together with the B-coefficients associated with the domain points $\{\xi^{T_{1,i}}_{3,0,2,4}, \xi^{T_{1,i}}_{2,2,1,4}\}_{i=1}^3$ in (M8$^2_{T^1_{WF}}$), we have uniquely and stably determined all B-coefficients corresponding to the domain points in the minimal determining set from Lemma A.2 of a bivariate polynomial in $\mathcal{P}^2_5$ defined on F^4_{CT}. Thus, the remaining undetermined B-coefficients of g_4 are uniquely and stably determined.

Now, following Lemma 2.33 for $j = 0$, we can uniquely and stably compute the remaining undetermined B-coefficients of s corresponding to domain points in the three tubes $E_5(\langle v_1, v_2\rangle)$, $E_5(\langle v_2, v_3\rangle)$, and $E_5(\langle v_3, v_1\rangle)$ using the C^1, C^2, and C^3 smoothness conditions at the interior faces $\langle v_1, v_2, v_T\rangle$, $\langle v_2, v_3, v_T\rangle$, and $\langle v_3, v_1, v_T\rangle$ of T^1_{WF}.

Then, we examine the B-coefficients of $s|_{F^5}$, where the triangle F^5, which is subjected to a Clough-Tocher split, is formed by the domain points in the shells $R^{T_{1,1}}_4(v_T)$, $R^{T_{1,2}}_4(v_T)$, and $R^{T_{1,3}}_4(v_T)$. Since the spline s has C^7 supersmoothness at the edge $\langle v_F, v_T\rangle$, we can consider the B-coefficients of $s|_{F^5}$ as those of a bivariate polynomial g_5 defined on F^5. Considering the polynomial g_5, it can be seen that the B-coefficients corresponding to the domain points on the edges of F^5 are already uniquely and stably determined. Then, using the C^1 smoothness conditions at the vertices of F^5, we can uniquely and stably compute the remaining undetermined B-coefficients of g_5 corresponding to the domain points with a distance of one from these vertices. At this point, together with the B-coefficients associated with the domain points $\{\xi^{T_{1,i}}_{2,0,2,5}\}_{i=1}^3$ in (M8$^2_{T^1_{WF}}$), we have uniquely and stably determined all B-coefficients corresponding to the domain points in the minimal determining set from Lemma A.1 of a bivariate polynomial in $\mathcal{P}^2_4$ defined on F^5_{CT}. Hence, we can uniquely and stably compute the remaining undetermined B-coefficients of $s|_{F^5}$ in the way described in the proof of Lemma A.1.

Now, we consider the last undetermined B-coefficients of s. These are associated with domain points in the ball $D_7(v_T)$. Due to the C^7 supersmoothness of s at the split point v_T, we can consider the B-coefficients corresponding to the domain points in $D_7(v_T)$ as those of a trivariate polynomial g defined on the tetrahedron $\widetilde{T}$, which is formed by the domain points in $D_7(v_T)$. Note that $\widetilde{T}$ is also subjected to a first order partial Worsey-Farin split. Examining g, we see that the B-coefficients associated with the domain points on the non-split faces of $\widetilde{T}$ and the B-coefficients corresponding to the domain points within a distance of three from the face that has been refined with the Clough-Tocher split are already uniquely and stably determined. Then, setting the B-coefficient of g associated with the domain point $\xi^{T_{1,1}}_{0,0,0,9}$ is equal to setting the function value of g at v_T. Thus, following Lemma 2.42, the remaining undetermined B-coefficients of g can be uniquely and stably determined from smoothness conditions. Therefore, we have uniquely and stably determined all B-coefficients of s.

Now, since $\mathcal{M}_1^2$ is a minimal determining set for $\widetilde{\mathcal{S}}_9^2(T^1_{WF})$, it follows that $\dim \widetilde{\mathcal{S}}_9^2(T^1_{WF}) = \#\mathcal{M}_1^2 = 259$. □

Lemma 3.5:
Let Δ be a tetrahedral partition of a polyhedral domain $\Omega \subset \mathbb{R}^3$ and $T := \langle v_1, v_2, v_3, v_4 \rangle \in \Delta$ a tetrahedron, that shares exactly the face $\langle v_1, v_2, v_3 \rangle$ with the tetrahedron $\widetilde{T}$ and the opposite vertex v_4 with the tetrahedron $\hat{T}$, with $\widetilde{T}, \hat{T} \in \Delta$. Moreover, T and $\widetilde{T}$ are refined with a first order partial Worsey-Farin split at the split points v_T and $v_{\widetilde{T}}$ in the interior of T and $\widetilde{T}$, respectively, and the split point v_F in the interior of the face $\langle v_1, v_2, v_3 \rangle$. Then let $\widetilde{T}_i$ be a subtetrahedron of $\widetilde{T}^1_{WF}$ that contains the edge $\langle v_i, v_{i+1} \rangle$, for $i = 2, 3, 4$, where $v_4 := v_1$ and $v_5 := v_2$.

Let $\widetilde{\mathcal{S}}_9^2(T^1_{WF})$ be the superspline space defined in Theorem 3.4. Then, for any spline $s \in \mathcal{S}_9^{2,4,3}(\Delta \setminus T)$, there is a spline $g \in \widetilde{\mathcal{S}}_9^2(T^1_{WF})$ such that

$$\widetilde{s} := \begin{cases} s, & \text{on } \Delta \setminus T \\ g, & \text{on } T, \end{cases}$$

belongs to $\mathcal{S}_9^{2,4,3}(\Delta)$. Moreover, the spline g is uniquely and stably determined by the value of $g(v_T)$ and the B-coefficients of $\widetilde{s}|_{\Delta \setminus T}$ by the following additional smoothness conditions across the edges of T:

$$\tau^{\widetilde{T}_j, T_j}_{4-i,1+i,4,0} \widetilde{s} = 0, \; i = 0, \dots, 3, \; j = 2,3,4 \tag{3.2}$$

$$\tau^{\widetilde{T}_j, T_j}_{4-i,0,3+i,2} \widetilde{s} = 0, \; i = 0,1, \; j = 2,3,4 \tag{3.3}$$

$$\tau^{\widetilde{T}_j, T_j}_{3,2,3,1} \widetilde{s} = \tau^{\widetilde{T}_j, T_j}_{2,3,3,1} \widetilde{s} = \tau^{\widetilde{T}_j, T_j}_{2,2,4,1} \widetilde{s} = 0, \; j = 2,3,4 \tag{3.4}$$

$$\tau^{\widetilde{T}_j, T_j}_{2,2,3,2} \widetilde{s} = \tau^{\widetilde{T}_j, T_j}_{3,2,2,2} \widetilde{s} = \tau^{\widetilde{T}_j, T_j}_{2,3,2,2} \widetilde{s} = 0, \; j = 2,3,4 \tag{3.5}$$

$$\begin{aligned}
&\tau_{4,0,2,3}^{\widetilde{T}_{3+j},T_{1,j}}\widetilde{s} = \tau_{3,0,3,3}^{\widetilde{T}_{3+j},T_{1,j}}\widetilde{s} = \tau_{3,1,2,3}^{\widetilde{T}_{3+j},T_{1,j}}\widetilde{s} = \\
&\tau_{3,0,2,4}^{\widetilde{T}_{3+j},T_{1,j}}\widetilde{s} = \tau_{2,2,1,4}^{\widetilde{T}_{3+j},T_{1,j}}\widetilde{s} = \tau_{2,0,2,5}^{\widetilde{T}_{3+j},T_{1,j}}\widetilde{s} = 0, \qquad (3.6)\\
&j = 1,2,3, \text{ with } \widetilde{T}_5 := \widetilde{T}_2 \text{ and } \widetilde{T}_6 := \widetilde{T}_3
\end{aligned}$$

$$\tau_{2,0,4,3}^{\widetilde{T}_4,T_{1,1}}\widetilde{s} = 0 \qquad (3.7)$$

Proof:
Using the C^2 smoothness conditions across the face $\langle v_1, v_2, v_3\rangle$, the C^4 supersmoothness conditions at the vertices of T, and the C^3 supersmoothness conditions at the edges of T, we can uniquely and stably determine the B-coefficients of g corresponding to the domain points in $(M1^2_{T^1_{WF}})$, $(M2^2_{T^1_{WF}})$, and $(M3^2_{T^1_{WF}})$ from Theorem 3.4.

From the additional smoothness conditions in (3.2), (3.2), (3.4), and (3.5), we can uniquely and stably determine the B-coefficients associated with the domain points in $(M4^2_{T^1_{WF}})$, $(M5^2_{T^1_{WF}})$, $(M6^2_{T^1_{WF}})$, and $(M7^2_{T^1_{WF}})$, respectively. Then, from the additional smoothness conditions in (3.6) and (3.7), we can uniquely and stably determine the B-coefficients of g corresponding to the domain points in $(M8^2_{T^1_{WF}})$, except for the one associated with $\xi^{T_{1,1}}_{0,0,0,9}$. Finally, this B-coefficient is equal to the value of $g(v_T)$. Thus, following Theorem 3.4, g is uniquely and stably determined. □

Lemma 3.6:
Let Δ be a tetrahedral partition of a polyhedral domain $\Omega \subset \mathbb{R}^3$ and $T := \langle v_1, v_2, v_3, v_4\rangle \in \Delta$ a tetrahedron, that shares exactly the face $\langle v_1, v_2, v_3\rangle$ with the tetrahedron $\widetilde{T}$, with $\widetilde{T} \in \Delta$. Moreover, T and $\widetilde{T}$ are refined with a first order partial Worsey-Farin split at the split points v_T and $v_{\widetilde{T}}$ in the interior of T and $\widetilde{T}$, respectively, and the split point v_F in the interior of the face $\langle v_1, v_2, v_3\rangle$. Then let $\widetilde{T}_i$ be a subtetrahedron of $\widetilde{T}^1_{WF}$ that contains the edge $\langle v_i, v_{i+1}\rangle$, for $i = 2,3,4$, where $v_4 := v_1$ and $v_5 := v_2$.

Let $\widetilde{\mathcal{S}}^2_9(T^1_{WF})$ be the superspline space defined in Theorem 3.4. Then, for any spline $s \in \mathcal{S}^{2,4,3}_9(\Delta \setminus T)$, there is a spline $g \in \widetilde{\mathcal{S}}^2_9(T^1_{WF})$ such that

$$\widetilde{s} := \begin{cases} s, & \text{on } \Delta \setminus T \\ g, & \text{on } T, \end{cases}$$

belongs to $\mathcal{S}_9^{2,4,3}(\Delta)$. Moreover, g is uniquely and stably determined by the values of $g(v_T)$ and $g(v_4)$, and the B-coefficients of $\widetilde{s}|_{\Delta\setminus T}$ by the following additional smoothness conditions across the edges of T:

$$\tau_{i,j,k,l}^{\widetilde{T}_2,T_2}\widetilde{s} = 0,\ 5 \le k \le 8,\ i+j+l = 9-k \tag{3.8}$$

$$\tau_{4-i,1+i,4,0}^{\widetilde{T}_j,T_j}\widetilde{s} = 0,\ i = 0,\dots,3,\ j = 2,3,4 \tag{3.9}$$

$$\tau_{4-i,0,3+i,2}^{\widetilde{T}_j,T_j}\widetilde{s} = 0,\ i = 0,1,\ j = 2,3,4 \tag{3.10}$$

$$\tau_{3,2,3,1}^{\widetilde{T}_j,T_j}\widetilde{s} = \tau_{2,3,3,1}^{\widetilde{T}_j,T_j}\widetilde{s} = \tau_{2,2,4,1}^{\widetilde{T}_j,T_j}\widetilde{s} = 0,\ j = 2,3,4 \tag{3.11}$$

$$\tau_{2,2,3,2}^{\widetilde{T}_j,T_j}\widetilde{s} = \tau_{3,2,2,2}^{\widetilde{T}_j,T_j}\widetilde{s} = \tau_{2,3,2,2}^{\widetilde{T}_j,T_j}\widetilde{s} = 0,\ j = 2,3,4 \tag{3.12}$$

$$\begin{aligned} &\tau_{4,0,2,3}^{\widetilde{T}_{3+j},T_{1,j}}\widetilde{s} = \tau_{3,0,3,3}^{\widetilde{T}_{3+j},T_{1,j}}\widetilde{s} = \tau_{3,1,2,3}^{\widetilde{T}_{3+j},T_{1,j}}\widetilde{s} = \\ &\tau_{3,0,2,4}^{\widetilde{T}_{3+j},T_{1,j}}\widetilde{s} = \tau_{2,2,1,4}^{\widetilde{T}_{3+j},T_{1,j}}\widetilde{s} = \tau_{2,0,2,5}^{\widetilde{T}_{3+j},T_{1,j}}\widetilde{s} = 0, \\ &j = 1,2,3,\ \text{with } \widetilde{T}_5 := \widetilde{T}_2 \text{ and } \widetilde{T}_6 := \widetilde{T}_3 \end{aligned} \tag{3.13}$$

$$\tau_{2,0,4,3}^{\widetilde{T}_4,T_{1,1}}\widetilde{s} = 0 \tag{3.14}$$

Proof:
The B-coefficient of g associated with the domain point $\xi_{0,0,9,0}^{T_4}$ is equal to the value of $g(v_4)$. Then the remaining B-coefficients corresponding to domain points in the disk $D_4^{T_4}(v_4)$ can be uniquely and stably computed using the C^4 supersmoothness conditions at v_4 and the additional smoothness conditions in (3.8).

The remaining B-coefficients corresponding to the domain points in the minimal determining set $\mathcal{M}_1^2$ from Theorem 3.4 can be determined in the same way as in the proof of Lemma 3.5 from the C^2 smoothness conditions across the face $\langle v_1, v_2, v_3\rangle$, the C^4 supersmoothness conditions at the remaining vertices of T, the C^3 supersmoothness conditions at the edges of T, the additional smoothness conditions in (3.9) - (3.14) and the value of $g(v_T)$. Therefore, all B-coefficients of g are uniquely and stably determined. □

Lemma 3.7:
Let Δ be a tetrahedral partition of a polyhedral domain $\Omega \subset \mathbb{R}^3$ and $T := \langle v_1, v_2, v_3, v_4 \rangle \in \Delta$ a tetrahedron, that shares exactly the three edges $\langle v_1, v_2 \rangle$, $\langle v_2, v_3 \rangle$, and $\langle v_3, v_1 \rangle$ with the tetrahedra $\widetilde{T}_4, \widetilde{T}_2$, and $\widetilde{T}_3$, respectively, and the opposite vertex v_4 with the tetrahedron $\hat{T}$, with $\widetilde{T}_j, \hat{T} \in \Delta$, $j = 2,3,4$. Moreover, T is refined with a first order partial Worsey-Farin split at the split point v_T in the interior of T, and the split point v_F in the interior of the face $\langle v_1, v_2, v_3 \rangle$.

Let $\widetilde{\mathcal{S}}_9^2(T_{WF}^1)$ be the superspline space defined in Theorem 3.4. Then, for any spline $s \in \mathcal{S}_9^{2,4,3}(\Delta \setminus T)$, there is a spline $g \in \widetilde{\mathcal{S}}_9^2(T_{WF}^1)$ such that

$$\widetilde{s} := \begin{cases} s, & \text{on } \Delta \setminus T \\ g, & \text{on } T, \end{cases}$$

belongs to $\mathcal{S}_9^{2,4,3}(\Delta)$. Moreover, g is uniquely and stably determined by the value of $g(v_T)$ and the B-coefficients of $\widetilde{s}|_{\Delta \setminus T}$ by the following additional smoothness conditions across the edges of T:

$$\tau_{3,0,4,2}^{\widetilde{T}_{3+j}, T_{1,j}} \widetilde{s} = 0, \ j = 1,2,3 \tag{3.15}$$

$$\tau_{4-i,1+i,4,0}^{\widetilde{T}_j, T_j} \widetilde{s} = 0, \ i = 0,\dots,3, \ j = 2,3,4 \tag{3.16}$$

$$\tau_{4-i,0,3+i,2}^{\widetilde{T}_j, T_j} \widetilde{s} = 0, \ i = 0,1, \ j = 2,3,4 \tag{3.17}$$

$$\tau_{3,2,3,1}^{\widetilde{T}_j, T_j} \widetilde{s} = \tau_{2,3,3,1}^{\widetilde{T}_j, T_j} \widetilde{s} = \tau_{2,2,4,1}^{\widetilde{T}_j, T_j} \widetilde{s} = 0, \ j = 2,3,4 \tag{3.18}$$

$$\tau_{2,2,3,2}^{\widetilde{T}_j, T_j} \widetilde{s} = \tau_{3,2,2,2}^{\widetilde{T}_j, T_j} \widetilde{s} = \tau_{2,3,2,2}^{\widetilde{T}_j, T_j} \widetilde{s} = 0, \ j = 2,3,4 \tag{3.19}$$

$$\begin{aligned} &\tau_{4,0,2,3}^{\widetilde{T}_{3+j}, T_{1,j}} \widetilde{s} = \tau_{3,0,3,3}^{\widetilde{T}_{3+j}, T_{1,j}} \widetilde{s} = \tau_{3,1,2,3}^{\widetilde{T}_{3+j}, T_{1,j}} \widetilde{s} = \\ &\tau_{3,0,2,4}^{\widetilde{T}_{3+j}, T_{1,j}} \widetilde{s} = \tau_{2,2,1,4}^{\widetilde{T}_{3+j}, T_{1,j}} \widetilde{s} = \tau_{2,0,2,5}^{\widetilde{T}_{3+j}, T_{1,j}} \widetilde{s} = 0, \\ &j = 1,2,3, \text{ with } \widetilde{T}_5 := \widetilde{T}_2 \text{ and } \widetilde{T}_6 := \widetilde{T}_3 \end{aligned} \tag{3.20}$$

$$\tau_{2,0,4,3}^{\widetilde{T}_4, T_{1,1}} \widetilde{s} = 0 \tag{3.21}$$

Proof:
Using the additional smoothness conditions in (3.15), the B-coefficients of g associated with the domain points in $(\mathrm{M3}^2_{T_{WF}^1})$ from Theorem 3.4 can be uniquely and stably determined.

The remaining B-coefficients associated with the domain points in the minimal determining set $\mathcal{M}_1^2$ from Theorem 3.4 can be determined in the same way as in the proof of Lemma 3.5 from the C^4 supersmoothness conditions at the vertices of T, the C^3 supersmoothness conditions at the edges of T, the additional smoothness conditions in (3.16) - (3.21) and the value of $g(v_T)$. Therefore, all B-coefficients of g are uniquely and stably determined. □

Lemma 3.8:
Let Δ be a tetrahedral partition of a polyhedral domain $\Omega \subset \mathbb{R}^3$ and $T := \langle v_1, v_2, v_3, v_4 \rangle \in \Delta$ a tetrahedron, that shares exactly the three edges $\langle v_1, v_2 \rangle$, $\langle v_2, v_3 \rangle$, and $\langle v_3, v_1 \rangle$ with the tetrahedra $\widetilde{T}_4$, $\widetilde{T}_2$, and $\widetilde{T}_3$, respectively, with $\widetilde{T}_j \in \Delta$, $j = 2,3,4$. Moreover, T is refined with a first order partial Worsey-Farin split at the split point v_T in the interior of T, and the split point v_F in the interior of the face $\langle v_1, v_2, v_3 \rangle$.

Let $\widetilde{\mathcal{S}}_9^2(T_{WF}^1)$ be the superspline space defined in Theorem 3.4. Then, for any spline $s \in \mathcal{S}_9^{2,4,3}(\Delta \setminus T)$, there is a spline $g \in \widetilde{\mathcal{S}}_9^2(T_{WF}^1)$ such that

$$\widetilde{s} := \begin{cases} s, & \text{on } \Delta \setminus T \\ g, & \text{on } T, \end{cases}$$

belongs to $\mathcal{S}_9^{2,4,3}(\Delta)$. Moreover, g is uniquely and stably determined by the values of $g(v_T)$ and $g(v_4)$, and the B-coefficients of $\widetilde{s}|_{\Delta \setminus T}$ by the following additional smoothness conditions across the edges of T:

$$\tau_{i,j,k,l}^{\widetilde{T}_2,T_2} \widetilde{s} = 0,\ 5 \le k \le 8,\ i+j+l = 9-k \tag{3.22}$$

$$\tau_{3,0,4,2}^{\widetilde{T}_{3+j},T_{1,j}} \widetilde{s} = 0,\ j = 1,2,3 \tag{3.23}$$

$$\tau_{4-i,1+i,4,0}^{\widetilde{T}_j,T_j} \widetilde{s} = 0,\ i = 0,\dots,3,\ j = 2,3,4 \tag{3.24}$$

$$\tau_{4-i,0,3+i,2}^{\widetilde{T}_j,T_j} \widetilde{s} = 0,\ i = 0,1,\ j = 2,3,4 \tag{3.25}$$

$$\tau_{3,2,3,1}^{\widetilde{T}_j,T_j} \widetilde{s} = \tau_{2,3,3,1}^{\widetilde{T}_j,T_j} \widetilde{s} = \tau_{2,2,4,1}^{\widetilde{T}_j,T_j} \widetilde{s} = 0,\ j = 2,3,4 \tag{3.26}$$

$$\tau_{2,2,3,2}^{\widetilde{T}_j,T_j} \widetilde{s} = \tau_{3,2,2,2}^{\widetilde{T}_j,T_j} \widetilde{s} = \tau_{2,3,2,2}^{\widetilde{T}_j,T_j} \widetilde{s} = 0,\ j = 2,3,4 \tag{3.27}$$

$$\begin{aligned} &\tau_{4,0,2,3}^{\widetilde{T}_{3+j},T_{1,j}} \widetilde{s} = \tau_{3,0,3,3}^{\widetilde{T}_{3+j},T_{1,j}} \widetilde{s} = \tau_{3,1,2,3}^{\widetilde{T}_{3+j},T_{1,j}} \widetilde{s} = \tau_{3,0,2,4}^{\widetilde{T}_{3+j},T_{1,j}} \widetilde{s} = \tau_{2,2,1,4}^{\widetilde{T}_{3+j},T_{1,j}} \widetilde{s} = \\ &\tau_{2,0,2,5}^{\widetilde{T}_{3+j},T_{1,j}} \widetilde{s} = 0,\ j = 1,2,3, \text{ with } \widetilde{T}_5 := \widetilde{T}_2 \text{ and } \widetilde{T}_6 := \widetilde{T}_3 \end{aligned} \tag{3.28}$$

$$\tau_{2,0,4,3}^{\widetilde{T}_4,T_{1,1}}\widetilde{s} = 0 \tag{3.29}$$

Proof:
The B-coefficient of g associated with the domain points in the disk $D_4^{T_4}(v_4)$ can be uniquely and stably computed in the same way as in the proof of Lemma 3.6 using the additional smoothness conditions (3.22) and the value of $g(v_4)$.

The remaining B-coefficients corresponding to the domain points in the minimal determining set $\mathcal{M}_1^2$ from Theorem 3.4 can be determined in the same way as in the proof of Lemma 3.5 and 3.7 from the C^4 supersmoothness conditions at the vertices of T, the C^3 supersmoothness conditions at the edges of T, the additional smoothness conditions in (3.23) - (3.29) and the value of $g(v_T)$. Thus, all B-coefficients of g are uniquely and stably determined. □

Theorem 3.9:
Let $\hat{\mathcal{M}}_1^2$ be the union of the following sets of domain points in $\mathcal{D}_{T_{WF}^1,9}$:

($\hat{\mathrm{M}}1^2_{T^1_{WF}}$) $D_4^{T_{v_i}}(v_i)$, $i = 1,\ldots,4$, with $v_i \in T_{v_i} \in T^1_{WF}$

($\hat{\mathrm{M}}2^2_{T^1_{WF}}$) $\bigcup_{i=1}^{2} E_3^{T_{\langle v_i,v_{i+1}\rangle}}(\langle v_i,v_{i+1}\rangle) \setminus (D_4^{T_{\langle v_i,v_{i+1}\rangle}}(v_i) \cup D_4^{T_{\langle v_i,v_{i+1}\rangle}}(v_{i+1}))$,
with $\langle v_i,v_{i+1}\rangle \in T_{\langle v_i,v_{i+1}\rangle}$

($\hat{\mathrm{M}}3^2_{T^1_{WF}}$) $\{\xi_{4,1,4,0}^{T_{1,1}}, \xi_{1,4,4,0}^{T_{1,2}}, \xi_{0,2,7,0}^{T_{1,1}}, \xi_{0,0,9,0}^{T_{1,1}}\}$

($\hat{\mathrm{M}}4^2_{T^1_{WF}}$) $\{\xi_{4,1,3,1}^{T_{1,1}}, \xi_{1,4,3,1}^{T_{1,2}}, \xi_{0,2,6,1}^{T_{1,1}}, \xi_{0,0,8,1}^{T_{1,1}}\}$

($\hat{\mathrm{M}}5^2_{T^1_{WF}}$) $\{\xi_{4,i,3-i,2}^{T_{1,1}}, \xi_{3,2,2,2}^{T_{1,1}}, \xi_{i,4,3-i,2}^{T_{1,2}}, \xi_{2,3,2,2}^{T_{1,2}}, \xi_{0,3,4,2}^{T_{1,1}}, \xi_{0,1,6,2}^{T_{1,1}}\}_{i=0}^{1}$

($\hat{\mathrm{M}}6^2_{T^1_{WF}}$) $\{\xi_{4-i,1+i,4,0}^{T_2}, \xi_{4-i,1+i,4,0}^{T_3}, \xi_{4-i,1+i,4,0}^{T_4}, \xi_{4-j,2+j,3,0}^{T_3}\}_{i,j=0}^{i=3,j=2}$

($\hat{\mathrm{M}}7^2_{T^1_{WF}}$) $\{\xi_{4-i,0,3+i,2}^{T_2}, \xi_{4-i,0,3+i,2}^{T_3}, \xi_{4-i,0,3+i,2}^{T_4}\}_{i=0}^{1}$

($\hat{\mathrm{M}}8^2_{T^1_{WF}}$) $\{\xi_{3,2,3,1}^{T_j}, \xi_{2,3,3,1}^{T_j}, \xi_{2,2,4,1}^{T_j}\}_{j=2}^{4}$

($\hat{\mathrm{M}}9^2_{T^1_{WF}}$) $\{\xi_{2,2,3,2}^{T_j}, \xi_{3,2,2,2}^{T_j}, \xi_{2,3,2,2}^{T_j}\}_{j=2}^{4}$

($\hat{\mathrm{M}}10^2_{T^1_{WF}}$) $\{\xi_{4,0,2,3}^{T_{1,i}}, \xi_{3,0,3,3}^{T_{1,i}}, \xi_{3,1,2,3}^{T_{1,i}}, \xi_{3,0,2,4}^{T_{1,i}}, \xi_{2,2,1,4}^{T_{1,i}}, \xi_{2,0,2,5}^{T_{1,i}}\}_{i=1}^{3} \cup \{\xi_{2,0,4,3}^{T_{1,1}}, \xi_{0,0,0,9}^{T_{1,1}}\}$

Then $\hat{\mathcal{M}}_1^2$ is a stable minimal determining set for

$$\begin{aligned}\hat{\mathcal{S}}_9^2(T_{WF}^1) := \{s \in C^2(T_{WF}^1) : s|_{\widetilde{T}} \in \mathcal{P}_9^3,\ \forall \widetilde{T} \in T_{WF}^1,\\ s \in C^4(v_i),\ i = 1,\dots,4,\\ s \in C^3(e), \text{ for all edges } e \in T,\\ s \in C^7(\langle v_F, v_T\rangle),\\ s \in C^7(v_T),\\ \tau_{3-i,5+i,1,0}^{T_{1,1},T_{1,2}} s = \tau_{4,4,1,0}^{T_{1,2},T_{1,3}} s = 0,\ i = 0,\dots,3,\\ \tau_{3-i,5+i,0,1}^{T_{1,1},T_{1,2}} s = \tau_{4,4,0,1}^{T_{1,2},T_{1,3}} s = 0,\ i = 0,\dots,3\},\end{aligned}$$

and the dimension of $\hat{\mathcal{S}}_9^2(T_{WF}^1)$ is equal to 255.

Proof:
First, we show that $\hat{\mathcal{M}}_1^2$ is a stable minimal determining set for $\hat{\mathcal{S}}_9^2(T_{WF}^1)$. To this end, we set the B-coefficients c_ξ of a spline $s \in \hat{\mathcal{S}}_9^2(T_{WF}^1)$ to arbitrary values for each domain point $\xi \in \hat{\mathcal{M}}_1^2$ and show that all other B-coefficients of s are uniquely and stably determined.

We first consider the B-coefficients of s corresponding to the domain points in the balls $D_4(v_i)$, $i = 1,\dots,4$. We can uniquely and stably determine all these B-coefficients from those corresponding to the domain points in ($\hat{M}1^2_{T_{WF}^1}$) using the C^4 supersmoothness conditions at the vertices of T.

Next, we uniquely and stably compute the remaining undetermined B-coefficients of s corresponding to domain points in the tubes $E_3(\langle v_1, v2\rangle)$ and $E_3(\langle v_2, v3\rangle)$ from the B-coefficients associated with the domain points in ($\hat{M}2^2_{T_{WF}^1}$) using the C^3 supersmoothness conditions at the corresponding two edges of T.

Now, we consider the B-coefficients associated with the domain points in the Clough-Tocher split face $F := \langle v_1, v_2, v_3\rangle$. Since they lie in one common plane, the three subtriangles forming this face are degenerated faces of the three subtetrahedra $T_{1,1}, T_{1,2}$, and $T_{1,3}$. Thus, we can regard the B-coefficients of $s|_F$ as those of a bivariate spline $g_0 \in \hat{\mathcal{S}}_9^2(F_{CT})$ defined in (A.22). Considering g_0, we see that the B-coefficients corresponding to the domain points within a distance of four from the vertices of F and corresponding to those within a distance of three from the two edges $\langle v_1, v_2\rangle$ and $\langle v_2, v_3\rangle$ of F are already uniquely and stably determined. Together with the

B-coefficients corresponding to the domain points in ($\hat{\text{M}}3^2_{T^1_{WF}}$), all B-coefficients associated with the domain points in the minimal determining set in Lemma A.14 are uniquely and stably determined. Thus all B-coefficients of g_0 can be uniquely and stably determined, and therefore also all B-coefficients of $s|_F$.

Next, we examine the B-coefficients of s corresponding to the domain points in the shells $R_8^{T_{1,1}}(v_T), R_8^{T_{1,2}}(v_T)$, and $R_8^{T_{1,3}}(v_T)$, which form a triangle F^1. Following the same arguments as above, these B-coefficients can also be regarded as those of a bivariate spline g_1 in the space $\hat{\mathcal{S}}_8^2(F^1_{CT})$ defined in (A.12). Considering this spline, we see that the B-coefficients corresponding to the domain points within a distance of three from the vertices and within a distance of two from the edges of F^1 lying in the shells $R_8^{T_{1,1}}(v_T)$ and $R_8^{T_{1,2}}(v_T)$ are already uniquely and stably determined. Together with the B-coefficients associated with the domain points in ($\hat{\text{M}}4^2_{T^1_{WF}}$), we have uniquely and stably determined all B-coefficients corresponding to the domain points in the minimal determining set in Lemma A.9. Thus, following this lemma, all B-coefficients of g_1 are uniquely and stably determined and thus also the B-coefficients of $s|_{F^1}$.

Subsequently, we investigate the B-coefficients of s associated with the domain points in the shells $R_7^{T_{1,1}}(v_T), R_7^{T_{1,2}}(v_T)$, and $R_7^{T_{1,3}}(v_T)$. These domain points form the triangle F^2. Again, the B-coefficients of $s|_{F^2}$ can be regarded as those of a bivariate spline $g_2 \in \hat{\mathcal{S}}_7^2(F^2_{CT})$ defined in (A.8). Considering this bivariate spline, it can be seen that the B-coefficients corresponding to the domain points within a distance of two from the vertices and within a distance of one from the two edges of F^2 that lie in the shells $R_7^{T_{1,1}}(v_T)$ and $R_7^{T_{1,2}}(v_T)$ are already uniquely and stably determined. Then, together with the B-coefficients associated with the domain points in (M$3^2_{T^1_{WF}}$), all B-coefficients corresponding to the domain points in the minimal determining set of Lemma A.7 are uniquely and stably determined. Therefore, all B-coefficients of g_2, and consequently all B-coefficients of $s|_{F^2}$, are uniquely and stably determined.

Now, we can use the C^2 smoothness conditions at the edge $\langle v_1, v_3\rangle$, in order to uniquely and stably compute the remaining undetermined B-coefficients of s corresponding to the domain points in $E_2(\langle v_1, v_3\rangle)$ from those already determined.

Next, we consider the other faces of T. At this point, all B-coefficients corresponding to the domain points on these faces, except those associated with the domain points in ($\hat{M}6^2_{T^1_{WF}}$), are uniquely and stably determined. Thus, the spline s is uniquely and stably determined on all faces of T.

Now, we can use Lemma 2.33 for the case $j = 0$, in order to uniquely and stably determine remaining B-coefficients of s corresponding to the domain points in $E_3(\langle v_1, v_3 \rangle)$ from those already determined using the C^3 supersmoothness conditions at $\langle v_1, v_3 \rangle$.

Then, the remaining undetermined B-coefficients of s can be computed in the same way as in the proof of Theorem 3.4.

Considering the dimension of $\hat{\mathcal{S}}^2_9(T^1_{WF})$, we can see that $\dim \widetilde{\mathcal{S}}^2_9(T^1_{WF}) = 255$, since $\hat{\mathcal{M}}^2_1$ is a minimal determining set for $\hat{\mathcal{S}}^2_9(T^1_{WF})$ and $\#\hat{\mathcal{M}}^2_1 = 255$.□

Lemma 3.10:
Let Δ be a tetrahedral partition of a polyhedral domain $\Omega \subset \mathbb{R}^3$ and $T := \langle v_1, v_2, v_3, v_4 \rangle \in \Delta$ a tetrahedron, that shares exactly the two edges $\langle v_1, v_2 \rangle$ and $\langle v_2, v_3 \rangle$ with the tetrahedra $\widetilde{T}_4$ and $\widetilde{T}_2$, respectively, and the opposite vertex v_4 with the tetrahedron $\hat{T}$, with $\widetilde{T}_2, \widetilde{T}_4, \hat{T} \in \Delta$. Moreover, T is refined with a first order partial Worsey-Farin split at the split point v_T in the interior of T, and the split point v_F in the interior of the face $\langle v_1, v_2, v_3 \rangle$.

Let $\hat{\mathcal{S}}^2_9(T^1_{WF})$ be the superspline space defined in Theorem 3.9. Then, for any spline $s \in \mathcal{S}^{2,4,3}_9(\Delta \setminus T)$, there is a spline $g \in \hat{\mathcal{S}}^2_9(T^1_{WF})$ such that

$$\widetilde{s} := \begin{cases} s, & \text{on } \Delta \setminus T \\ g, & \text{on } T, \end{cases}$$

belongs to $\mathcal{S}^{2,4,3}_9(\Delta)$. Moreover, g is uniquely and stably determined by the value of $g(v_T)$ and the B-coefficients of $\widetilde{s}|_{\Delta \setminus T}$ by the following additional smoothness conditions across the vertices and edges of T^1_{WF}:

$$\tau^{\widetilde{T}_4, T_{1,1}}_{4,1,4,0} \widetilde{s} = \tau^{\widetilde{T}_4, T_{1,1}}_{0,2,7,0} \widetilde{s} = \tau^{\widetilde{T}_4, T_{1,1}}_{0,0,9,0} \widetilde{s} = \tau^{\widetilde{T}_2, T_{1,2}}_{1,4,4,0} \widetilde{s} = 0 \tag{3.30}$$

$$\tau^{\widetilde{T}_4, T_{1,1}}_{4,1,3,1} \widetilde{s} = \tau^{\widetilde{T}_4, T_{1,1}}_{0,2,6,1} \widetilde{s} = \tau^{\widetilde{T}_4, T_{1,1}}_{0,0,8,1} \widetilde{s} = \tau^{\widetilde{T}_2, T_{1,2}}_{1,4,3,1} \widetilde{s} = 0 \tag{3.31}$$

$$\tau^{\widetilde{T}_4, T_{1,1}}_{4,i,3-i,2} \widetilde{s} = \tau^{\widetilde{T}_4, T_{1,1}}_{3,2,2,2} \widetilde{s} = \tau^{\widetilde{T}_4, T_{1,1}}_{0,3,4,2} \widetilde{s} = \tau^{\widetilde{T}_2, T_{1,2}}_{0,1,6,2} \widetilde{s} = \tau^{\widetilde{T}_2, T_{1,2}}_{i,4,3-i,2} \widetilde{s} = \tau^{\widetilde{T}_2, T_{1,2}}_{2,3,2,2} \widetilde{s} = 0 \tag{3.32}$$

$$\tau^{\widetilde{T}_j, T_j}_{4-i,1+i,4,0} \widetilde{s} = 0, \; i = 0, \ldots, 3, \; j = 2,4 \tag{3.33}$$

$$\tau^{\widetilde{T}_2, T_3}_{4-i,1+i,4,0} \widetilde{s} = \tau^{\widetilde{T}_4, T_3}_{2-i,3+i,4,0} \widetilde{s} = \tau^{\widetilde{T}_2, T_3}_{4-i,2+i,3,0} \widetilde{s} = \tau^{\widetilde{T}_4, T_3}_{2,4,3,0} \widetilde{s} = 0, \; i = 0,1 \tag{3.34}$$

$$\tau_{4-i,0,3+i,2}^{\widetilde{T}_j,T_j}\widetilde{s} = \tau_{4-i,0,3+i,2}^{\widetilde{T}_2,T_3}\widetilde{s} = 0,\ i = 0,1,\ j = 2,4 \tag{3.35}$$

$$\begin{aligned}&\tau_{3,2,3,1}^{\widetilde{T}_j,T_j}\widetilde{s} = \tau_{2,3,3,1}^{\widetilde{T}_j,T_j}\widetilde{s} = \tau_{2,2,4,1}^{\widetilde{T}_j,T_j}\widetilde{s} = \tau_{2,2,3,2}^{\widetilde{T}_j,T_j}\widetilde{s} = \\ &\tau_{3,2,2,2}^{\widetilde{T}_j,T_j}\widetilde{s} = \tau_{2,3,2,2}^{\widetilde{T}_j,T_j}\widetilde{s} = 0,\ j = 2,4\end{aligned} \tag{3.36}$$

$$\tau_{3,2,3,1}^{\widetilde{T}_2,T_3}\widetilde{s} = \tau_{2,3,3,1}^{\widetilde{T}_4,T_3}\widetilde{s} = \tau_{2,2,4,1}^{\widetilde{T}_2,T_3}\widetilde{s} = \tau_{2,2,3,2}^{\widetilde{T}_2,T_3}\widetilde{s} = \tau_{3,2,2,2}^{\widetilde{T}_2,T_3}\widetilde{s} = \tau_{2,3,2,2}^{\widetilde{T}_4,T_2}\widetilde{s} = 0 \tag{3.37}$$

$$\begin{aligned}&\tau_{4,0,2,3}^{\widetilde{T}_{3+j},T_{1,j}}\widetilde{s} = \tau_{3,0,3,3}^{\widetilde{T}_{3+j},T_{1,j}}\widetilde{s} = \tau_{3,1,2,3}^{\widetilde{T}_{3+j},T_{1,j}}\widetilde{s} = \tau_{3,0,2,4}^{\widetilde{T}_{3+j},T_{1,j}}\widetilde{s} = \\ &\tau_{2,2,1,4}^{\widetilde{T}_{3+j},T_{1,j}}\widetilde{s} = \tau_{2,0,2,5}^{\widetilde{T}_{3+j},T_{1,j}}\widetilde{s} = 0,\ j = 1,2,\ \text{with}\ \widetilde{T}_5 := \widetilde{T}_2\end{aligned} \tag{3.38}$$

$$\tau_{4,0,2,3}^{\widetilde{T}_2,T_{1,3}}\widetilde{s} = \tau_{3,0,3,3}^{\widetilde{T}_2,T_{1,3}}\widetilde{s} = \tau_{3,1,2,3}^{\widetilde{T}_2,T_{1,3}}\widetilde{s} = \tau_{3,0,2,4}^{\widetilde{T}_2,T_{1,3}}\widetilde{s} = \tau_{2,2,1,4}^{\widetilde{T}_2,T_{1,3}}\widetilde{s} = \tau_{2,0,2,5}^{\widetilde{T}_2,T_{1,3}}\widetilde{s} = 0 \tag{3.39}$$

$$\tau_{2,0,4,3}^{\widetilde{T}_4,T_{1,1}}\widetilde{s} = 0 \tag{3.40}$$

Proof:
Using the C^4 supersmoothness conditions at the vertices of T and the C^3 supersmoothness conditions at the edges of T, we can uniquely and stably determine the B-coefficients of g corresponding to the domain points in $(\hat{M}1^2_{T^1_{WF}})$ and $(\hat{M}2^2_{T^1_{WF}})$ from Theorem 3.9. From the additional smoothness conditions in (3.30), (3.31), and (3.32) we can uniquely and stably determine the B-coefficients associated with the domain points in $(\hat{M}3^2_{T^1_{WF}})$, $(\hat{M}4^2_{T^1_{WF}})$, and $(\hat{M}5^2_{T^1_{WF}})$, respectively. From the additional smoothness conditions in (3.33) - (3.37) we can uniquely and stably compute the B-coefficients corresponding to the domain points in $(\hat{M}6^2_{T^1_{WF}})$ - $(\hat{M}9^2_{T^1_{WF}})$. Finally, the B-coefficients associated with the remaining domain points in the minimal determining set $\hat{\mathcal{M}}^2_1$ from Theorem 3.9, those corresponding to the domain points in $(\hat{M}10^2_{T^1_{WF}})$, are uniquely and stably determined by the additional smoothness conditions in (3.38), (3.39), and (3.40), except for the B-coefficient associated with $\xi^{T_{1,1}}_{0,0,0,9}$. This B-coefficient is equal to the value of $g(v_T)$. Therefore, following Theorem 3.9, g is uniquely and stably determined. □

Lemma 3.11:
Let Δ be a tetrahedral partition of a polyhedral domain $\Omega \subset \mathbb{R}^3$ and $T := \langle v_1, v_2, v_3, v_4 \rangle \in \Delta$ a tetrahedron, that shares exactly the two edges $\langle v_1, v_2 \rangle$ and $\langle v_2, v_3 \rangle$ with the tetrahedra $\widetilde{T}_4$ and $\widetilde{T}_2$, respectively, with $\widetilde{T}_2, \widetilde{T}_4 \in \Delta$. Moreover, T is refined with a first order partial Worsey-Farin split at the split point v_T in the interior of T, and the split point v_F in the interior of the face $\langle v_1, v_2, v_3 \rangle$.

Let $\hat{\mathcal{S}}_9^2(T_{WF}^1)$ be the superspline space defined in Theorem 3.9. Then, for any spline $s \in \mathcal{S}_9^{2,4,3}(\Delta \setminus T)$, there is a spline $g \in \hat{\mathcal{S}}_9^2(T_{WF}^1)$ such that

$$\widetilde{s} := \begin{cases} s, & \text{on } \Delta \setminus T \\ g, & \text{on } T, \end{cases}$$

belongs to $\mathcal{S}_9^{2,4,3}(\Delta)$. Moreover, g is uniquely and stably determined by the values of $g(v_T)$ and $g(v_4)$, and the B-coefficients of $\widetilde{s}|_{\Delta \setminus T}$ by the following additional smoothness conditions across the vertices and edges of T_{WF}^1:

$$\tau_{i,j,k,l}^{\widetilde{T}_2,T_2} \widetilde{s} = 0,\ 5 \le k \le 8,\ i+j+l = 9-k \tag{3.41}$$

$$\tau_{4,1,4,0}^{\widetilde{T}_4,T_{1,1}} \widetilde{s} = \tau_{0,2,7,0}^{\widetilde{T}_4,T_{1,1}} \widetilde{s} = \tau_{0,0,9,0}^{\widetilde{T}_4,T_{1,1}} \widetilde{s} = \tau_{1,4,4,0}^{\widetilde{T}_2,T_{1,2}} \widetilde{s} = 0 \tag{3.42}$$

$$\tau_{4,1,3,1}^{\widetilde{T}_4,T_{1,1}} \widetilde{s} = \tau_{0,2,6,1}^{\widetilde{T}_4,T_{1,1}} \widetilde{s} = \tau_{0,0,8,1}^{\widetilde{T}_4,T_{1,1}} \widetilde{s} = \tau_{1,4,3,1}^{\widetilde{T}_2,T_{1,2}} \widetilde{s} = 0 \tag{3.43}$$

$$\tau_{4,i,3-i,2}^{\widetilde{T}_4,T_{1,1}} \widetilde{s} = \tau_{3,2,2,2}^{\widetilde{T}_4,T_{1,1}} \widetilde{s} = \tau_{0,3,4,2}^{\widetilde{T}_4,T_{1,1}} \widetilde{s} = \tau_{0,1,6,2}^{\widetilde{T}_2,T_{1,2}} \widetilde{s} = \tau_{i,4,3-i,2}^{\widetilde{T}_2,T_{1,2}} \widetilde{s} = \tau_{2,3,2,2}^{\widetilde{T}_2,T_{1,2}} \widetilde{s} = 0 \tag{3.44}$$

$$\tau_{4-i,1+i,4,0}^{\widetilde{T}_j,T_j} \widetilde{s} = 0,\ i = 0,\dots,3,\ j = 2,4 \tag{3.45}$$

$$\tau_{4-i,1+i,4,0}^{\widetilde{T}_2,T_3} \widetilde{s} = \tau_{2-i,3+i,4,0}^{\widetilde{T}_4,T_3} \widetilde{s} = \tau_{4-i,2+i,3,0}^{\widetilde{T}_2,T_3} \widetilde{s} = \tau_{2,4,3,0}^{\widetilde{T}_4,T_3} \widetilde{s} = 0,\ i = 0,1 \tag{3.46}$$

$$\tau_{4-i,0,3+i,2}^{\widetilde{T}_j,T_j} \widetilde{s} = \tau_{4-i,0,3+i,2}^{\widetilde{T}_2,T_3} \widetilde{s} = 0,\ i = 0,1,\ j = 2,4 \tag{3.47}$$

$$\begin{aligned} &\tau_{3,2,3,1}^{\widetilde{T}_j,T_j} \widetilde{s} = \tau_{2,3,3,1}^{\widetilde{T}_j,T_j} \widetilde{s} = \tau_{2,2,4,1}^{\widetilde{T}_j,T_j} \widetilde{s} = \tau_{3,2,3,1}^{\widetilde{T}_2,T_3} \widetilde{s} = \\ &\tau_{2,3,3,1}^{\widetilde{T}_4,T_3} \widetilde{s} = \tau_{2,2,4,1}^{\widetilde{T}_2,T_3} \widetilde{s} = 0,\ j = 2,4 \end{aligned} \tag{3.48}$$

$$\begin{aligned} &\tau_{2,2,3,2}^{\widetilde{T}_j,T_j} \widetilde{s} = \tau_{3,2,2,2}^{\widetilde{T}_j,T_j} \widetilde{s} = \tau_{2,3,2,2}^{\widetilde{T}_j,T_j} \widetilde{s} = \tau_{2,2,3,2}^{\widetilde{T}_2,T_3} \widetilde{s} = \\ &\tau_{3,2,2,2}^{\widetilde{T}_2,T_3} \widetilde{s} = \tau_{2,3,2,2}^{\widetilde{T}_4,T_2} \widetilde{s} = 0,\ j = 2,4 \end{aligned} \tag{3.49}$$

$$\begin{aligned}&\tau_{4,0,2,3}^{\widetilde{T}_{3+j},T_{1,j}}\widetilde{s} = \tau_{3,0,3,3}^{\widetilde{T}_{3+j},T_{1,j}}\widetilde{s} = \tau_{3,1,2,3}^{\widetilde{T}_{3+j},T_{1,j}}\widetilde{s} = \\ &\tau_{3,0,2,4}^{\widetilde{T}_{3+j},T_{1,j}}\widetilde{s} = \tau_{2,2,1,4}^{\widetilde{T}_{3+j},T_{1,j}}\widetilde{s} = \tau_{2,0,2,5}^{\widetilde{T}_{3+j},T_{1,j}}\widetilde{s} = 0,\end{aligned} \tag{3.50}$$

$j = 1,2$, with $\widetilde{T}_5 := \widetilde{T}_2$

$$\begin{aligned}&\tau_{4,0,2,3}^{\widetilde{T}_2,T_{1,3}}\widetilde{s} = \tau_{3,0,3,3}^{\widetilde{T}_2,T_{1,3}}\widetilde{s} = \tau_{3,1,2,3}^{\widetilde{T}_2,T_{1,3}}\widetilde{s} = \\ &\tau_{3,0,2,4}^{\widetilde{T}_2,T_{1,3}}\widetilde{s} = \tau_{2,2,1,4}^{\widetilde{T}_2,T_{1,3}}\widetilde{s} = \tau_{2,0,2,5}^{\widetilde{T}_2,T_{1,3}}\widetilde{s} = 0\end{aligned} \tag{3.51}$$

$$\tau_{2,0,4,3}^{\widetilde{T}_4,T_{1,1}}\widetilde{s} = 0 \tag{3.52}$$

Proof:
We can uniquely and stably determine the B-coefficient of g associated with the domain points in the disk $D_4^{T_4}(v_4)$ in the same way as in the proof of Lemma 3.6 using the additional smoothness conditions (3.41) and the value of $g(v_4)$.

The remaining B-coefficients associated with the domain points in the minimal determining set $\hat{\mathcal{M}}_1^2$ from Theorem 3.9 can be determined in the same way as in the proof of Lemma 3.10 from the C^4 supersmoothness conditions at the vertices of T, the C^3 supersmoothness conditions at the edges of T, the additional smoothness conditions in (3.42) - (3.52) and the value of $g(v_T)$. Thus, all B-coefficients of g are uniquely and stably determined.□

3.2.2 Minimal determining sets for C^2 supersplines based on second order partial Worsey-Farin splits

In this subsection, we investigate minimal determining sets for C^2 supersplines on second order partial Worsey-Farin splits of a tetrahedron $T := \langle v_1, v_2, v_3, v_4 \rangle$. To this end, let T_{WF}^2 be the second order partial Worsey-Farin split of T at the split point v_T in the interior of T, and let the faces $\langle v_1, v_2, v_3 \rangle$ and $\langle v_2, v_3, v_4 \rangle$ be refined with a Clough-Tocher split at the split points v_{F_1} and v_{F_2}, respectively. The corresponding subtetrahedra are denoted by

$$\begin{aligned} T_{1,1} &:= \langle v_1, v_2, v_{F_1}, v_T \rangle, & T_{1,2} &:= \langle v_2, v_3, v_{F_1}, v_T \rangle, \\ T_{1,3} &:= \langle v_3, v_1, v_{F_1}, v_T \rangle, & T_{2,1} &:= \langle v_2, v_3, v_{F_2}, v_T \rangle, \\ T_{2,2} &:= \langle v_3, v_4, v_{F_2}, v_T \rangle, & T_{2,3} &:= \langle v_4, v_2, v_{F_2}, v_T \rangle, \\ T_3 &:= \langle v_3, v_1, v_4, v_T \rangle, & T_4 &:= \langle v_1, v_2, v_4, v_T \rangle. \end{aligned} \tag{3.53}$$

Theorem 3.12:
Let $\mathcal{M}_2^2$ be the union of the following sets of domain points in $\mathcal{D}_{T_{WF}^2,9}$:

($\mathrm{M1}^2_{T^2_{WF}}$) $D_4^{T_{v_i}}(v_i)$, $i=1,\dots,4$, with $v_i \in T_{v_i} \in T^2_{WF}$

($\mathrm{M2}^2_{T^2_{WF}}$) $\bigcup_{i=1}^{3} E_3^{T_{\langle v_i,v_{i+1}\rangle}}(\langle v_i,v_{i+1}\rangle) \setminus (D_4^{T_{\langle v_i,v_{i+1}\rangle}}(v_i) \cup D_4^{T_{\langle v_i,v_{i+1}\rangle}}(v_{i+1}))$,
with $\langle v_i,v_{i+1}\rangle \in T_{\langle v_i,v_{i+1}\rangle}$

($\mathrm{M3}^2_{T^2_{WF}}$) $\{\xi^{T_{i,1}}_{4,1,4,0}, \xi^{T_{i,2}}_{1,4,4,0}, \xi^{T_{i,1}}_{0,2,7,0}\xi^{T_{i,1}}_{0,0,9,0}\}_{i=1}^{2}$

($\mathrm{M4}^2_{T^2_{WF}}$) $\{\xi^{T_3}_{4-i,1+i,4,0}, \xi^{T_4}_{4,4-i,1+i,0}\}_{i=0}^{3}$

($\mathrm{M5}^2_{T^2_{WF}}$) $\{\xi^{T_{i,1}}_{4,1,3,1}, \xi^{T_{i,2}}_{1,4,3,1}, \xi^{T_{i,1}}_{0,2,6,1}, \xi^{T_{i,1}}_{0,0,8,1}\}_{i=1}^{2}$

($\mathrm{M6}^2_{T^2_{WF}}$) $\{\xi^{T_4}_{4,0,3,2}, \xi^{T_4}_{3,0,4,2}\}$

($\mathrm{M7}^2_{T^2_{WF}}$) $\{\xi^{T_3}_{3,3,2,1}, \xi^{T_3}_{2,3,3,1}, \xi^{T_3}_{2,4,2,1}, \xi^{T_4}_{3,2,3,1}, \xi^{T_4}_{2,3,3,1}, \xi^{T_4}_{2,2,4,1}\}$

($\mathrm{M8}^2_{T^2_{WF}}$) $\{\xi^{T_{j,i}}_{3,0,4,2}\}$, $i=1,2,3$, $j=1,2$

($\mathrm{M9}^2_{T^2_{WF}}$) $\{\xi^{T_j}_{3,2,2,2}, \xi^{T_j}_{2,3,2,2}, \xi^{T_j}_{2,2,3,2}\}_{j=3}^{4}$

($\mathrm{M10}^2_{T^2_{WF}}$) $\{\xi^{T_{1,2}}_{4-i,1+i,0,4}\}_{i=0}^{3}$

($\mathrm{M11}^2_{T^2_{WF}}$) $\{\xi^{T_{1+i,2-i}}_{3,0,3,3}, \xi^{T_{1+i,2-i}}_{0,3,3,3}, \xi^{T_{1+i,2-i}}_{2,2,2,3}, \xi^{T_{1+i,2-i}}_{1,1,4,3}, \xi^{T_{1+i,2-i}}_{2,0,4,3}, \xi^{T_{1+i,2-i}}_{0,2,4,3}\}_{i=0}^{1}$

($\mathrm{M12}^2_{T^2_{WF}}$) $\{\xi^{T_{1,2}}_{2,1,2,4}, \xi^{T_{1,2}}_{1,2,2,4}, \xi^{T_{1,2}}_{1,1,3,4}\}$

($\mathrm{M13}^2_{T^2_{WF}}$) $\{\xi^{T_{1,1}}_{0,0,0,9}\}$

Then $\mathcal{M}_2^2$ is a stable minimal determining set for

$$
\begin{aligned}
\widetilde{\mathcal{S}}_9^2(T^2_{WF}) := \{s \in C^2(T^2_{WF}) : {}& s|_{\widetilde{T}} \in \mathcal{P}_9^3,\ \forall \widetilde{T} \in T^2_{WF},\\
& s \in C^4(v_i),\ i=1,\dots,4,\\
& s \in C^3(e), \text{ for all edges } e \in T,\\
& s \in C^7(\langle v_{F_i}, v_T\rangle),\ i=1,2\\
& s \in C^7(v_T),\\
& \tau^{T_{j,1},T_{j,2}}_{3-i,5+i,1,0}s = \tau^{T_{j,2},T_{j,3}}_{4,4,1,0}s = 0,\ i=0,\dots,3,\ j=1,2,\\
& \tau^{T_{j,1},T_{j,2}}_{3-i,5+i,0,1}s = 0,\ i=0,\dots,3,\ j=1,2\},
\end{aligned}
$$

and the dimension of $\widetilde{\mathcal{S}}_9^2(T^2_{WF})$ is equal to 264.

Proof:
It is first shown that $\mathcal{M}_2^2$ is a stable minimal determining set for $\widetilde{\mathcal{S}}_9^2(T_{WF}^2)$. To this end we set the B-coefficients c_ξ of a spline $s \in \widetilde{\mathcal{S}}_9^2(T_{WF}^2)$ to arbitrary values for each domain point $\xi \in \mathcal{M}_2^2$. Then we show that all other B-coefficients of s are uniquely and stably determined.

The B-coefficients of s associated with the domain points in the balls $D_4(v_i)$, $i = 1,\dots,4$, are uniquely and stably determined from the B-coefficients corresponding to the domain points in $(\mathrm{M1}^2_{T^2_{WF}})$ by the C^4 supersmoothness conditions at the vertices of T.

Now, we can uniquely and stably determine the remaining B-coefficients of s corresponding to the domain points in the tubes $E_3(\langle v_i, v_{i+1}\rangle)$, $i = 1,2,3$, from the B-coefficients associated with the domain points in $(\mathrm{M2}^2_{T^2_{WF}})$ using the C^3 supersmoothness conditions at the corresponding three edges of T.

Next, we consider the B-coefficients corresponding to the domain points in the Clough-Tocher split faces $F_1 := \langle v_1, v_2, v_3\rangle$ and $F_2 := \langle v_2, v_3, v_4\rangle$. The subtriangles forming these faces are degenerated faces of the subtetrahedra $T_{j,1}, T_{j,2}$, and $T_{j,3}, j = 1,2$, since they lie in one common plane, respectively. Therefore, we can consider the B-coefficients of $s|_{F_j}$, $j = 1,2$, as those of a bivariate spline $g_0^{F_j} \in \hat{\mathcal{S}}_9^2(F_{j,CT})$, $j = 1,2$, defined in (A.22). Examining $g_0^{F_j}$, we can see that the B-coefficients corresponding to the domain points within a distance of four from the vertices of F_j, $j = 1,2$, and corresponding to those within a distance of three from two edges of F_j, $j = 1,2$, are already uniquely and stably determined. Then, together with the B-coefficients associated with the domain points in $(\mathrm{M3}^2_{T^2_{WF}})$, all B-coefficients associated with the domain points in the minimal determining set in Lemma A.14 are uniquely and stably determined. Thus all B-coefficients of $g_0^{F_j}$, $j = 1,2$, are uniquely and stably determined, and therefore also all B-coefficients of $s|_{F_j}$, $j = 1,2$.

Now, we consider the B-coefficients of s restricted to the two non-split faces $\langle v_3, v_1, v_4\rangle$ and $\langle v_1, v_2, v_4\rangle$ of T. The only undetermined B-coefficients of s corresponding to domain points on these faces are associated with the domain points in $(\mathrm{M4}^2_{T^2_{WF}})$. Thus, s restricted to these faces is uniquely and stably determined.

Now, following Lemma 2.33 for the case $j = 0$, we can use the C^1 smoothness conditions at the edges $\langle v_1, v_3\rangle$, $\langle v_1, v_4\rangle$, and $\langle v_2, v_4\rangle$, to uniquely and stably determined the remaining B-coefficients of s corresponding to the domain points within a distance of one from these edges.

Next, we investigate the B-coefficients of s associated with the domain points in the shells $R_8^{T_{j,1}}(v_T), R_8^{T_{j,2}}(v_T)$, and $R_8^{T_{j,3}}(v_T)$, $j = 1,2$, which form a triangle F_j^1, respectively. Following the same arguments as above, these B-coefficients can also be regarded as those of a bivariate spline $g_1^{F_j^1}$ in the space $\bar{\mathcal{S}}_8^2(F_{j,CT}^1)$, $j = 1,2$, defined in (A.14). Examining these splines, we see that the B-coefficients corresponding to the domain points within a distance of three from the vertices and within a distance of two from two of the edges of F_j^1, $j = 1,2$, are already uniquely and stably determined. Together with the B-coefficients associated with the domain points in (M5$^2_{T^2_{WF}}$), all B-coefficients corresponding to the domain points in the minimal determining set of Lemma A.10 are uniquely and stably determined. Therefore, all B-coefficients of $s|_{F_j^1}$, $j = 1,2$, are uniquely and stably determined.

Now, we consider the B-coefficients of s corresponding to the domain points within a distance of two from the edge $\langle v_1, v_4 \rangle$. Together with the B-coefficients of s associated with the domain points in (M6$^2_{T^2_{WF}}$), we have already uniquely and stably determined all B-coefficients of s corresponding to the domain points on the interior face $\langle v_1, v_4, v_T \rangle$ within a distance of two from this edge. Furthermore, the B-coefficients of s restricted to the faces of T are already uniquely and stably determined. Therefore, the only undetermined B-coefficients of s considered here are associated with domain points in the shell $R_8(v_T)$ with a distance of one from the interior face $\langle v_1, v_4, v_T \rangle$. Thus, using the C^1 and C^2 smoothness conditions across this face and following Lemma 2.33 for the case $j = 1$, we can uniquely and stably determine the last undetermined B-coefficients of $E_2(\langle v_1, v_4 \rangle)$.

Next, following Lemma 2.33 for the case $j = 0$, we can use the C^1 and C^2 smoothness conditions at the three interior faces $\langle v_1, v_3, v_T \rangle$, $\langle v_1, v_4, v_T \rangle$, and $\langle v_2, v_4, v_T \rangle$ of T^2_{WF} to uniquely and stably determined the remaining B-coefficients of s corresponding to the domain points within a distance of two from the edges $\langle v_1, v_3 \rangle$, $\langle v_1, v_4 \rangle$, and $\langle v_2, v_4 \rangle$.

Subsequently, we examine the B-coefficients of s corresponding to the domain points with a distance of one from the two faces $\langle v_3, v_1, v_4 \rangle$ and $\langle v_1, v_2, v_4 \rangle$. The only undetermined B-coefficients of s associated with the domain points in the shells $R_8^{T_3}(v_T)$ and $R_8^{T_4}(v_T)$ correspond to the domain points in (M7$^2_{T^2_{WF}}$). Thus, since these domain points are contained in the minimal determining set, s is uniquely and stably determine on these two shells.

Following Lemma 2.33 for the case $j=0$, we can use the C^3 supersmoothness conditions at the edges $\langle v_1,v_3\rangle$, $\langle v_1,v_4\rangle$, and $\langle v_2,v_4\rangle$ of T^2_{WF} to uniquely and stably determined the remaining B-coefficients of s corresponding to the domain points within a distance of three from these edges.

Now, we consider the B-coefficients of s associated with the domain points in the shells $R_7^{T_{j,1}}(v_T), R_7^{T_{j,2}}(v_T)$, and $R_7^{T_{j,3}}(v_T)$, $j=1,2$, which form a triangle F_j^2, respectively. Following the same arguments as above, these B-coefficients can be regarded as those of a bivariate spline $g_2^{F_j^2}$ in the space $\mathcal{S}_7^2(F^2_{j,CT})$, $j=1,2$, defined in (A.6). Considering these splines, we see that the B-coefficients corresponding to the domain points within a distance of two from the vertices and within a distance of one from the edges of F_j^1, $j=1,2$, are already uniquely and stably determined. Together with the B-coefficients associated with the domain points in (M8$^2_{T^2_{WF}}$), all B-coefficients corresponding to the domain points in the minimal determining set in Lemma A.6 are uniquely and stably determined. Thus, following Lemma A.6, all B-coefficients of $s|_{F_j^2}$, $j=1,2$, are uniquely and stably determined.

Next, we examine the B-coefficients of s corresponding to the domain points with a distance of two from the faces $\langle v_3,v_1,v_4\rangle$ and $\langle v_1,v_2,v_4\rangle$ of T. Here, the only undetermined B-coefficients of s associated with the domain points in the shells $R_7^{T_3}(v_T)$ and $R_7^{T_4}(v_T)$ correspond to the domain points in (M9$^2_{T^2_{WF}}$). Thus, s is uniquely and stably determine on these shell.

Following Lemma 2.33 for the case $j=1$, we can use the C^1 and C^2 smoothness conditions at the face $\langle v_2,v_3,v_T\rangle$ to uniquely and stably compute the remaining B-coefficients corresponding to domain points within a distance of four from the edge $\langle v_2,v_3\rangle$ from the already determined B-coefficients and those associated with the domain points in (M10$^2_{T^2_{WF}}$).

Now, we consider the B-coefficients of s associated with the domain points in the shells $R_6^{T_{j,1}}(v_T), R_6^{T_{j,2}}(v_T)$, and $R_6^{T_{j,3}}(v_T)$, $j=1,2$, which form a triangle F_j^3, respectively. These B-coefficients can be regarded as those of a bivariate polynomial $g_3^{F_j^3}$ defined on F_j^3, $j=1,2$. Examining these polynomials, we see that the B-coefficients corresponding to the domain points within a distance of one from the vertices, on two edges and within a distance of one from one edge of F_j^3, $j = 1,2$, are already uniquely and stably determined. Together with the B-coefficients associated with the domain points in (M11$^2_{T^2_{WF}}$), all B-coefficients corresponding to the domain points in the minimal determining set in Lemma A.4 are uniquely and stably determined. Therefore, all B-coefficients of $s|_{F_j^3}$, $j = 1,2$, are uniquely and stably determined.

Following Lemma 2.33 for the case $j = 0$, we can use the C^1, C^2, and C^3 smoothness conditions at the interior face $\langle v_2, v_3, v_T \rangle$, which are imposed by the C^7 supersmoothness at v_T, to uniquely and stably compute the remaining B-coefficients of s corresponding to the domain points within a distance of five from the edge $\langle v_2, v_3 \rangle$ from the already determined B-coefficients. In the same way, we can uniquely and stably compute the remaining undetermined B-coefficients of s corresponding to the domain points within a distance of four from the edges $\langle v_1, v_2 \rangle$ and $\langle v_3, v_1 \rangle$ from those already determined by using the C^1 and C^2 smoothness conditions at the faces $\langle v_1, v_2, v_T \rangle$ and $\langle v_3, v_1, v_T \rangle$.

Next, we investigate the B-coefficients of s associated with the domain points in the shells $R_5^{T_{1,1}}(v_T), R_5^{T_{1,2}}(v_T)$, and $R_5^{T_{1,3}}(v_T)$, which form a triangle F_1^4. Following the arguments from above, these B-coefficients can be considered as those of a bivariate polynomial $g_4^{F_1^4}$ of degree five defined on F_1^4. Examining this polynomial, it can be seen that the B-coefficients corresponding to the domain points within a distance of one from the vertices, on two edges and within a distance of one from one edge of F_1^4 are already uniquely and stably determined. Together with the B-coefficients corresponding to the domain points in (M12$^2_{T^2_{WF}}$), all B-coefficients corresponding to the domain points in the minimal determining set in Lemma A.3 are uniquely and stably determined. Thus, all B-coefficients of $s|_{F_1^4}$ are uniquely and stably determined.

Finally, we consider the last undetermined B-coefficients of s. These are associated with domain points in the ball $D_7(v_T)$. Due to the C^7 supersmoothness of s at the split point v_T, we can consider the B-coefficients associated with the domain points in $D_7(v_T)$ as those of a trivariate polynomial g defined on the tetrahedron $\widetilde{T}$, which is formed by the domain points in $D_7(v_T)$. Note that $\widetilde{T}$ is also subjected to a second order partial Worsey-Farin split. Examining g, we see that the B-coefficients associated with the domain points on the non-split faces of $\widetilde{T}$, corresponding to the domain points within a distance of one from one of the faces that has been refined with the Clough-Tocher split, and within a distance of two from the second refined face are already uniquely and stably determined. Now, setting the B-coefficient of g associated with the domain point $\xi^{T_{1,1}}_{0,0,0,9}$ is equal to setting the function value of g at v_T. Thus, following Lemma 2.42, the remaining undetermined B-coefficients of g can be uniquely and stably determined from smoothness conditions. Therefore, we have uniquely and stably determined all B-coefficients of s.

Since the set of domain points $\mathcal{M}^2_2$ is a minimal determining set for $\widetilde{\mathcal{S}}^2_9(T^2_{WF})$, it follows that $\dim \widetilde{\mathcal{S}}^2_9(T^2_{WF}) = \#\mathcal{M}^2_2 = 264$. □

Lemma 3.13:
Let Δ be a tetrahedral partition of a polyhedral domain $\Omega \subset \mathbb{R}^3$ and $T := \langle v_1, v_2, v_3, v_4 \rangle \in \Delta$ a tetrahedron, that shares exactly the three edges $\langle v_1, v_2 \rangle$, $\langle v_2, v_3 \rangle$, and $\langle v_3, v_4 \rangle$ with the tetrahedra $\widetilde{T}_1, \widetilde{T}_2$, and $\widetilde{T}_3$, respectively, with $\widetilde{T}_j \in \Delta$, $j = 1,2,3$. Moreover, T is refined with a second order partial Worsey-Farin split at the split point v_T in the interior of T, and the split points v_{F_1} and v_{F_2} in the interior of the faces $\langle v_1, v_2, v_3 \rangle$ and $\langle v_2, v_3, v_4 \rangle$, respectively.

Let $\widetilde{\mathcal{S}}^2_9(T^2_{WF})$ be the superspline space defined in Theorem 3.12. Then, for any spline $s \in \mathcal{S}^{2,4,3}_9(\Delta \setminus T)$, there is a spline $g \in \widetilde{\mathcal{S}}^2_9(T^2_{WF})$ such that

$$\widetilde{s} := \begin{cases} s, & \text{on } \Delta \setminus T \\ g, & \text{on } T, \end{cases}$$

belongs to $\mathcal{S}^{2,4,3}_9(\Delta)$. Moreover, g is uniquely and stably determined by the value of $g(v_T)$ and the B-coefficients of $\widetilde{s}|_{\Delta \setminus T}$ by the following additional smoothness conditions across the edges of T:

$$\tau^{\widetilde{T}_j, T_{j,1}}_{4,1,4,0}\widetilde{s} = \tau^{\widetilde{T}_j, T_{j,1}}_{0,2,7,0}\widetilde{s} = \tau^{\widetilde{T}_j, T_{j,1}}_{0,0,9,0}\widetilde{s} = \tau^{\widetilde{T}_{j+1}, T_{j,2}}_{1,4,4,0}\widetilde{s} = 0,\ j = 1,2 \tag{3.54}$$

$$\tau^{\widetilde{T}_3, T_3}_{4-i,1+i,4,0}\widetilde{s} = \tau^{\widetilde{T}_1, T_4}_{4,4-i,1+i,0}\widetilde{s} = 0,\ i = 0,\dots,3 \tag{3.55}$$

$$\tau^{\widetilde{T}_j, T_{j,1}}_{4,1,3,1}\widetilde{s} = \tau^{\widetilde{T}_j, T_{j,1}}_{0,2,6,1}\widetilde{s} = \tau^{\widetilde{T}_j, T_{j,1}}_{0,0,8,1}\widetilde{s} = \tau^{\widetilde{T}_{j+1}, T_{j,2}}_{1,4,3,1}\widetilde{s} = 0,\ i = 0,\dots,3,\ j = 1,2 \tag{3.56}$$

$$\tau^{\widetilde{T}_1, T_4}_{4,0,3,2}\widetilde{s} = \tau^{\widetilde{T}_1, T_4}_{3,0,4,2}\widetilde{s} = 0 \tag{3.57}$$

$$\tau^{\widetilde{T}_3, T_3}_{3,3,2,1}\widetilde{s} = \tau^{\widetilde{T}_3, T_3}_{2,3,3,1}\widetilde{s} = \tau^{\widetilde{T}_3, T_3}_{2,4,2,1}\widetilde{s} = \tau^{\widetilde{T}_1, T_4}_{3,2,3,1}\widetilde{s} = \tau^{\widetilde{T}_1, T_4}_{2,3,3,1}\widetilde{s} = \tau^{\widetilde{T}_1, T_4}_{2,2,4,1}\widetilde{s} = 0 \tag{3.58}$$

$$\tau^{\widetilde{T}_j, T_{j,1}}_{3,0,4,2}\widetilde{s} = \tau^{\widetilde{T}_{1+j}, T_{j,2}}_{3,0,4,2}\widetilde{s} = \tau^{\widetilde{T}_{1+j}, T_{j,3}}_{3,0,4,2}\widetilde{s} = 0,\ j = 1,2 \tag{3.59}$$

$$\tau_{3,2,2,2}^{\widetilde{T}_j,T_j}\widetilde{s} = \tau_{2,3,2,2}^{\widetilde{T}_j,T_j}\widetilde{s} = \tau_{2,2,3,2}^{\widetilde{T}_j,T_j}\widetilde{s} = 0,\ j = 2,3,4,\ \text{with}\ \widetilde{T}_4 := \widetilde{T}_1 \tag{3.60}$$

$$\tau_{4-i,1+1,0,4}^{\widetilde{T}_2,T_{1,2}}\widetilde{s} = 0,\ i = 0,\dots,3 \tag{3.61}$$

$$\begin{aligned}&\tau_{3,0,3,3}^{\widetilde{T}_2,T_{1+i,2-i}}\widetilde{s} = \tau_{0,3,3,3}^{\widetilde{T}_2,T_{1+i,2-i}}\widetilde{s} = \tau_{2,2,2,3}^{\widetilde{T}_2,T_{1+i,2-i}}\widetilde{s} = \tau_{1,1,4,3}^{\widetilde{T}_2,T_{1+i,2-i}}\widetilde{s}\\ &= \tau_{2,0,4,3}^{\widetilde{T}_2,T_{1+i,2-i}}\widetilde{s} = \tau_{0,2,4,3}^{\widetilde{T}_2,T_{1+i,2-i}}\widetilde{s} = 0,\ i = 0,1\end{aligned} \tag{3.62}$$

$$\tau_{2,1,2,4}^{\widetilde{T}_2,T_{1,2}}\widetilde{s} = \tau_{1,2,2,4}^{\widetilde{T}_2,T_{1,2}}\widetilde{s} = \tau_{1,1,3,4}^{\widetilde{T}_2,T_{1,2}}\widetilde{s} = 0 \tag{3.63}$$

Proof:
Using the C^4 supersmoothness conditions at the vertices of T and the C^3 supersmoothness conditions at the edges $\langle v_i, v_{i+1}\rangle$, $i = 1,2,3$, of T we can uniquely and stably determine the B-coefficients of g corresponding to the domain points in $(M1^2_{T^2_{WF}})$ and $(M2^2_{T^2_{WF}})$ from Theorem 3.12. From the additional smoothness conditions in (3.54) we can uniquely and stably determine the B-coefficients of g corresponding to the domain points in $(M3^2_{T^2_{WF}})$. Then, from the additional smoothness conditions in (3.55) - (3.63), we can uniquely and stably determine the B-coefficients of g corresponding to the domain points in $(M4^2_{T^2_{WF}})$ - $(M12^2_{T^2_{WF}})$. Finally, the B-coefficient associated with the domain point in $(M13^2_{T^2_{WF}})$ is equal to the value of $g(v_T)$. Thus, following Theorem 3.12, g is uniquely and stably determined. □

Theorem 3.14:
Let $\widetilde{\mathcal{M}}^2_2$ be the union of the following sets of domain points in $\mathcal{D}_{T^2_{WF},9}$:

$(\widetilde{M}1^2_{T^2_{WF}})$ $D_4^{T_{v_i}}(v_i)$, $i = 1,\dots,4$, with $v_i \in T_{v_i} \in T^2_{WF}$

$(\widetilde{M}2^2_{T^2_{WF}})$ $\{\xi^{T_4}_{i,j,k,l}\}_{l=0}^2$, $i,j,k \le 4$, $i+j+k = 9-l$

$(\widetilde{M}3^2_{T^2_{WF}})$ $\{\xi^{T_4}_{4-i,2+i,0,3}, \xi^{T_4}_{4-i,0,2+i,3}, \xi^{T_4}_{0,4-i,2+i,3}\}_{i=0}^2$

$(\widetilde{M}4^2_{T^2_{WF}})$ $E_3^{T_{1,2}}(\langle v_2, v_3\rangle) \setminus (D_4^{T_{1,2}}(v_2) \cup D_4^{T_{1,2}}(v_3))$

$(\widetilde{M}5^2_{T^2_{WF}})$ $\{\xi^{T_{i,1}}_{4,1,4,0}, \xi^{T_{i,1+i}}_{1,4,4,0}, \xi^{T_{i,1}}_{0,2,7,0}\xi^{T_{i,1}}_{0,0,9,0}\}_{i=1}^2$

$(\widetilde{M}6^2_{T^2_{WF}})$ $\{\xi^{T_3}_{4,4-i,1+i,0}\}_{i=0}^3$

$(\widetilde{M}7^2_{T^2_{WF}})$ $\{\xi^{T_{i,1}}_{4,1,3,1}, \xi^{T_{i,1+i}}_{1,4,3,1}, \xi^{T_{i,1}}_{0,2,6,1}, \xi^{T_{i,1}}_{0,0,8,1}\}_{i=1}^2$

$(\widetilde{M}8^2_{T^2_{WF}})$ $\{\xi^{T_3}_{4,2,2,1}, \xi^{T_3}_{3,3,2,1}, \xi^{T_3}_{3,2,3,1}\}$

$(\widetilde{\mathrm{M}}9^2_{T^2_{WF}})$ $\{\xi^{T_{j,i}}_{3,0,4,2}\}$, $i = 1,2,3$, $j = 1,2$
$(\widetilde{\mathrm{M}}10^2_{T^2_{WF}})$ $\{\xi^{T_3}_{3,2,2,2}, \xi^{T_3}_{2,3,2,2}, \xi^{T_3}_{2,2,3,2}\}$
$(\widetilde{\mathrm{M}}11^2_{T^2_{WF}})$ $\{\xi^{T_{1,2}}_{4-i,1+i,0,4}\}_{i=0}^{3}$
$(\widetilde{\mathrm{M}}12^2_{T^2_{WF}})$ $\{\xi^{T_{1+i,2-i}}_{3,0,3,3}, \xi^{T_{1+i,2-i}}_{0,3,3,3}, \xi^{T_{1+i,2-i}}_{2,2,2,3}, \xi^{T_{1+i,2-i}}_{1,1,4,3}, \xi^{T_{1+i,2-i}}_{2,0,4,3}, \xi^{T_{1+i,2-i}}_{0,2,4,3}\}_{i=0}^{1}$
$(\widetilde{\mathrm{M}}13^2_{T^2_{WF}})$ $\{\xi^{T_4}_{3,1,2,3}, \xi^{T_4}_{2,2,2,3}, \xi^{T_4}_{2,1,3,3}\}$
$(\widetilde{\mathrm{M}}14^2_{T^2_{WF}})$ $\{\xi^{T_{1,1}}_{0,0,0,9}\}$

Then $\widetilde{\mathcal{M}}^2_2$ is a stable minimal determining set for $\widetilde{\mathcal{S}}^2_9(T^2_{WF})$.

Proof:
In order to show that $\widetilde{\mathcal{M}}^2_2$ is a stable minimal determining set for $\widetilde{\mathcal{S}}^2_9(T^2_{WF})$ we set the B-coefficients c_ξ of a spline $s \in \widetilde{\mathcal{S}}^2_9(T^2_{WF})$ to arbitrary values for each domain point $\xi \in \widetilde{\mathcal{M}}^2_2$ and show that all other B-coefficients of s are uniquely and stably determined.

We can uniquely and stably compute the undetermined B-coefficients of s corresponding to the domain points in the balls $D_4(v_i)$, $i = 1,\ldots,4$, from those associated with the domain points in $(\widetilde{\mathrm{M}}1^2_{T^2_{WF}})$ using the C^4 supersmoothness conditions at the vertices of T.

Then, together with the B-coefficients associated with the domain points in $(\widetilde{\mathrm{M}}2^2_{T^2_{WF}})$ and $(\widetilde{\mathrm{M}}3^2_{T^2_{WF}})$, all B-coefficients corresponding to the domain points within a distance of two from the face $\langle v_1, v_2, v_4\rangle$ and within a distance of three from the corresponding edges are uniquely and stably determined.

Now, using the C^3 supersmoothness conditions at the edges of T and the B-coefficients associated with the domain points in $(\widetilde{\mathrm{M}}4^2_{T^2_{WF}})$, we can uniquely and stably compute all B-coefficients of s corresponding to the domain points within a distance of three from the edges $\langle v_1, v_2\rangle$, $\langle v_2, v_4\rangle$, $\langle v_4, v_1\rangle$, and $\langle v_2, v_3\rangle$.

Next, we examine the B-coefficients associated with the domain points in the Clough-Tocher split faces $F_1 := \langle v_1, v_2, v_3\rangle$ and $F_2 := \langle v_2, v_3, v_4\rangle$. These can be determined in the same way as in the proof of Theorem 3.12 from the already determined B-coefficients and those corresponding to the domain points in $(\widetilde{\mathrm{M}}5^2_{T^2_{WF}})$.

Since the remaining undetermined B-coefficients of s restricted to the face $\langle v_3, v_1, v_4 \rangle$ are associated with the domain points in $(\widetilde{M}6^2_{T^2_{WF}})$, the spline s is uniquely and stably determined on this face.

Subsequently, following Lemma 2.33 for the case $j = 0$, we can uniquely and stably determine the remaining undetermined B-coefficients with a distance of one from the edges $\langle v_1, v_3 \rangle$ and $\langle v_3, v_4 \rangle$ using the C^1 smoothness conditions at these edges.

Now, the B-coefficients of s corresponding to the domain points in the shells $R_8^{T_{j,i}}(v_T)$, $i = 1,2,3$, $j = 1,2$, are considered. Again, the remaining undetermined B-coefficients of s restricted to these shells can be computed in the same way as in the proof of Theorem 3.12 from the B-coefficients associated with the domain points in $(\widetilde{M}7^2_{T^2_{WF}})$ and those already determined.

Next, following Lemma 2.33 for the case $j = 0$, we can use the C^1 and C^2 smoothness conditions at the interior faces $\langle v_1, v_3, v_T \rangle$ and $\langle v_3, v_4, v_T \rangle$ of T^2_{WF} in order to uniquely and stably compute the remaining undetermined B-coefficients of s corresponding to the domain points within a distance of two from the edges $\langle v_1, v_3 \rangle$ and $\langle v_3, v_4 \rangle$, respectively.

Then, since the only undetermined B-coefficients of s restricted to the domain points in the shell $R_8^{T_3}(v_T)$ correspond to the domain points in $(\widetilde{M}8^2_{T^2_{WF}})$, all B-coefficients of s within a distance of one from the faces of T are uniquely and stably determined.

Again, following Lemma 2.33 for the case $j = 0$, we can uniquely and stably compute the remaining undetermined B-coefficients of s corresponding to the domain points within a distance of three from the edges $\langle v_1, v_3 \rangle$ and $\langle v_3, v_4 \rangle$ using the C^3 supersmoothness conditions at these edges.

Then, the last undetermined B-coefficients of s corresponding to the domain points in the shells $R_7^{T_{j,i}}(v_T)$, $i = 1,2,3$, $j = 1,2$, can be computed in the same way as in the proof of Theorem 3.12 from the already determined B-coefficients and those associated with the domain points in $(\widetilde{M}9^2_{T^2_{WF}})$.

The remaining undetermined B-coefficients of s restricted to the shell $R_7^{T_3}(v_T)$ are associated with the domain points in $(\widetilde{M}10^2_{T^2_{WF}})$. Thus, all B-coefficients of s corresponding to the domain points within a distance of two from the faces of T are uniquely and stably determined.

Now, following Lemma 2.33 for the case $j = 1$, we can uniquely and stably compute the remaining undetermined B-coefficients of s corresponding to the domain points within a distance of four from the edge $\langle v_2, v_3 \rangle$

from the already determined B-coefficients and those associated with the domain points in ($\widetilde{\mathrm{M}}11^2_{T^2_{WF}}$) using the C^1 and C^2 smoothness conditions at the interior face $\langle v_2, v_3, v_T \rangle$.

Then the remaining unknown B-coefficients of s associated with domain points in the shells $R_6^{T_{j,i}}(v_T)$, $i = 1,2,3$, $j = 1,2$, can be determined in the same way as in the proof of Theorem 3.12 from the already determined B-coefficients and those associated with the domain points in ($\widetilde{\mathrm{M}}12^2_{T^2_{WF}}$).

Next, following Lemma 2.33 for the case $j = 0$, we can uniquely and stably compute the remaining undetermined B-coefficients of s corresponding to the domain points within a distance of four from the edges $\langle v_1, v_2 \rangle$ and $\langle v_2, v_4 \rangle$ using the C^1 and C^2 smoothness conditions at the faces $\langle v_1, v_2, v_T \rangle$ and $\langle v_2, v_4, v_T \rangle$, respectively.

Then, the last undetermined B-coefficients of s restricted to the shell $R_6^{T_4}(v_T)$ are associated with the domain points in ($\widetilde{\mathrm{M}}13^2_{T^2_{WF}}$). Thus, all B-coefficients of s restricted to $R_6^{T_4}(v_T)$ are uniquely and stably determined.

Finally, the last undetermined B-coefficients of s are considered. These are associated with domain points in the ball $D_7(v_T)$. Since s has C^7 supersmoothness at the split point v_T, we can consider the corresponding B-coefficients as those of a trivariate polynomial g of degree seven that is defined on the tetrahedron $\widetilde{T}$, which is bounded by the domain points in $R_7(v_T)$. Due to the partial Worsey-Farin split of T, the tetrahedron $\widetilde{T}$ is also subjected to a second order partial Worsey-Farin split. Investigating g it can be seen that the B-coefficients associated with the domain points within a distance of one from the two refined faces, within a distance of one from a non-split face, and on the last non-split face are already uniquely and stably determined. Then, setting the B-coefficient associated with the domain point $\xi_{0,0,0,9}^{T_{1,1}}$ is equal to setting the value of g at the point v_T. Therefore, following Lemma 2.42, the remaining undetermined B-coefficients of g can be uniquely and stably determined from smoothness conditions. Thus, we have uniquely and stably determined the spline s. □

Lemma 3.15:
Let Δ be a tetrahedral partition of a polyhedral domain $\Omega \subset \mathbb{R}^3$ and $T := \langle v_1, v_2, v_3, v_4 \rangle \in \Delta$ a tetrahedron, that shares exactly one face $\langle v_1, v_2, v_4 \rangle$ and the edge $\langle v_2, v_3 \rangle$ with the tetrahedra $\widetilde{T}_1$ and $\widetilde{T}_2$, respectively, with $\widetilde{T}_1, \widetilde{T}_2 \in \Delta$. Moreover, T is refined with a second order partial Worsey-Farin split at the split point v_T in the interior of T, and the split points v_{F_1} and v_{F_2} in the interior of the faces $\langle v_1, v_2, v_3 \rangle$ and $\langle v_2, v_3, v_4 \rangle$, respectively.

Let $\widetilde{\mathcal{S}}_9^2(T_{WF}^2)$ be the superspline space defined in Theorem 3.12. Then, for any spline $s \in \mathcal{S}_9^{2,4,3}(\Delta \setminus T)$, there is a spline $g \in \widetilde{\mathcal{S}}_9^2(T_{WF}^2)$ such that

$$\widetilde{s} := \begin{cases} s, & \text{on } \Delta \setminus T \\ g, & \text{on } T, \end{cases}$$

belongs to $\mathcal{S}_9^{2,4,3}(\Delta)$. Moreover, the spline g is uniquely and stably determined by the value of $g(v_T)$ and the B-coefficients of $\widetilde{s}|_{\Delta \setminus T}$ by the following additional smoothness conditions across the edges and faces of T:

$$\tau_{4,1,4,0}^{\widetilde{T}_j,T_{j,1}}\widetilde{s} = \tau_{0,2,7,0}^{\widetilde{T}_j,T_{j,1}}\widetilde{s} = \tau_{0,0,9,0}^{\widetilde{T}_j,T_{j,1}}\widetilde{s} = \tau_{1,4,4,0}^{\widetilde{T}_{j+1},T_{j,2}}\widetilde{s} = 0,\ j = 1,2, \text{ with } \widetilde{T}_3 := \widetilde{T}_1 \tag{3.64}$$

$$\tau_{4,4-i,1+i,0}^{\widetilde{T}_1,T_3}\widetilde{s} = 0,\ i = 0,\ldots,3 \tag{3.65}$$

$$\begin{aligned} &\tau_{4,1,3,1}^{\widetilde{T}_j,T_{j,1}}\widetilde{s} = \tau_{0,2,6,1}^{\widetilde{T}_j,T_{j,1}}\widetilde{s} = \tau_{0,0,8,1}^{\widetilde{T}_j,T_{j,1}}\widetilde{s} = \tau_{1,4,3,1}^{\widetilde{T}_{j+1},T_{j,2}}\widetilde{s} = 0, \\ &i = 0,\ldots,3,\ j = 1,2, \text{ with } \widetilde{T}_3 := \widetilde{T}_1 \end{aligned} \tag{3.66}$$

$$\tau_{4,2,2,1}^{\widetilde{T}_1,T_3}\widetilde{s} = \tau_{3,3,2,1}^{\widetilde{T}_1,T_3}\widetilde{s} = \tau_{3,2,3,1}^{\widetilde{T}_1,T_3}\widetilde{s} = 0 \tag{3.67}$$

$$\tau_{3,0,4,2}^{\widetilde{T}_1,T_{i,j}}\widetilde{s} = 0,\ i = 1,2,3,\ j = 1,2 \tag{3.68}$$

$$\tau_{3,2,2,2}^{\widetilde{T}_1,T_3}\widetilde{s} = \tau_{2,3,2,2}^{\widetilde{T}_1,T_3}\widetilde{s} = \tau_{2,2,3,2}^{\widetilde{T}_1,T_3}\widetilde{s} = 0 \tag{3.69}$$

$$\tau_{4-i,1+i,0,4}^{\widetilde{T}_2,T_{1,2}}\widetilde{s} = 0,\ i = 0,\ldots,3 \tag{3.70}$$

$$\begin{aligned} &\tau_{3,0,3,3}^{\widetilde{T}_2,T_{1+i,2-i}}\widetilde{s} = \tau_{0,3,3,3}^{\widetilde{T}_2,T_{1+i,2-i}}\widetilde{s} = \tau_{2,2,2,3}^{\widetilde{T}_2,T_{1+i,2-i}}\widetilde{s} = \tau_{1,1,4,3}^{\widetilde{T}_2,T_{1+i,2-i}}\widetilde{s} \\ &= \tau_{2,0,4,3}^{\widetilde{T}_2,T_{1+i,2-i}}\widetilde{s} = \tau_{0,2,4,3}^{\widetilde{T}_2,T_{1+i,2-i}}\widetilde{s} = 0 \end{aligned} \tag{3.71}$$

$$\tau_{3,1,2,3}^{\widetilde{T}_1,T_4}\widetilde{s} = \tau_{2,2,2,3}^{\widetilde{T}_1,T_4}\widetilde{s} = \tau_{2,1,3,3}^{\widetilde{T}_1,T_4}\widetilde{s} = 0 \tag{3.72}$$

Proof:
Using the C^4 supersmoothness conditions at the vertices of T, the C^3 supersmoothness conditions at the edges $\langle v_1, v_2 \rangle$, $\langle v_2, v_3 \rangle$, $\langle v_2, v_4 \rangle$, and $\langle v_1, v_4 \rangle$, and the C^2 smoothness conditions at the face $\langle v_1, v_2, v_4 \rangle$ we can uniquely and stably determine the B-coefficients of g corresponding to the domain points in $(\widetilde{M}1^2_{T_{WF}^2})$, $(\widetilde{M}2^2_{T_{WF}^2})$, $(\widetilde{M}3^2_{T_{WF}^2})$, and $(\widetilde{M}4^2_{T_{WF}^2})$ from Theorem 3.14. From the additional smoothness conditions in (3.64) we can uniquely and stably determine the B-coefficients of g corresponding to the domain points

in ($\widetilde{\text{M}}5^2_{T^2_{WF}}$). Then, from the additional smoothness conditions in (3.65) - (3.72), the B-coefficients of g corresponding to the domain points in ($\widetilde{\text{M}}6^2_{T^2_{WF}}$) - ($\widetilde{\text{M}}13^2_{T^2_{WF}}$) can be uniquely and stably determined. The B-coefficients associated with the last domain point in the minimal determining set $\widetilde{\mathcal{M}}^2_2$, the one in ($\widetilde{\text{M}}14^2_{T^2_{WF}}$), is uniquely determined by the value of $g(v_T)$. Thus, following Theorem 3.14, g is uniquely and stably determined. □

Theorem 3.16:
Let $\hat{\mathcal{M}}^2_2$ be the union of the following sets of domain points in $\mathcal{D}_{T^2_{WF},9}$:

($\hat{\text{M}}1^2_{T^2_{WF}}$) $D_4^{T_{v_i}}(v_i)$, $i = 1, \ldots, 4$, with $v_i \in T_{v_i} \in T^2_{WF}$

($\hat{\text{M}}2^2_{T^2_{WF}}$) $\{\xi^{T_4}_{i,j,k,l}\}^2_{l=0}$, $i,j,k \leq 4$, $i+j+k = 9-l$

($\hat{\text{M}}3^2_{T^2_{WF}}$) $\{\xi^{T_j}_{4-i,2+i,0,3}, \xi^{T_j}_{4-i,0,2+i,3}, \xi^{T_4}_{0,4-i,2+i,3}\}^2_{i=0}$, $j = 3,4$

($\hat{\text{M}}4^2_{T^2_{WF}}$) $\{\xi^{T_3}_{4,4-i,1+i,0}, \xi^{T_3}_{3,4-i,1+i,1}, \xi^{T_3}_{2,4-i,1+i,2}, \xi^{T_3}_{4,3-i,i,2}\}^3_{i=0}$

($\hat{\text{M}}5^2_{T^2_{WF}}$) $\{\xi^{T_3}_{4,4-i,i,1}, \xi^{T_3}_{3,4-i,i,2}\}^4_{i=0}$

($\hat{\text{M}}6^2_{T^2_{WF}}$) $\{\xi^{T_{1,1}}_{1,4,4,0}, \xi^{T_{1,3}}_{4,1,4,0}, \xi^{T_{1,1}}_{2,0,7,0}\xi^{T_{1,1}}_{0,0,9,0}, \xi^{T_{2,2}}_{4,1,4,0}, \xi^{T_{2,3}}_{1,4,4,0}, \xi^{T_{2,2}}_{0,2,7,0}\xi^{T_{2,2}}_{0,0,9,0}\}^2_{i=1}$

($\hat{\text{M}}7^2_{T^2_{WF}}$) $\{\xi^{T_{1,2}}_{4,3,0,2}, \xi^{T_{1,2}}_{3,4,0,2}\}$

($\hat{\text{M}}8^2_{T^2_{WF}}$) $\{\xi^{T_{1,1}}_{1,4,3,1}, \xi^{T_{1,3}}_{4,1,3,1}, \xi^{T_{1,1}}_{2,0,6,1}, \xi^{T_{2,2}}_{4,1,3,1}, \xi^{T_{2,3}}_{1,4,3,1}, \xi^{T_{2,2}}_{0,2,6,1}\}$

($\hat{\text{M}}9^2_{T^2_{WF}}$) $\{\xi^{T_{j,i}}_{3,0,4,2}\}$, $i = 1,2,3$, $j = 1,2$

($\hat{\text{M}}10^2_{T^2_{WF}}$) $\{\xi^{T_{1,2}}_{4-i,1+i,0,4}\}^3_{i=0}$

($\hat{\text{M}}11^2_{T^2_{WF}}$) $\{\xi^{T_{1+i,2-i}}_{3,0,3,3}, \xi^{T_{1+i,2-i}}_{0,3,3,3}, \xi^{T_{1+i,2-i}}_{2,2,2,3}, \xi^{T_{1+i,2-i}}_{1,1,4,3}, \xi^{T_{1+i,2-i}}_{2,0,4,3}, \xi^{T_{1+i,2-i}}_{0,2,4,3}\}^1_{i=0}$

($\hat{\text{M}}12^2_{T^2_{WF}}$) $\{\xi^{T_4}_{3,1,2,3}, \xi^{T_4}_{2,2,2,3}, \xi^{T_4}_{2,1,3,3}\}$

($\hat{\text{M}}13^2_{T^2_{WF}}$) $\{\xi^{T_{1,1}}_{0,0,0,9}\}$

Then $\hat{\mathcal{M}}_2^2$ is a stable minimal determining set for

$$\begin{aligned}\hat{\mathcal{S}}_9^2(T_{WF}^2) := \{s \in C^2(T_{WF}^2) : s|_{\widetilde{T}} \in \mathcal{P}_9^3,\ \forall \widetilde{T} \in T_{WF}^2,\\
& s \in C^4(v_i),\ i = 1,\dots,4,\\
& s \in C^3(e),\ \text{for all edges } e \in T,\\
& s \in C^7(\langle v_{F_i}, v_T\rangle),\ i = 1,2\\
& s \in C^7(v_T),\\
& \tau_{3-i,5+i,1,0}^{T_{j,j},T_{j,3}} s = \tau_{4,4,1,0}^{T_{j,3},T_{j,3-j}} s = 0,\ i = 0,\dots,3,\ j = 1,2,\\
& \tau_{3-i,5+i,0,1}^{T_{j,1},T_{j,3}} s = 0,\ i = 0,\dots,2,\ j = 1,2\},\end{aligned}$$

and the dimension of $\hat{\mathcal{S}}_9^2(T_{WF}^2)$ is equal to 266.

Proof:
First, we show that $\hat{\mathcal{M}}_2^2$ is a stable minimal determining set for $\hat{\mathcal{S}}_9^2(T_{WF}^2)$. To this end we set the B-coefficients c_ξ of a spline $s \in \hat{\mathcal{S}}_9^2(T_{WF}^2)$ to arbitrary values for each domain point $\xi \in \hat{\mathcal{M}}_2^2$ and show that all other B-coefficients of s are uniquely and stably determined.

From the B-coefficients associated with the domain points in $(\hat{M}1_{T_{WF}^2}^2)$, we can uniquely and stably determine the B-coefficients of s corresponding to the domain points in the balls $D_4(v_i)$, $i = 1,\dots,4$, using the C^4 supersmoothness conditions at the vertices of T.

Subsequently, together with the B-coefficients associated with the domain points in $(\hat{M}2_{T_{WF}^2}^2)$ and some in $(\hat{M}3_{T_{WF}^2}^2)$, all B-coefficients of $s|_{T_4}$ corresponding to the domain points within a distance of two from the face $\langle v_1, v_2, v_4\rangle$ and within a distance of three from the corresponding edges are uniquely and stably determined.

Now, using the C^3 supersmoothness conditions at the edges of $\langle v_1, v_2, v_4\rangle$, we can uniquely and stably determine the remaining B-coefficients corresponding to domain points within a distance of three from these edges.

Then, together with the remaining B-coefficients associated with the domain points in $(\hat{M}3_{T_{WF}^2}^2)$ and those in $(\hat{M}4_{T_{WF}^2}^2)$ and $(\hat{M}5_{T_{WF}^2}^2)$, all B-coefficients of $s|_{T_3}$ corresponding to the domain points within a distance of two from the face $\langle v_3, v_1, v_4\rangle$ and within a distance of three from the corresponding edges are uniquely and stably determined.

Next, using the C^3 supersmoothness conditions at the edges of $\langle v_3, v_1, v_4 \rangle$ we can uniquely and stably compute the remaining B-coefficients of s corresponding to domain points within a distance of three from the edges $\langle v_3, v_1 \rangle$ and $\langle v_4, v_3 \rangle$.

Subsequently, we consider the B-coefficients associated with the domain points in the Clough-Tocher split faces $F_1 := \langle v_1, v_2, v_3 \rangle$ and $F_2 := \langle v_2, v_3, v_4 \rangle$ of T_{WF}^2. These can be determined in the same way as in the proof of Theorem 3.12 from the already determined B-coefficients and those corresponding to the domain points in ($\hat{\text{M}}6^2_{T^2_{WF}}$).

Then, following Lemma 2.33 for the case $j = 0$, we can use a C^1 smoothness condition at the edge $\langle v_2, v_3 \rangle$ in order to compute the last undetermined B-coefficient of s corresponding to a domain point within a distance of one from this edge.

Now, using the C^1 and C^2 smoothness conditions at the edge $\langle v_2, v_3 \rangle$ we can uniquely and stably compute the B-coefficients corresponding to the domain points within a distance of two from this edge from the already determined B-coefficients and those associated with the domain points in ($\hat{\text{M}}7^2_{T^2_{WF}}$), by Lemma 2.33 for the case $j = 1$.

Next, we examine the B-coefficients of s associated with the domain points in the shells $R_8^{T_{j,1}}(v_T), R_8^{T_{j,2}}(v_T)$, and $R_8^{T_{j,3}}(v_T)$, $j = 1,2$, which form a triangle F_j^1, respectively. Since the three subtriangles forming the faces $\langle v_1, v_2, v_3 \rangle$ and $\langle v_2, v_3, v_4 \rangle$ are degenerated faces of the subtetrahedra $T_{j,1}$, $T_{j,2}$, and $T_{j,3}, j = 1,2$, since they lie in one common plane, respectively, we can consider the B-coefficients of $s|_{F_j^1}$, $j = 1,2$, as those of a bivariate spline $g_1^{F_j^1} \in \check{\mathcal{S}}_8^2(F^1_{j,CT})$, $j = 1,2$, defined in (A.16). Considering these bivariate splines, we see that the B-coefficients corresponding to the domain points within a distance of three from the vertices and within a distance of two from two of the edges and within a distance of one from the third edge of F_j^1, $j = 1,2$, are already uniquely and stably determined. Thus, together with the B-coefficients associated with the domain points in ($\hat{\text{M}}8^2_{T^2_{WF}}$), all B-coefficients corresponding to the domain points in the minimal determining set in Lemma A.11 are uniquely and stably determined. Therefore, all B-coefficients of $s|_{F_j^1}$, $j = 1,2$, are uniquely and stably determined.

Subsequently, following Lemma 2.33 for the case $j = 0$, we can uniquely and stably compute the remaining undetermined B-coefficients of s corresponding to the domain points within a distance of three from the edge $\langle v_2, v_3 \rangle$ using the C^3 supersmoothness conditions at these edges.

Now, the remaining B-coefficients of s associated with the domain points in the shells $R_7^{T_{j,i}}(v_T)$, $i=1,2,3$, $j=1,2$, can be determined in the same way as in the proof of Theorem 3.12 from the already determined B-coefficients and those associated with the domain points in ($\hat{\text{M}}9^2_{T^2_{WF}}$).

Then, all other undetermined B-coefficients of s can be uniquely and stably computed in the same way as in the proof of Theorem 3.14 from those already determined and the ones corresponding to the domain points in ($\hat{\text{M}}10^2_{T^2_{WF}}$) - ($\hat{\text{M}}13^2_{T^2_{WF}}$).

Then, considering the dimension of $\hat{\mathcal{S}}^2_9(T^2_{WF})$ we see that $\dim \hat{\mathcal{S}}^2_9(T^2_{WF}) = 266 = \#\hat{\mathcal{M}}^2_2$, since $\hat{\mathcal{M}}^2_2$ is a minimal determining set for this superspline space. □

Lemma 3.17:
Let Δ be a tetrahedral partition of a polyhedral domain $\Omega \subset \mathbb{R}^3$ and $T := \langle v_1, v_2, v_3, v_4 \rangle \in \Delta$ a tetrahedron, that shares exactly the two faces $\langle v_1, v_2, v_4 \rangle$ and $\langle v_3, v_1, v_4 \rangle$ with the tetrahedra $\widetilde{T}_1$ and $\widetilde{T}_2$, respectively, with $\widetilde{T}_1, \widetilde{T}_2 \in \Delta$. Moreover, T is refined with a second order partial Worsey-Farin split at the split point v_T in the interior of T, and the split points v_{F_1} and v_{F_2} in the interior of the faces $\langle v_1, v_2, v_3 \rangle$ and $\langle v_2, v_3, v_4 \rangle$, respectively.

Let $\hat{\mathcal{S}}^2_9(T^2_{WF})$ be the superspline space defined in Theorem 3.16. Then, for any spline $s \in \mathcal{S}^{2,4,3}_9(\Delta \setminus T)$, there is a spline $g \in \hat{\mathcal{S}}^2_9(T^2_{WF})$ such that

$$\widetilde{s} := \begin{cases} s, & \text{on } \Delta \setminus T \\ g, & \text{on } T, \end{cases}$$

belongs to $\mathcal{S}^{2,4,3}_9(\Delta)$. Moreover, the spline g is uniquely and stably determined by the value of $g(v_T)$ and the B-coefficients of $\widetilde{s}|_{\Delta \setminus T}$ by the following additional smoothness conditions across the vertices, edges and faces of T:

$$\begin{aligned} &\tau^{\widetilde{T}_1,T_{1,1}}_{1,4,4,0}\widetilde{s} = \tau^{\widetilde{T}_1,T_{1,1}}_{2,0,7,0}\widetilde{s} = \tau^{\widetilde{T}_1,T_{1,1}}_{0,0,9,0}\widetilde{s} = \tau^{\widetilde{T}_2,T_{1,3}}_{4,1,4,0}\widetilde{s} \\ &= \tau^{\widetilde{T}_2,T_{2,2}}_{4,1,4,0}\widetilde{s} = \tau^{\widetilde{T}_2,T_{2,2}}_{0,2,7,0}\widetilde{s} = \tau^{\widetilde{T}_2,T_{2,2}}_{0,0,9,0}\widetilde{s} = \tau^{\widetilde{T}_1,T_{2,3}}_{1,4,4,0}\widetilde{s} = 0 \end{aligned} \tag{3.73}$$

$$\tau^{\widetilde{T}_1,T_{1,2}}_{4,3,0,2}\widetilde{s} = \tau^{\widetilde{T}_1,T_{1,2}}_{3,4,0,2}\widetilde{s} = 0 \tag{3.74}$$

$$\tau^{\widetilde{T}_1,T_{1,1}}_{1,4,3,1}\widetilde{s} = \tau^{\widetilde{T}_1,T_{1,1}}_{2,0,6,1}\widetilde{s} = \tau^{\widetilde{T}_2,T_{1,3}}_{4,1,3,1}\widetilde{s} = \tau^{\widetilde{T}_2,T_{2,2}}_{4,1,3,1}\widetilde{s} = \tau^{\widetilde{T}_2,T_{2,2}}_{0,2,6,1}\widetilde{s} = \tau^{\widetilde{T}_1,T_{2,3}}_{1,4,3,1}\widetilde{s} = 0 \tag{3.75}$$

$$\tau^{\widetilde{T}_j,T_{i,j}}_{3,0,4,2}\widetilde{s} = \tau^{\widetilde{T}_2,T_{1,3}}_{3,0,4,2}\widetilde{s} = \tau^{\widetilde{T}_1,T_{2,3}}_{3,0,4,2}\widetilde{s} = 0, \; i,j = 1,2 \tag{3.76}$$

$$\tau_{4-i,1+i,0,4}^{\widetilde{T}_1,T_{1,2}}\widetilde{s} = 0,\ i = 0,\dots,3 \tag{3.77}$$

$$\begin{aligned}&\tau_{3,0,3,3}^{\widetilde{T}_{1+i},T_{1+i,2-i}}\widetilde{s} = \tau_{0,3,3,3}^{\widetilde{T}_{1+i},T_{1+i,2-i}}\widetilde{s} = \tau_{2,2,2,3}^{\widetilde{T}_{1+i},T_{1+i,2-i}}\widetilde{s} = \tau_{1,1,4,3}^{\widetilde{T}_{1+i},T_{1+i,2-i}}\widetilde{s}\\ &= \tau_{2,0,4,3}^{\widetilde{T}_{1+i},T_{1+i,2-i}}\widetilde{s} = \tau_{0,2,4,3}^{\widetilde{T}_{1+i},T_{1+i,2-i}}\widetilde{s} = 0,\ i = 0,1\end{aligned} \tag{3.78}$$

$$\tau_{3,1,2,3}^{\widetilde{T}_1,T_4}\widetilde{s} = \tau_{2,2,2,3}^{\widetilde{T}_1,T_4}\widetilde{s} = \tau_{2,1,3,3}^{\widetilde{T}_1,T_4}\widetilde{s} = 0 \tag{3.79}$$

Proof:
Using the C^4 supersmoothness conditions at the vertices of T, the C^3 supersmoothness conditions at the five edges $\langle v_1,v_2\rangle$, $\langle v_1,v_3\rangle$, $\langle v_1,v_4\rangle$, $\langle v_2,v_4\rangle$, and $\langle v_3,v_4\rangle$, and the C^2 smoothness conditions at the faces $\langle v_1,v_2,v_4\rangle$ and $\langle v_3,v_1,v_4\rangle$ we can uniquely and stably determine the B-coefficients of g corresponding to the domain points in ($\hat{\mathrm{M}}1^2_{T^2_{WF}}$), ($\hat{\mathrm{M}}2^2_{T^2_{WF}}$), ($\hat{\mathrm{M}}3^2_{T^2_{WF}}$), ($\hat{\mathrm{M}}4^2_{T^2_{WF}}$), and ($\hat{\mathrm{M}}5^2_{T^2_{WF}}$) from Theorem 3.16. From the additional smoothness conditions in (3.73) we can uniquely and stably determine the B-coefficients of g corresponding to the domain points in ($\hat{\mathrm{M}}6^2_{T^2_{WF}}$). Now, using the additional smoothness conditions in (3.74) - (3.79) we can uniquely and stably determine the B-coefficients associated with the domain points in ($\hat{\mathrm{M}}6^2_{T^2_{WF}}$) - ($\hat{\mathrm{M}}12^2_{T^2_{WF}}$). The last undetermined B-coefficient corresponding to a domain point in the minimal determining set $\hat{\mathcal{M}}^2_2$ is uniquely determined by the value of $g(v_T)$. Thus, following Theorem 3.16, g is uniquely and stably determined. □

Theorem 3.18:
Let $\bar{\mathcal{M}}^2_2$ be the union of the following sets of domain points in $\mathcal{D}_{T^2_{WF},9}$:

($\bar{\mathrm{M}}1^2_{T^2_{WF}}$) $D_4^{T_{v_i}}(v_i)$, $i = 1,\dots,4$, with $v_i \in T_{v_i} \in T^2_{WF}$

($\bar{\mathrm{M}}2^2_{T^2_{WF}}$) $\{\xi^{T_4}_{i,j,k,l}\}_{l=0}^{2}$, $i,j,k \le 4$, $i+j+k = 9-l$

($\bar{\mathrm{M}}3^2_{T^2_{WF}}$) $\{\xi^{T_j}_{4-i,2+i,0,3}, \xi^{T_j}_{4-i,0,2+i,3}, \xi^{T_4}_{0,4-i,2+i,3}\}_{i=0}^{2}$, $j = 3,4$

($\bar{\mathrm{M}}4^2_{T^2_{WF}}$) $\{\xi^{T_3}_{4,4-i,1+i,0}, \xi^{T_3}_{3,4-i,1+i,1}, \xi^{T_3}_{2,4-i,1+i,2}, \xi^{T_3}_{4,3-i,i,2}\}_{i=0}^{3}$

($\bar{\mathrm{M}}5^2_{T^2_{WF}}$) $\{\xi^{T_3}_{4,4-i,i,1}, \xi^{T_3}_{3,4-i,i,2}\}_{i=0}^{4}$

($\bar{\mathrm{M}}6^2_{T^2_{WF}}$) $E_3^{T_{1,2}}(\langle v_2,v_3\rangle) \setminus (D_4^{T_{1,2}}(v_2) \cup D_4^{T_{1,2}}(v_3))$

($\bar{\mathrm{M}}7^2_{T^2_{WF}}$) $\{\xi^{T_{j,i}}_{3,0,4,2}\}$, $i = 1,2,3$, $j = 1,2$

($\bar{M}8^2_{T^2_{WF}}$) $\{\xi^{T_{1,2}}_{4-i,1+i,0,4}\}_{i=0}^{3}$
($\bar{M}9^2_{T^2_{WF}}$) $\{\xi^{T_{1+i,2-i}}_{3,0,3,3}, \xi^{T_{1+i,2-i}}_{0,3,3,3}, \xi^{T_{1+i,2-i}}_{2,2,2,3}, \xi^{T_{1+i,2-i}}_{1,1,4,3}, \xi^{T_{1+i,2-i}}_{2,0,4,3}, \xi^{T_{1+i,2-i}}_{0,2,4,3}\}_{i=0}^{1}$
($\bar{M}10^2_{T^2_{WF}}$) $\{\xi^{T_4}_{3,1,2,3}, \xi^{T_4}_{2,2,2,3}, \xi^{T_4}_{2,1,3,3}\}$
($\bar{M}11^2_{T^2_{WF}}$) $\{\xi^{T_{1,1}}_{0,0,0,9}\}$
Then $\bar{\mathcal{M}}^2_2$ is a stable minimal determining set for

$$\begin{aligned}\bar{\mathcal{S}}^2_9(T^2_{WF}) := \{s \in C^2(T^2_{WF}) : s|_{\widetilde{T}} \in \mathcal{P}^3_9,\ \forall \widetilde{T} \in T^2_{WF},\\ s \in C^4(v_i),\ i = 1,\dots,4,\\ s \in C^3(e),\ \text{for all edges } e \in T,\\ s \in C^7(\langle v_{F_i}, v_T\rangle),\ i = 1,2\\ s \in C^7(v_T),\\ \tau^{T_{j,i},T_{j,i+1}}_{3,5,1,0} s = \tau^{T_{j,i},T_{j,i+1}}_{3,5,0,1} s = 0,\\ i = 1,2,3,\ j = 1,2,\ \text{with } T_{j,4} := T_{j,1}\},\end{aligned}$$

and the dimension of $\bar{\mathcal{S}}^2_9(T^2_{WF})$ is equal to 270.

Proof:
We first show that $\bar{\mathcal{M}}^2_2$ is a stable minimal determining set for $\bar{\mathcal{S}}^2_9(T^2_{WF})$. Therefore, we set the B-coefficients c_ξ of a spline $s \in \bar{\mathcal{S}}^2_9(T^2_{WF})$ to arbitrary values for each domain point $\xi \in \bar{\mathcal{M}}^2_2$ and show that all other B-coefficients of s can be uniquely and stably computed from those.

Considering the B-coefficients associated with the domain points in the balls $D_4(v_i)$, $i = 1,\dots,4$, we see that these are uniquely and stably determined from those corresponding to the domain points in ($\bar{M}1^2_{T^2_{WF}}$) by the C^4 supersmoothness conditions at the four vertices of T.

Then, the B-coefficients corresponding to the domain points within a distance of two from the faces $\langle v_1, v_2, v_4\rangle$ and $\langle v_3, v_1, v_4\rangle$, and within a distance of three from the corresponding five edges, are uniquely and stably determined from the B-coefficients associated with the domain points in ($\bar{M}2^2_{T^2_{WF}}$) - ($\bar{M}5^2_{T^2_{WF}}$) in the same way as in the proof of Theorem 3.16.

Now, we can uniquely and stably compute all remaining B-coefficients corresponding to domain points within a distance of three from the edge $\langle v_2, v_3\rangle$ from those associated with the domain points in ($\bar{M}6^2_{T^2_{WF}}$) .

Next, we examine the B-coefficients in the shells $R_9(v_T)^{T_{j,i}}$ and $R_8(v_T)^{T_{j,i}}$, for $i = 1,2,3$ and $j = 1,2$. These can be uniquely and stably computed in the same way as those corresponding to the domain points in the shells $R_9(v_T)^{T_{1,i}}$ and $R_8(v_T)^{T_{1,i}}$, $i = 1,2,3$, in the proof of Theorem 3.4 from the already determined B-coefficients by applying Lemma A.13 and Lemma A.8.

Then, the remaining undetermined B-coefficients of s can be uniquely and stably determined in the same way as in the proof of Theorem 3.16 from the already determined B-coefficients and those associated with the domain points in ($\bar{\mathrm{M}}7^2_{T^2_{WF}}$) - ($\bar{\mathrm{M}}11^2_{T^2_{WF}}$).

Finally, considering the dimension of the space $\bar{\mathcal{S}}^2_9(T^2_{WF})$ we can state that $\dim \bar{\mathcal{S}}^2_9(T^2_{WF}) = 270$, since $\bar{\mathcal{M}}^2_2$ is a minimal determining set for $\bar{\mathcal{S}}^2_9(T^2_{WF})$ and the cardinality of this set is equal to 270. □

Lemma 3.19:
Let Δ be a tetrahedral partition of a polyhedral domain $\Omega \subset \mathbb{R}^3$ and $T := \langle v_1, v_2, v_3, v_4 \rangle \in \Delta$ a tetrahedron, that shares exactly the two faces $\langle v_1, v_2, v_4 \rangle$ and $\langle v_3, v_1, v_4 \rangle$ with the tetrahedra $\widetilde{T}_1$ and $\widetilde{T}_2$, respectively, and the edge $\langle v_2, v_3 \rangle$ with the tetrahedron $\hat{T}$, with $\widetilde{T}_1, \widetilde{T}_2, \hat{T} \in \Delta$. Moreover, T is refined with a second order partial Worsey-Farin split at the split point v_T in the interior of T, and the split points v_{F_1} and v_{F_2} in the interior of the faces $\langle v_1, v_2, v_3 \rangle$ and $\langle v_2, v_3, v_4 \rangle$, respectively.

Let $\bar{\mathcal{S}}^2_9(T^2_{WF})$ be the superspline space defined in Theorem 3.18. Then, for any spline $s \in \mathcal{S}^{2,4,3}_9(\Delta \setminus T)$, there is a spline $g \in \bar{\mathcal{S}}^2_9(T^2_{WF})$ such that

$$\widetilde{s} := \begin{cases} s, & \text{on } \Delta \setminus T \\ g, & \text{on } T, \end{cases}$$

belongs to $\mathcal{S}^{2,4,3}_9(\Delta)$. Moreover, g is uniquely and stably determined by the value of $g(v_T)$ and the B-coefficients of $\widetilde{s}|_{\Delta \setminus T}$ by the following additional smoothness conditions across the edges and faces of T:

$$\tau^{\widetilde{T}_i, T_{i,i}}_{3,0,4,2}\widetilde{s} = \tau^{\widetilde{T}_{3-i}, T_{i,3}}_{3,0,4,2}\widetilde{s} = \tau^{\hat{T}, T_{1,2}}_{3,0,4,2}\widetilde{s} = \tau^{\hat{T}, T_{2,1}}_{3,0,4,2}\widetilde{s} = 0, \; i = 1,2 \tag{3.80}$$

$$\tau^{\hat{T}, T_{1,2}}_{4-i,1+i,0,4}\widetilde{s} = 0, \; i = 0, \ldots, 3 \tag{3.81}$$

$$\begin{aligned} \tau^{\hat{T}, T_{1+i,2-i}}_{3,0,3,3}\widetilde{s} &= \tau^{\hat{T}, T_{1+i,2-i}}_{0,3,3,3}\widetilde{s} = \tau^{\hat{T}, T_{1+i,2-i}}_{2,2,2,3}\widetilde{s} = \tau^{\hat{T}, T_{1+i,2-i}}_{1,1,4,3}\widetilde{s} = \tau^{\hat{T}, T_{1+i,2-i}}_{2,0,4,3}\widetilde{s} \\ &= \tau^{\hat{T}, T_{1+i,2-i}}_{0,2,4,3}\widetilde{s} = 0, \; i = 0,1 \end{aligned} \tag{3.82}$$

$$\tau_{3,1,2,3}^{\widetilde{T}_1,T_4}\widetilde{s} = \tau_{2,2,2,3}^{\widetilde{T}_1,T_4}\widetilde{s} = \tau_{2,1,3,3}^{\widetilde{T}_1,T_4}\widetilde{s} = 0 \tag{3.83}$$

Proof:
Using the C^4 supersmoothness conditions at the vertices of T, the C^3 supersmoothness conditions at the edges of T, and the C^2 smoothness conditions at the faces $\langle v_1,v_2,v_4\rangle$ and $\langle v_3,v_1,v_4\rangle$ we can uniquely and stably determine the B-coefficients of g corresponding to the domain points in ($\bar{M}1^2_{T^2_{WF}}$) - ($\bar{M}6^2_{T^2_{WF}}$) from Theorem 3.18. Then, from the additional smoothness conditions in (3.80) - (3.83) we can uniquely and stably determine the B-coefficients associated with the domain points in ($\bar{M}7^2_{T^2_{WF}}$) - ($\bar{M}10^2_{T^2_{WF}}$). The B-coefficients associated with the last domain point in the minimal determining set $\bar{\mathcal{M}}^2_2$ from Theorem 3.18 is equal to the value of $g(v_T)$. Thus, following Theorem 3.18, g is uniquely and stably determined. □

Theorem 3.20:
Let $\check{\mathcal{M}}^2_2$ be the union of the following sets of domain points in $\mathcal{D}_{T^2_{WF},9}$:

($\check{M}1^2_{T^2_{WF}}$) $D_4^{T_{v_i}}(v_i)$, $i = 1,\dots,4$, with $v_i \in T_{v_i} \in T^2_{WF}$

($\check{M}2^2_{T^2_{WF}}$) $\{\xi^{T_4}_{i,j,k,l}\}_{l=0}^{2}$, $i,j,k \leq 4$, $i+j+k = 9-l$

($\check{M}3^2_{T^2_{WF}}$) $\{\xi^{T_j}_{4-i,2+i,0,3}, \xi^{T_j}_{4-i,0,2+i,3}, \xi^{T_4}_{0,4-i,2+i,3}\}_{i=0}^{2}$, $j = 3,4$

($\check{M}4^2_{T^2_{WF}}$) $\{\xi^{T_3}_{4,4-i,1+i,0}, \xi^{T_3}_{3,4-i,1+i,1}, \xi^{T_3}_{2,4-i,1+i,2}, \xi^{T_3}_{4,3-i,i,2}\}_{i=0}^{3}$

($\check{M}5^2_{T^2_{WF}}$) $\{\xi^{T_3}_{4,4-i,i,1}, \xi^{T_3}_{3,4-i,i,2}\}_{i=0}^{4}$

($\check{M}6^2_{T^2_{WF}}$) $E_3^{T_{1,2}}(\langle v_2,v_3\rangle) \setminus (D_4^{T_{1,2}}(v_2) \cup D_4^{T_{1,2}}(v_3))$

($\check{M}7^2_{T^2_{WF}}$) $\{\xi^{T_{1,1}}_{1,0,8-i,i}, \xi^{T_{1,1}}_{0,1,8-i,i}, \xi^{T_{1,1}}_{0,0,9-i,i}\}_{i=0}^{1}$

($\check{M}8^2_{T^2_{WF}}$) $\{\xi^{T_{j,i}}_{3,0,4,2}\}$, $i = 1,2,3$, $j = 1,2$

($\check{M}9^2_{T^2_{WF}}$) $\{\xi^{T_{1,2}}_{4-i,1+i,0,4}\}_{i=0}^{3}$

($\check{M}10^2_{T^2_{WF}}$) $\{\xi^{T_{1+i,2-i}}_{3,0,3,3}, \xi^{T_{1+i,2-i}}_{0,3,3,3}, \xi^{T_{1+i,2-i}}_{2,2,2,3}, \xi^{T_{1+i,2-i}}_{1,1,4,3}, \xi^{T_{1+i,2-i}}_{2,0,4,3}, \xi^{T_{1+i,2-i}}_{0,2,4,3}\}_{i=0}^{1}$

($\check{M}11^2_{T^2_{WF}}$) $\{\xi^{T_4}_{3,1,2,3}, \xi^{T_4}_{2,2,2,3}, \xi^{T_4}_{2,1,3,3}\}$

($\check{M}12^2_{T^2_{WF}}$) $\{\xi^{T_{1,1}}_{0,0,0,9}\}$

Then $\check{\mathcal{M}}_2^2$ is a stable minimal determining set for

$$\begin{aligned}\check{\mathcal{S}}_9^2(T_{WF}^2) := \{s \in C^2(T_{WF}^2) : s|_{\widetilde{T}} \in \mathcal{P}_9^3,\ \forall \widetilde{T} \in T_{WF}^2,\\ s \in C^4(v_i),\ i = 1,\dots,4,\\ s \in C^3(e),\ \text{for all edges } e \in T,\\ s \in C^7(\langle v_{F_i}, v_T\rangle),\ i = 1,2\\ s \in C^7(v_T),\\ \tau_{3,5,1,0}^{T_{2,i},T_{2,i+1}} s = 0,\ i = 1,2,3, \text{ with } T_{2,4} := T_{2,1},\\ \tau_{3,5,0,1}^{T_{2,i},T_{2,i+1}} s = 0,\ i = 1,2,3, \text{ with } T_{2,4} := T_{2,1}\},\end{aligned}$$

and the dimension of $\check{\mathcal{S}}_9^2(T_{WF}^2)$ is equal to 276.

Proof:
First, we show that $\check{\mathcal{M}}_2^2$ is a stable minimal determining set for the superspline space $\check{\mathcal{S}}_9^2(T_{WF}^2)$. To this end we set the B-coefficients c_ξ of a spline $s \in \check{\mathcal{S}}_9^2(T_{WF}^2)$ to arbitrary values for each domain point $\xi \in \check{\mathcal{M}}_2^2$ and show that all other B-coefficients of s are uniquely and stably determined.

The B-coefficients corresponding to the domain points within a distance of four from the vertices of T, within a distance of three from the edges of T, and within a distance of two from the faces $\langle v_1, v_2, v_4\rangle$ and $\langle v_3, v_1, v_4\rangle$ can be uniquely and stably determined from those associated with the domain points in ($\check{\text{M}}1^2_{T^2_{WF}}$) -($\check{\text{M}}6^2_{T^2_{WF}}$) in the same way as in the proof of Theorem 3.18.

Now, we consider the B-coefficients associated with the domain points in the Clough-Tocher split face $F := \langle v_1, v_2, v_3\rangle$. The three subtriangles forming this face are degenerated faces of the three subtetrahedra $T_{1,1}, T_{1,2}$, and $T_{1,3}$, since they lie in one common plane. Thus, we can regard the B-coefficients of $s|_F$ as those of a bivariate spline $g_0 \in \check{\mathcal{S}}_9^2(F_{CT})$ defined in (A.24). Investigating g_0, it can be seen that the B-coefficients corresponding to the domain points within a distance of four from the vertices of F and within a distance of three from the edges of F are already uniquely and stably determined. Together with the B-coefficients associated with the domain points in ($\check{\text{M}}7^2_{T^2_{WF}}$) with $i = 0$, we have determined all B-coefficients corresponding to the domain points in the minimal determining set in Lemma A.15. Therefore, following Lemma A.15, all B-coefficients of g_0 are uniquely and stably determined, and thus also all B-coefficients of $s|_F$.

Next, we investigate the B-coefficients of s corresponding to the domain points in the shells $R_8^{T_{1,1}}(v_T), R_8^{T_{1,2}}(v_T)$, and $R_8^{T_{1,3}}(v_T)$, which form a triangle F^1. Following the same arguments as above, these B-coefficients can be regarded as those of a bivariate spline g_1 in the space $\dot{\mathcal{S}}_8^2(F^1_{CT})$ defined in (A.18). Examining this spline, we see that the B-coefficients corresponding to the domain points within a distance of three from the vertices and within a distance of two from the edges of F^1 are already uniquely and stably determined. Then, together with the B-coefficients associated with the domain points in ($\check{\text{M}}7^2_{T^2_{WF}}$) with $i = 1$, all B-coefficients corresponding to the domain points in the minimal determining set in Lemma A.12 are uniquely and stably determined. Thus, by Lemma A.12 the remaining B-coefficients of g_1 are uniquely and stably determined, and therefore also all B-coefficients of $s|_{F^1}$.

Then, the remaining undetermined B-coefficients of s can be computed from those already determined and the ones associated with the domain points in ($\check{\text{M}}8^2_{T^2_{WF}}$) - ($\check{\text{M}}12^2_{T^2_{WF}}$) in the same way as in the proof of Theorem 3.16.

Now, considering the dimension of the space $\check{\mathcal{S}}_9^2(T^2_{WF})$ we can see that $\dim \check{\mathcal{S}}_9^2(T^2_{WF}) = 276$, since the cardinality of the minimal determining set $\check{\mathcal{M}}_2^2$ is equal to 276. □

Lemma 3.21:
Let Δ be a tetrahedral partition of a polyhedral domain $\Omega \subset \mathbb{R}^3$ and $T := \langle v_1, v_2, v_3, v_4 \rangle \in \Delta$ a tetrahedron, that shares exactly the three faces $\langle v_1, v_2, v_4 \rangle$, $\langle v_3, v_1, v_4 \rangle$, and $\langle v_1, v_2, v_3 \rangle$ with the tetrahedra $\widetilde{T}_1$, $\widetilde{T}_2$, and $\widetilde{T}_3$, respectively, with $\widetilde{T}_1, \widetilde{T}_2, \widetilde{T}_3 \in \Delta$. Moreover, T is refined with a second order partial Worsey-Farin split at the split point v_T in the interior of T, and the split points v_{F_1} and v_{F_2} in the interior of the faces $\langle v_1, v_2, v_3 \rangle$ and $\langle v_2, v_3, v_4 \rangle$, respectively. The tetrahedron $\widetilde{T}_3$ is also subjected to a partial Worsey-Farin split, such that the common face with T is refined with a Clough-Tocher split. The three subtetrahedra are denoted by $\widetilde{T}_{3,i}$, $i = 1,2,3$, where $\widetilde{T}_{3,i}$ has a common face with $T_{1,i}$.

Let $\check{\mathcal{S}}_9^2(T^2_{WF})$ be the superspline space defined in Theorem 3.20. Then, for any spline $s \in \mathcal{S}_9^{2,4,3}(\Delta \setminus T)$, there is a spline $g \in \check{\mathcal{S}}_9^2(T^2_{WF})$ such that

$$\widetilde{s} := \begin{cases} s, & \text{on } \Delta \setminus T \\ g, & \text{on } T, \end{cases}$$

belongs to $\mathcal{S}_9^{2,4,3}(\Delta)$. Moreover, the spline g is uniquely and stably determined by the value of $g(v_T)$ and the B-coefficients of $\widetilde{s}|_{\Delta\setminus T}$ by the following additional smoothness conditions across the edges and faces of T:

$$\tau_{3,0,4,2}^{\widetilde{T}_{3,2},T_{2,1}}\widetilde{s} = \tau_{3,0,4,2}^{\widetilde{T}_2,T_{2,2}}\widetilde{s} = \tau_{3,0,4,2}^{\widetilde{T}_1,T_{2,3}}\widetilde{s} = 0 \tag{3.84}$$

$$\tau_{4-i,1+i,0,4}^{\widetilde{T}_{3,2},T_{1,2}}\widetilde{s} = 0,\ i = 0,\dots,3 \tag{3.85}$$

$$\begin{aligned}&\tau_{3,0,3,3}^{\widetilde{T}_{3,2},T_{1+i,2-i}}\widetilde{s} = \tau_{0,3,3,3}^{\widetilde{T}_{3,2},T_{1+i,2-i}}\widetilde{s} = \tau_{2,2,2,3}^{\widetilde{T}_{3,2},T_{1+i,2-i}}\widetilde{s} = \tau_{1,1,4,3}^{\widetilde{T}_{3,2},T_{1+i,2-i}}\widetilde{s}\\ &= \tau_{2,0,4,3}^{\widetilde{T}_{3,2},T_{1+i,2-i}}\widetilde{s} = \tau_{0,2,4,3}^{\widetilde{T}_{3,2},T_{1+i,2-i}}\widetilde{s} = 0,\ i = 0,1\end{aligned} \tag{3.86}$$

$$\tau_{3,1,2,3}^{\widetilde{T}_1,T_4}\widetilde{s} = \tau_{2,2,2,3}^{\widetilde{T}_1,T_4}\widetilde{s} = \tau_{2,1,3,3}^{\widetilde{T}_1,T_4}\widetilde{s} = 0 \tag{3.87}$$

Proof:
Using the C^4 supersmoothness conditions at the vertices of T, the C^3 supersmoothness conditions at the edges of T, and the C^2 smoothness conditions at the faces $\langle v_1,v_2,v_4\rangle$, $\langle v_3,v_1,v_4\rangle$, and $\langle v_1,v_2,v_3\rangle$ we can uniquely and stably determine the B-coefficients of g corresponding to the domain points in ($\check{\mathrm{M}}1^2_{T^2_{WF}}$) - ($\check{\mathrm{M}}7^2_{T^2_{WF}}$), and in ($\check{\mathrm{M}}8^2_{T^2_{WF}}$) for $j = 1$, from Theorem 3.20. Then, using the additional smoothness conditions in (3.84) - (3.87) we can uniquely and stably determine the B-coefficients associated with the remaining domain points in ($\check{\mathrm{M}}8^2_{T^2_{WF}}$), and those in ($\check{\mathrm{M}}9^2_{T^2_{WF}}$) - ($\check{\mathrm{M}}11^2_{T^2_{WF}}$). The last B-coefficients corresponding to the domain point $\xi_{0,0,0,9}^{T_{1,1}}$ in $\check{\mathcal{M}}_2^2$ is equal to the value of $g(v_T)$. Thus, following Theorem 3.20, g is uniquely and stably determined. □

Theorem 3.22:
Let $\dot{\mathcal{M}}_2^2$ be the union of the following sets of domain points in $\mathcal{D}_{T^2_{WF},9}$:

($\dot{\mathrm{M}}1^2_{T^2_{WF}}$) $D_4^{T_{v_i}}(v_i)$, $i = 1,\dots,4$, with $v_i \in T_{v_i} \in T^2_{WF}$

($\dot{\mathrm{M}}2^2_{T^2_{WF}}$) $\{\xi_{i,j,k,l}^{T_4}\}_{l=0}^2$, $i,j,k \le 4$, $i+j+k = 9-l$

($\dot{\mathrm{M}}3^2_{T^2_{WF}}$) $\{\xi_{4-i,2+i,0,3}^{T_j}, \xi_{4-i,0,2+i,3}^{T_j}, \xi_{0,4-i,2+i,3}^{T_4}\}_{i=0}^2$, $j = 3,4$

($\dot{\mathrm{M}}4^2_{T^2_{WF}}$) $\{\xi_{4,4-i,1+i,0}^{T_3}, \xi_{3,4-i,1+i,1}^{T_3}, \xi_{2,4-i,1+i,2}^{T_3}, \xi_{4,3-i,i,2}^{T_3}\}_{i=0}^3$

($\dot{\mathrm{M}}5^2_{T^2_{WF}}$) $\{\xi_{4,4-i,i,1}^{T_3}, \xi_{3,4-i,i,2}^{T_3}\}_{i=0}^4$

($\dot{\mathrm{M}}6^2_{T^2_{WF}}$) $E_3^{T_{1,2}}(\langle v_2,v_3\rangle) \setminus (D_4^{T_{1,2}}(v_2) \cup D_4^{T_{1,2}}(v_3))$

($\dot{M}7^2_{T^2_{WF}}$) $\{\xi^{T_{j,1}}_{1,0,8-i,i}, \xi^{T_{j,1}}_{0,1,8-i,i}, \xi^{T_{j,1}}_{0,0,9-i,i}\}_{i=0}^{1}$, $j = 1,2$
($\dot{M}8^2_{T^2_{WF}}$) $\{\xi^{T_{j,i}}_{3,0,4,2}\}$, $i = 1,2,3$, $j = 1,2$
($\dot{M}9^2_{T^2_{WF}}$) $\{\xi^{T_{1,2}}_{4-i,1+i,0,4}\}_{i=0}^{3}$
($\dot{M}10^2_{T^2_{WF}}$) $\{\xi^{T_{1+i,2-i}}_{3,0,3,3}, \xi^{T_{1+i,2-i}}_{0,3,3,3}, \xi^{T_{1+i,2-i}}_{2,2,2,3}, \xi^{T_{1+i,2-i}}_{1,1,4,3}, \xi^{T_{1+i,2-i}}_{2,0,4,3}, \xi^{T_{1+i,2-i}}_{0,2,4,3}\}_{i=0}^{1}$
($\dot{M}11^2_{T^2_{WF}}$) $\{\xi^{T_4}_{3,1,2,3}, \xi^{T_4}_{2,2,2,3}, \xi^{T_4}_{2,1,3,3}\}$
($\dot{M}12^2_{T^2_{WF}}$) $\{\xi^{T_{1,1}}_{0,0,0,9}\}$

Then $\dot{\mathcal{M}}^2_2$ is a stable minimal determining set for

$$\begin{aligned}\dot{\mathcal{S}}^2_9(T^2_{WF}) := \{s \in C^2(T^2_{WF}) : s|_{\widetilde{T}} \in \mathcal{P}^3_9, \forall \widetilde{T} \in T^2_{WF},\\ s \in C^4(v_i),\ i = 1,\dots,4,\\ s \in C^3(e), \text{ for all edges } e \in T,\\ s \in C^7(\langle v_{F_i}, v_T\rangle),\ i = 1,2\\ s \in C^7(v_T)\},\end{aligned}$$

and the dimension of $\dot{\mathcal{S}}^2_9(T^2_{WF})$ is equal to 282.

Proof:
We first show that $\dot{\mathcal{M}}^2_2$ is a stable minimal determining set for the superspline space $\dot{\mathcal{S}}^2_9(T^2_{WF})$. Therefore, we set the B-coefficients c_ξ of a spline $s \in \dot{\mathcal{S}}^2_9(T^2_{WF})$ to arbitrary values for each domain point $\xi \in \dot{\mathcal{M}}^2_2$ and show that all other B-coefficients of s are uniquely and stably determined.

The B-coefficients corresponding to the domain points within a distance of two from the faces $\langle v_1, v_2, v_3\rangle$, $\langle v_3, v_1, v_4\rangle$, and $\langle v_1, v_2, v_4\rangle$, within a distance of three from the edges of T, and within a distance of four from the vertices of T are uniquely and stably determined in the same way as in the proof of Theorem 3.20 from the B-coefficients associated with the domain points in ($\dot{M}1^2_{T^2_{WF}}$) - ($\dot{M}6^2_{T^2_{WF}}$), and those in ($\dot{M}7^2_{T^2_{WF}}$) with $j = 1$.

Now, the B-coefficients of s associated with the domain points in the shells $R^{T_{2,i}}_9(v_T)$ and $R^{T_{2,i}}_8(v_T)$, $i = 1,2,3$, can be uniquely and stably computed from the already determined B-coefficients and those corresponding to the domain points in ($\dot{M}7^2_{T^2_{WF}}$) with $j = 2$ in the same way as the B-coefficients associated with the domain points in the rings $R^{T_{1,i}}_9(v_T)$ and $R^{T_{1,i}}_8(v_T)$, $i = 1,2,3$, in Theorem 3.20.

The remaining undetermined B-coefficients of s can also be determined in the same way as in the proof of Theorem 3.20 from the B-coefficients corresponding to the domain points in $(\dot{M}8^2_{T^2_{WF}})$ - $(\dot{M}12^2_{T^2_{WF}})$.

Considering the dimension of $\dot{\mathcal{S}}^2_9(T^2_{WF})$ we can see that $\dim \dot{\mathcal{S}}^2_9(T^2_{WF}) = 282 = \#\dot{\mathcal{M}}^2_2$, since $\dot{\mathcal{M}}^2_2$ is a minimal determining set for $\dot{\mathcal{S}}^2_9(T^2_{WF})$. □

Lemma 3.23:
Let Δ be a tetrahedral partition of a polyhedral domain $\Omega \subset \mathbb{R}^3$ and $T := \langle v_1, v_2, v_3, v_4 \rangle \in \Delta$ a tetrahedron, that shares the four faces $\langle v_1, v_2, v_4 \rangle$, $\langle v_3, v_1, v_4 \rangle$, $\langle v_1, v_2, v_3 \rangle$, and $\langle v_2, v_3, v_4 \rangle$ with the tetrahedra $\widetilde{T}_1$, $\widetilde{T}_2$, $\widetilde{T}_3$, and $\widetilde{T}_4$, respectively, with $\widetilde{T}_i \in \Delta$, $i = 1, \ldots, 4$. Moreover, T is refined with a second order partial Worsey-Farin split at the split point v_T in the interior of T, and the split points v_{F_1} and v_{F_2} in the interior of the faces $\langle v_1, v_2, v_3 \rangle$ and $\langle v_2, v_3, v_4 \rangle$, respectively. The tetrahedra $\widetilde{T}_3$ and $\widetilde{T}_3$ are also subjected to a partial Worsey-Farin split, such that the common face with T is refined with a Clough-Tocher split, respectively. The three subtetrahedra are denoted by $\widetilde{T}_{j,i}$, $i = 1,2,3$, $j = 3,4$, respectively, where $\widetilde{T}_{j,i}$ has a common face with $T_{j-2,i}$.

Let $\dot{\mathcal{S}}^2_9(T^2_{WF})$ be the superspline space defined in Theorem 3.22. Then, for any spline $s \in \mathcal{S}^{2,4,3}_9(\Delta \setminus T)$, there is a spline $g \in \dot{\mathcal{S}}^2_9(T^2_{WF})$ such that

$$\widetilde{s} := \begin{cases} s, & \text{on } \Delta \setminus T \\ g, & \text{on } T, \end{cases}$$

belongs to $\mathcal{S}^{2,4,3}_9(\Delta)$. Moreover, g is uniquely and stably determined by the value of $g(v_T)$ and the B-coefficients of $\widetilde{s}|_{\Delta \setminus T}$ by the following additional smoothness conditions across the faces of T:

$$\tau^{\widetilde{T}_{3,2}, T_{1,2}}_{4-i,1+i,0,4} \widetilde{s} = 0, \; i = 0, \ldots, 3 \tag{3.88}$$

$$\begin{aligned} &\tau^{\widetilde{T}_{3+i,2-i}, T_{1+i,2-i}}_{3,0,3,3} \widetilde{s} = \tau^{\widetilde{T}_{3+i,2-i}, T_{1+i,2-i}}_{0,3,3,3} \widetilde{s} = \tau^{\widetilde{T}_{3+i,2-i}, T_{1+i,2-i}}_{2,2,2,3} \widetilde{s} \\ &= \tau^{\widetilde{T}_{3+i,2-i}, T_{1+i,2-i}}_{1,1,4,3} \widetilde{s} = \tau^{\widetilde{T}_{3+i,2-i}, T_{1+i,2-i}}_{2,0,4,3} \widetilde{s} = \tau^{\widetilde{T}_{3+i,2-i}, T_{1+i,2-i}}_{0,2,4,3} \widetilde{s} = 0, \; i = 0,1 \end{aligned} \tag{3.89}$$

$$\tau^{\widetilde{T}_1, T_4}_{3,1,2,3} \widetilde{s} = \tau^{\widetilde{T}_1, T_4}_{2,2,2,3} \widetilde{s} = \tau^{\widetilde{T}_1, T_4}_{2,1,3,3} \widetilde{s} = 0 \tag{3.90}$$

Proof:
Using the C^4 supersmoothness conditions at the vertices of T, the C^3 supersmoothness conditions at the edges of T, and the C^2 smoothness conditions

at the faces of T, we can uniquely and stably determine the B-coefficients of g corresponding to the domain points in $(\dot{M}1^2_{T^2_{WF}})$ - $(\dot{M}8^2_{T^2_{WF}})$, from Theorem 3.22.

From the additional smoothness conditions in (3.88) - (3.90) we can stably and uniquely compute the B-coefficients associated with the domain points in $(\dot{M}9^2_{T^2_{WF}})$ - $(\dot{M}11^2_{T^2_{WF}})$. Since the B-coefficient associated with the domain point in $(\dot{M}12^2_{T^2_{WF}})$ is equal to the value of $g(v_T)$, following Theorem 3.22, g is uniquely and stably determined. □

3.2.3 Minimal determining sets for C^2 supersplines based on third order partial Worsey-Farin splits

In this subsection, we examine minimal determining sets for C^2 splines on third order partial Worsey-Farin splits of a tetrahedron $T := \langle v_1, v_2, v_3, v_4 \rangle$. Therefore, let T^3_{WF} be the third order partial Worsey-Farin split of T at the split point v_T in the interior of T, and let the faces $\langle v_1, v_2, v_3 \rangle$, $\langle v_2, v_3, v_4 \rangle$, and $\langle v_3, v_1, v_4 \rangle$ be refined with a Clough-Tocher split at the split point v_{F_1}, v_{F_2}, and v_{F_3}, respectively. The corresponding subtetrahedra are denoted by

$$\begin{aligned} T_{1,1} &:= \langle v_1, v_2, v_{F_1}, v_T \rangle, & T_{1,2} &:= \langle v_2, v_3, v_{F_1}, v_T \rangle, \\ T_{1,3} &:= \langle v_3, v_1, v_{F_1}, v_T \rangle, & T_{2,1} &:= \langle v_2, v_3, v_{F_2}, v_T \rangle, \\ T_{2,2} &:= \langle v_3, v_4, v_{F_2}, v_T \rangle, & T_{2,3} &:= \langle v_4, v_2, v_{F_2}, v_T \rangle, \\ T_{3,1} &:= \langle v_3, v_1, v_{F_3}, v_T \rangle, & T_{3,2} &:= \langle v_1, v_4, v_{F_3}, v_T \rangle, \\ T_{3,3} &:= \langle v_4, v_3, v_{F_3}, v_T \rangle, & T_4 &:= \langle v_1, v_2, v_4, v_T \rangle. \end{aligned} \tag{3.91}$$

Theorem 3.24:
Let $\mathcal{M}^2_3$ be the union of the following sets of domain points in $\mathcal{D}_{T^3_{WF},9}$:

$(M1^2_{T^3_{WF}})$ $D^{T_{v_i}}_4(v_i)$, $i = 1, \dots, 4$, with $v_i \in T_{v_i} \in T^3_{WF}$

$(M2^2_{T^3_{WF}})$ $\left(\bigcup_{i=1}^{2} E^{T_{1,i}}_3(\langle v_i, v_{1+i} \rangle) \cup E^{T_{1,3}}_3(\langle v_1, v_3 \rangle) \cup E^{T_{2,2}}_3(\langle v_4, v_3 \rangle) \right) \setminus \bigcup_{i=1}^{4} (D_4(v_i))$

$(M3^2_{T^3_{WF}})$ $\{\xi^{T_4}_{4-i,1+i,4,0}, \xi^{T_4}_{4-i,1+i,3,1}, \xi^{T_4}_{3-i,i,4,2}, \xi^{T_4}_{4-i,1+i,2,2}\}_{i=0}^{3}$

$(M4^2_{T^3_{WF}})$ $\{\xi^{T_4}_{4-i,i,4,1}, \xi^{T_4}_{4-i,i,3,2}\}_{i=0}^{4}$

$(M5^2_{T^3_{WF}})$ $\{\xi^{T_4}_{4-i,0,2+i,3}, \xi^{T_4}_{0,4-i,2+i,3}\}_{i=0}^{2}$

$(M6^2_{T^3_{WF}})$ $\{\xi^{T_{j,i}}_{3,0,4,2}\}$, $i = 1,2,3$, $j = 1,2,3$

$(M7^2_{T^3_{WF}})$ $\{\xi^{T_4}_{i,j,k,l}\}$, $i,j,k \geq 1$, $l = 3,4,5$, $i+j+k = 9-l$

$(M8^2_{T^3_{WF}})$ $\{\xi^{T_{1,1}}_{0,0,0,9}\}$

Then $\mathcal{M}^2_3$ is a stable minimal determining set for

$$\begin{aligned}\widetilde{\mathcal{S}}^2_9(T^3_{WF}) := \{s \in C^2(T^3_{WF}) : s|_{\widetilde{T}} \in \mathcal{P}^3_9,\ \forall \widetilde{T} \in T^3_{WF},\\ s \in C^4(v_i),\ i = 1,\dots,4,\\ s \in C^3(e), \text{ for all edges } e \in T,\\ s \in C^7(\langle v_{F_i}, v_T\rangle),\ i = 1,2\\ s \in C^7(v_T),\\ \tau^{T_{j,i},T_{j,i+1}}_{3,5,1,0} s = 0,\ i,j = 1,2,3, \text{ with } T_{j,4} := T_{j,1},\\ \tau^{T_{j,i},T_{j,i+1}}_{3,5,0,1} s = 0,\ i,j = 1,2,3, \text{ with } T_{j,4} := T_{j,1}\},\end{aligned}$$

and the dimension of $\widetilde{\mathcal{S}}^2_9(T^3_{WF})$ is equal to 281.

Proof:
We first show that $\mathcal{M}^2_3$ is a stable minimal determining set for $\widetilde{\mathcal{S}}^2_9(T^3_{WF})$. To this end, we set the B-coefficients c_ξ of a spline $s \in \widetilde{\mathcal{S}}^2_9(T^3_{WF})$ to arbitrary values for each domain point $\xi \in \mathcal{M}^2_3$. Then we show that all other B-coefficients of s are uniquely and stably determined.

From the B-coefficients associated with the domain points in $(M1^2_{T^3_{WF}})$ we can uniquely and stably determine the remaining B-coefficients corresponding to the domain points in the balls $D_4(v_i)$, $i = 1,\dots,4$, using the C^4 supersmoothness conditions at the vertices of T.

Then, the remaining undetermined B-coefficients of s corresponding to domain points within a distance of three from the edges $\langle v_1, v_3\rangle$, $\langle v_2, v_3\rangle$, $\langle v_4, v_3\rangle$, and $\langle v_1, v_2\rangle$ can be uniquely and stably computed from the already determined B-coefficients and those associated with the domain points in $(M2^2_{T^3_{WF}})$ using the C^3 supersmoothness conditions at the corresponding edges.

Now, the only undetermined B-coefficients of s corresponding to domain points within a distance of two from the face $\langle v_1, v_2, v_4\rangle$ are associated with the domain points in $(M3^2_{T^3_{WF}})$ and $(M4^2_{T^3_{WF}})$. Thus, the B-coefficients of s associated with the domain points in the shells $R^{T_4}_{9-i}(v_T)$, $i = 0,1,2$, are uniquely and stably determined.

Together with the B-coefficients associated with the domain points in the set ($M5^2_{T^3_{WF}}$) and the already determined ones, we can uniquely and stably compute the remaining undetermined B-coefficients of s corresponding to the domain points within a distance of three from the two edges $\langle v_2, v_4 \rangle$ and $\langle v_4, v_1 \rangle$ using the corresponding C^3 supersmoothness conditions.

Subsequently, we consider the B-coefficients associated with the domain points in the shells $R_9^{T_{j,i}}(v_T), R_8^{T_{j,i}}(v_T)$, and $R_7^{T_{j,i}}(v_T)$, $i = 1,2,3$, $j = 1,2,3$. These can be uniquely and stably computed from those already determined and the ones associated with the domain points in ($M6^2_{T^3_{WF}}$) in the same way as those associated with the domain points in the shells $R_9^{T_{1,i}}(v_T)$, $R_8^{T_{1,i}}(v_T)$, and $R_7^{T_{1,i}}(v_T)$, $i = 1,2,3$, in the proof of Theorem 3.4.

Now, together with the B-coefficients associated with the domain points in ($M7^2_{T^3_{WF}}$) with $l = 3$, all B-coefficients of s corresponding to the domain points in the shell $R_6^{T_4}(v_T)$ are uniquely and stably determined.

Then, following Lemma 2.33 for the case $j = 0$, we can uniquely and stably compute the remaining undetermined B-coefficients of s corresponding to the domain points within a distance of four from the edges $\langle v_1, v_2 \rangle$, $\langle v_2, v_4 \rangle$, and $\langle v_4, v_1 \rangle$ of T from the already determined B-coefficients by using the C^2 smoothness conditions at the corresponding interior faces of T^3_{WF}.

Thus, together with the B-coefficients corresponding to those domain points in ($M7^2_{T^3_{WF}}$) with $l = 4$, the B-coefficients of s associated with the domain points in the shell $R_5^{T_4}(v_T)$ are uniquely and stably determined.

Again, following Lemma 2.33 for the case $j = 0$, the remaining undetermined B-coefficients of s corresponding to the domain points within a distance of five from the edges $\langle v_1, v_2 \rangle$, $\langle v_2, v_4 \rangle$, and $\langle v_4, v_1 \rangle$ can be uniquely and stably computed from the ones already determined by using the C^3 supersmoothness conditions at the corresponding interior faces of T^3_{WF}, which are contained in the C^7 supersmoothness conditions at the split point v_T.

Now, we examine the last undetermined B-coefficients of s. These correspond to domain points in the ball $D_7(v_T)$. Since s has C^7 supersmoothness at v_T, we can regard the corresponding B-coefficients as those of a trivariate polynomial g defined on the tetrahedron $\widetilde{T}$, which is bounded by the domain points in $R_7(v_T)$ and also subjected to a third order partial Worsey-Farin split at v_T. Considering g, we see that the B-coefficients associated with the domain points on the three Clough-Tocher split faces of $\widetilde{T}$ and the B-coefficients corresponding to the domain points within a distance of four from the non-split face are already uniquely and stably determined. Now, setting the B-coefficient of g associated with the domain point $\xi_{0,0,0,9}^{T_{1,1}}$ is equivalent to setting the function value of g at v_T. Thus, following Lemma 2.42, the remaining undetermined B-coefficients of g can be uniquely and stably determined from smoothness conditions. Therefore, we have uniquely and stably determined all B-coefficients of s.

Finally, considering the dimension of the space $\widetilde{\mathcal{S}}_9^2(T_{WF}^3)$, we can see that $\dim \widetilde{\mathcal{S}}_9^2(T_{WF}^3) = \#\mathcal{M}_3^2 = 281$, since the set $\mathcal{M}_3^2$ is a minimal determining set for this superspline space. □

Lemma 3.25:
Let Δ be a tetrahedral partition of a polyhedral domain $\Omega \subset \mathbb{R}^3$ and $T := \langle v_1, v_2, v_3, v_4 \rangle \in \Delta$ a tetrahedron, that shares exactly the three edges $\langle v_1, v_3 \rangle$, $\langle v_2, v_3 \rangle$, and $\langle v_4, v_3 \rangle$ with the tetrahedra $\widetilde{T}_1, \widetilde{T}_2$, and $\widetilde{T}_3$, respectively, with $\widetilde{T}_j \in \Delta$, $j = 1,2,3$. Moreover, T is refined with a third order partial Worsey-Farin split at the split point v_T in the interior of T, and the split points v_{F_1}, v_{F_2}, and v_{F_3} in the interior of the faces $\langle v_1, v_2, v_3 \rangle$, $\langle v_2, v_3, v_4 \rangle$, and $\langle v_1, v_3, v_4 \rangle$, respectively.

Let $\widetilde{\mathcal{S}}_9^2(T_{WF}^3)$ be the superspline space defined in Theorem 3.24. Then, for any spline $s \in \mathcal{S}_9^{2,4,3}(\Delta \setminus T)$, there is a spline $g \in \widetilde{\mathcal{S}}_9^2(T_{WF}^3)$ such that

$$\widetilde{s} := \begin{cases} s, & \text{on } \Delta \setminus T \\ g, & \text{on } T, \end{cases}$$

belongs to $\mathcal{S}_9^{2,4,3}(\Delta)$. Moreover, g is uniquely and stably determined by the value of $g(v_T)$ and the B-coefficients of $\widetilde{s}|_{\Delta \setminus T}$ by the following additional smoothness conditions across the vertices and edges of T:

$$\tau_{i,j,k,l}^{\widetilde{T}_1, T_{1,1}} \widetilde{s} = 0, \; i,j < 5, \; k+l \leq 3, \; i+j+k+l = 9 \tag{3.92}$$

$$\tau_{4-i,1+i,4,0}^{\widetilde{T}_3, T_4} \widetilde{s} = \tau_{4-i,1+i,3,1}^{\widetilde{T}_3, T_4} \widetilde{s} = \tau_{3-i,i,4,2}^{\widetilde{T}_3, T_4} \widetilde{s} = \tau_{4-i,1+i,2,2}^{\widetilde{T}_3, T_4} \widetilde{s} = 0, \; i = 0, \ldots, 3 \tag{3.93}$$

$$\tau_{4-i,i,4,1}^{\widetilde{T}_3, T_4} \widetilde{s} = \tau_{4-i,i,3,2}^{\widetilde{T}_3, T_4} \widetilde{s} = 0, \; i = 0, \ldots, 4 \tag{3.94}$$

$$\tau_{4-i,0,2+i,3}^{\widetilde{T}_1, T_4} \widetilde{s} = \tau_{0,4-i,2+i,3}^{\widetilde{T}_2, T_4} \widetilde{s} = 0, \; i = 0,1,2 \tag{3.95}$$

$$\tau_{3,0,4,2}^{\widetilde{T}_j,T_{j,i}}\widetilde{s} = \tau_{3,0,4,2}^{\widetilde{T}_{j+1},T_{j,2}}\widetilde{s} = \tau_{3,0,4,2}^{\widetilde{T}_3,T_{3,j+1}}\widetilde{s} = \tau_{3,0,4,2}^{\widetilde{T}_1,T_{3,1}}\widetilde{s} = 0,\ i = 1,3,\ j = 1,2 \tag{3.96}$$

$$\tau_{i,j,k,l}^{\widetilde{T}_1,T_4}\widetilde{s} = 0,\ i,j,k \geq 1,\ l = 3,4,5,\ i+j+k = 9-l \tag{3.97}$$

Proof:
Using the C^4 supersmoothness conditions at the vertices of T and the C^3 supersmoothness conditions at the three common edges of T with Δ, we can uniquely and stably compute the B-coefficients of g associated with the domain points in (M$1^2_{T^3_{WF}}$) and (M$2^2_{T^3_{WF}}$), except for those in $E_3^{T_{1,1}}(\langle v_1,v_2\rangle) \setminus (D_4(v_1) \cup D_4(v_2))$, from Theorem 3.24.

From the additional smoothness conditions in (3.92) - (3.97) we can stably and uniquely determine the B-coefficients associated with the remaining domain points in (M$2^2_{T^3_{WF}}$) and with those in (M$3^2_{T^3_{WF}}$) - (M$7^2_{T^3_{WF}}$). Finally, the last undetermined B-coefficient corresponding to a domain point in the minimal determining set $\mathcal{M}_3^2$ from Theorem 3.24 is equal to the value of $g(v_T)$. Thus, following Theorem 3.24, g is uniquely and stably determined. □

Lemma 3.26:
Let Δ be a tetrahedral partition of a polyhedral domain $\Omega \subset \mathbb{R}^3$ and $T := \langle v_1,v_2,v_3,v_4\rangle \in \Delta$ a tetrahedron, that shares exactly the four edges $\langle v_1,v_3\rangle$, $\langle v_2,v_3\rangle$, $\langle v_4,v_3\rangle$, and $\langle v_1,v_2\rangle$ with the tetrahedra $\widetilde{T}_1$, $\widetilde{T}_2$, $\widetilde{T}_3$, and $\widetilde{T}_4$, respectively, with $\widetilde{T}_j \in \Delta$, $j = 1,\dots,4$. Moreover, T is refined with a third order partial Worsey-Farin split at the split point v_T in the interior of T, and the split points v_{F_1}, v_{F_2}, and v_{F_3} in the interior of the faces $\langle v_1,v_2,v_3\rangle$, $\langle v_2,v_3,v_4\rangle$, and $\langle v_1,v_3,v_4\rangle$, respectively.

Let $\widetilde{\mathcal{S}}_9^2(T^3_{WF})$ be the superspline space defined in Theorem 3.24. Then, for any spline $s \in \mathcal{S}_9^{2,4,3}(\Delta \setminus T)$, there is a spline $g \in \widetilde{\mathcal{S}}_9^2(T^3_{WF})$ such that

$$\widetilde{s} := \begin{cases} s, & \text{on } \Delta \setminus T \\ g, & \text{on } T, \end{cases}$$

belongs to $\mathcal{S}_9^{2,4,3}(\Delta)$. Moreover, the spline g is uniquely and stably determined by the value of $g(v_T)$ and the B-coefficients of $\widetilde{s}|_{\Delta \setminus T}$ by the following additional smoothness conditions across the vertices and edges of T:

$$\tau_{4-i,1+i,4,0}^{\widetilde{T}_3,T_4}\widetilde{s} = \tau_{4-i,1+i,3,1}^{\widetilde{T}_3,T_4}\widetilde{s} = \tau_{3-i,i,4,2}^{\widetilde{T}_3,T_4}\widetilde{s} = \tau_{4-i,1+i,2,2}^{\widetilde{T}_3,T_4}\widetilde{s} = 0,\ i = 0,\dots,3 \tag{3.98}$$

$$\tau_{4-i,i,4,1}^{\widetilde{T}_3,T_4}\widetilde{s} = \tau_{4-i,i,3,2}^{\widetilde{T}_3,T_4}\widetilde{s} = 0,\ i = 0,\dots,4 \tag{3.99}$$

$$\tau_{4-i,0,2+i,3}^{\widetilde{T}_1,T_4}\widetilde{s} = \tau_{0,4-i,2+i,3}^{\widetilde{T}_2,T_4}\widetilde{s} = 0,\ i = 0,1,2 \tag{3.100}$$

$$\tau_{3,0,4,2}^{\widetilde{T}_j,T_{j,i}}\widetilde{s} = \tau_{3,0,4,2}^{\widetilde{T}_{j+1},T_{j,2}}\widetilde{s} = \tau_{3,0,4,2}^{\widetilde{T}_3,T_{3,j+1}}\widetilde{s} = \tau_{3,0,4,2}^{\widetilde{T}_1,T_{3,1}}\widetilde{s} = 0,\ i = 1,3,\ j = 1,2 \tag{3.101}$$

$$\tau_{i,j,k,l}^{\widetilde{T}_1,T_4}\widetilde{s} = 0,\ i,j,k \geq 1,\ l = 3,4,5,\ i+j+k = 9-l \tag{3.102}$$

Proof:
We can uniquely and stably determine the B-coefficients of g associated with the domain points in (M$1^2_{T^3_{WF}}$) and (M$2^2_{T^3_{WF}}$) from Theorem 3.24 by using the C^4 supersmoothness conditions at the vertices of T and the C^3 supersmoothness conditions at the edges $\langle v_1,v_3\rangle, \langle v_2,v_3\rangle, \langle v_4,v_3\rangle$, and $\langle v_1,v_2\rangle$.

Using the additional smoothness conditions of g in (3.98) - (3.102), we can uniquely and stably compute the B-coefficients corresponding to the domain points in (M$3^2_{T^3_{WF}}$) - (M$7^2_{T^3_{WF}}$). Finally, the B-coefficient corresponding to the domain point in (M$8^2_{T^3_{WF}}$) is equal to the value of $g(v_T)$. Thus, following Theorem 3.24, g is uniquely and stably determined. □

Lemma 3.27:
Let Δ be a tetrahedral partition of a polyhedral domain $\Omega \subset \mathbb{R}^3$ and $T := \langle v_1,v_2,v_3,v_4\rangle \in \Delta$ a tetrahedron, that shares exactly the face $\langle v_1,v_2,v_4\rangle$ with the tetrahedron $\hat{T}$, and the edges $\langle v_1,v_3\rangle$ and $\langle v_2,v_3\rangle$ with the tetrahedra $\widetilde{T}_1$ and $\widetilde{T}_2$, respectively, with $\hat{T},\widetilde{T}_1,\widetilde{T}_2 \in \Delta$. Moreover, T is refined with a third order partial Worsey-Farin split at the split point v_T in the interior of T, and the split points v_{F_1}, v_{F_2}, and v_{F_3} in the interior of the faces $\langle v_1,v_2,v_3\rangle$, $\langle v_2,v_3,v_4\rangle$, and $\langle v_1,v_3,v_4\rangle$, respectively.

Let $\widetilde{\mathcal{S}}^2_9(T^3_{WF})$ be the superspline space defined in Theorem 3.24. Then, for any spline $s \in \mathcal{S}^{2,4,3}_9(\Delta \setminus T)$, there is a spline $g \in \widetilde{\mathcal{S}}^2_9(T^3_{WF})$ such that

$$\widetilde{s} := \begin{cases} s, & \text{on } \Delta \setminus T \\ g, & \text{on } T, \end{cases}$$

belongs to $\mathcal{S}^{2,4,3}_9(\Delta)$. Moreover, the spline g is uniquely and stably determined by the value of $g(v_T)$ and the B-coefficients of $\widetilde{s}|_{\Delta\setminus T}$ by the following additional smoothness conditions across the vertices, edges, and faces of T:

$$\tau_{i,j,k,l}^{\hat{T}_1,T_{2,2}}\widetilde{s} = 0,\ i,j < 5,\ k+l \leq 3,\ i+j+k+l = 9 \tag{3.103}$$

$$
\begin{aligned}
&\tau_{3,0,4,2}^{\hat{T}_j,T_{i,i}}\widetilde{s} = \tau_{3,0,4,2}^{\hat{T},T_{2+j,3-j}}\widetilde{s} = \tau_{3,0,4,2}^{\widetilde{T}_2,T_{1+j,2-j}}\widetilde{s} \\
&= \tau_{3,0,4,2}^{\widetilde{T}_1,T_{1+2j,3-2j}}\widetilde{s} = 0,\ i = 1,2,3,\ j = 0,1
\end{aligned} \tag{3.104}
$$

$$
\tau_{i,j,k,l}^{\hat{T},T_4}\widetilde{s} = 0,\ i,j,k \geq 1,\ l = 3,4,5,\ i+j+k = 9-l \tag{3.105}
$$

Proof:
Using the C^4 supersmoothness conditions at the vertices of T, the C^3 supersmoothness conditions at the edges of T, except for $\langle v_3, v_4\rangle$, and the C^2 smoothness conditions at the face $\langle v_1, v_2, v_4\rangle$, we can uniquely and stably compute the B-coefficients of g associated with the domain points in (M1$^2_{T^3_{WF}}$), (M2$^2_{T^3_{WF}}$), except for those in $E_3^{T_{2,2}}(\langle v_4, v_3\rangle) \setminus (D_4(v_3) \cup D_4(v_4))$, (M3$^2_{T^3_{WF}}$), (M4$^2_{T^3_{WF}}$), and (M5$^2_{T^3_{WF}}$) from Theorem 3.24.

From the additional smoothness conditions of g in (3.103) - (3.105), we can uniquely and stably determine the B-coefficients associated with the remaining domain points in (M2$^2_{T^3_{WF}}$), and the ones associated with the domain points in (M6$^2_{T^3_{WF}}$) and (M7$^2_{T^3_{WF}}$). The last undetermined B-coefficient corresponding to a domain point in the minimal determining set $\mathcal{M}_3^2$ is equal to the value of $g(v_T)$. Therefore, following Theorem 3.24, g is uniquely and stably determined. □

Lemma 3.28:
Let Δ be a tetrahedral partition of a polyhedral domain $\Omega \subset \mathbb{R}^3$ and $T := \langle v_1, v_2, v_3, v_4\rangle \in \Delta$ a tetrahedron, that shares exactly the face $\langle v_1, v_2, v_4\rangle$ with the tetrahedron $\hat{T}$, and the edges $\langle v_1, v_3\rangle$, $\langle v_2, v_3\rangle$, and $\langle v_4, v_3\rangle$ with the tetrahedra $\widetilde{T}_1$, $\widetilde{T}_2$, and $\widetilde{T}_3$ respectively, with $\hat{T}, \widetilde{T}_j \in \Delta$, $j = 1,2,3$. Moreover, T is refined with a third order partial Worsey-Farin split at the split point v_T in the interior of T, and the split points v_{F_1}, v_{F_2}, and v_{F_3} in the interior of the faces $\langle v_1, v_2, v_3\rangle$, $\langle v_2, v_3, v_4\rangle$, and $\langle v_1, v_3, v_4\rangle$, respectively.

Let $\widetilde{\mathcal{S}}_9^2(T^3_{WF})$ be the superspline space defined in Theorem 3.24. Then, for any spline $s \in \mathcal{S}_9^{2,4,3}(\Delta \setminus T)$, there is a spline $g \in \widetilde{\mathcal{S}}_9^2(T^3_{WF})$ such that

$$
\widetilde{s} := \begin{cases} s, & \text{on } \Delta \setminus T \\ g, & \text{on } T, \end{cases}
$$

belongs to $\mathcal{S}_9^{2,4,3}(\Delta)$. Moreover, the spline g is uniquely and stably determined by the value of $g(v_T)$ and the B-coefficients of $\widetilde{s}|_{\Delta \setminus T}$ by the following additional smoothness conditions across the edges and faces of T:

$$\tau_{3,0,4,2}^{\hat{T},T_{1,1}}\widetilde{s} = \tau_{3,0,4,2}^{\hat{T},T_{2+j,3-j}}\widetilde{s} = \tau_{3,0,4,2}^{\widetilde{T}_2,T_{1+j,2-j}}\widetilde{s}$$
$$= \tau_{3,0,4,2}^{\widetilde{T}_1,T_{1+2j,3-2j}}\widetilde{s} = \tau_{3,0,4,2}^{\widetilde{T}_{3-j},T_{2+j,2+j}}\widetilde{s} = 0,\ j = 0,1 \tag{3.106}$$
$$\tau_{i,j,k,l}^{\hat{T},T_4}\widetilde{s} = 0,\ i,j,k \geq 1,\ l = 3,4,5,\ i+j+k = 9-l \tag{3.107}$$

Proof:
From the C^4 supersmoothness conditions at the vertices of T, the C^3 supersmoothness conditions at the edges of T, and the C^2 smoothness conditions at the face $\langle v_1, v_2, v_4 \rangle$, we can uniquely and stably determine the B-coefficients of g associated with the domain points in (M$1^2_{T^3_{WF}}$) - (M$5^2_{T^3_{WF}}$) from Theorem 3.24.

Using the additional smoothness conditions of g in (3.106) - (3.107), we can uniquely and stably compute the B-coefficients corresponding to the domain points in (M$6^2_{T^3_{WF}}$) and (M$7^2_{T^3_{WF}}$). Since the value of $g(v_T)$ is equal to the B-coefficient corresponding to the domain point in (M$8^2_{T^3_{WF}}$), following Theorem 3.24, g is uniquely and stably determined. □

Theorem 3.29:
Let $\hat{\mathcal{M}}^2_3$ be the union of the following sets of domain points in $\mathcal{D}_{T^3_{WF},9}$:

($\hat{\text{M}}1^2_{T^3_{WF}}$) $D_4^{T_{v_i}}(v_i)$, $i = 1,\dots,4$, with $v_i \in T_{v_i} \in T^3_{WF}$

($\hat{\text{M}}2^2_{T^3_{WF}}$) $\left(\bigcup_{i=1}^{2} E_3^{T_{1,i}}(\langle v_i, v_{1+i}\rangle) \cup E_3^{T_{1,3}}(\langle v_1, v_3\rangle) \cup E_3^{T_{2,2}}(\langle v_4, v_3\rangle)\right) \setminus \bigcup_{i=1}^{4}(D_4(v_i))$

($\hat{\text{M}}3^2_{T^3_{WF}}$) $\{\xi^{T_4}_{4-i,1+i,4,0}, \xi^{T_4}_{4-i,1+i,3,1}, \xi^{T_4}_{3-i,i,4,2}, \xi^{T_4}_{4-i,1+i,2,2}\}_{i=0}^{3}$

($\hat{\text{M}}4^2_{T^3_{WF}}$) $\{\xi^{T_4}_{4-i,i,4,1}, \xi^{T_4}_{4-i,i,3,2}\}_{i=0}^{4}$

($\hat{\text{M}}5^2_{T^3_{WF}}$) $\{\xi^{T_4}_{4-i,0,2+i,3}, \xi^{T_4}_{0,4-i,2+i,3}\}_{i=0}^{2}$

($\hat{\text{M}}6^2_{T^3_{WF}}$) $\{\xi^{T_{1,1}}_{1,0,8-i,i}, \xi^{T_{1,1}}_{0,1,8-i,i}, \xi^{T_{1,1}}_{0,0,9-i,i}\}_{i=0}^{1}$

($\hat{\text{M}}7^2_{T^3_{WF}}$) $\{\xi^{T_{j,i}}_{3,0,4,2}\}$, $i = 1,2,3$, $j = 1,2,3$

($\hat{\text{M}}8^2_{T^3_{WF}}$) $\{\xi^{T_4}_{i,j,k,l}\}$, $i,j,k \geq 1$, $l = 3,4,5$, $i+j+k = 9-l$

($\hat{\text{M}}9^2_{T^3_{WF}}$) $\{\xi^{T_{1,1}}_{0,0,0,9}\}$

Then $\hat{\mathcal{M}}_3^2$ is a stable minimal determining set for

$$\begin{aligned}\hat{\mathcal{S}}_9^2(T_{WF}^3) := \{s \in C^2(T_{WF}^3) : s|_{\widetilde{T}} \in \mathcal{P}_9^3,\ \forall \widetilde{T} \in T_{WF}^3,\\ s \in C^4(v_i),\ i = 1,\dots,4,\\ s \in C^3(e),\ \text{for all edges } e \in T,\\ s \in C^7(\langle v_{F_i}, v_T\rangle),\ i = 1,2\\ s \in C^7(v_T),\\ \tau_{3,5,1,0}^{T_{j,i},T_{j,i+1}} s = \tau_{3,5,0,1}^{T_{j,i},T_{j,i+1}} s = 0,\\ i = 1,2,3,\ j = 2,3,\ \text{with } T_{j,4} := T_{j,1}\},\end{aligned}$$

and the dimension of $\hat{\mathcal{S}}_9^2(T_{WF}^3)$ is equal to 287.

Proof:
First, we show that $\hat{\mathcal{M}}_3^2$ is a stable minimal determining set for $\hat{\mathcal{S}}_9^2(T_{WF}^3)$. To this end, we set the B-coefficients c_ξ of a spline $s \in \hat{\mathcal{S}}_9^2(T_{WF}^3)$ to arbitrary values for each domain point $\xi \in \hat{\mathcal{M}}_3^2$. Then we show that all other B-coefficients of s are uniquely and stably determined.

The B-coefficients corresponding to the domain points within a distance of four from the vertices of T, within a distance of three from the edges of T, and within a distance of two from the face $\langle v_1, v_2, v_4\rangle$ can be uniquely and stably determined in the same way as in the proof of Theorem 3.24 from those in ($\hat{\text{M}}1^2_{T_{WF}^3}$) - ($\hat{\text{M}}5^2_{T_{WF}^3}$).

Then, the B-coefficients associated with the domain points in the shells $R_9^{T_{1,i}}(v_T)$ and $R_8^{T_{1,i}}(v_T)$, $i = 1,2,3$, can be uniquely and stably computed from the already determined ones and those associated with the domain points in ($\widetilde{\text{M}}6^2_{T_{WF}^2}$) in the same way as the corresponding B-coefficients are determined in the proof of Theorem 3.20.

Now, the remaining undetermined B-coefficients can be uniquely and stably computed from those corresponding to the domain points in ($\hat{\text{M}}7^2_{T_{WF}^3}$) - ($\hat{\text{M}}9^2_{T_{WF}^3}$) and the already determined ones in the same way as in the proof of Theorem 3.24.

Considering the dimension of the spline space $\hat{\mathcal{S}}_9^2(T_{WF}^3)$, we can state that $\dim \hat{\mathcal{S}}_9^2(T_{WF}^3) = 287$, which is equal to the cardinality of the minimal determining set $\hat{\mathcal{M}}_3^2$. □

Lemma 3.30:
Let Δ be a tetrahedral partition of a polyhedral domain $\Omega \subset \mathbb{R}^3$ and $T := \langle v_1, v_2, v_3, v_4 \rangle \in \Delta$ a tetrahedron, that shares exactly the face $\langle v_1, v_2, v_3 \rangle$ with the tetrahedron $\hat{T}$, and the edge $\langle v_4, v_3 \rangle$ with the tetrahedra $\widetilde{T}$, with $\hat{T}, \widetilde{T} \in \Delta$. Moreover, T is refined with a third order partial Worsey-Farin split at the split point v_T in the interior of T, and the split points v_{F_1}, v_{F_2}, and v_{F_3} in the interior of the faces $\langle v_1, v_2, v_3 \rangle$, $\langle v_2, v_3, v_4 \rangle$, and $\langle v_1, v_3, v_4 \rangle$, respectively. The tetrahedron $\hat{T}$ is also subjected to a partial Worsey-Farin split, such that the common face with T is refined with a Clough-Tocher split. The three corresponding subtetrahedra are denoted by $\hat{T}_i$, $i = 1,2,3$, respectively, where $\hat{T}_i$ has a common face with $T_{1,i}$.

Let $\hat{\mathcal{S}}_9^2(T_{WF}^3)$ be the superspline space defined in Theorem 3.29. Then, for any spline $s \in \mathcal{S}_9^{2,4,3}(\Delta \setminus T)$, there is a spline $g \in \hat{\mathcal{S}}_9^2(T_{WF}^3)$ such that

$$\widetilde{s} := \begin{cases} s, & \text{on } \Delta \setminus T \\ g, & \text{on } T, \end{cases}$$

belongs to $\mathcal{S}_9^{2,4,3}(\Delta)$. Moreover, the spline g is uniquely and stably determined by the value of $g(v_T)$ and the B-coefficients of $\widetilde{s}|_{\Delta \setminus T}$ by the following additional smoothness conditions across the vertices, edges and faces of T:

$$\tau_{4-i,1+i,4,0}^{\hat{T}_1,T_4}\widetilde{s} = \tau_{4-i,1+i,3,1}^{\hat{T}_1,T_4}\widetilde{s} = \tau_{3-i,i,4,2}^{\hat{T}_1,T_4}\widetilde{s} = \tau_{4-i,1+i,2,2}^{\hat{T}_1,T_4}\widetilde{s} = 0,\ i = 0,\dots,3 \quad (3.108)$$

$$\tau_{4-i,i,4,1}^{\hat{T}_1,T_4}\widetilde{s} = \tau_{4-i,i,3,2}^{\hat{T}_1,T_4}\widetilde{s} = 0,\ i = 0,\dots,4 \quad (3.109)$$

$$\tau_{4-i,0,2+i,3}^{\hat{T}_1,T_4}\widetilde{s} = \tau_{0,4-i,2+i,3}^{\hat{T}_1,T_4}\widetilde{s} = 0,\ i = 0,1,2 \quad (3.110)$$

$$\tau_{3,0,4,2}^{\widetilde{T},T_{j,j}}\widetilde{s} = \tau_{3,0,4,2}^{\hat{T}_1,T_{2+i,3-i}}\widetilde{s} = \tau_{3,0,4,2}^{\hat{T}_2,T_{2,1}}\widetilde{s} = \tau_{3,0,4,2}^{\hat{T}_3,T_{3,1}}\widetilde{s} = 0,\ i = 0,1,\ j = 2,3 \quad (3.111)$$

$$\tau_{i,j,k,l}^{\hat{T}_1,T_4}\widetilde{s} = 0,\ i,j,k \geq 1,\ l = 3,4,5,\ i+j+k = 9-l \quad (3.112)$$

Proof:
From the C^4 supersmoothness conditions at the vertices of T, the C^3 supersmoothness conditions at the edges $\langle v_1, v_2 \rangle$, $\langle v_1, v_3 \rangle$, $\langle v_2, v_3 \rangle$, and $\langle v_4, v_3 \rangle$ of T, and the C^2 smoothness conditions at the face $\langle v_1, v_2, v_3 \rangle$, we can uniquely and stably compute the B-coefficients of g associated with the domain points in $(\hat{M}1^2_{T^3_{WF}})$, $(\hat{M}2^2_{T^3_{WF}})$, $(\hat{M}6^2_{T^3_{WF}})$, and those in $(\hat{M}7^2_{T^3_{WF}})$ with $j = 1$ from Theorem 3.29.

Using the additional smoothness conditions of g in (3.108) - (3.112), we can uniquely and stably determine the B-coefficients associated with the remaining domain points in $(\hat{M}7^2_{T^3_{WF}})$, the ones associated with the domain points in $(\hat{M}3^2_{T^3_{WF}})$ - $(\hat{M}5^2_{T^3_{WF}})$, and $(\hat{M}8^2_{T^3_{WF}})$. Since the B-coefficient corresponding to the domain point in $(\hat{M}9^2_{T^3_{WF}})$ is equal to the value of $g(v_T)$, following Theorem 3.29, g is uniquely and stably determined. □

Lemma 3.31:
Let Δ be a tetrahedral partition of a polyhedral domain $\Omega \subset \mathbb{R}^3$ and $T := \langle v_1,v_2,v_3,v_4\rangle \in \Delta$ a tetrahedron, that shares exactly the two faces $\langle v_1,v_2,v_3\rangle$ and $\langle v_1,v_2,v_4\rangle$ with the tetrahedra $\hat{T}$ and $\widetilde{T}$, respectively, with $\hat{T},\widetilde{T} \in \Delta$. Moreover, T is refined with a third order partial Worsey-Farin split at the split point v_T in the interior of T, and the split points v_{F_1}, v_{F_2}, and v_{F_3} in the interior of the faces $\langle v_1,v_2,v_3\rangle$, $\langle v_2,v_3,v_4\rangle$, and $\langle v_1,v_3,v_4\rangle$, respectively. The tetrahedron $\hat{T}$ is also subjected to a partial Worsey-Farin split, such that the common face with T is refined with a Clough-Tocher split. The three corresponding subtetrahedra are denoted by $\hat{T}_i$, $i = 1,2,3$, respectively, where $\hat{T}_i$ has a common face with $T_{1,i}$.

Let $\hat{\mathcal{S}}^2_9(T^3_{WF})$ be the superspline space defined in Theorem 3.29. Then, for any spline $s \in \mathcal{S}^{2,4,3}_9(\Delta \setminus T)$, there is a spline $g \in \hat{\mathcal{S}}^2_9(T^3_{WF})$ such that

$$\widetilde{s} := \begin{cases} s, & \text{on } \Delta \setminus T \\ g, & \text{on } T, \end{cases}$$

belongs to $\mathcal{S}^{2,4,3}_9(\Delta)$. Moreover, the spline g is uniquely and stably determined by the value of $g(v_T)$ and the B-coefficients of $\widetilde{s}|_{\Delta\setminus T}$ by the following additional smoothness conditions across the vertices, edges and faces of T:

$$\tau^{\widetilde{T},T_{2,2}}_{i,j,k,l}\widetilde{s} = 0,\ i,j < 5,\ k+l \leq 3,\ i+j+k+l = 9 \tag{3.113}$$

$$\tau^{\widetilde{T},T_{j,j}}_{3,0,4,2}\widetilde{s} = \tau^{\widetilde{T},T_{2+i,3-i}}_{3,0,4,2}\widetilde{s} = \tau^{\hat{T}_1,T_{2,1}}_{3,0,4,2}\widetilde{s} = \tau^{\hat{T}_3,T_{3,1}}_{3,0,4,2}\widetilde{s} = 0,\ j = 2,3 \tag{3.114}$$

$$\tau^{\widetilde{T},T_4}_{i,j,k,l}\widetilde{s} = 0,\ i,j,k \geq 1,\ l = 3,4,5,\ i+j+k = 9-l \tag{3.115}$$

Proof:
Using the C^4 supersmoothness conditions at the vertices of T, the C^3 supersmoothness conditions at the edges of T, except for $\langle v_4,v_3\rangle$, and the C^2 smoothness conditions at the faces $\langle v_1,v_2,v_3\rangle$ and $\langle v_1,v_2,v_4\rangle$, we can

uniquely and stably determine the B-coefficients of g corresponding to the domain points in $(\hat{M}1^2_{T^3_{WF}})$, $(\hat{M}3^2_{T^3_{WF}})$ - $(\hat{M}6^2_{T^3_{WF}})$, $(\hat{M}2^2_{T^3_{WF}})$, except for those corresponding to the domain points in $E_3^{T_{2,2}}(\langle v_4,v_3\rangle) \setminus (D_4(v_3) \cup D_4(v_4))$, and those in $(\hat{M}7^2_{T^3_{WF}})$ with $j=1$ from Theorem 3.29.

From the additional smoothness conditions in (3.113), (3.114), and (3.105) we can uniquely and stably compute the B-coefficients associated with the remaining domain points in $(\hat{M}2^2_{T^3_{WF}})$ and $(\hat{M}7^2_{T^3_{WF}})$, and those corresponding to the domain points in $(\hat{M}8^2_{T^3_{WF}})$. The last B-coefficients associated with a domain point in $\hat{\mathcal{M}}^2_3$ is equal to the value of $g(v_T)$. Thus, following Theorem 3.29, g is uniquely and stably determined. □

Lemma 3.32:
Let Δ be a tetrahedral partition of a polyhedral domain $\Omega \subset \mathbb{R}^3$ and $T := \langle v_1,v_2,v_3,v_4\rangle \in \Delta$ a tetrahedron, that shares exactly the two faces $\langle v_1,v_2,v_3\rangle$ and $\langle v_1,v_2,v_4\rangle$ with the tetrahedra $\hat{T}_1$ and $\hat{T}_2$, respectively, and the edge $\langle v_4,v_3\rangle$ with the tetrahedron $\widetilde{T}$, with $\hat{T}_1,\hat{T}_2,\widetilde{T} \in \Delta$. Moreover, T is refined with a third order partial Worsey-Farin split at the split point v_T in the interior of T, and the split points v_{F_1}, v_{F_2}, and v_{F_3} in the interior of the faces $\langle v_1,v_2,v_3\rangle$, $\langle v_2,v_3,v_4\rangle$, and $\langle v_1,v_3,v_4\rangle$, respectively. The tetrahedron $\hat{T}_1$ is also subjected to a partial Worsey-Farin split, such that the common face with T is refined with a Clough-Tocher split. The three corresponding subtetrahedra are denoted by $\hat{T}_{1,i}$, $i=1,2,3$, respectively, where $\hat{T}_{1,i}$ has a common face with $T_{1,i}$.

Let $\hat{\mathcal{S}}^2_9(T^3_{WF})$ be the superspline space defined in Theorem 3.29. Then, for any spline $s \in \mathcal{S}^{2,4,3}_9(\Delta \setminus T)$, there is a spline $g \in \hat{\mathcal{S}}^2_9(T^3_{WF})$ such that

$$\widetilde{s} := \begin{cases} s, & \text{on } \Delta \setminus T \\ g, & \text{on } T, \end{cases}$$

belongs to $\mathcal{S}^{2,4,3}_9(\Delta)$. Moreover, the spline g is uniquely and stably determined by the value of $g(v_T)$ and the B-coefficients of $\widetilde{s}|_{\Delta \setminus T}$ by the following additional smoothness conditions across the edges and faces of T:

$$\tau_{3,0,4,2}^{\widetilde{T},T_{j,j}}\widetilde{s} = \tau_{3,0,4,2}^{\hat{T}_2,T_{2+i,3-i}}\widetilde{s} = \tau_{3,0,4,2}^{\hat{T}_{1,j},T_{j,1}}\widetilde{s} = 0,\ i=0,1,\ j=2,3 \tag{3.116}$$

$$\tau_{i,j,k,l}^{\hat{T}_2,T_4}\widetilde{s} = 0,\ i,j,k \geq 1,\ l=3,4,5,\ i+j+k=9-l \tag{3.117}$$

Proof:
From the C^4 supersmoothness conditions at the vertices of T, the C^3 supersmoothness conditions at the edges of T, and the C^2 smoothness conditions at the faces $\langle v_1,v_2,v_3\rangle$ and $\langle v_1,v_2,v_4\rangle$, we can uniquely and stably determine the B-coefficients of g corresponding to the domain points in ($\hat{\text{M}}1^2_{T^3_{WF}}$) - ($\hat{\text{M}}6^2_{T^3_{WF}}$), and those in ($\hat{\text{M}}7^2_{T^3_{WF}}$) with $j=1$ from Theorem 3.29.

From the additional smoothness conditions in (3.116) and (3.117) we can uniquely and stably determine the B-coefficients corresponding to the remaining domain points in ($\hat{\text{M}}7^2_{T^3_{WF}}$), and to those in ($\hat{\text{M}}8^2_{T^3_{WF}}$). Since the value of $g(v_T)$ is equal to the B-coefficient associated with the domain point in ($\hat{\text{M}}9^2_{T^3_{WF}}$), following Theorem 3.29, g is uniquely and stably determined.□

Theorem 3.33:
Let $\check{\mathcal{M}}^2_3$ be the union of the following sets of domain points in $\mathcal{D}_{T^3_{WF},9}$:

($\check{\text{M}}1^2_{T^3_{WF}}$) $D_4^{T_{v_i}}(v_i)$, $i=1,\ldots,4$, with $v_i \in T_{v_i} \in T^3_{WF}$

($\check{\text{M}}2^2_{T^3_{WF}}$) $\left(\bigcup_{i=1}^{2} E_3^{T_{1,i}}(\langle v_i,v_{1+i}\rangle) \cup E_3^{T_{1,3}}(\langle v_1,v_3\rangle) \cup E_3^{T_{2,2}}(\langle v_4,v_3\rangle)\right) \setminus \bigcup_{i=1}^{4}(D_4(v_i))$

($\check{\text{M}}3^2_{T^3_{WF}}$) $\{\xi^{T_4}_{4-i,1+i,4,0}, \xi^{T_4}_{4-i,1+i,3,1}, \xi^{T_4}_{3-i,i,4,2}, \xi^{T_4}_{4-i,1+i,2,2}\}_{i=0}^{3}$

($\check{\text{M}}4^2_{T^3_{WF}}$) $\{\xi^{T_4}_{4-i,i,4,1}, \xi^{T_4}_{4-i,i,3,2}\}_{i=0}^{4}$

($\check{\text{M}}5^2_{T^3_{WF}}$) $\{\xi^{T_4}_{4-i,0,2+i,3}, \xi^{T_4}_{0,4-i,2+i,3}\}_{i=0}^{2}$

($\check{\text{M}}6^2_{T^3_{WF}}$) $\{\xi^{T_{j,1}}_{1,0,8-i,i}, \xi^{T_{j,1}}_{0,1,8-i,i}, \xi^{T_{j,1}}_{0,0,9-i,i}\}_{i=0}^{1}$, $j=1,2$

($\check{\text{M}}7^2_{T^3_{WF}}$) $\{\xi^{T_{j,i}}_{3,0,4,2}\}$, $i=1,2,3$, $j=1,2,3$

($\check{\text{M}}8^2_{T^3_{WF}}$) $\{\xi^{T_4}_{i,j,k,l}\}$, $i,j,k \geq 1$, $l=3,4,5$, $i+j+k=9-l$

($\check{\text{M}}9^2_{T^3_{WF}}$) $\{\xi^{T_{1,1}}_{0,0,0,9}\}$

Then $\check{\mathcal{M}}_3^2$ is a stable minimal determining set for

$$
\begin{aligned}
\check{\mathcal{S}}_9^2(T_{WF}^3) := \{s \in C^2(T_{WF}^3) : s|_{\widetilde{T}} \in \mathcal{P}_9^3,\ \forall \widetilde{T} \in T_{WF}^3,\\
s \in C^4(v_i),\ i = 1,\dots,4,\\
s \in C^3(e),\ \text{for all edges } e \in T,\\
s \in C^7(\langle v_{F_i}, v_T \rangle),\ i = 1,2\\
s \in C^7(v_T),\\
\tau_{3,5,1,0}^{T_{3,i},T_{3,i+1}} s = 0,\ i = 1,2,3,\ \text{with } T_{3,4} := T_{3,1},\\
\tau_{3,5,0,1}^{T_{3,i},T_{3,i+1}} s = 0,\ i = 1,2,3,\ \text{with } T_{3,4} := T_{3,1}\},
\end{aligned}
$$

and the dimension of $\check{\mathcal{S}}_9^2(T_{WF}^3)$ is equal to 293.

Proof:
We first show that $\check{\mathcal{M}}_3^2$ is a stable minimal determining set for $\check{\mathcal{S}}_9^2(T_{WF}^3)$. Therefore, we set the B-coefficients c_ξ of a spline $s \in \check{\mathcal{S}}_9^2(T_{WF}^3)$ to arbitrary values for each domain point $\xi \in \check{\mathcal{M}}_3^2$ and show that all other B-coefficients of s are uniquely and stably determined.

The B-coefficients corresponding to the domain points within a distance of four from the vertices of T, within a distance of three from the edges of T, and within a distance of two from the faces $\langle v_1, v_2, v_3 \rangle$ and $\langle v_1, v_2, v_4 \rangle$ can be uniquely and stably determined in the same way as in the proof of Theorem 3.29 from the B-coefficients associated with the domain points in ($\check{\text{M}}1^2_{T^3_{WF}}$) - ($\check{\text{M}}5^2_{T^3_{WF}}$) and those corresponding to the domain points in ($\check{\text{M}}6^2_{T^2_{WF}}$) with $j = 1$.

Now, the B-coefficients corresponding to the domain points in the shells $R_9^{T_{2,i}}(v_T)$ and $R_8^{T_{2,i}}(v_T)$, $i = 1,2,3$, can be uniquely and stably computed from those already determined and the ones associated with the domain points in ($\check{\text{M}}6^2_{T^2_{WF}}$) with $j = 2$ in the same way as the B-coefficients in the shells $R_9^{T_{1,i}}(v_T)$ and $R_8^{T_{1,i}}(v_T)$, $i = 1,2,3$, are determined in the proof of Theorem 3.20.

Then, the remaining B-coefficients of s can be uniquely and stably determined from those corresponding to the domain points in ($\check{\text{M}}7^2_{T^3_{WF}}$) - ($\check{\text{M}}8^2_{T^3_{WF}}$) and the already determined B-coefficients in the same way as in the proof of Theorem 3.24.

Finally, examining the dimension of the space $\check{\mathcal{S}}_9^2(T_{WF}^3)$, we can see that $\dim \check{\mathcal{S}}_9^2(T_{WF}^3) = 293 = \#\check{\mathcal{M}}_3^2$, since the set $\check{\mathcal{M}}_3^2$ is a minimal determining set for the superspline space $\check{\mathcal{S}}_9^2(T_{WF}^3)$. □

Lemma 3.34:
Let Δ be a tetrahedral partition of a polyhedral domain $\Omega \subset \mathbb{R}^3$ and $T := \langle v_1, v_2, v_3, v_4 \rangle \in \Delta$ a tetrahedron, that shares exactly the three faces $\langle v_1, v_2, v_3 \rangle$, $\langle v_2, v_3, v_4 \rangle$, and $\langle v_1, v_2, v_4 \rangle$ with the tetrahedra $\hat{T}_1$, $\hat{T}_2$, and $\widetilde{T}$, respectively, with $\hat{T}_1, \hat{T}_2, \widetilde{T} \in \Delta$. Moreover, T is refined with a third order partial Worsey-Farin split at the split point v_T in the interior of T, and the split points v_{F_1}, v_{F_2}, and v_{F_3} in the interior of the faces $\langle v_1, v_2, v_3 \rangle$, $\langle v_2, v_3, v_4 \rangle$, and $\langle v_1, v_3, v_4 \rangle$, respectively. The tetrahedra $\hat{T}_1$ and $\hat{T}_2$ are also subjected to a partial Worsey-Farin split, such that the common faces with T are refined with a Clough-Tocher split, respectively. The subtetrahedra are denoted by $\hat{T}_{j,i}$, $i = 1,2,3$, $j = 1,2$, respectively, where $\hat{T}_{j,i}$ has a common face with $T_{j,i}$.

Let $\check{\mathcal{S}}_9^2(T_{WF}^3)$ be the superspline space defined in Theorem 3.33. Then, for any spline $s \in \mathcal{S}_9^{2,4,3}(\Delta \setminus T)$, there is a spline $g \in \check{\mathcal{S}}_9^2(T_{WF}^3)$ such that

$$\widetilde{s} := \begin{cases} s, & \text{on } \Delta \setminus T \\ g, & \text{on } T, \end{cases}$$

belongs to $\mathcal{S}_9^{2,4,3}(\Delta)$. Moreover, the spline g is uniquely and stably determined by the value of $g(v_T)$ and the B-coefficients of $\widetilde{s}|_{\Delta \setminus T}$ by the following additional smoothness conditions across the edges and faces of T:

$$\tau_{3,0,4,2}^{\hat{T}_{1,3}, T_{3,1}} \widetilde{s} = \tau_{3,0,4,2}^{\widetilde{T}, T_{3,2}} \widetilde{s} = \tau_{3,0,4,2}^{\hat{T}_{2,2}, T_{3,3}} \widetilde{s} = 0 \tag{3.118}$$

$$\tau_{i,j,k,l}^{\widetilde{T}, T_4} \widetilde{s} = 0,\ i,j,k \geq 1,\ l = 3,4,5,\ i+j+k = 9-l \tag{3.119}$$

Proof:
Using the C^4 supersmoothness conditions at the vertices of T, the C^3 supersmoothness conditions at the edges of T, and the C^2 smoothness conditions at the faces $\langle v_1, v_2, v_3 \rangle$, $\langle v_2, v_3, v_4 \rangle$, and $\langle v_1, v_2, v_4 \rangle$, we can uniquely and stably compute the B-coefficients of g corresponding to the domain points in ($\check{M}1^2_{T^3_{WF}}$) - ($\check{M}6^2_{T^3_{WF}}$), and those in ($\check{M}7^2_{T^3_{WF}}$) with $j = 1$ and $j = 2$ from Theorem 3.33.

From the additional smoothness conditions in (3.118) and (3.119) we can uniquely and stably determine the B-coefficients associated with the remaining domain points in ($\check{M}7^2_{T^3_{WF}}$), and with those in ($\check{M}8^2_{T^3_{WF}}$). The B-coefficient associated with the last domain point in $\check{\mathcal{M}}^2_3$, the one in ($\check{M}9^2_{T^3_{WF}}$), is equal to the value of $g(v_T)$. Thus, following Theorem 3.33, g is uniquely and stably determined. □

Theorem 3.35:
Let $\dot{\mathcal{M}}^2_3$ be the union of the following sets of domain points in $\mathcal{D}_{T^3_{WF},9}$:

($\dot{M}1^2_{T^3_{WF}}$) $D_4^{T_{v_i}}(v_i)$, $i = 1,\dots,4$, with $v_i \in T_{v_i} \in T^3_{WF}$

($\dot{M}2^2_{T^3_{WF}}$) $\left(\bigcup_{i=1}^{2} E_3^{T_{1,i}}(\langle v_i, v_{1+i}\rangle) \cup E_3^{T_{1,3}}(\langle v_1, v_3\rangle) \cup E_3^{T_{2,2}}(\langle v_4, v_3\rangle)\right) \setminus \bigcup_{i=1}^{4}(D_4(v_i))$

($\dot{M}3^2_{T^3_{WF}}$) $\{\xi^{T_4}_{4-i,1+i,4,0}, \xi^{T_4}_{4-i,1+i,3,1}, \xi^{T_4}_{3-i,i,4,2}, \xi^{T_4}_{4-i,1+i,2,2}\}_{i=0}^{3}$

($\dot{M}4^2_{T^3_{WF}}$) $\{\xi^{T_4}_{4-i,i,4,1}, \xi^{T_4}_{4-i,i,3,2}\}_{i=0}^{4}$

($\dot{M}5^2_{T^3_{WF}}$) $\{\xi^{T_4}_{4-i,0,2+i,3}, \xi^{T_4}_{0,4-i,2+i,3}\}_{i=0}^{2}$

($\dot{M}6^2_{T^3_{WF}}$) $\{\xi^{T_{j,1}}_{1,0,8-i,i}, \xi^{T_{j,1}}_{0,1,8-i,i}, \xi^{T_{j,1}}_{0,0,9-i,i}\}_{i=0}^{1}$, $j = 1,2,3$

($\dot{M}7^2_{T^3_{WF}}$) $\{\xi^{T_{j,i}}_{3,0,4,2}\}$, $i = 1,2,3$, $j = 1,2,3$

($\dot{M}8^2_{T^3_{WF}}$) $\{\xi^{T_4}_{i,j,k,l}\}$, $i,j,k \geq 1$, $l = 3,4,5$, $i+j+k = 9-l$

($\dot{M}9^2_{T^3_{WF}}$) $\{\xi^{T_{1,1}}_{0,0,0,9}\}$

Then $\dot{\mathcal{M}}^2_3$ is a stable minimal determining set for

$$\begin{aligned}\dot{\mathcal{S}}^2_9(T^3_{WF}) := \{s \in C^2(T^3_{WF}) : s|_{\widetilde{T}} \in \mathcal{P}^3_9,\ \forall \widetilde{T} \in T^3_{WF},\\ s \in C^4(v_i),\ i = 1,\dots,4,\\ s \in C^3(e), \text{ for all edges } e \in T,\\ s \in C^7(\langle v_{F_i}, v_T\rangle),\ i = 1,2\\ s \in C^7(v_T)\},\end{aligned}$$

and the dimension of $\dot{\mathcal{S}}^2_9(T^3_{WF})$ is equal to 299.

Proof:
First, we show that the set $\dot{\mathcal{M}}^2_3$ is a stable minimal determining set for the spline space $\dot{\mathcal{S}}^2_9(T^3_{WF})$. To this end, we set the B-coefficients c_ξ of a spline $s \in \dot{\mathcal{S}}^2_9(T^3_{WF})$ to arbitrary values for each domain point $\xi \in \dot{\mathcal{M}}^2_3$. Then, we show that all other B-coefficients of s are uniquely and stably determined.

The B-coefficients corresponding to the domain points within a distance of four from the vertices of T, within a distance of three from the edges of T, and within a distance of two from the faces $\langle v_1, v_2, v_3 \rangle$, $\langle v_2, v_3, v_4 \rangle$, and $\langle v_1, v_2, v_4 \rangle$ can be uniquely and stably determined in the same way as in the proof of Theorem 3.33 from the B-coefficients associated with the domain points in ($\dot{\text{M}}1^2_{T^3_{WF}}$) - ($\dot{\text{M}}5^2_{T^3_{WF}}$) and the ones associated with the domain points in ($\dot{\text{M}}6^2_{T^3_{WF}}$) with $j = 1$ and $j = 2$.

Then, we can uniquely and stably compute the B-coefficients corresponding to the domain points in the shells $R_9^{T_{3,i}}(v_T)$ and $R_8^{T_{3,i}}(v_T)$, $i = 1,2,3$, from the already determined B-coefficients and those associated with the domain points in ($\widetilde{\text{M}}6^2_{T^2_{WF}}$) with $j = 3$ in the same way as the B-coefficients corresponding to the domain points in the shells $R_9^{T_{1,i}}(v_T)$ and $R_8^{T_{1,i}}(v_T)$, $i = 1,2,3$, are determined in the proof of Theorem 3.20.

The remaining B-coefficients of s can be uniquely and stably computed from those corresponding to the domain points in ($\dot{\text{M}}7^2_{T^3_{WF}}$) - ($\dot{\text{M}}9^2_{T^3_{WF}}$) and the already determined ones in the same way as in the proof of Theorem 3.24.

At last, considering the dimension of the superspline space $\dot{\mathcal{S}}_9^2(T^3_{WF})$, we obtain that $\dim \dot{\mathcal{S}}_9^2(T^3_{WF}) = \#\dot{\mathcal{M}}_3^2 = 299$, since the set $\dot{\mathcal{M}}_3^2$ is a minimal determining set for the space of supersplines $\dot{\mathcal{S}}_9^2(T^3_{WF})$. □

Lemma 3.36:
Let Δ be a tetrahedral partition of a polyhedral domain $\Omega \subset \mathbb{R}^3$ and $T := \langle v_1, v_2, v_3, v_4 \rangle \in \Delta$ a tetrahedron, that shares the four faces $\langle v_1, v_2, v_3 \rangle$, $\langle v_2, v_3, v_4 \rangle$, $\langle v_1, v_3, v_4 \rangle$, and $\langle v_1, v_2, v_4 \rangle$ with the tetrahedra $\hat{T}_1$, $\hat{T}_2$, $\hat{T}_3$, and $\widetilde{T}$, respectively, with $\hat{T}_1, \hat{T}_2, \hat{T}_3, \widetilde{T} \in \Delta$. Moreover, T is refined with a third order partial Worsey-Farin split at the split point v_T in the interior of T, and the split points v_{F_1}, v_{F_2}, and v_{F_3} in the interior of the faces $\langle v_1, v_2, v_3 \rangle$, $\langle v_2, v_3, v_4 \rangle$, and $\langle v_1, v_3, v_4 \rangle$, respectively. The tetrahedra $\hat{T}_1$, $\hat{T}_2$, and $\hat{T}_3$ are also subjected to a partial Worsey-Farin split, such that the common faces with T are refined with a Clough-Tocher split, respectively. The subtetrahedra are denoted by $\hat{T}_{j,i}$, $i = 1,2,3$, $j = 1,2,3$ respectively, where $\hat{T}_{j,i}$ has a common face with $T_{j,i}$.

Let $\dot{\mathcal{S}}_9^2(T_{WF}^3)$ be the superspline space defined in Theorem 3.35. Then, for any spline $s \in \mathcal{S}_9^{2,4,3}(\Delta \setminus T)$, there is a spline $g \in \dot{\mathcal{S}}_9^2(T_{WF}^3)$ such that

$$\widetilde{s} := \begin{cases} s, & \text{on } \Delta \setminus T \\ g, & \text{on } T, \end{cases}$$

belongs to $\mathcal{S}_9^{2,4,3}(\Delta)$. Moreover, the spline g is uniquely and stably determined by the value of $g(v_T)$ and the B-coefficients of $\widetilde{s}|_{\Delta \setminus T}$ by the following additional smoothness conditions across the face $\langle v_1, v_2, v_4 \rangle$ of T:

$$\tau_{i,j,k,l}^{\widetilde{T},T_4} \widetilde{s} = 0,\ i,j,k \geq 1,\ l = 3,4,5,\ i+j+k = 9-l \tag{3.120}$$

Proof:
From the C^4 supersmoothness conditions at the vertices of T, the C^3 supersmoothness conditions at the edges of T, and the C^2 smoothness conditions at the faces of T, we can uniquely and stably compute the B-coefficients of g corresponding to the domain points in ($\check{M}1^2_{T^3_{WF}}$) - ($\check{M}7^2_{T^3_{WF}}$) from Theorem 3.35.

From the additional smoothness conditions in (3.120) we can uniquely and stably compute the B-coefficients corresponding to the domain points in the set ($\check{M}8^2_{T^3_{WF}}$). Since, the B-coefficient associated with the domain point in ($\dot{M}9^2_{T^3_{WF}}$) is equal to the value of $g(v_T)$, following Theorem 3.35, g is uniquely and stably determined. □

3.2.4 Minimal determining sets for C^2 supersplines based on fourth order partial Worsey-Farin splits

In this subsection, we construct minimal determining sets for C^2 splines of degree nine based on the fourth order partial Worsey-Farin split of a tetrahedron $T := \langle v_1, v_2, v_3, v_4 \rangle$. Therefore, let T_{WF}^4 be the fourth order partial Worsey-Farin split of T at the split point v_T in the interior of T, and let the four faces $\langle v_1, v_2, v_3 \rangle$, $\langle v_2, v_3, v_4 \rangle$, $\langle v_3, v_1, v_4 \rangle$, and $\langle v_1, v_2, v_4 \rangle$ be refined with a Clough-Tocher split at the split point v_{F_1}, v_{F_2}, v_{F_3}, and v_{F_4}, respectively.

The corresponding subtetrahedra are denoted by

$$\begin{aligned}
T_{1,1} &:= \langle v_1, v_2, v_{F_1}, v_T \rangle, & T_{1,2} &:= \langle v_2, v_3, v_{F_1}, v_T \rangle, \\
T_{1,3} &:= \langle v_3, v_1, v_{F_1}, v_T \rangle, & T_{2,1} &:= \langle v_2, v_3, v_{F_2}, v_T \rangle, \\
T_{2,2} &:= \langle v_3, v_4, v_{F_2}, v_T \rangle, & T_{2,3} &:= \langle v_4, v_2, v_{F_2}, v_T \rangle, \\
T_{3,1} &:= \langle v_3, v_1, v_{F_3}, v_T \rangle, & T_{3,2} &:= \langle v_1, v_4, v_{F_3}, v_T \rangle, \\
T_{3,3} &:= \langle v_4, v_3, v_{F_3}, v_T \rangle, & T_{4,1} &:= \langle v_1, v_2, v_{F_4}, v_T \rangle, \\
T_{4,2} &:= \langle v_2, v_4, v_{F_4}, v_T \rangle, & T_{4,3} &:= \langle v_4, v_1, v_{F_4}, v_T \rangle.
\end{aligned} \tag{3.121}$$

Theorem 3.37:
Let $\mathcal{M}_4^2$ be the union of the following sets of domain points in $\mathcal{D}_{T_{WF}^4,9}$:

(M1$^2_{T^4_{WF}}$) $D_4^{T_{v_i}}(v_i)$, $i = 1,\ldots,4$, with $v_i \in T_{v_i} \in T_{WF}^4$

(M2$^2_{T^4_{WF}}$) $\big(E_3^{T_{1,1}}(\langle v_1,v_2\rangle) \cup E_3^{T_{1,2}}(\langle v_2,v_3\rangle) \cup E_3^{T_{1,3}}(\langle v_3,v_1\rangle) \cup E_3^{T_{2,2}}(\langle v_3,v_4\rangle) \cup$

$$\cup E_3^{T_{2,3}}(\langle v_4,v_2\rangle) \cup E_3^{T_{3,1}}(\langle v_1,v_4\rangle)\big) \setminus \bigcup_{i=1}^{4} (D_4(v_i))$$

(M3$^2_{T^4_{WF}}$) $\{\xi_{3,0,4,2}^{T_{j,i}}\}$, $i = 1,2,3$, $j = 1,\ldots,4$

(M4$^2_{T^4_{WF}}$) $\{\xi_{4,0,2,3}^{T_{1,i}}, \xi_{3,0,3,3}^{T_{1,i}}, \xi_{3,1,2,3}^{T_{1,i}}, \xi_{3,0,2,4}^{T_{1,i}}, \xi_{2,2,1,4}^{T_{1,i}}, \xi_{2,0,2,5}^{T_{1,i}}\}_{i=1}^{3}$

(M5$^2_{T^4_{WF}}$) $\{\xi_{2,0,4,3}^{T_{1,1}}\}$

(M6$^2_{T^4_{WF}}$) $\{\xi_{0,0,0,9}^{T_{1,1}}\}$

Then $\mathcal{M}_4^2$ is a stable minimal determining set for

$$\begin{aligned}
\widetilde{\mathcal{S}}_9^2(T_{WF}^4) := \{ s \in C^2(T_{WF}^4) : {}& s|_{\widetilde{T}} \in \mathcal{P}_9^3,\ \forall \widetilde{T} \in T_{WF}^4, \\
& s \in C^4(v_i),\ i = 1,\ldots,4, \\
& s \in C^3(e), \text{ for all edges } e \in T, \\
& s \in C^7(\langle v_{F_i}, v_T\rangle),\ i = 1,2 \\
& s \in C^7(v_T), \\
& \tau_{3,5,1,0}^{T_{j,i},T_{j,i+1}} s = \tau_{3,5,0,1}^{T_{j,i},T_{j,i+1}} s = 0, \\
& i = 1,2,3,\ j = 1,\ldots,4, \text{ with } T_{j,4} := T_{j,1} \},
\end{aligned}$$

and the dimension of $\widetilde{\mathcal{S}}_9^2(T_{WF}^4)$ is equal to 292.

Proof:
First, we show that $\mathcal{M}_4^2$ is a stable minimal determining set for the super-spline space $\widetilde{\mathcal{S}}_9^2(T_{WF}^4)$. To this end we set the B-coefficients c_ξ of a spline $s \in \widetilde{\mathcal{S}}_9^2(T_{WF}^4)$ to arbitrary values for each domain point $\xi \in \mathcal{M}_4^2$ and show that all other B-coefficients of s are uniquely and stably determined.

We can uniquely and stably determine the B-coefficients of s associated with the domain points in the balls $D_4(v_i)$, $i = 1,\dots,4$, from the B-coefficients corresponding to the domain points in $(\mathrm{M1}^2_{T_{WF}^4})$ by using the C^4 supersmoothness conditions at the vertices of T.

Then, we can use the C^3 supersmoothness conditions at the edges of T to uniquely and stably determine the remaining B-coefficients of s corresponding to the domain points within a distance of three from the edges of T from the already determined B-coefficients and those associated with the domain points in $(\mathrm{M2}^2_{T_{WF}^4})$.

The B-coefficients associated with the domain points within a distance of two from the faces of T can be uniquely and stably computed from the already determined B-coefficients and those corresponding to the domain points in the $(\mathrm{M3}^2_{T_{WF}^4})$ in the same way as the B-coefficients associated with the domain points in the shells $R_9^{T_{1,i}}(v_T)$, $R_8^{T_{1,i}}(v_T)$, and $R_7^{T_{1,i}}(v_T)$, $i = 1,2,3$, are computed in Theorem 3.4.

Now, we consider the B-coefficients corresponding to the domain points in the shells $R_6^{T_{1,i}}(v_T)$, $R_5^{T_{1,i}}(v_T)$, and $R_4^{T_{1,i}}(v_T)$, $i = 1,2,3$. The B-coefficients of s restricted to these shells can be uniquely and stably determined from the B-coefficients associated with the domain points in $(\mathrm{M4}^2_{T_{WF}^4})$, $(\mathrm{M5}^2_{T_{WF}^4})$ and those already determined in the same way as the B-coefficients of s restricted to these shells in Theorem 3.4.

At this point, we can consider the last undetermined B-coefficients of s, which are associated with domain points in the ball $D_7(v_T)$. Since s has C^7 supersmoothness at the split point v_T, we can regard the B-coefficients corresponding to the domain points in $D_7(v_T)$ as those of a trivariate polynomial g defined on the tetrahedron $\widetilde{T}$, bounded by the domain points in $R_7(v_T)$, which is refined with a fourth order partial Worsey-Farin split at v_T. Considering the polynomial g, we can see that the B-coefficients associated with the domain points on three of the faces of $\widetilde{T}$ and those corresponding to the domain points within a distance of three from the fourth face are already uniquely and stably determined. Since the setting of the B-coefficient of g associated with the domain point $\xi_{0,0,0,9}^{T_{1,1}}$ in (M6$^2_{T^4_{WF}}$) is equal to setting the function value of g at v_T, we can apply Lemma 2.42. Thus, the remaining undetermined B-coefficients of g can be uniquely and stably computed by the C^7 smoothness conditions at v_T. Therefore, we have also uniquely and stably determined all B-coefficients of s.

Since $\mathcal{M}_4^2$ is a minimal determining set for the spline space $\widetilde{\mathcal{S}}_9^2(T^4_{WF})$, we obtain that $\widetilde{\mathcal{S}}_9^2(T^4_{WF}) = \#\mathcal{M}_4^2 = 292$. □

Lemma 3.38:
Let Δ be a tetrahedral partition of a polyhedral domain $\Omega \subset \mathbb{R}^3$ and $T := \langle v_1, v_2, v_3, v_4 \rangle \in \Delta$ a tetrahedron, that shares exactly the four edges $\langle v_1, v_3 \rangle$, $\langle v_1, v_4 \rangle$, $\langle v_2, v_3 \rangle$, and $\langle v_2, v_4 \rangle$ with the tetrahedra $\widetilde{T}_1, \widetilde{T}_2, \widetilde{T}_3$, and $\widetilde{T}_4$, respectively, with $\widetilde{T}_j \in \Delta$, $j = 1, \ldots, 4$. Moreover, T is refined with a fourth order partial Worsey-Farin split at the split point v_T in the interior of T, and the split points $v_{F_1}, v_{F_2}, v_{F_3}$, and v_{F_4} in the interior of the faces $\langle v_1, v_2, v_3 \rangle$, $\langle v_2, v_3, v_4 \rangle$, $\langle v_1, v_3, v_4 \rangle$, and $\langle v_1, v_2, v_4 \rangle$, respectively.

Let $\widetilde{\mathcal{S}}_9^2(T^4_{WF})$ be the superspline space defined in Theorem 3.37. Then, for any spline $s \in \mathcal{S}_9^{2,4,3}(\Delta \setminus T)$, there is a spline $g \in \widetilde{\mathcal{S}}_9^2(T^4_{WF})$ such that

$$\widetilde{s} := \begin{cases} s, & \text{on } \Delta \setminus T \\ g, & \text{on } T, \end{cases}$$

belongs to $\mathcal{S}_9^{2,4,3}(\Delta)$. Moreover, the spline g is uniquely and stably determined by the value of $g(v_T)$ and the B-coefficients of $\widetilde{s}|_{\Delta \setminus T}$ by the following additional smoothness conditions across the vertices and edges of T:

$$\tau_{i,j,k,l}^{\widetilde{T}_1, T_{1,1}} \widetilde{s} = \tau_{i,j,k,l}^{\widetilde{T}_2, T_{2,2}} \widetilde{s} = 0, \; i,j < 5, \; k+l \leq 3, \; i+j+k+l = 9 \tag{3.122}$$

$$\begin{aligned} \tau_{3,0,4,2}^{\widetilde{T}_1, T_{j,1}} \widetilde{s} &= \tau_{3,0,4,2}^{\widetilde{T}_1, T_{1,3}} \widetilde{s} = \tau_{3,0,4,2}^{\widetilde{T}_2, T_{2+i,2+i}} \widetilde{s} = \tau_{3,0,4,2}^{\widetilde{T}_2, T_{3+i,2+i}} \widetilde{s} = \tau_{3,0,4,2}^{\widetilde{T}_3, T_{1+i,2-i}} \widetilde{s} \\ &= \tau_{3,0,4,2}^{\widetilde{T}_4, T_{2+2i,3-i}} \widetilde{s} = 0, \; i = 0,1, \; j = 1,3,4 \end{aligned} \tag{3.123}$$

$$\begin{aligned} \tau_{4,0,2,3}^{\widetilde{T}_1, T_{1,i}} \widetilde{s} &= \tau_{4,0,2,3}^{\widetilde{T}_3, T_{1,2}} \widetilde{s} = \tau_{3,0,3,3}^{\widetilde{T}_1, T_{1,i}} \widetilde{s} = \tau_{3,0,3,3}^{\widetilde{T}_3, T_{1,2}} \widetilde{s} = \tau_{3,1,2,3}^{\widetilde{T}_1, T_{1,i}} \widetilde{s} \\ &= \tau_{3,1,2,3}^{\widetilde{T}_3, T_{1,2}} \widetilde{s} = \tau_{3,0,2,4}^{\widetilde{T}_1, T_{1,i}} \widetilde{s} = \tau_{3,0,2,4}^{\widetilde{T}_3, T_{1,2}} \widetilde{s} = \tau_{2,2,1,4}^{\widetilde{T}_1, T_{1,i}} \widetilde{s} = \tau_{2,2,1,4}^{\widetilde{T}_3, T_{1,2}} \widetilde{s} \\ &= \tau_{2,0,2,5}^{\widetilde{T}_1, T_{1,i}} \widetilde{s} = \tau_{2,0,2,5}^{\widetilde{T}_3, T_{1,2}} \widetilde{s} = 0, \; i = 1,3 \end{aligned} \tag{3.124}$$

$$\tau_{2,0,4,3}^{\widetilde{T}_1, T_{1,1}} \widetilde{s} = 0 \tag{3.125}$$

Proof:
From the C^4 supersmoothness conditions at the vertices and the C^3 supersmoothness conditions at the edges $\langle v_1, v_3\rangle$, $\langle v_1, v_4\rangle$, $\langle v_2, v_3\rangle$, and $\langle v_2, v_4\rangle$ of T, we can uniquely and stably compute the B-coefficients of g corresponding to the domain points in (M$1^2_{T^4_{WF}}$) and (M$2^2_{T^4_{WF}}$), except for those associated with the domain points $\left(E_3^{T_{1,1}}(\langle v_1, v_2\rangle) \cup E_3^{T_{2,2}}(\langle v_3, v_4\rangle)\right) \setminus \bigcup_{i=1}^4 D_4(v_i)$, from Theorem 3.37.

From the additional smoothness conditions in (3.122) we can uniquely and stably compute the remaining B-coefficients associated with domain points in (M$2^2_{T^4_{WF}}$). Using the additional smoothness conditions in (3.123), (3.124), and (3.125) the B-coefficients corresponding to the domain points in (M$3^2_{T^4_{WF}}$) - (M$5^2_{T^4_{WF}}$) can be uniquely and stably determined. The B-coefficient associated with the last domain point $\xi^{T_{1,1}}_{0,0,0,9}$ in $\mathcal{M}^2_4$ is equal to the value of $g(v_T)$. Thus, following Theorem 3.37, g is uniquely and stably determined. □

Lemma 3.39:
Let Δ be a tetrahedral partition of a polyhedral domain $\Omega \subset \mathbb{R}^3$ and $T := \langle v_1, v_2, v_3, v_4\rangle \in \Delta$ a tetrahedron, that shares exactly the five edges $\langle v_1, v_3\rangle$, $\langle v_1, v_4\rangle$, $\langle v_2, v_3\rangle$, $\langle v_2, v_4\rangle$, and $\langle v_1, v_2\rangle$ with the tetrahedra $\widetilde{T}_1, \widetilde{T}_2, \widetilde{T}_3$, $\widetilde{T}_4$, and $\widetilde{T}_5$, respectively, with $\widetilde{T}_j \in \Delta$, $j = 1,\ldots,5$. Moreover, T is refined with a fourth order partial Worsey-Farin split at the split point v_T in the interior of T, and the split points $v_{F_1}, v_{F_2}, v_{F_3}$, and v_{F_4} in the interior of the faces $\langle v_1, v_2, v_3\rangle$, $\langle v_2, v_3, v_4\rangle$, $\langle v_1, v_3, v_4\rangle$, and $\langle v_1, v_2, v_4\rangle$, respectively.

Let $\widetilde{\mathcal{S}}^2_9(T^4_{WF})$ be the superspline space defined in Theorem 3.37. Then, for any spline $s \in \mathcal{S}^{2,4,3}_9(\Delta \setminus T)$, there is a spline $g \in \widetilde{\mathcal{S}}^2_9(T^4_{WF})$ such that

$$\widetilde{s} := \begin{cases} s, & \text{on } \Delta \setminus T \\ g, & \text{on } T, \end{cases}$$

belongs to $\mathcal{S}^{2,4,3}_9(\Delta)$. Moreover, the spline g is uniquely and stably determined by the value of $g(v_T)$ and the B-coefficients of $\widetilde{s}|_{\Delta \setminus T}$ by the following additional smoothness conditions across the vertices and edges of T:

$$\tau_{i,j,k,l}^{\widetilde{T}_2,T_{2,2}}\widetilde{s}=0,\ i,j<5,\ k+l\leq 3,\ i+j+k+l=9 \tag{3.126}$$

$$\begin{aligned}&\tau_{3,0,4,2}^{\widetilde{T}_1,T_{1+2i,3-2i}}\widetilde{s}=\tau_{3,0,4,2}^{\widetilde{T}_2,T_{2+i,2+i}}\widetilde{s}=\tau_{3,0,4,2}^{\widetilde{T}_2,T_{3+i,2+i}}\widetilde{s}=\tau_{3,0,4,2}^{\widetilde{T}_3,T_{1+i,2-i}}\widetilde{s}\\&=\tau_{3,0,4,2}^{\widetilde{T}_4,T_{2+2i,3-i}}\widetilde{s}=\tau_{3,0,4,2}^{\widetilde{T}_5,T_{1+3i,1}}\widetilde{s}=0,\ i=0,1\end{aligned} \tag{3.127}$$

$$\begin{aligned}&\tau_{4,0,2,3}^{\widetilde{T}_{5-2i},T_{1,1+i}}\widetilde{s}=\tau_{3,0,3,3}^{\widetilde{T}_{5-2i},T_{1,1+i}}\widetilde{s}=\tau_{3,1,2,3}^{\widetilde{T}_{5-2i},T_{1,1+i}}\widetilde{s}\\&=\tau_{3,0,2,4}^{\widetilde{T}_{5-2i},T_{1,1+i}}\widetilde{s}=\tau_{2,2,1,4}^{\widetilde{T}_{5-2i},T_{1,1+i}}\widetilde{s}=\tau_{2,0,2,5}^{\widetilde{T}_{5-2i},T_{1,1+i}}\widetilde{s}=0,\ i=0,1,2\end{aligned} \tag{3.128}$$

$$\tau_{2,0,4,3}^{\widetilde{T}_5,T_{1,1}}\widetilde{s}=0 \tag{3.129}$$

Proof:
From the C^4 supersmoothness conditions at the vertices of T and the C^3 supersmoothness conditions at the edges $\langle v_1,v_3\rangle$, $\langle v_1,v_4\rangle$, $\langle v_2,v_3\rangle$, $\langle v_2,v_4\rangle$, and $\langle v_1,v_2\rangle$ of T, we can uniquely and stably compute the B-coefficients of g corresponding to the domain points in $(M1^2_{T^4_{WF}})$ and $(M2^2_{T^4_{WF}})$, except for those associated with the domain points $E_3^{T_{2,2}}(\langle v_3,v_4\rangle)\setminus(D_4(v_3)\cup(D_4(v_4))$, from Theorem 3.37.

Using the additional smoothness conditions in (3.126) we can uniquely and stably determine the B-coefficients corresponding to the remaining domain points in $(M2^2_{T^4_{WF}})$. From the additional smoothness conditions in (3.127), (3.128), and (3.129) we can uniquely and stably compute the B-coefficients corresponding to the domain points in $(M3^2_{T^4_{WF}})$ - $(M5^2_{T^4_{WF}})$. Since the B-coefficient associated with the domain point in $(M6^2_{T^4_{WF}})$ is equal to the value of $g(v_T)$, following Theorem 3.37, g is uniquely and stably determined. □

Lemma 3.40:
Let Δ be a tetrahedral partition of a polyhedral domain $\Omega\subset\mathbb{R}^3$ and $T:=\langle v_1,v_2,v_3,v_4\rangle\in\Delta$ a tetrahedron, that shares the six edges $\langle v_1,v_3\rangle$, $\langle v_1,v_4\rangle$, $\langle v_2,v_3\rangle$, $\langle v_2,v_4\rangle$, $\langle v_1,v_2\rangle$, and $\langle v_3,v_4\rangle$ with the tetrahedra $\widetilde{T}_1,\widetilde{T}_2,\widetilde{T}_3$, $\widetilde{T}_4$, $\widetilde{T}_5$, and $\widetilde{T}_6$, respectively, with $\widetilde{T}_j\in\Delta$, $j=1,\ldots,6$. Moreover, T is refined with a fourth order partial Worsey-Farin split at the split point v_T in the interior of T, and the split points v_{F_1},v_{F_2},v_{F_3}, and v_{F_4} in the interior of the faces $\langle v_1,v_2,v_3\rangle$, $\langle v_2,v_3,v_4\rangle$, $\langle v_1,v_3,v_4\rangle$, and $\langle v_1,v_2,v_4\rangle$, respectively.

Let $\widetilde{\mathcal{S}}_9^2(T_{WF}^4)$ be the superspline space defined in Theorem 3.37. Then, for any spline $s \in \mathcal{S}_9^{2,4,3}(\Delta \setminus T)$, there is a spline $g \in \widetilde{\mathcal{S}}_9^2(T_{WF}^4)$ such that

$$\widetilde{s} := \begin{cases} s, & \text{on } \Delta \setminus T \\ g, & \text{on } T, \end{cases}$$

belongs to $\mathcal{S}_9^{2,4,3}(\Delta)$. Moreover, the spline g is uniquely and stably determined by the value of $g(v_T)$ and the B-coefficients of $\widetilde{s}|_{\Delta \setminus T}$ by the following additional smoothness conditions across the vertices and edges of T:

$$\begin{aligned} &\tau_{3,0,4,2}^{\widetilde{T}_1,T_{1+2i,3-2i}}\widetilde{s} = \tau_{3,0,4,2}^{\widetilde{T}_2,T_{3+i,2+i}}\widetilde{s} = \tau_{3,0,4,2}^{\widetilde{T}_3,T_{1+i,2-i}}\widetilde{s} = \tau_{3,0,4,2}^{\widetilde{T}_4,T_{2+2i,3-i}}\widetilde{s} \\ &= \tau_{3,0,4,2}^{\widetilde{T}_5,T_{1+3i,1}}\widetilde{s} = \tau_{3,0,4,2}^{\widetilde{T}_6,T_{2+i,2+i}}\widetilde{s} = 0, \; i = 0,1 \end{aligned} \tag{3.130}$$

$$\begin{aligned} &\tau_{4,0,2,3}^{\widetilde{T}_{5-2i},T_{1,1+i}}\widetilde{s} = \tau_{3,0,3,3}^{\widetilde{T}_{5-2i},T_{1,1+i}}\widetilde{s} = \tau_{3,1,2,3}^{\widetilde{T}_{5-2i},T_{1,1+i}}\widetilde{s} \\ &= \tau_{3,0,2,4}^{\widetilde{T}_{5-2i},T_{1,1+i}}\widetilde{s} = \tau_{2,2,1,4}^{\widetilde{T}_{5-2i},T_{1,1+i}}\widetilde{s} = \tau_{2,0,2,5}^{\widetilde{T}_{5-2i},T_{1,1+i}}\widetilde{s} = 0, \; i = 0,1,2 \end{aligned} \tag{3.131}$$

$$\tau_{2,0,4,3}^{\widetilde{T}_5,T_{1,1}}\widetilde{s} = 0 \tag{3.132}$$

Proof:
From the C^4 supersmoothness conditions at the vertices and the C^3 supersmoothness conditions at the edges of T, we can uniquely and stably compute the B-coefficients of g corresponding to the domain points in (M$1^2_{T_{WF}^4}$) and (M$2^2_{T_{WF}^4}$) from Theorem 3.37.

From the additional smoothness conditions in (3.130), (3.131), and (3.132) we can uniquely and stably determine the B-coefficients corresponding to the domain points in (M$3^2_{T_{WF}^4}$) - (M$5^2_{T_{WF}^4}$). The last undetermined B-coefficient corresponding to a domain point in $\mathcal{M}_4^2$ is equal to the value of $g(v_T)$. Thus, following Theorem 3.37, g is uniquely and stably determined. □

Theorem 3.41:
Let $\hat{\mathcal{M}}_4^2$ be the union of the following sets of domain points in $\mathcal{D}_{T_{WF}^4,9}$:

($\hat{\text{M}}1^2_{T_{WF}^4}$) $D_4^{T_{v_i}}(v_i)$, $i = 1,\dots,4$, with $v_i \in T_{v_i} \in T_{WF}^4$

($\hat{\text{M}}2^2_{T_{WF}^4}$) $(E_3^{T_{1,1}}(\langle v_1,v_2\rangle) \cup E_3^{T_{1,2}}(\langle v_2,v_3\rangle) \cup E_3^{T_{1,3}}(\langle v_3,v_1\rangle) \cup E_3^{T_{2,2}}(\langle v_3,v_4\rangle) \cup$
$\cup E_3^{T_{2,3}}(\langle v_4,v_2\rangle) \cup E_3^{T_{3,1}}(\langle v_1,v_4\rangle)) \setminus \bigcup_{i=1}^{4}(D_4(v_i))$

($\hat{\text{M}}3^2_{T^4_{WF}}$) $\{\xi^{T_{1,1}}_{1,0,8-i,i}, \xi^{T_{1,1}}_{0,1,8-i,i}, \xi^{T_{1,1}}_{0,0,9-i,i}\}_{i=0}^{1}$
($\hat{\text{M}}4^2_{T^4_{WF}}$) $\{\xi^{T_{j,i}}_{3,0,4,2}\}$, $i = 1,2,3$, $j = 1,\ldots,4$
($\hat{\text{M}}5^2_{T^4_{WF}}$) $\{\xi^{T_{1,i}}_{4,0,2,3}, \xi^{T_{1,i}}_{3,0,3,3}, \xi^{T_{1,i}}_{3,1,2,3}, \xi^{T_{1,i}}_{3,0,2,4}, \xi^{T_{1,i}}_{2,2,1,4}, \xi^{T_{1,i}}_{2,0,2,5}\}_{i=1}^{3}$
($\hat{\text{M}}6^2_{T^4_{WF}}$) $\{\xi^{T_{1,1}}_{2,0,4,3}\}$
($\hat{\text{M}}7^2_{T^4_{WF}}$) $\{\xi^{T_{1,1}}_{0,0,0,9}\}$

Then $\hat{\mathcal{M}}^2_4$ is a stable minimal determining set for

$$\begin{aligned}\hat{\mathcal{S}}^2_9(T^4_{WF}) := \{s \in C^2(T^4_{WF}) : s|_{\widetilde{T}} \in \mathcal{P}^3_9,\ \forall \widetilde{T} \in T^4_{WF},\\ s \in C^4(v_i),\ i = 1,\ldots,4,\\ s \in C^3(e),\ \text{for all edges } e \in T,\\ s \in C^7(\langle v_{F_i}, v_T\rangle),\ i = 1,2\\ s \in C^7(v_T),\\ \tau^{T_{j,i},T_{j,i+1}}_{3,5,1,0} s = \tau^{T_{j,i},T_{j,i+1}}_{3,5,0,1} s = 0,\\ i = 1,2,3,\ j = 2,3,4,\ \text{with } T_{j,4} := T_{j,1}\},\end{aligned}$$

and the dimension of $\hat{\mathcal{S}}^2_9(T^4_{WF})$ is equal to 298.

Proof:
We first show that the set $\hat{\mathcal{M}}^2_4$ is a stable minimal determining set for the superspline space $\hat{\mathcal{S}}^2_9(T^4_{WF})$. Therefore, we set the B-coefficients c_ξ of a spline $s \in \hat{\mathcal{S}}^2_9(T^4_{WF})$ to arbitrary values for each domain point $\xi \in \hat{\mathcal{M}}^2_4$ and show that all other B-coefficients of s are uniquely and stably determined.

The B-coefficients of s corresponding to the domain points within a distance of four from the vertices of T, within a distance of three from the edges of T and within a distance of two from the faces $\langle v_2, v_3, v_4\rangle$, $\langle v_3, v_1, v_4\rangle$, and $\langle v_1, v_2, v_4\rangle$ can be uniquely and stably determined from those corresponding to the domain points in the sets ($\hat{\text{M}}1^2_{T^4_{WF}}$), ($\hat{\text{M}}2^2_{T^4_{WF}}$), and ($\hat{\text{M}}4^2_{T^4_{WF}}$) with $j = 2,3,4$, in the same way as in the proof of Theorem 3.37.

We can uniquely and stably determine the B-coefficients of s associated with the domain points in the shells $R^{T_{1,i}}_9(v_T)$, $R^{T_{1,i}}_8(v_T)$, and $R^{T_{1,i}}_7(v_T)$, $i = 1,2,3$, from the B-coefficients corresponding to the domain points in

$(\hat{M}3^2_{T^4_{WF}})$ and those in $(\hat{M}4^2_{T^4_{WF}})$ with $j = 1$ in the same way as in the proof of Theorem 3.20.

Then, the remaining undetermined B-coefficients can be uniquely and stably computed from those already determined and the ones associated with the domain points in $(\hat{M}5^2_{T^4_{WF}})$, $(\hat{M}6^2_{T^4_{WF}})$, and $(\hat{M}7^2_{T^4_{WF}})$ in the same way as in the proof of Theorem 3.37.

Considering the dimension of the space $\hat{\mathcal{S}}^2_9(T^4_{WF})$, we can see that $\hat{\mathcal{S}}^2_9(T^4_{WF}) = 298 = \#\hat{\mathcal{M}}^2_4$, since $\hat{\mathcal{M}}^2_4$ is a minimal determining set for this superspline space. □

Lemma 3.42:
Let Δ be a tetrahedral partition of a polyhedral domain $\Omega \subset \mathbb{R}^3$ and $T := \langle v_1, v_2, v_3, v_4 \rangle \in \Delta$ a tetrahedron, that shares exactly the face $\langle v_1, v_2, v_3 \rangle$ with the tetrahedron $\hat{T}$, and the edges $\langle v_1, v_4 \rangle$ and $\langle v_2, v_4 \rangle$ with the tetrahedra $\widetilde{T}_1$ and $\widetilde{T}_2$, respectively, with $\hat{T}, \widetilde{T}_1, \widetilde{T}_2 \in \Delta$. Moreover, T is refined with a fourth order partial Worsey-Farin split at the split point v_T in the interior of T, and the split points $v_{F_1}, v_{F_2}, v_{F_3}$, and v_{F_4} in the interior of the faces $\langle v_1, v_2, v_3 \rangle$, $\langle v_2, v_3, v_4 \rangle$, $\langle v_1, v_3, v_4 \rangle$, and $\langle v_1, v_2, v_4 \rangle$, respectively. The tetrahedron $\hat{T}$ is also subjected to a partial Worsey-Farin split, such that the common face with T is refined with a Clough-Tocher split. The three corresponding subtetrahedra are denoted by $\hat{T}_i$, $i = 1,2,3$, respectively, where $\hat{T}_i$ has a common face with $T_{1,i}$.

Let $\hat{\mathcal{S}}^2_9(T^4_{WF})$ be the superspline space defined in Theorem 3.41. Then, for any spline $s \in \mathcal{S}^{2,4,3}_9(\Delta \setminus T)$, there is a spline $g \in \hat{\mathcal{S}}^2_9(T^4_{WF})$ such that

$$\widetilde{s} := \begin{cases} s, & \text{on } \Delta \setminus T \\ g, & \text{on } T, \end{cases}$$

belongs to $\mathcal{S}^{2,4,3}_9(\Delta)$. Moreover, the spline g is uniquely and stably determined by the value of $g(v_T)$ and the B-coefficients of $\widetilde{s}|_{\Delta \setminus T}$ by the following additional smoothness conditions across the vertices, edges and faces of T:

$$\tau^{\hat{T}_2, T_{2,2}}_{i,j,k,l} \widetilde{s} = 0, \; i,j < 5, \; k+l \leq 3, \; i+j+k+l = 9 \tag{3.133}$$

$$\begin{aligned} \tau^{\hat{T}_1, T_{4,1}}_{3,0,4,2} \widetilde{s} &= \tau^{\hat{T}_2, T_{2+i,2+i}}_{3,0,4,2} \widetilde{s} = \tau^{\hat{T}_2, T_{2,1}}_{3,0,4,2} \widetilde{s} = \tau^{\hat{T}_3, T_{3,1}}_{3,0,4,2} \widetilde{s} = \tau^{\widetilde{T}_1, T_{3+i,2+i}}_{3,0,4,2} \widetilde{s} \\ &= \tau^{\widetilde{T}_2, T_{2+2i,3-i}}_{3,0,4,2} \widetilde{s} = 0, \; i = 0,1 \end{aligned} \tag{3.134}$$

$$\begin{aligned}
&\tau_{4,0,2,3}^{\hat{T}_i,T_{1,i}}\widetilde{s} = \tau_{3,0,3,3}^{\hat{T}_i,T_{1,i}}\widetilde{s} = \tau_{3,1,2,3}^{\hat{T}_i,T_{1,i}}\widetilde{s} \\
&= \tau_{3,0,2,4}^{\hat{T}_i,T_{1,i}}\widetilde{s} = \tau_{2,2,1,4}^{\hat{T}_i,T_{1,i}}\widetilde{s} = \tau_{2,0,2,5}^{\hat{T}_i,T_{1,i}}\widetilde{s} = 0, \; i = 1,2,3
\end{aligned} \tag{3.135}$$

$$\tau_{2,0,4,3}^{\hat{T}_1,T_{1,1}}\widetilde{s} = 0 \tag{3.136}$$

Proof:
From the C^4 supersmoothness conditions of g at the vertices of T, the C^3 supersmoothness conditions at the five edges $\langle v_1,v_2\rangle$, $\langle v_2,v_3\rangle$, $\langle v_3,v_1\rangle$, $\langle v_1,v_4\rangle$, and $\langle v_3,v_4\rangle$, and the C^2 smoothness conditions at the face $\langle v_1,v_2,v_3\rangle$ we can uniquely and stably compute the B-coefficients of g corresponding to the domain points in $(\hat{\mathrm{M}}1^2_{T^4_{WF}})$ and $(\hat{\mathrm{M}}2^2_{T^4_{WF}})$, except for those associated with the domain points $E_3^{T_{2,2}}(\langle v_3,v_4\rangle) \setminus (D_4(v_3) \cup (D_4(v_4))$, and the B-coefficients associated with the domain points in $(\hat{\mathrm{M}}3^2_{T^4_{WF}})$ and $(\hat{\mathrm{M}}4^2_{T^4_{WF}})$ with $j = 1$ from Theorem 3.41.

Using the additional smoothness conditions of g in (3.133) and (3.134), the remaining B-coefficients associated with the domain points in $(\hat{\mathrm{M}}2^2_{T^4_{WF}})$ and $(\hat{\mathrm{M}}4^2_{T^4_{WF}})$ can be uniquely and stably determined. Using the additional smoothness conditions in (3.135) and (3.136) we can uniquely and stably compute the B-coefficients corresponding to the domain points in $(\hat{\mathrm{M}}5^2_{T^4_{WF}})$ and $(\hat{\mathrm{M}}6^2_{T^4_{WF}})$. The last undetermined B-coefficient corresponding to a domain point in $\hat{\mathcal{M}}_4^2$ is equal to the value of $g(v_T)$. Thus, following Theorem 3.41, g is uniquely and stably determined. □

Lemma 3.43:
Let Δ be a tetrahedral partition of a polyhedral domain $\Omega \subset \mathbb{R}^3$ and $T := \langle v_1,v_2,v_3,v_4\rangle \in \Delta$ a tetrahedron, that shares exactly the face $\langle v_1,v_2,v_3\rangle$ with the tetrahedron $\hat{T}$, and the edges $\langle v_1,v_4\rangle$, $\langle v_2,v_4\rangle$, and $\langle v_3,v_4\rangle$ with the tetrahedra $\widetilde{T}_1$, $\widetilde{T}_2$, and $\widetilde{T}_3$, respectively, with $\hat{T},\widetilde{T}_j \in \Delta$, $j = 1,2,3$. Moreover, T is refined with a fourth order partial Worsey-Farin split at the split point v_T in the interior of T, and the split points v_{F_1}, v_{F_2}, v_{F_3}, and v_{F_4} in the interior of the faces $\langle v_1,v_2,v_3\rangle$, $\langle v_2,v_3,v_4\rangle$, $\langle v_1,v_3,v_4\rangle$, and $\langle v_1,v_2,v_4\rangle$, respectively. The tetrahedron $\hat{T}$ is also subjected to a partial Worsey-Farin split, such that the common face with T is refined with a Clough-Tocher split. The three corresponding subtetrahedra are denoted by $\hat{T}_i$, $i = 1,2,3$, respectively, where $\hat{T}_i$ has a common face with $T_{1,i}$.

Let $\hat{\mathcal{S}}_9^2(T_{WF}^4)$ be the superspline space defined in Theorem 3.41. Then, for any spline $s \in \mathcal{S}_9^{2,4,3}(\Delta \setminus T)$, there is a spline $g \in \hat{\mathcal{S}}_9^2(T_{WF}^4)$ such that

$$\widetilde{s} := \begin{cases} s, & \text{on } \Delta \setminus T \\ g, & \text{on } T, \end{cases}$$

belongs to $\mathcal{S}_9^{2,4,3}(\Delta)$. Moreover, the spline g is uniquely and stably determined by the value of $g(v_T)$ and the B-coefficients of $\widetilde{s}|_{\Delta \setminus T}$ by the following additional smoothness conditions across the edges and faces of T:

$$\begin{aligned} &\tau_{3,0,4,2}^{\hat{T}_1,T_{4,1}}\widetilde{s} = \tau_{3,0,4,2}^{\hat{T}_2,T_{2,1}}\widetilde{s} = \tau_{3,0,4,2}^{\hat{T}_3,T_{3,1}}\widetilde{s} = \tau_{3,0,4,2}^{\widetilde{T}_1,T_{3+i,2+i}}\widetilde{s} = \tau_{3,0,4,2}^{\widetilde{T}_2,T_{2+2i,3-i}}\widetilde{s} \\ &= \tau_{3,0,4,2}^{\widetilde{T}_3,T_{2+i,2+i}}\widetilde{s} = 0, \; i = 0,1 \end{aligned} \tag{3.137}$$

$$\begin{aligned} &\tau_{4,0,2,3}^{\hat{T}_i,T_{1,i}}\widetilde{s} = \tau_{3,0,3,3}^{\hat{T}_i,T_{1,i}}\widetilde{s} = \tau_{3,1,2,3}^{\hat{T}_i,T_{1,i}}\widetilde{s} \\ &= \tau_{3,0,2,4}^{\hat{T}_i,T_{1,i}}\widetilde{s} = \tau_{2,2,1,4}^{\hat{T}_i,T_{1,i}}\widetilde{s} = \tau_{2,0,2,5}^{\hat{T}_i,T_{1,i}}\widetilde{s} = 0, \; i = 1,2,3 \end{aligned} \tag{3.138}$$

$$\tau_{2,0,4,3}^{\hat{T}_1,T_{1,1}}\widetilde{s} = 0 \tag{3.139}$$

Proof:
From the C^4 supersmoothness conditions at the vertices of T, the C^3 supersmoothness conditions at the edges of T, and the C^2 smoothness conditions at the face $\langle v_1, v_2, v_3 \rangle$ we can uniquely and stably compute the B-coefficients of g corresponding to the domain points in $(\hat{M}1^2_{T_{WF}^4})$, $(\hat{M}2^2_{T_{WF}^4})$, $(\hat{M}3^2_{T_{WF}^4})$ and $(\hat{M}4^2_{T_{WF}^4})$ with $j = 1$ from Theorem 3.41.

From the additional smoothness conditions in (3.137) we can uniquely and stably compute the remaining B-coefficients associated with the domain points in $(\hat{M}4^2_{T_{WF}^4})$. Then, from the additional smoothness conditions in (3.138) and (3.139) we can uniquely and stably determine the B-coefficients corresponding to the domain points in $(\hat{M}5^2_{T_{WF}^4})$ and $(\hat{M}6^2_{T_{WF}^4})$. Finally, the B-coefficient corresponding to the domain point $\xi_{0,0,0,9}^{T_{1,1}}$ in $(\hat{M}7^2_{T_{WF}^4})$ is equal to the value of $g(v_T)$. Thus, following Theorem 3.41, g is uniquely and stably determined. □

Theorem 3.44:
Let $\check{\mathcal{M}}_4^2$ be the union of the following sets of domain points in $\mathcal{D}_{T_{WF}^4,9}$:
($\check{\mathrm{M}}1^2_{T^4_{WF}}$) $D_4^{T_{v_i}}(v_i)$, $i=1,\dots,4$, with $v_i \in T_{v_i} \in T_{WF}^4$
($\check{\mathrm{M}}2^2_{T^4_{WF}}$) $\big(E_3^{T_{1,1}}(\langle v_1,v_2\rangle) \cup E_3^{T_{1,2}}(\langle v_2,v_3\rangle) \cup E_3^{T_{1,3}}(\langle v_3,v_1\rangle) \cup E_3^{T_{2,2}}(\langle v_3,v_4\rangle) \cup$

$\cup E_3^{T_{2,3}}(\langle v_4,v_2\rangle) \cup E_3^{T_{3,1}}(\langle v_1,v_4\rangle)\big) \setminus \bigcup_{i=1}^{4}(D_4(v_i))$
($\check{\mathrm{M}}3^2_{T^4_{WF}}$) $\{\xi^{T_{j,1}}_{1,0,8-i,i}, \xi^{T_{j,1}}_{0,1,8-i,i}, \xi^{T_{j,1}}_{0,0,9-i,i}\}_{i=0}^{1}$, $j=1,2$
($\check{\mathrm{M}}4^2_{T^4_{WF}}$) $\{\xi^{T_{j,i}}_{3,0,4,2}\}$, $i=1,2,3$, $j=1,\dots,4$
($\check{\mathrm{M}}5^2_{T^4_{WF}}$) $\{\xi^{T_{1,i}}_{4,0,2,3}, \xi^{T_{1,i}}_{3,0,3,3}, \xi^{T_{1,i}}_{3,1,2,3}, \xi^{T_{1,i}}_{3,0,2,4}, \xi^{T_{1,i}}_{2,2,1,4}, \xi^{T_{1,i}}_{2,0,2,5}\}_{i=1}^{3}$
($\check{\mathrm{M}}6^2_{T^4_{WF}}$) $\{\xi^{T_{1,1}}_{2,0,4,3}\}$
($\check{\mathrm{M}}7^2_{T^4_{WF}}$) $\{\xi^{T_{1,1}}_{0,0,0,9}\}$
Then $\check{\mathcal{M}}_4^2$ is a stable minimal determining set for

$$\begin{aligned}\check{\mathcal{S}}_9^2(T_{WF}^4) := \{s \in C^2(T_{WF}^4) : s|_{\widetilde{T}} \in \mathcal{P}_9^3,\ \forall \widetilde{T} \in T_{WF}^4,\\ s \in C^4(v_i),\ i=1,\dots,4,\\ s \in C^3(e), \text{ for all edges } e \in T,\\ s \in C^7(\langle v_{F_i}, v_T\rangle),\ i=1,2\\ s \in C^7(v_T),\\ \tau^{T_{j,i},T_{j,i+1}}_{3,5,1,0} s = \tau^{T_{j,i},T_{j,i+1}}_{3,5,0,1} s = 0,\\ i=1,2,3,\ j=3,4, \text{ with } T_{j,4} := T_{j,1}\},\end{aligned}$$

and the dimension of $\check{\mathcal{S}}_9^2(T_{WF}^4)$ is equal to 304.

Proof:
First, we show that $\check{\mathcal{M}}_4^2$ is a stable minimal determining set for the superspline space $\check{\mathcal{S}}_9^2(T_{WF}^4)$. To this end we set the B-coefficients c_ξ of a spline $s \in \check{\mathcal{S}}_9^2(T_{WF}^4)$ to arbitrary values for each domain point $\xi \in \check{\mathcal{M}}_4^2$. Then we show that all other B-coefficients of s are uniquely and stably determined.

The B-coefficients of s corresponding to the domain points within a distance of four from the vertices of T, within a distance of three from the edges of T and within a distance of two from the faces $\langle v_1,v_2,v_3\rangle$, $\langle v_3,v_1,v_4\rangle$, and $\langle v_1,v_2,v_4\rangle$ can be uniquely and stably determined from those corresponding to the domain points in $(\check{M}1^2_{T^4_{WF}})$, $(\check{M}2^2_{T^4_{WF}})$, $(\check{M}3^2_{T^4_{WF}})$ with $j = 1$, and $(\check{M}4^2_{T^4_{WF}})$ with $j = 1,3,4$, in the same way as in the proof of Theorem 3.41.

The B-coefficients of s corresponding to the domain points in the shells $R_9^{T_{2,i}}(v_T)$, $R_8^{T_{2,i}}(v_T)$, and $R_7^{T_{2,i}}(v_T)$, $i = 1,2,3$, can be uniquely and stably determined from the already determined B-coefficients and those corresponding to the domain points in $(\check{M}3^2_{T^4_{WF}})$ with $j = 2$ and those in $(\check{M}4^2_{T^4_{WF}})$ with $j = 2$ in the same way as the ones associated with the domain points $R_{9-j}^{T_{2,i}}(v_T)$, $i = 1,2,3$, $j = 0,1,2$, in the proof of Theorem 3.20.

The remaining B-coefficients of s can be uniquely and stably computed from the ones already determined and those associated with the domain points in $(\check{M}5^2_{T^4_{WF}})$, $(\check{M}6^2_{T^4_{WF}})$, and $(\check{M}7^2_{T^4_{WF}})$ in the same way as in the proof of Theorem 3.37.

Finally, since the set $\check{\mathcal{M}}_4^2$ is a minimal determining set for the superspline space $\check{\mathcal{S}}_9^2(T^4_{WF})$, we can state that $\check{\mathcal{S}}_9^2(T^4_{WF}) = \#\check{\mathcal{M}}_4^2 = 304$. □

Lemma 3.45:
Let Δ be a tetrahedral partition of a polyhedral domain $\Omega \subset \mathbb{R}^3$ and $T := \langle v_1, v_2, v_3, v_4 \rangle \in \Delta$ a tetrahedron, that shares exactly the two faces $\langle v_1, v_2, v_3 \rangle$ and $\langle v_2, v_3, v_4 \rangle$ with the tetrahedra $\hat{T}_1$ and $\hat{T}_2$, respectively, with $\hat{T}_1, \hat{T}_2 \in \Delta$. Moreover, T is refined with a fourth order partial Worsey-Farin split at the split point v_T in the interior of T, and the split points v_{F_1}, v_{F_2}, v_{F_3}, and v_{F_4} in the interior of the faces $\langle v_1, v_2, v_3 \rangle$, $\langle v_2, v_3, v_4 \rangle$, $\langle v_1, v_3, v_4 \rangle$, and $\langle v_1, v_2, v_4 \rangle$, respectively. The tetrahedra $\hat{T}_1$ and $\hat{T}_2$ are also subjected to partial Worsey-Farin splits, such that the common faces with T are refined with a Clough-Tocher split. The corresponding subtetrahedra are denoted by $\hat{T}_{j,i}$, $i = 1,2,3$, $j = 1,2$, respectively, where $\hat{T}_{j,i}$ has a common face with $T_{j,i}$.

Let $\check{\mathcal{S}}_9^2(T^4_{WF})$ be the superspline space defined in Theorem 3.44. Then, for any spline $s \in \mathcal{S}_9^{2,4,3}(\Delta \setminus T)$, there is a spline $g \in \check{\mathcal{S}}_9^2(T^4_{WF})$ such that

$$\widetilde{s} := \begin{cases} s, & \text{on } \Delta \setminus T \\ g, & \text{on } T, \end{cases}$$

belongs to $\mathcal{S}_9^{2,4,3}(\Delta)$. Moreover, the spline g is uniquely and stably determined by the value of $g(v_T)$ and the B-coefficients of $\widetilde{s}|_{\Delta \setminus T}$ by the following additional smoothness conditions across the vertices, edges and faces of T:

$$\tau_{i,j,k,l}^{\hat{T}_{1,3},T_{3,1}} \widetilde{s} = 0,\ i,j < 5,\ k+l \leq 3,\ i+j+k+l = 9 \tag{3.140}$$

$$\begin{aligned} &\tau_{3,0,4,2}^{\hat{T}_{1,1},T_{3,2}} \widetilde{s} = \tau_{3,0,4,2}^{\hat{T}_{1,1},T_{4,3}} \widetilde{s} = \tau_{3,0,4,2}^{\hat{T}_{1,1},T_{4,1}} \widetilde{s} = \tau_{3,0,4,2}^{\hat{T}_{1,3},T_{3,1}} \widetilde{s} = \tau_{3,0,4,2}^{\hat{T}_{2,2},T_{3,3}} \widetilde{s} \\ &= \tau_{3,0,4,2}^{\hat{T}_{2,3},T_{4,2}} \widetilde{s} = 0,\ i = 0,1 \end{aligned} \tag{3.141}$$

$$\begin{aligned} &\tau_{4,0,2,3}^{\hat{T}_{1,i},T_{1,i}} \widetilde{s} = \tau_{3,0,3,3}^{\hat{T}_{1,i},T_{1,i}} \widetilde{s} = \tau_{3,1,2,3}^{\hat{T}_{1,i},T_{1,i}} \widetilde{s} \\ &= \tau_{3,0,2,4}^{\hat{T}_{1,i},T_{1,i}} \widetilde{s} = \tau_{2,2,1,4}^{\hat{T}_{1,i},T_{1,i}} \widetilde{s} = \tau_{2,0,2,5}^{\hat{T}_{1,i},T_{1,i}} \widetilde{s} = 0,\ i = 1,2,3 \end{aligned} \tag{3.142}$$

$$\tau_{2,0,4,3}^{\hat{T}_{1,1},T_{1,1}} \widetilde{s} = 0 \tag{3.143}$$

Proof:
Using the C^4 supersmoothness conditions of g at the vertices of T, the C^3 supersmoothness conditions at the edges $\langle v_1, v_2 \rangle$, $\langle v_1, v_3 \rangle$, $\langle v_2, v_3 \rangle$, $\langle v_2, v_4 \rangle$, and $\langle v_3, v_4 \rangle$, and the C^2 smoothness conditions at the faces $\langle v_1, v_2, v_3 \rangle$ and $\langle v_2, v_3, v_4 \rangle$ we can uniquely and stably determine the B-coefficients of g corresponding to the domain points in ($\check{M}1^2_{T^4_{WF}}$) and ($\check{M}2^2_{T^4_{WF}}$), except for those associated with the domain points $E_3^{T_{3,1}}(\langle v_1, v_4 \rangle) \setminus (D_4(v_1) \cup (D_4(v_4))$, and the B-coefficients associated with the domain points in ($\check{M}3^2_{T^4_{WF}}$) and ($\check{M}4^2_{T^4_{WF}}$) with $j = 1$ and $j = 2$ from Theorem 3.44.

From the additional smoothness conditions in (3.140) and (3.141) we can uniquely and stably determine the remaining B-coefficients associated with the domain points in ($\check{M}2^2_{T^4_{WF}}$) and ($\check{M}4^2_{T^4_{WF}}$). From the additional smoothness conditions in (3.142) and (3.143) we can uniquely and stably compute the B-coefficients corresponding to the domain points in ($\check{M}5^2_{T^4_{WF}}$) and ($\check{M}6^2_{T^4_{WF}}$). Finally, the last B-coefficient corresponding to a domain point in $\mathcal{M}_4^2$ is equal to the value of $g(v_T)$. Thus, following Theorem 3.41, g is uniquely and stably determined. □

Lemma 3.46:
Let Δ be a tetrahedral partition of a polyhedral domain $\Omega \subset \mathbb{R}^3$ and $T := \langle v_1, v_2, v_3, v_4 \rangle \in \Delta$ a tetrahedron, that shares exactly the two faces $\langle v_1, v_2, v_3 \rangle$ and $\langle v_2, v_3, v_4 \rangle$ with the tetrahedra $\hat{T}_1$ and $\hat{T}_2$, respectively, and the edge $\langle v_1, v_4 \rangle$ with the tetrahedron $\widetilde{T}$, with $\hat{T}_1, \hat{T}_2, \widetilde{T} \in \Delta$. Moreover, T is refined with a fourth order partial Worsey-Farin split at the split point v_T in the interior of T, and the split points $v_{F_1}, v_{F_2}, v_{F_3}$, and v_{F_4} in the interior

of the faces $\langle v_1,v_2,v_3\rangle$, $\langle v_2,v_3,v_4\rangle$, $\langle v_1,v_3,v_4\rangle$, and $\langle v_1,v_2,v_4\rangle$, respectively. The tetrahedra $\hat{T}_1$ and $\hat{T}_2$ are also subjected to partial Worsey-Farin splits, such that the common faces with T are refined with a Clough-Tocher split. The corresponding subtetrahedra are denoted by $\hat{T}_{j,i}$, $i=1,2,3$, $j=1,2$, respectively, where $\hat{T}_{j,i}$ has a common face with $T_{j,i}$.

Let $\check{\mathcal{S}}_9^2(T_{WF}^4)$ be the superspline space defined in Theorem 3.44. Then, for any spline $s \in \mathcal{S}_9^{2,4,3}(\Delta \setminus T)$, there is a spline $g \in \check{\mathcal{S}}_9^2(T_{WF}^4)$ such that

$$\widetilde{s} := \begin{cases} s, & \text{on } \Delta \setminus T \\ g, & \text{on } T, \end{cases}$$

belongs to $\mathcal{S}_9^{2,4,3}(\Delta)$. Moreover, the spline g is uniquely and stably determined by the value of $g(v_T)$ and the B-coefficients of $\widetilde{s}|_{\Delta \setminus T}$ by the following additional smoothness conditions across the vertices, edges and faces of T:

$$\begin{aligned} &\tau_{3,0,4,2}^{\widetilde{T},T_{3+i,2+i}}\widetilde{s} = \tau_{3,0,4,2}^{\hat{T}_{1,1},T_{4,1}}\widetilde{s} = \tau_{3,0,4,2}^{\hat{T}_{1,3},T_{3,1}}\widetilde{s} = \\ &\tau_{3,0,4,2}^{\hat{T}_{2,2},T_{3,3}}\widetilde{s} = \tau_{3,0,4,2}^{\hat{T}_{2,3},T_{4,2}}\widetilde{s} = 0,\ i=0,1 \end{aligned} \tag{3.144}$$

$$\begin{aligned} &\tau_{4,0,2,3}^{\hat{T}_{1,i},T_{1,i}}\widetilde{s} = \tau_{3,0,3,3}^{\hat{T}_{1,i},T_{1,i}}\widetilde{s} = \tau_{3,1,2,3}^{\hat{T}_{1,i},T_{1,i}}\widetilde{s} \\ &= \tau_{3,0,2,4}^{\hat{T}_{1,i},T_{1,i}}\widetilde{s} = \tau_{2,2,1,4}^{\hat{T}_{1,i},T_{1,i}}\widetilde{s} = \tau_{2,0,2,5}^{\hat{T}_{1,i},T_{1,i}}\widetilde{s} = 0,\ i=1,2,3 \end{aligned} \tag{3.145}$$

$$\tau_{2,0,4,3}^{\hat{T}_{1,1},T_{1,1}}\widetilde{s} = 0 \tag{3.146}$$

Proof:
Using the C^4 supersmoothness conditions at the vertices of T, the C^3 supersmoothness conditions at the edges of T, and the C^2 smoothness conditions at the faces $\langle v_1,v_2,v_3\rangle$ and $\langle v_2,v_3,v_4\rangle$ we can uniquely and stably determine the B-coefficients of g corresponding to the domain points in ($\check{M}1^2_{T^4_{WF}}$), ($\check{M}2^2_{T^4_{WF}}$), ($\check{M}3^2_{T^4_{WF}}$), and ($\check{M}4^2_{T^4_{WF}}$) with $j=1$ and $j=2$ from Theorem 3.44.

Using the additional smoothness conditions in (3.144) the remaining B-coefficients associated with the domain points in ($\check{M}4^2_{T^4_{WF}}$) can be uniquely and stably determined. Then, using the additional smoothness conditions in (3.145) and (3.146) we can uniquely and stably compute the B-coefficients corresponding to the domain points in ($\check{M}5^2_{T^4_{WF}}$) and ($\check{M}6^2_{T^4_{WF}}$). Since the B-coefficient associated with the domain point in ($\check{M}7^2_{T^4_{WF}}$) is equal to the value of $g(v_T)$, following Theorem 3.44, g is uniquely and stably determined. □

Theorem 3.47:
Let $\dot{\mathcal{M}}^2_4$ be the union of the following sets of domain points in $\mathcal{D}_{T^4_{WF},9}$:

($\dot{M}1^2_{T^4_{WF}}$) $D_4^{T_{v_i}}(v_i)$, $i = 1,\ldots,4$, with $v_i \in T_{v_i} \in T^4_{WF}$

($\dot{M}2^2_{T^4_{WF}}$) $\big(E_3^{T_{1,1}}(\langle v_1,v_2\rangle) \cup E_3^{T_{1,2}}(\langle v_2,v_3\rangle) \cup E_3^{T_{1,3}}(\langle v_3,v_1\rangle) \cup E_3^{T_{2,2}}(\langle v_3,v_4\rangle) \cup$
$\cup E_3^{T_{2,3}}(\langle v_4,v_2\rangle) \cup E_3^{T_{3,1}}(\langle v_1,v_4\rangle)\big) \setminus \bigcup_{i=1}^{4} (D_4(v_i))$

($\dot{M}3^2_{T^4_{WF}}$) $\{\xi^{T_{j,1}}_{1,0,8-i,i}, \xi^{T_{j,1}}_{0,1,8-i,i}, \xi^{T_{j,1}}_{0,0,9-i,i}\}_{i=0}^{1}$, $j = 1,2,3$

($\dot{M}4^2_{T^4_{WF}}$) $\{\xi^{T_{j,i}}_{3,0,4,2}\}$, $i = 1,2,3$, $j = 1,\ldots,4$

($\dot{M}5^2_{T^4_{WF}}$) $\{\xi^{T_{1,i}}_{4,0,2,3}, \xi^{T_{1,i}}_{3,0,3,3}, \xi^{T_{1,i}}_{3,1,2,3}, \xi^{T_{1,i}}_{3,0,2,4}, \xi^{T_{1,i}}_{2,2,1,4}, \xi^{T_{1,i}}_{2,0,2,5}\}_{i=1}^{3}$

($\dot{M}6^2_{T^4_{WF}}$) $\{\xi^{T_{1,1}}_{2,0,4,3}\}$

($\dot{M}7^2_{T^4_{WF}}$) $\{\xi^{T_{1,1}}_{0,0,0,9}\}$

Then $\dot{\mathcal{M}}^2_4$ is a stable minimal determining set for

$$\begin{aligned}\dot{\mathcal{S}}^2_9(T^4_{WF}) := \{s \in C^2(T^4_{WF}) : s|_{\widetilde{T}} \in \mathcal{P}^3_9,\ \forall \widetilde{T} \in T^4_{WF},\\ s \in C^4(v_i),\ i = 1,\ldots,4,\\ s \in C^3(e), \text{ for all edges } e \in T,\\ s \in C^7(\langle v_{F_i}, v_T\rangle),\ i = 1,2\\ s \in C^7(v_T),\\ \tau^{T_{4,i},T_{4,i+1}}_{3,5,1,0} s = 0,\ i = 1,2,3, \text{ with } T_{4,4} := T_{4,1},\\ \tau^{T_{4,i},T_{4,i+1}}_{3,5,0,1} s = 0,\ i = 1,2,3, \text{ with } T_{4,4} := T_{4,1}\},\end{aligned}$$

and the dimension of $\dot{\mathcal{S}}^2_9(T^4_{WF})$ is equal to 310.

Proof:
We first show that the set $\dot{\mathcal{M}}_4^2$ is a stable minimal determining set for the space $\dot{\mathcal{S}}_9^2(T_{WF}^4)$ of supersplines. Therefore, we set the B-coefficients c_ξ of a spline $s \in \dot{\mathcal{S}}_9^2(T_{WF}^4)$ to arbitrary values for each domain point $\xi \in \dot{\mathcal{M}}_4^2$ and show that all other B-coefficients of s are uniquely and stably determined.

The B-coefficients of s corresponding to the domain points within a distance of four from the vertices of T, within a distance of three from the edges of T and within a distance of two from the faces $\langle v_1, v_2, v_3 \rangle$, $\langle v_2, v_3, v_4 \rangle$, and $\langle v_1, v_2, v_4 \rangle$ can be uniquely and stably determined from the B-coefficients associated with the domain points in ($\dot{\text{M}}1^2_{T_{WF}^4}$), ($\dot{\text{M}}2^2_{T_{WF}^4}$), ($\dot{\text{M}}3^2_{T_{WF}^4}$) with $j = 1$ and $j = 2$, and ($\dot{\text{M}}4^2_{T_{WF}^4}$) with $j = 1,2,4$, in the same way as in the proof of Theorem 3.44.

Then, we can uniquely and stably compute the B-coefficients associated with the domain points in the shells $R_9^{T_{3,i}}(v_T)$, $R_8^{T_{3,i}}(v_T)$, and $R_7^{T_{3,i}}(v_T)$, $i = 1,2,3$, from the already determined B-coefficients and the ones corresponding to the domain points in ($\dot{\text{M}}3^2_{T_{WF}^4}$) with $j = 3$ and those in ($\dot{\text{M}}4^2_{T_{WF}^4}$) with $j = 3$ in the same way as the ones associated with the domain points $R_{9-j}^{T_{2,i}}(v_T)$, $i = 1,2,3$, $j = 0,1,2$, in the proof of Theorem 3.20.

We can uniquely and stably compute the remaining B-coefficients of s from those already determined and the B-coefficients associated with the domain points in ($\dot{\text{M}}5^2_{T_{WF}^4}$), ($\dot{\text{M}}6^2_{T_{WF}^4}$), and ($\dot{\text{M}}7^2_{T_{WF}^4}$) in the same way as in the proof of Theorem 3.37.

Examining the dimension of $\dot{\mathcal{S}}_9^2(T_{WF}^4)$, we see that $\dot{\mathcal{S}}_9^2(T_{WF}^4) = \#\dot{\mathcal{M}}_4^2 = 310$, since $\dot{\mathcal{M}}_4^2$ is a minimal determining set for the spline space $\dot{\mathcal{S}}_9^2(T_{WF}^4)$.□

Lemma 3.48:
Let Δ be a tetrahedral partition of a polyhedral domain $\Omega \subset \mathbb{R}^3$ and $T := \langle v_1, v_2, v_3, v_4 \rangle \in \Delta$ a tetrahedron, that shares exactly the three faces $\langle v_1, v_2, v_3 \rangle$, $\langle v_2, v_3, v_4 \rangle$, and $\langle v_3, v_1, v_4 \rangle$ with the tetrahedra $\hat{T}_1$, $\hat{T}_2$, and $\hat{T}_3$, respectively, with $\hat{T}_1, \hat{T}_2, \hat{T}_3 \in \Delta$. Moreover, T is refined with a fourth order partial Worsey-Farin split at the split point v_T in the interior of T, and the split points v_{F_1}, v_{F_2}, v_{F_3}, and v_{F_4} in the interior of the faces $\langle v_1, v_2, v_3 \rangle$, $\langle v_2, v_3, v_4 \rangle$, $\langle v_1, v_3, v_4 \rangle$, and $\langle v_1, v_2, v_4 \rangle$, respectively. The tetrahedra $\hat{T}_1$, $\hat{T}_2$, and $\hat{T}_3$ are also subjected to partial Worsey-Farin splits, such that the common faces with T are refined with a Clough-Tocher split. The corresponding subtetrahedra are denoted by $\hat{T}_{j,i}$, $i = 1,2,3$, $j = 1,2,3$ respectively, where $\hat{T}_{j,i}$ has a common face with $T_{j,i}$.

Let $\dot{\mathcal{S}}_9^2(T_{WF}^4)$ be the superspline space defined in Theorem 3.47. Then, for any spline $s \in \mathcal{S}_9^{2,4,3}(\Delta \setminus T)$, there is a spline $g \in \dot{\mathcal{S}}_9^2(T_{WF}^4)$ such that

$$\widetilde{s} := \begin{cases} s, & \text{on } \Delta \setminus T \\ g, & \text{on } T, \end{cases}$$

belongs to $\mathcal{S}_9^{2,4,3}(\Delta)$. Moreover, the spline g is uniquely and stably determined by the value of $g(v_T)$ and the B-coefficients of $\widetilde{s}|_{\Delta \setminus T}$ by the following additional smoothness conditions across the edges and faces of T:

$$\tau_{3,0,4,2}^{\hat{T}_{1,1},T_{4,1}}\widetilde{s} = \tau_{3,0,4,2}^{\hat{T}_{2,3},T_{4,2}}\widetilde{s} = \tau_{3,0,4,2}^{\hat{T}_{3,2},T_{4,3}}\widetilde{s} = 0 \tag{3.147}$$

$$\begin{aligned} &\tau_{4,0,2,3}^{\hat{T}_{1,i},T_{1,i}}\widetilde{s} = \tau_{3,0,3,3}^{\hat{T}_{1,i},T_{1,i}}\widetilde{s} = \tau_{3,1,2,3}^{\hat{T}_{1,i},T_{1,i}}\widetilde{s} \\ &= \tau_{3,0,2,4}^{\hat{T}_{1,i},T_{1,i}}\widetilde{s} = \tau_{2,2,1,4}^{\hat{T}_{1,i},T_{1,i}}\widetilde{s} = \tau_{2,0,2,5}^{\hat{T}_{1,i},T_{1,i}}\widetilde{s} = 0, \; i = 1,2,3 \end{aligned} \tag{3.148}$$

$$\tau_{2,0,4,3}^{\hat{T}_{1,1},T_{1,1}}\widetilde{s} = 0 \tag{3.149}$$

Proof:
Using the C^4 supersmoothness conditions at the vertices of T, the C^3 supersmoothness conditions at the edges of T, and the C^2 smoothness conditions at the faces $\langle v_1, v_2, v_3 \rangle$, $\langle v_2, v_3, v_4 \rangle$, and $\langle v_3, v_1, v_4 \rangle$ we can uniquely and stably determine the B-coefficients of g corresponding to the domain points in ($\dot{M}1^2_{T_{WF}^4}$), ($\dot{M}2^2_{T_{WF}^4}$), ($\dot{M}3^2_{T_{WF}^4}$), and ($\dot{M}4^2_{T_{WF}^4}$) with $j = 1,2,3$ from Theorem 3.47.

From the additional smoothness conditions in (3.147) the remaining B-coefficients associated with the domain points in ($\dot{M}4^2_{T_{WF}^4}$) can be uniquely and stably computed. Using the additional smoothness conditions in (3.148) and (3.149) we can uniquely and stably determine the B-coefficients corresponding to the domain points in ($\dot{M}5^2_{T_{WF}^4}$) and ($\dot{M}6^2_{T_{WF}^4}$). Moreover, the B-coefficient associated with the domain point in ($\dot{M}7^2_{T_{WF}^4}$) is equal to the value of $g(v_T)$. Thus, following Theorem 3.47, g is uniquely and stably determined. □

Theorem 3.49:
Let $\bar{\mathcal{M}}_4^2$ be the union of the following sets of domain points in $\mathcal{D}_{T^4_{WF},9}$:

($\bar{\text{M}}1^2_{T^4_{WF}}$) $D_4^{T_{v_i}}(v_i)$, $i = 1,\dots,4$, with $v_i \in T_{v_i} \in T^4_{WF}$

($\bar{\text{M}}2^2_{T^4_{WF}}$) $\big(E_3^{T_{1,1}}(\langle v_1,v_2\rangle) \cup E_3^{T_{1,2}}(\langle v_2,v_3\rangle) \cup E_3^{T_{1,3}}(\langle v_3,v_1\rangle) \cup E_3^{T_{2,2}}(\langle v_3,v_4\rangle) \cup$

$\cup E_3^{T_{2,3}}(\langle v_4,v_2\rangle) \cup E_3^{T_{3,1}}(\langle v_1,v_4\rangle)\big) \setminus \bigcup_{i=1}^{4}(D_4(v_i))$

($\bar{\text{M}}3^2_{T^4_{WF}}$) $\{\xi^{T_{j,1}}_{1,0,8-i,i}, \xi^{T_{j,1}}_{0,1,8-i,i}, \xi^{T_{j,1}}_{0,0,9-i,i}\}_{i=0}^{1}$, $j = 1,\dots,4$

($\bar{\text{M}}4^2_{T^4_{WF}}$) $\{\xi^{T_{j,i}}_{3,0,4,2}\}$, $i = 1,2,3$, $j = 1,\dots,4$

($\bar{\text{M}}5^2_{T^4_{WF}}$) $\{\xi^{T_{1,i}}_{4,0,2,3}, \xi^{T_{1,i}}_{3,0,3,3}, \xi^{T_{1,i}}_{3,1,2,3}, \xi^{T_{1,i}}_{3,0,2,4}, \xi^{T_{1,i}}_{2,2,1,4}, \xi^{T_{1,i}}_{2,0,2,5}\}_{i=1}^{3}$

($\bar{\text{M}}6^2_{T^4_{WF}}$) $\{\xi^{T_{1,1}}_{2,0,4,3}\}$

($\bar{\text{M}}7^2_{T^4_{WF}}$) $\{\xi^{T_{1,1}}_{0,0,0,9}\}$

Then $\bar{\mathcal{M}}_4^2$ is a stable minimal determining set for

$$\begin{aligned}\bar{\mathcal{S}}_9^2(T^4_{WF}) := \{s \in C^2(T^4_{WF}) : s|_{\widetilde{T}} \in \mathcal{P}_9^3,\ \forall \widetilde{T} \in T^4_{WF},\\ s \in C^4(v_i),\ i = 1,\dots,4,\\ s \in C^3(e),\ \text{for all edges } e \in T,\\ s \in C^7(\langle v_{F_i}, v_T\rangle),\ i = 1,2\\ s \in C^7(v_T)\},\end{aligned}$$

and the dimension of $\bar{\mathcal{S}}_9^2(T^4_{WF})$ is equal to 316.

Proof:
First, we show that the set $\bar{\mathcal{M}}_4^2$ is a stable minimal determining set for the superspline space $\bar{\mathcal{S}}_9^2(T^4_{WF})$. To this end we set the B-coefficients c_ξ of a spline $s \in \bar{\mathcal{S}}_9^2(T^4_{WF})$ to arbitrary values for each domain point $\xi \in \bar{\mathcal{M}}_4^2$ and show that all other B-coefficients of s are uniquely and stably determined.

We can uniquely and stably determine the B-coefficients of s corresponding to the domain points within a distance of four from the vertices of T, within a distance of three from the edges of T and within a distance of two from the faces $\langle v_1,v_2,v_3\rangle$, $\langle v_2,v_3,v_4\rangle$, and $\langle v_3,v_1,v_4\rangle$ from the B-coefficients associated with the domain points in ($\bar{\text{M}}1^2_{T^4_{WF}}$), ($\bar{\text{M}}2^2_{T^4_{WF}}$), ($\bar{\text{M}}3^2_{T^4_{WF}}$) with $j = 1,2,3$, and ($\bar{\text{M}}4^2_{T^4_{WF}}$) with $j = 1,2,3$, in the same way as in the proof of Theorem 3.47.

Now, in the same way as the B-coefficients associated with the domain points in $R_{9-j}^{T_{2,i}}(v_T)$, $i = 1,2,3$, $j = 0,1,2$, are determined in the proof of Theorem 3.20, we can uniquely and stably compute the B-coefficients of s corresponding to the domain points in the shells $R_9^{T_{4,i}}(v_T)$, $R_8^{T_{4,i}}(v_T)$, and $R_7^{T_{4,i}}(v_T)$, $i = 1,2,3$, from the already determined B-coefficients and those corresponding to the domain points in $(\bar{\text{M}}3^2_{T^4_{WF}})$ with $j = 4$ and in $(\bar{\text{M}}4^2_{T^4_{WF}})$ with $j = 3$.

The remaining B-coefficients of s can be uniquely and stably determined from those already determined and the B-coefficients associated with the domain points in $(\bar{\text{M}}5^2_{T^4_{WF}})$, $(\bar{\text{M}}6^2_{T^4_{WF}})$, and $(\bar{\text{M}}7^2_{T^4_{WF}})$ in the same way as in the proof of Theorem 3.37.

Now, considering the dimension of the space $\bar{\mathcal{S}}_9^2(T^4_{WF})$, we can see that $\bar{\mathcal{S}}_9^2(T^4_{WF}) = 316$, since the cardinality of the minimal determining set $\bar{\mathcal{M}}_4^2$ is equal to 316. □

Lemma 3.50:
Let Δ be a tetrahedral partition of a polyhedral domain $\Omega \subset \mathbb{R}^3$ and $T := \langle v_1, v_2, v_3, v_4 \rangle \in \Delta$ a tetrahedron, that shares the faces $\langle v_1, v_2, v_3 \rangle$, $\langle v_2, v_3, v_4 \rangle$, $\langle v_3, v_1, v_4 \rangle$, and $\langle v_1, v_2, v_4 \rangle$ with the tetrahedra $\hat{T}_1$, $\hat{T}_2$, $\hat{T}_3$, and $\hat{T}_4$, respectively, with $\hat{T}_j \in \Delta$, $j = 1, \ldots, 4$. Moreover, T is refined with a fourth order partial Worsey-Farin split at the split point v_T in the interior of T, and the split points $v_{F_1}, v_{F_2}, v_{F_3}$, and v_{F_4} in the interior of the faces $\langle v_1, v_2, v_3 \rangle$, $\langle v_2, v_3, v_4 \rangle$, $\langle v_1, v_3, v_4 \rangle$, and $\langle v_1, v_2, v_4 \rangle$, respectively. The tetrahedra $\hat{T}_j$, $j = 1, \ldots, 4$, are also subjected to partial Worsey-Farin splits, such that the common faces with T are refined with a Clough-Tocher split. The corresponding subtetrahedra are denoted by $\hat{T}_{j,i}$, $i = 1,2,3$, $j = 1, \ldots, 4$ respectively, where $\hat{T}_{j,i}$ has a common face with $T_{j,i}$.

Let $\bar{\mathcal{S}}_9^2(T^4_{WF})$ be the superspline space defined in Theorem 3.49. Then, for any spline $s \in \mathcal{S}_9^{2,4,3}(\Delta \setminus T)$, there is a spline $g \in \bar{\mathcal{S}}_9^2(T^4_{WF})$ such that

$$\widetilde{s} := \begin{cases} s, & \text{on } \Delta \setminus T \\ g, & \text{on } T, \end{cases}$$

belongs to $\mathcal{S}_9^{2,4,3}(\Delta)$. Moreover, the spline g is uniquely and stably determined by the value of $g(v_T)$ and the B-coefficients of $\widetilde{s}|_{\Delta \setminus T}$ by the following additional smoothness conditions across the faces of T:

$$
\begin{aligned}
&\tau_{4,0,2,3}^{\hat{T}_{1,i},T_{1,i}}\widetilde{s} = \tau_{3,0,3,3}^{\hat{T}_{1,i},T_{1,i}}\widetilde{s} = \tau_{3,1,2,3}^{\hat{T}_{1,i},T_{1,i}}\widetilde{s} \\
&= \tau_{3,0,2,4}^{\hat{T}_{1,i},T_{1,i}}\widetilde{s} = \tau_{2,2,1,4}^{\hat{T}_{1,i},T_{1,i}}\widetilde{s} = \tau_{2,0,2,5}^{\hat{T}_{1,i},T_{1,i}}\widetilde{s} = 0,\ i = 1,2,3
\end{aligned} \tag{3.150}
$$

$$
\tau_{2,0,4,3}^{\hat{T}_{1,1},T_{1,1}}\widetilde{s} = 0 \tag{3.151}
$$

Proof:
From the C^4 supersmoothness conditions at the vertices, the C^3 supersmoothness conditions at the edges, and the C^2 smoothness conditions at the faces of T we can uniquely and stably determine the B-coefficients of g corresponding to the domain points in $(\bar{\text{M}}1^2_{T^4_{WF}})$, $(\bar{\text{M}}2^2_{T^4_{WF}})$, $(\bar{\text{M}}3^2_{T^4_{WF}})$, and $(\bar{\text{M}}4^2_{T^4_{WF}})$ from Theorem 3.49.

Then, from the additional smoothness conditions in (3.150) and (3.151) we can uniquely and stably determine the B-coefficients corresponding to the domain points in $(\bar{\text{M}}5^2_{T^4_{WF}})$ and $(\bar{\text{M}}6^2_{T^4_{WF}})$. The last undetermined B-coefficient associated with a domain point in the minimal determining set $\bar{\mathcal{M}}^2_4$ is equal to the value of $g(v_T)$. Thus, following Theorem 3.49, g is uniquely and stably determined. □

Remark 3.51:
The space $\widetilde{\mathcal{S}}^2_9(T^4_{WF})$ of supersplines has already been studied in [9], though a different minimal determining set was constructed there. For the space $\bar{\mathcal{S}}^2_9(T^4_{WF})$ of C^2 splines based on the fourth order partial Worsey-Farin split, which is also just called Worsey-Farin split, another minimal determining set was constructed in [66], which is also considered in chapter 8.

4 Local Lagrange interpolation by cubic C^1 splines on type-4 cube partitions

In this chapter, the Lagrange interpolation method on type-4 cube partitions constructed by Matt and Nürnberger [67] is introduced. In section 4.1, we consider cube partitions and their classification. Subsequently, in section 4.2, we examine the type-4 partition, which is obtained by dividing each cube of a partition into five tetrahedra, according to the classification of the previous section. In section 4.3, the tetrahedra of a type-4 partition are refined with partial Worsey-Farin splits based on the classification of the cubes and the location of the tetrahedra in these. Moreover, interpolation points for $\mathcal{S}_3^1(\Delta_4^*)$ are chosen. In section 4.4, we consider the local Lagrange interpolation with $\mathcal{S}_3^1(\Delta_4^*)$. Furthermore, we examine a nodal minimal determining set corresponding to this spline space. In the next section, bounds on the error of the interpolant are examined, which shows that the interpolation method yields optimal approximation order. Finally, in section 4.6, numerical tests and visualizations are shown. These also verify the optimal approximation order of the method.

4.1 Cube partitions

In this section cube partitions are considered. Therefore, first cube partitions are defined and afterwards the cubes of a partition are divided into different classes. This classification is obtained by considering the indices of the different cubes.

Now, let n be an odd integer and let $\Omega := [0,n] \times [0,n] \times [0,n] \subseteq \mathbb{R}^3$. Then the cube partition

$$\diamond := \{Q_{i,j,k} : Q_{i,j,k} = [i,i+1] \times [j,j+1] \times [k,k+1],\ i,j,k = 0,\ldots,n-1\}$$

of Ω is obtained by intersecting Ω with n parallel planes in each of the three space dimensions. The partition $\diamond$ consists of n^3 cubes.

According to their indices, the cubes $Q_{i,j,k}$, $i,j,k = 0,\ldots,n-1$, can be divided into the five classes $\mathcal{K}_0,\ldots,\mathcal{K}_4$. Following Table 4.1, one of the possible classifications of the cubes can be obtained, where all other cubes are put in class $\mathcal{K}_4$. The properties of the indices of cubes in class $\mathcal{K}_4$ are shown in Table 4.2. Note that this is the partition used throughout this chapter. An example for $n = 5$ of the classification of $\diamond$ can be found in Figure 4.1.

	i	j	k
$\mathcal{K}_0$	even	even	even
$\mathcal{K}_1$	odd	odd	even
$\mathcal{K}_2$	odd	even	odd
$\mathcal{K}_3$	even	odd	odd

Table 4.1: Classification of the cubes in $\diamond$ for the classes $\mathcal{K}_0,\ldots,\mathcal{K}_3$.

	i	j	k
$\mathcal{K}_4$	even	even	odd
$\mathcal{K}_4$	even	odd	even
$\mathcal{K}_4$	odd	even	even
$\mathcal{K}_4$	odd	odd	odd

Table 4.2: Possible cases for the indices of cubes in class $\mathcal{K}_4$.

For a better understanding of the structure of the classification, see Figure 4.2, where two layers of a cube partition are shown, with some cubes left out.

In the following lemma, the relations between cubes in the different classes $\mathcal{K}_0,\ldots,\mathcal{K}_4$ are considered. These play an important role for the local Lagrange interpolation method on the type-4 partition, constructed later in this chapter.

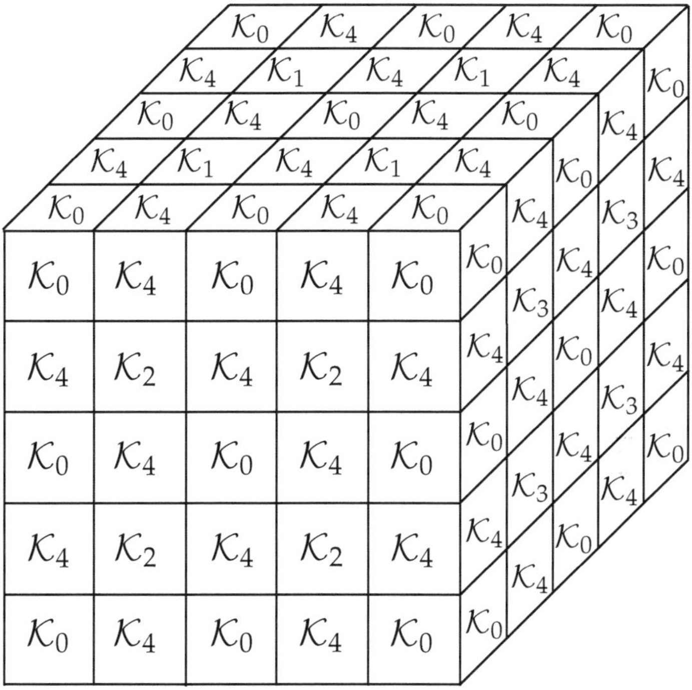

Figure 4.1: Classification of a cube partition $\diamond$.

Lemma 4.1:

(i) Two different cubes in the same class $\mathcal{K}_i$, $i = 0,\ldots,3$, are disjoint.

(ii) Each cube Q in class $\mathcal{K}_i$, $i = 1,2,3$, shares common edges with at most four cubes in class $\mathcal{K}_j$, $j = 0,\ldots,i-1$, respectively (see Figure 4.3).

(iii) Two different cubes in class $\mathcal{K}_4$ share at most one common edge.

(iv) If two different cubes in class $\mathcal{K}_4$ share a common edge e, then e is also an edge of two cubes in the classes $\mathcal{K}_i$ and $\mathcal{K}_j$, with $i,j = 0,\ldots,3$, and $i \neq j$ (see Figure 4.4).

(v) Each vertex of $\diamond$ is a vertex of a cube in class $\mathcal{K}_0$.

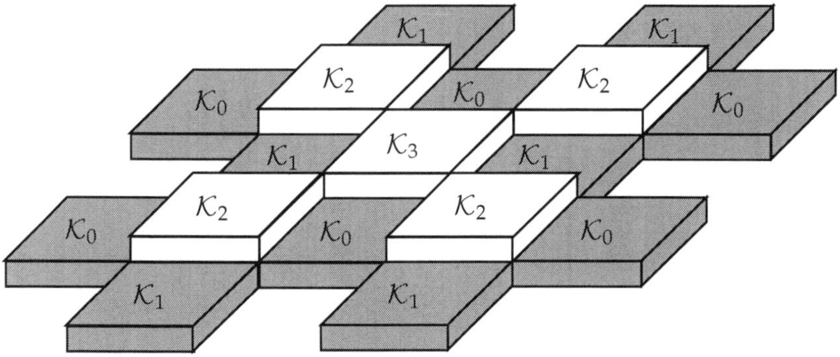

Figure 4.2: Two layers of a classified cube partition.

Proof:
First, statement (i) is considered. Therefore, let $Q := Q_{i,j,k}$ be a cube in class $\mathcal{K}_0$. Thus, i, j, and k must be even. Now, let $\widetilde{Q}$ be another cube sharing a common vertex, edge, or face with Q, respectively. Then, the triple of indices of $\widetilde{Q}$ must be of the form $(i+\alpha, j+\beta, k+\gamma)$, with $\alpha, \beta, \gamma \in \{-1, 0, 1\}$ and $(\alpha, \beta, \gamma) \neq 0$. Thus, at least one of the indices of $\widetilde{Q}$ must be odd, which implies, that $\widetilde{Q} \notin \mathcal{K}_0$.

The proof is analogue for the cubes in class $\mathcal{K}_1, \mathcal{K}_2$, and $\mathcal{K}_3$.

Next, assertion (ii) is verified. Therefore, let $Q := Q_{i,j,k}$ be a cube in class $\mathcal{K}_3$ in the interior of $\diamond$. Since $Q \in \mathcal{K}_3$, the index i has to be even and the indices j and k must be odd. Now, we consider the 18 cubes, that share at least one edge with Q. The four cubes $Q_{i,j-1,k-1}, Q_{i,j-1,k+1}, Q_{i,j+1,k-1}$, and $Q_{i,j+1,k+1}$ have a common edge with Q, respectively. Moreover, since i is even and j and k are odd, the indices of these four cubes are all even. Thus, following Table 4.1,the four cubes are in class $\mathcal{K}_0$. Next, the four cubes $Q_{i-1,j,k-1}, Q_{i-1,j,k+1}, Q_{i+1,j,k-1}$, and $Q_{i+1,j,k+1}$ are considered. Each of them has exactly one common edge with Q and their first two indices are odd and the last one is even. Therefore, these cubes are in class $\mathcal{K}_1$. Subsequently, the cubes $Q_{i-1,j-1,k}, Q_{i-1,j+1,k}, Q_{i+1,j-1,k}$, and $Q_{i+1,j+1,k}$, which share a common edge with Q, respectively, are regarded. For these cubes, the first and the last index is odd, respectively, and the second one is even. Therefore, they are in class $\mathcal{K}_2$. The six other cubes sharing common edges with Q, must touch Q at a common face. Thus, for their indices holds one of the combinations shown in Table 4.2, respectively.

A cube $\widetilde{Q}$ in class $\mathcal{K}_3$ at the boundary of $\diamond$ has less surrounding cubes. Thus, the proof is similar for this case, except that there can be less than four cubes in class $\mathcal{K}_i$, $i = 0,1,2$, sharing edges with $\widetilde{Q}$.

The proof is analog for cubes in the classes $\mathcal{K}_i$, $i = 0,1,2$.

In the following, assertion (iii) is considered. Therefore, let $Q := Q_{i,j,k}$ be a cube in class $\mathcal{K}_4$, where the indices i and j are even and k is odd. Then, the six cubes sharing a common face with Q are $Q_{i,j,k-1}$, $Q_{i,j,k+1}$, $Q_{i,j-1,k}$, $Q_{i,j+1,k}$, $Q_{i-1,j,k}$, and $Q_{i+1,j,k}$. The indices of the two cubes $Q_{i,j,k-1}$ and $Q_{i,j,k+1}$ are all even. Thus, these two cubes are in class $\mathcal{K}_0$. Considering the two cubes $Q_{i,j-1,k}$ and $Q_{i,j+1,k}$, it can be seen that their first index is even and the last two are odd, respectively. Then, following Table 4.1, these two cubes must be in class $\mathcal{K}_3$. For the last two cubes, $Q_{i-1,j,k}$ and $Q_{i+1,j,k}$, only the second index is even, the other two are odd, respectively. Therefore, these two cubes are in class $\mathcal{K}_2$.

In the same way it can be shown for the other three cases for the indices of Q shown in Table 4.2, that no two different cubes in class $\mathcal{K}_4$ share a common face.

Next, assertion (iv) is proved. Therefore, let Q and $\widetilde{Q}$ be two cubes in class $\mathcal{K}_4$ sharing the common edge e. Following statement (iii), Q and $\widetilde{Q}$ can not share a common face. Thus, e must be an interior edge of $\diamond$. Due to this reason, there must be two other cubes in $\diamond$ sharing e and touching Q and $\widetilde{Q}$ at common faces, respectively. Again, following assertion (iii), these two cubes can not be in class $\mathcal{K}_4$. Thus, these two cubes have to be in one of the classes $\mathcal{K}_0, \ldots, \mathcal{K}_3$.

Finally, statement (v) is verified. The partition $\diamond$ contains a total number of $(n+1)^3$ vertices. The cardinality of the set of even numbers from 0 to n is equal to $\frac{n+1}{2}$. Now, since three indices of all cubes in class $\mathcal{K}_0$ are even, the partition $\diamond$ contains exactly $\left(\frac{n+1}{2}\right) = \frac{(n+1)^3}{8}$ cubes that are in class $\mathcal{K}_0$. Due to the fact that each cube has a total of eight vertices and that the cubes in class $\mathcal{K}_0$ are disjoint, following statement (i), each vertex of $\diamond$ must be a vertex of a cube in class $\mathcal{K}_0$. □

Remark 4.2:

It is also possible to create a cube partition $\diamond$ and a corresponding classification for $\Omega := [0,n] \times [0,n] \times [0,n] \subseteq \mathbb{R}^3$, where n is even. However, then the property of the classification that all vertices of $\diamond$ are vertices of cubes in a class $\mathcal{K}_0$ can not be obtained, though this property is quite important for the Lagrange interpolation method considered in this chapter.

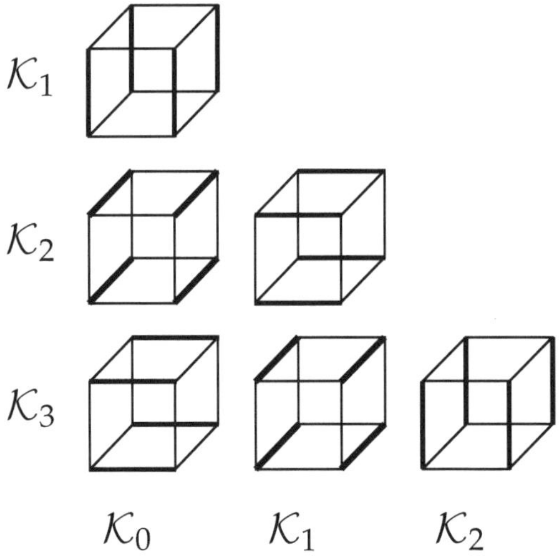

Figure 4.3: Possible common edges of cubes in the classes $\mathcal{K}_i$, $i = 1,2,3$, with cubes in the lower classes $\mathcal{K}_j$, $j = 0,1,2$, respectively. Common edges are marked with thick lines, the classes $\mathcal{K}_i$ are shown left, and the classes $\mathcal{K}_j$ are shown at the bottom.

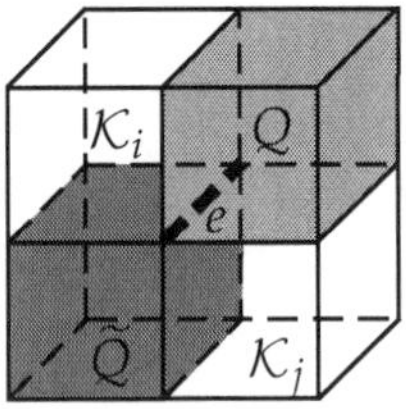

Figure 4.4: Two cubes Q and $\widetilde{Q}$ in class $\mathcal{K}_4$ with a common edge e and two cubes in the classes $\mathcal{K}_i$ and $\mathcal{K}_j$, $i,j = 0,\ldots,3$, and $i \neq j$, respectively.

Remark 4.3:
It is also possible to create other classifications of the cubes in $\diamond$, where the properties in Lemma 4.1 still hold (cf. [42]). However, the classification used in [42] is just a rotation of the one used here.

4.2 Type-4 partition

In this section the type-4 partition is described, which is indicated by Δ_4 throughout this chapter. In order to construct a type-4 partition, each cube of a cube partition $\diamond$ is divided into exactly five tetrahedra. Therefore, the set $\mathcal{V}$ of vertices of $\diamond$ is split up into the two sets $\mathcal{V}_1$ and $\mathcal{V}_2$. Then it is possible to divide each cube into five tetrahedra according to the two sets $\mathcal{V}_1$ and $\mathcal{V}_2$ (cf. [95]). The type-4 partition is the basis for the Lagrange interpolation method investigated in this chapter.

At first, the vertices of a cube partition $\diamond$ are divided into the two sets $\mathcal{V}_1$ and $\mathcal{V}_2$. Therefore, let $\mathcal{V} := \{v_{i,j,k} := (i,j,k)\}_{i,j,k=0}^{n}$ be the set of $(n+1)^3$ vertices of $\diamond$.

Lemma 4.4:
The set $\mathcal{V}$ of vertices of a cube partition $\diamond$ can be divided into two sets $\mathcal{V}_1$ and $\mathcal{V}_2$, such that for each vertex $v \in \mathcal{V}_\nu$ all other vertices in $\mathcal{V}$ sharing an edge with v are contained in the set $\mathcal{V}_\mu$, with $\nu, \mu = 1,2$ and $\nu \neq \mu$.

Proof:
The sets $\mathcal{V}_1$ and $\mathcal{V}_2$ can be chosen as follows:

$$\mathcal{V}_1 := \{v_{i,j,k} \in \mathcal{V} : \; i+j+k \text{ is even}\}$$
$$\mathcal{V}_2 := \{v_{i,j,k} \in \mathcal{V} : \; i+j+k \text{ is odd}\}$$

Now, let $v := v_{i,j,k}$ be a vertex contained in the set $\mathcal{V}_1$. According to the construction of $\diamond$, the vertex v shares exactly six edges with the vertices $\{v_{i,j,k-1}, v_{i,j,k+1}, v_{i,j-1,k}, v_{i,j+1,k}, v_{i-1,j,k}, v_{i+1,j,k}\} \subset \mathcal{V}$, respectively. In case v is located at the boundary of $\diamond$, there are less common edges with other vertices in $\mathcal{V}$. Now, considering the sums of the indices of the six vertices sharing an edge with v, it can be seen that these are equal to $i+j+k+1$ or $i+j+k-1$, respectively. Since $v \in \mathcal{V}_1$, the sum $i+j+k$ is even. Thus, the sums $i+j+k+1$ and $i+j+k-1$ are odd, which implies, that the vertices sharing a common edge with v are contained in the set $\mathcal{V}_2$. The proof is analog for the vertices in $\mathcal{V}_2$. □

The division of $\mathcal{V}$ into the two sets $\mathcal{V}_1$ and $\mathcal{V}_2$ considered in Lemma 4.4 is not unique, since it depends on the choice of the first vertex. Without loss of generality, it can be assumed that $v_{0,0,0} \in \mathcal{V}_1$. Thus, the two sets $\mathcal{V}_1$ and $\mathcal{V}_2$ are uniquely defined and they can be written as

$$\mathcal{V}_1 := \{v_{i,j,k} \in \mathcal{V} : \; i+j+k \text{ is even}\}, \text{ and}$$
$$\mathcal{V}_2 := \{v_{i,j,k} \in \mathcal{V} : \; i+j+k \text{ is odd}\},$$

as in the proof on Lemma 4.4. In the following, the vertices in the set $\mathcal{V}_1$ are called type-1 vertices and those in $\mathcal{V}_2$ are type-2 vertices.

Now, it is possible to define the type-4 partition of a cube partition $\diamond$.

Definition 4.5:
Let $\diamond$ be a cube partition in $\mathbb{R}^3$. The tetrahedral partition Δ_4 obtained by splitting each cube in $\diamond$ into five tetrahedra by connecting the four type-2 vertices in each cube is called type-4 partition.

The five tetrahedra in each cube are named following Notation 4.6.

Notation 4.6:
The individual tetrahedra in a cube $Q_{i,j,k}$ contained in one of the classes $\mathcal{K}_0, \mathcal{K}_1, \mathcal{K}_2$, or $\mathcal{K}_3$ are denoted by

$$\begin{aligned}
T_{i,j,k}^1 &:= \langle v_{i,j,k}, v_{i,j,k+1}, v_{i,j+1,k}, v_{i+1,j,k} \rangle, \\
T_{i,j,k}^2 &:= \langle v_{i,j,k+1}, v_{i,j+1,k}, v_{i,j+1,k+1}, v_{i+1,j+1,k+1} \rangle, \\
T_{i,j,k}^3 &:= \langle v_{i,j+1,k}, v_{i+1,j,k}, v_{i+1,j+1,k}, v_{i+1,j+1,k+1} \rangle, \\
T_{i,j,k}^4 &:= \langle v_{i,j,k+1}, v_{i+1,j,k}, v_{i+1,j,k+1}, v_{i+1,j+1,k+1} \rangle, \\
T_{i,j,k}^5 &:= \langle v_{i,j,k+1}, v_{i,j+1,k}, v_{i+1,j,k}, v_{i+1,j+1,k+1} \rangle
\end{aligned}$$

see Figure 4.5.

The individual tetrahedra in a cube $Q_{i,j,k} \in \mathcal{K}_4$ are denoted by

$$\begin{aligned}
T_{i,j,k}^1 &:= \langle v_{i,j,k}, v_{i,j,k+1}, v_{i,j+1,k+1}, v_{i+1,j,k+1} \rangle, \\
T_{i,j,k}^2 &:= \langle v_{i,j,k}, v_{i,j+1,k}, v_{i,j+1,k+1}, v_{i+1,j+1,k} \rangle, \\
T_{i,j,k}^3 &:= \langle v_{i,j+1,k+1}, v_{i+1,j,k+1}, v_{i+1,j+1,k}, v_{i+1,j+1,k+1} \rangle, \\
T_{i,j,k}^4 &:= \langle v_{i,j,k}, v_{i+1,j,k}, v_{i+1,j,k+1}, v_{i+1,j+1,k} \rangle, \\
T_{i,j,k}^5 &:= \langle v_{i,j,k}, v_{i,j+1,k+1}, v_{i+1,j,k+1}, v_{i+1,j+1,k} \rangle,
\end{aligned}$$

see Figure 4.6.

In Figure 4.7 the cube partition from Figure 4.1 has been refined to a type-4 partition. The inserted edges connecting the type-2 vertices are marked in gray.

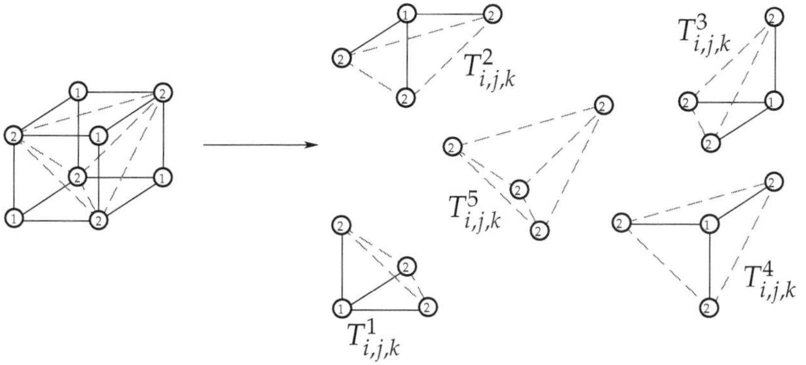

Figure 4.5: Division of a cube in class $\mathcal{K}_0, \mathcal{K}_1, \mathcal{K}_2$, or $\mathcal{K}_3$, with marked type-1 and type-2 vertices, into five tetrahedra.

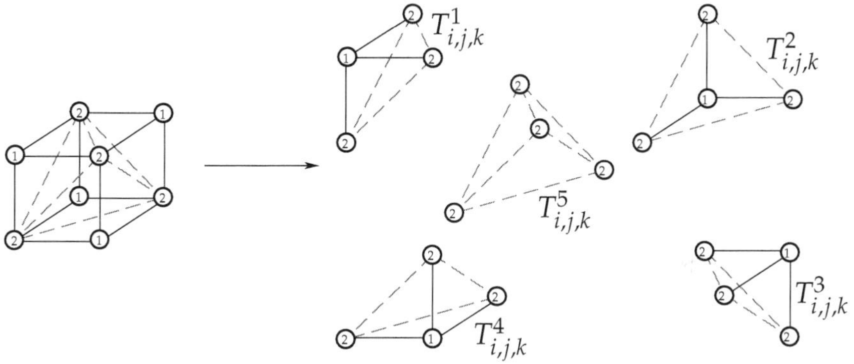

Figure 4.6: Division of a cube in class $\mathcal{K}_4$, with marked type-1 and type-2 vertices, into five tetrahedra.

Remark 4.7:
Analogue to bivariate crosscut partitions, where a domain is cut by straight lines, the type-4 partition can also be obtained by cutting a cube partition $\diamond$ with planes. A cube Q has to be cut with exactly four planes, in order to get the type-4 partition of Q.

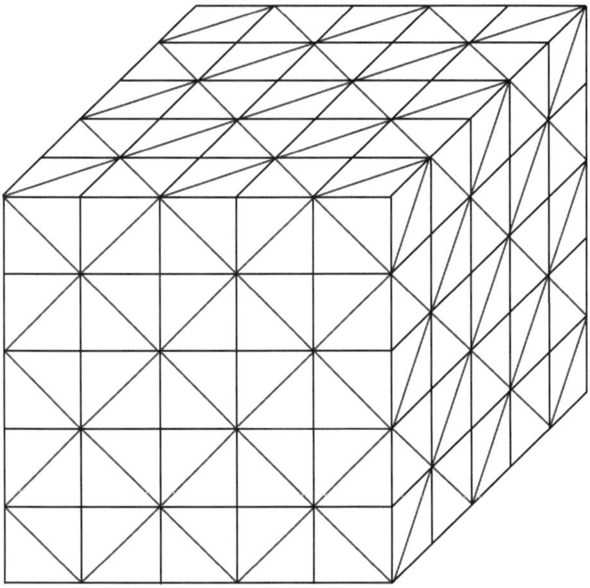

Figure 4.7: Type-4 partition of a cube partition $\diamond$.

4.3 Selection of interpolation points and refinement of Δ_4

In this section the selection of the Lagrange interpolation points as well as the refinement of Δ_4 to Δ_4^* for the local Lagrange interpolation with $\mathcal{S}_3^1(\Delta_4^*)$ is considered. In order to refine the partition Δ_4 some of the tetrahedra of the type-4 partition are divided with a partial Worsey-Farin split (see Definition 2.8).

Let $\diamond$ be a cube partition in $\mathbb{R}^3$ and Δ_4 the corresponding type-4 partition. Moreover, let the cubes in $\diamond$ be classified into the classes $\mathcal{K}_i$, $i = 0,\ldots,4$, as in section 4.1 and the tetrahedra be identified as in Notation 4.6.

Algorithm 4.8:
Step 1: For each cube $Q_{i,j,k} \in \mathcal{K}_0$,

a) choose the 20 points $\mathcal{D}_{T^1_{i,j,k},3}$,

b) choose the 10 points $\mathcal{D}_{T^2_{i,j,k},3} \setminus E_1^{T^2_{i,j,k}}(\langle v_{i,j,k+1}, v_{i,j+1,k}\rangle)$,

c) choose the 4 points $D_1^{T^3_{i,j,k}}(v_{i+1,j,k+1})$, $v_F \in \langle v_{i+1,j,k}, v_{i+1,j,k+1}, v_{i+1,j+1,k+1}\rangle$, and $v_{\widetilde{F}} \in \langle v_{i,j,k+1}, v_{i+1,j,k}, v_{i+1,j+1,k+1}\rangle$, and split $T^3_{i,j,k}$ with a first order partial Worsey-Farin split at $\langle v_{i,j+1,k}, v_{i+1,j,k}, v_{i+1,j+1,k+1}\rangle$,

d) choose the 4 points $D_1^{T^4_{i,j,k}}(v_{i+1,j+1,k})$ and split $T^4_{i,j,k}$ with a first order partial Worsey-Farin split at $\langle v_{i,j+1,k}, v_{i+1,j,k}, v_{i+1,j+1,k+1}\rangle$.

Step 2: Define all edges of $\Delta_4 \setminus \mathcal{K}_0$ as "unmarked" and all edges of the cubes in class $\mathcal{K}_0$ as "marked".
Step 3: For each cube $Q_{i,j,k}$ in $\mathcal{K}_l$, $l = 1, \ldots, 4$,
for each tetrahedron $T^h_{i,j,k}$, $h = 1, \ldots, 4$, in $Q_{i,j,k}$,

a) if $T := T^h_{i,j,k}$ has $m > 0$ faces with two or three marked edges, then split these faces with a Clough-Tocher split, T with a m-th order partial Worsey-Farin split and replace T in Δ_4 by the resulting subtetrahedra.

b) if a face $\langle v_1, v_2, v_3\rangle$ of T has none or two marked edges, choose the point v_F.

c) mark all edges of T.

Step 4: Split each tetrahedron $T^5_{i,j,k}$ in Δ_4 with a 4-th order partial Worsey-Farin split.

In Figure 4.8 the selection of the Lagrange interpolation points in the interior of the faces of the tetrahedra in Δ_4 and the corresponding refinements are shown. Therefore, "marked" edges are pointed out as thicker lines. Interpolation points are indicated with $\bullet$.

The set of interpolation points chosen in Algorithm 4.8 is denoted by $\mathcal{L}_4$. The refined tetrahedral partition, obtained by dividing some of the tetrahedra of Δ_4 with partial Worsey-Farin splits, is called Δ_4^*.

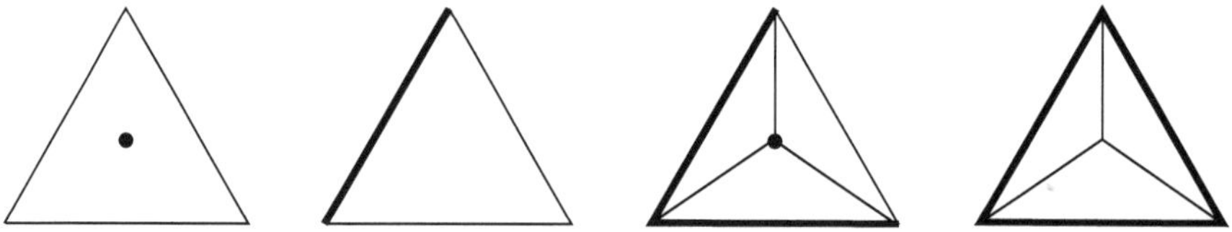

Figure 4.8: Selection of interpolation points in the interior of the faces of tetrahedra in Δ_4 and refinements with Clough-Tocher splits.

In order to facilitate the understanding of Algorithm 4.8, the order of the splits that may be applied to the different tetrahedra, in the various classes of cubes, are shown in Table 4.3, where m-WF stands for the m-th order partial Worsey-Farin split.

The symbol "-" indicates, that the corresponding split does not occur for a specified tetrahedron. With "o" the cases are identified, where the corresponding split is used because the cube containing a considered tetrahedron is on the boundary of $\diamond$. The refinements for the cases where the cubes containing the regarded tetrahedron are in the interior of $\diamond$ are indicated by the symbol "x".

4.4 Lagrange interpolation with $\mathcal{S}_3^1(\Delta_4^*)$

In this section the main theorem of this chapter is stated. It is shown that the set $\mathcal{L}_4$, constructed in section 4.3, is a Lagrange interpolation set for the spline space $\mathcal{S}_3^1(\Delta_4^*)$. In the course of the proof some of the tetrahedra in Δ_4^* have to be refined even further, since Δ_4^* is not a proper tetrahedral partition, as defined in section 2.1. There are some tetrahedra sharing common face, where the corresponding face is split in one and non-split in the other tetrahedron. Though a spline can be uniquely determined on the tetrahedra which are not split enough, according to their neighbors, and the corresponding tetrahedra are subdivided later on to determine the spline on the other tetrahedra. This leads to smaller systems of linear equations, since there are less subtetrahedra at the time the spline is determined.

	no split	1-WF	2-WF	3-WF	4-WF
$\mathcal{K}_0 T^1_{i,j,k}$	x	-	-	-	-
$\mathcal{K}_0 T^2_{i,j,k}$	x	-	-	-	-
$\mathcal{K}_0 T^3_{i,j,k}$	-	x	-	-	-
$\mathcal{K}_0 T^4_{i,j,k}$	-	x	-	-	-
$\mathcal{K}_0 T^5_{i,j,k}$	-	-	-	-	x
$\mathcal{K}_1 T^1_{i,j,k}$	x	-	-	-	-
$\mathcal{K}_1 T^2_{i,j,k}$	-	x	-	-	-
$\mathcal{K}_1 T^3_{i,j,k}$	-	-	x	-	-
$\mathcal{K}_1 T^4_{i,j,k}$	-	-	-	x	-
$\mathcal{K}_1 T^5_{i,j,k}$	-	-	-	-	x
$\mathcal{K}_2 T^1_{i,j,k}$	o	x	-	-	-
$\mathcal{K}_2 T^2_{i,j,k}$	-	o	x	-	-
$\mathcal{K}_2 T^3_{i,j,k}$	-	-	o	x	-
$\mathcal{K}_2 T^4_{i,j,k}$	-	-	-	o	x
$\mathcal{K}_2 T^5_{i,j,k}$	-	-	-	-	x
$\mathcal{K}_3 T^1_{i,j,k}$	o	-	-	x	-
$\mathcal{K}_3 T^2_{i,j,k}$	o	-	-	x	-
$\mathcal{K}_3 T^3_{i,j,k}$	-	-	-	o	x
$\mathcal{K}_3 T^4_{i,j,k}$	-	-	-	o	x
$\mathcal{K}_3 T^5_{i,j,k}$	-	-	-	-	x
$\mathcal{K}_4 T^1_{i,j,k}$	-	o	-	-	x
$\mathcal{K}_4 T^2_{i,j,k}$	-	o	o	-	x
$\mathcal{K}_4 T^3_{i,j,k}$	-	-	-	o	x
$\mathcal{K}_4 T^4_{i,j,k}$	-	-	-	-	x
$\mathcal{K}_4 T^5_{i,j,k}$	-	-	-	-	x

Table 4.3: Possible refinements of the various tetrahedra in Δ_4.

Theorem 4.9:
Let $\mathcal{L}_4$ and Δ_4^* be the set of interpolation points and the refined tetrahedral partition of a type-4 partition based on $\diamond$ obtained by Algorithm 4.8, respectively. Then $\mathcal{L}_4$ is a local and stable Lagrange interpolation set for $\mathcal{S}_3^1(\Delta_4^*)$.

Proof:
In order to show that $\mathcal{L}_4$ is a Lagrange interpolation set for $\mathcal{S}_3^1(\Delta_4^*)$, the values $\{z_\kappa\}_{\kappa \in \mathcal{L}_4}$ are fixed arbitrarily for a spline $s \in \mathcal{S}_3^1(\Delta_4^*)$. Then it is shown that s is locally, stably, and uniquely determined. Therefore, three steps have to be performed. In the first step s is determined on the cubes in class $\mathcal{K}_0$. Subsequently, s is computed restricted to the cubes in $\mathcal{K}_1, \mathcal{K}_2$, and $\mathcal{K}_3$. In the final step, the spline s is determined on the cubes in class $\mathcal{K}_4$. Thus, we can distinguish the following cases:

Step 1: $s|_{\mathcal{K}_0}$
Following statement (i) in Lemma 4.1, the cubes in class $\mathcal{K}_0$ are disjoint. Thus, it suffices to consider only one cube in this class, on the other cubes in class $\mathcal{K}_0$ the spline s can be determined analogously. Therefore, let $Q_{i,j,k}$ be a cube in class $\mathcal{K}_0$ and $T^l := T^l_{i,j,k}$, $l = 1, \ldots, 5$ the five tetrahedra in $Q_{i,j,k}$ (cf. Notation 4.6).

First, $s|_{T^1}$ is considered. Following Algorithm 4.8, the set $\mathcal{L}_4$ contains the domain points $\mathcal{D}_{T^1,3}$, all in the tetrahedron T^1. Thus, $s|_{T^1}$ can be uniquely and stably determined from the values $\{z_\kappa\}_{\kappa \in (\mathcal{L}_4 \cap T^1)}$, with $\mathcal{L}_4 \cap T^1 = \mathcal{D}_{T^1,3}$. Since the spline restricted to T^1 depends only on the values of the interpolation points contained in T^1, the computation of $s|_{T^1}$ is also local.

Since all B-coefficients of s corresponding to the domain points in T^1 are uniquely determined, all B-coefficients associated with domain points within a distance of one from the vertices and edges of T^1 can be uniquely and stably computed from the C^1 smoothness conditions at these edges and vertices. These B-coefficients of s are also determined locally, since they only depend on the values $\{z_\kappa\}_{\kappa \in (\mathcal{L}_4 \cap T^1)}$.

Next, the spline s restricted to the tetrahedron T^2 is examined. The tetrahedron T^2 shares the common edge $\langle v_{i,j,k+1}, v_{i,j+1,k} \rangle$ with the tetrahedron T^1. Thus, the B-coefficients of s corresponding to the domain points in the tube $E_1^{T_2}(\langle v_{i,j,k+1}, v_{i,j+1,k} \rangle)$ are already uniquely determined. Since the domain points associated with the remaining undetermined B-coefficients of $s|_{T^2}$ are contained in the set $\mathcal{L}_4$, these B-coefficients of $s|_{T^2}$ can be uniquely and stably computed from the values $\{z_\kappa\}_{\kappa\in(\mathcal{L}_4\cap T^2)}$. The computation of all B-coefficients of s restricted to the tetrahedron T^2 is local, since they only depend on the values $\{z_\kappa\}_{\kappa\in(\mathcal{L}_4\cap(T^1\cup T^2))}$.

Now, using the C^1 smoothness conditions at the vertices and edges of T^2, all B-coefficients of s corresponding to domain points within a distance of one from the vertices and edges of T^2 can be uniquely and stably determined. Their computation is local, since the only depend on the values $\{z_\kappa\}_{\kappa\in(\mathcal{L}_4\cap(T^1\cup T^2))}$.

Subsequently, $s|_{T^3}$ is investigated. Following Algorithm 4.8, the tetrahedron T^3 has been divided with a first order partial Worsey-Farin split at the face $\langle v_{i,j+1,k}, v_{i+1,j+1,k}, v_{i+1,j+1,k+1}\rangle$ (see also Table 4.3). Moreover, since T^3 shares the two edges $\langle v_{i,j+1,k}, v_{i+1,j,k}\rangle$ and $\langle v_{i,j+1,k}, v_{i+1,j+1,k+1}\rangle$ with the tetrahedra T^1 and T^2, respectively, the B-coefficients of s associated with the domain points in the two tubes $E_1^{T_3}(\langle v_{i,j+1,k}, v_{i+1,j,k}\rangle)$ and $E_1^{T_3}(\langle v_{i,j+1,k}, v_{i+1,j+1,k+1}\rangle)$ are already uniquely and stably determined. The B-coefficients associated with the domain points in $D_1(v_{i+1,j+1,k}) \cap T^3$ can be uniquely and stably determined from the values $\{z_\kappa\}_{\kappa\in D_1^{T^3}(v_{i+1,j+1,k})\subset\mathcal{L}_4}$. Therefore, the remaining undetermined B-coefficients corresponding to domain points on the edges of T^3 are uniquely and stably determined from these values and the already determined B-coefficients on these edges by solving three univariate interpolation problems, since s restricted to each of three edges at $v_{i+1,j+1,k}$ is just a univariate spline. Then, using the C^1 smoothness conditions at $v_{i+1,j+1,k}$, the remaining undetermined B-coefficients corresponding to domain points in $D_1(v_{i+1,j+1,k}) \cap T^3$ can be uniquely and stably determined (T^3 has been subjected to a partial Worsey-Farin split, thus there are three subtetrahedra sharing $v_{i+1,j+1,k}$). Then, the spline s restricted to the two faces $F_1 := \langle v_{i+1,j,k}, v_{i+1,j+1,k}, v_{i+1,j+1,k+1}\rangle$ and $F_2 := \langle v_{i,j+1,k}, v_{i+1,j,k}, v_{i+1,j+1,k+1}\rangle$ of T^3 is considered. Following Lemma 2.42 for $n = 3$ and Lemma 2.45, the remaining undetermined B-coefficients of $s|_{F_1}$ and $s|_{F_2}$ can be uniquely and stably determined from the values at the two interpolation points v_{F_1} and v_{F_2}, respectively. Then, all B-coefficients of s corresponding to the domain points in the minimal determining set $\mathcal{M}_1^1$ from Theorem 3.3 are determined. Thus, all other B-coefficients of $s|_{T^3}$ can be uniquely and stably computed from those already determined. Moreover, the computation of the B-coefficients of $s|_{T^3}$ is local, since they only depend on the values $\{z_\kappa\}_{\kappa\in\mathcal{L}_4\cap(T^1\cup T^2\cup T^3)}$.

Since the B-coefficients of $s|_{T^3}$ are uniquely computed, all undetermined B-coefficients of s corresponding to domain points within a distance of one from the edges and vertices of T^3 can be uniquely and stably determined using the C^1 smoothness conditions at the edges and vertices of T^3. The computation of these B-coefficients is also local, since they only depend on the values corresponding to the interpolation points in $\mathcal{L}_4 \cap (T^1 \cup T^2 \cup T^3)$.

Now, the spline s restricted to the tetrahedron T^4 is considered. Following Algorithm 4.8, T^4 has been subjected to a first order partial Worsey-Farin split, where the face $\langle v_{i,j,k+1}, v_{i+1,j,k}, v_{i+1,j+1,k+1}\rangle$ has been refined with a Clough-Tocher split. Since T^4 shares the edges $\langle v_{i,j,k+1}, v_{i+1,j,k}\rangle$, $\langle v_{i,j,k+1}, v_{i+1,j+1,k+1}\rangle$, and $\langle v_{i+1,j,k}, v_{i+1,j+1,k+1}\rangle$ with the tetrahedra T^1, T^2, and T^3, respectively, the B-coefficients of $s|_{T^4}$ corresponding to the domain points in $(E_1(\langle v_{i,j,k+1}, v_{i+1,j,k}\rangle) \cup E_1(\langle v_{i,j,k+1}, v_{i+1,j+1,k+1}\rangle) \cup E_1(\langle v_{i+1,j,k}, v_{i+1,j+1,k+1}\rangle)) \cap T^4$ are already uniquely and stably determined. From the values corresponding to the four interpolation points in the disk $D_1^{T^4}(v_{i+1,j,k+1})$, the B-coefficients corresponding to the domain points in $D_1(v_{i+1,j+1,k}) \cap T^4$ can be uniquely and stably determined. Therefore, three univariate interpolation problems have to be solved first, and afterwards the C^1 smoothness conditions at $v_{i+1,j,k+1}$ have to be used. These univariate interpolation problems correspond to the three edges of T^4 containing $v_{i+1,j,k+1}$ and can be uniquely and stably solved. Since T^4 has been subjected to a partial Worsey-Farin split, the C^1 smoothness conditions have to be used in order to computed the last undetermined B-coefficients of $D_1(v_{i+1,j+1,k}) \cap T^4$ associated with the domain point on the interior edge in T^4 with a distance of one from the vertex $v_{i+1,j,k+1}$. This B-coefficient is also uniquely and stably determined. Now, all B-coefficients of s corresponding to the minimal determining set $\mathcal{M}_1^1$ from Theorem 3.3 for the tetrahedron T^4 are determined. Thus, all remaining undetermined B-coefficients of $s|_{T^4}$ can be uniquely and stably computed. It is easy to see, that the computation of the B-coefficients of $s|_{T^4}$ is local, since they only depend on the values associated with the interpolation points in $\mathcal{L}_4 \cap (T^1 \cup T^2 \cup T^3 \cup T^4)$.

Now, since $s|_{T^4}$ is uniquely determined, the remaining unknown B-coefficients of s corresponding to domain points within a distance of one from the edges and vertices of T^4 can be uniquely and stably determined using the C^1 smoothness conditions. The computation of these B-coefficients is local, since they only depend on the values associated with the interpolation points $\mathcal{L}_4 \cap (T^1 \cup T^2 \cup T^3 \cup T^4)$.

Now, s restricted to T^5, the last tetrahedron in $Q_{i,j,k}$, is examined. Following Algorithm 4.8 T^5 has been subjected to a fourth order partial Worsey-Farin split, also just called Worsey-Farin split. From the construction of the type-4 partition it can be seen that T^5 shares exactly one common face with each of the tetrahedra T^1, T^2, T^3, and T^4 (see also Notation 4.6 and Figure 4.5). Since T^5 is subjected to a fourth order partial Worsey-Farin split, each face of T^5 is divided with a Clough-Tocher split. Though at this time only the corresponding faces in the tetrahedra T^3 and T^4 are split. The two faces of T^5 which are in common with T^1 and T^2 are not split in these tetrahedra. Thus, since the spline s restricted to these two tetrahedra is already known, the de Casteljau algorithm can be used to subdivide the spline on these tetrahedra with a first order partial Worsey-Farin split. Then, using the C^1 smoothness conditions at the faces of T^5, all B-coefficients of $s|_{T^5}$ within a distance of one from the faces of T^5 can be uniquely and stably determined. Therefore, all B-coefficients corresponding to the minimal determining set $\mathcal{M}_4^1$ from Theorem 3.3 for the tetrahedron T^5 are uniquely and stably determined. Then, all other B-coefficients of $s|_{T^5}$ can be uniquely and stably computed from smoothness conditions. The computation of the B-coefficients of s restricted to T^5 is local, since the B-coefficients of $s|_{T^5}$ only depend on the values $\{z_\kappa\}_{\kappa \in \mathcal{L}_4 \cap (T^1 \cup T^2 \cup T^3 \cup T^4)}$ (in Algorithm 4.8 no interpolation points are chosen in T^5).

Now, the spline s is uniquely and stably determined on all cubes in class $\mathcal{K}_0$. Since the B-coefficients of s corresponding to the domain points in the disks with radius one around the vertices of each cube in class $\mathcal{K}_0$ are also already determined, and since each vertex of $\diamond$ is a vertex of a cube in class $\mathcal{K}_0$ (cf. statement (v) in Lemma 4.1), s is already uniquely and stably determined on all edges of Δ_4^*. Thus, to compute the remaining undetermined B-coefficients of s, it suffices to consider the faces of the tetrahedra in Δ_4.

Step 2: $s|_{\mathcal{K}_1 \cup \mathcal{K}_2 \cup \mathcal{K}_3}$

Now, s restricted to the tetrahedra in the cubes in class $\mathcal{K}_1$, $\mathcal{K}_2$, and $\mathcal{K}_3$ is considered. The computation of the B-coefficients of s on these cubes is quite similar to the way the B-coefficients of s restricted to the cubes in class $\mathcal{K}_0$ are determined. The spline is first considered on the cubes in class $\mathcal{K}_1$, then on those in class $\mathcal{K}_2$ and at last on those in class $\mathcal{K}_3$. Following statement (i) in Lemma 4.1, two cubes in class $\mathcal{K}_i$, $i = 1,2,3$, are disjoint. Thus, it is possible to determine s in the same way on each of the cubes in class $\mathcal{K}_i$, $i = 1,2,3$, respectively. Moreover, since s is already uniquely and stably determined on the edges of Δ_4^*, it suffices to examine the faces of the tetrahedra in the cubes in class $\mathcal{K}_1$, $\mathcal{K}_2$, and $\mathcal{K}_3$, which can be performed in the same way for each of these classes. Therefore, let $\widetilde{Q}_{i,j,k}$ be a cube in one of the classes $\mathcal{K}_1$, $\mathcal{K}_2$, or $\mathcal{K}_3$ and let $\widetilde{T}^l := \widetilde{T}^l_{i,j,k}$, $l = 1,\dots,5$, be the five tetrahedra in $\widetilde{Q}_{i,j,k}$ (cf. Notation 4.6).

First, s is determined on the tetrahedron $\widetilde{T}^1$. Therefore, for each face $\widetilde{F}$ of $\widetilde{T}^1$ four distinct cases have to be considered:

Case 1: The face $\widetilde{F}$ has no marked edges at the time it is considered in Algorithm 4.8. Then the point $v_{\widetilde{F}}$ is contained in the set $\mathcal{L}_4$, and thus, following Lemma 2.42 for $n = 3$, the remaining undetermined B-coefficients of $s|_{\widetilde{F}}$ can be uniquely and stably determined.

Case 2: The face $\widetilde{F}$ has exactly one marked edge at the time it is considered in Algorithm 4.8. Then the B-coefficients of $s|_{\widetilde{F}}$ corresponding to the domain points within a distance of one from the vertices of $\widetilde{F}$ and within a distance of one from the marked edge of $\widetilde{F}$ are already uniquely and stably determined by C^1 smoothness conditions, and thus $s|_{\widetilde{F}}$ is determined.

Case 3: The face $\widetilde{F}$ has exactly two marked edges at the time it is considered in Algorithm 4.8. Then $\widetilde{F}$ has been subjected to a Clough-Tocher split, and the B-coefficients of $s|_{\widetilde{F}}$ corresponding to the domain points within a distance of one from the vertices of $\widetilde{F}$ and within a distance of one from the two marked edges of $\widetilde{F}$ are already uniquely and stably determined by C^1 smoothness conditions. Moreover, the set $\mathcal{L}_4$ contains the split point $v_{\widetilde{F}}$ of $\widetilde{F}$. Thus, following Lemma 2.45, the remaining undetermined B-coefficients of $s|_{\widetilde{F}}$ can be uniquely and stably determined.

Case 4: The face $\widetilde{F}$ has exactly three marked edges at the time it is considered in Algorithm 4.8. In this situation, the triangle $\widetilde{F}$ has been subjected to a Clough-Tocher split, and the B-coefficients of $s|_{\widetilde{F}}$ corresponding to the domain points within a distance of one from the vertices of $\widetilde{F}$ and within a distance of one from the edges of $\widetilde{F}$ are already uniquely and stably determined by C^1 smoothness conditions. Then, the remaining B-coefficients of $s|_{\widetilde{F}}$ can be uniquely and stably determined from the C^1 smoothness conditions across the interior edges of $\widetilde{F}_{CT}$ (cf. Theorem 6.6 and [6, 59]).

If none of the faces of $\widetilde{T}^1$ are refined with a Clough-Tocher split, the tetrahedron $\widetilde{T}^1$ is not divided with partial Worsey-Farin split. Thus $s|_{\widetilde{T}^1}$ is already uniquely and stably determined. In case a total number of $m > 0$ faces of $\widetilde{T}^1$ are refined with a Clough-Tocher split, the tetrahedron $\widetilde{T}^1$ is subjected to an m-th order partial Worsey-Farin split. Then, following Theorem 3.3, the remaining undetermined B-coefficients of $s|_{\widetilde{T}^1}$ can be uniquely and stably computed from those already determined by using smoothness conditions, since all B-coefficients of $s|_{\widetilde{T}^1}$ corresponding to the minimal determining set $\mathcal{M}_m^1$ from Theorem 3.3 are known already.

The computation of the B-coefficients of $s|_{\widetilde{T}^1}$ is local, since $s|_{\widetilde{T}^1}$ is uniquely determined by the values $\{z_\kappa\}_{\kappa \in \mathcal{L}_4 \cap \widetilde{T}^1}$ and the B-coefficients of s restricted to the at most three cubes in class $\mathcal{K}_0$ sharing common vertices with $\widetilde{T}^1$ and the at most three cubes in the classes $\mathcal{K}_0, \ldots, \mathcal{K}_l$ sharing a common edge with $\widetilde{T}^1$, with $\widetilde{T}^1 \subset \widetilde{Q}_{i,j,k} \in \mathcal{K}_l$ (cf. Lemma 4.1 and Figures 4.3, 4.1, and 4.2). Thus, if $\widetilde{Q}_{i,j,k}$ is a cube in class $\mathcal{K}_1$, the B-coefficients of $s|_{\widetilde{T}^1}$ only depend on the values associated with the interpolation points of the set $\mathcal{L}_4$ in the tetrahedron $\widetilde{T}^1$ and in the three cubes $\widetilde{Q}_{i-1,j-1,k}$, $\widetilde{Q}_{i+1,j-1,k}$, and $\widetilde{Q}_{i-1,j+1,k}$, within a distance of one from $\widetilde{Q}_{i,j,k}$. For the case $\widetilde{T}^1 \subset \widetilde{Q}_{i,j,k} \in \mathcal{K}_2$ is in the interior of $\diamond$, the spline s restricted to the tetrahedron $\widetilde{T}^1$ depends on the values at the interpolation points of the set $\mathcal{L}_4$ contained in the tetrahedron $\widetilde{T}^1$ and in six cubes in lower classes within a distance of two from $\widetilde{Q}_{i,j,k}$. If $\widetilde{T}^1 \subset \widetilde{Q}_{i,j,k} \in \mathcal{K}_2$ lies on the boundary of Δ_4^* the spline $s|_{\widetilde{T}^1}$ only depends on the values corresponding to the interpolation points of the set $\mathcal{L}_4$ contained in the tetrahedron $\widetilde{T}^1$ and in the three cubes $\widetilde{Q}_{i-1,j,k-1}$, $\widetilde{Q}_{i-1,j,k+1}$, and $\widetilde{Q}_{i+1,j,k-1}$, within a distance of one from $\widetilde{Q}_{i,j,k}$. For the case that $\widetilde{T}^1 \subset \widetilde{Q}_{i,j,k} \in \mathcal{K}_3$ is in the interior of $\diamond$, the B-coefficients of $s|_{\widetilde{T}^1}$ depend on the values corresponding to the interpolation points of the set $\mathcal{L}_4$ contained in the tetrahedron $\widetilde{T}^1$ and in 18 cubes in lower classes within a distance of three from $\widetilde{Q}_{i,j,k}$. If $\widetilde{T}^1 \subset \widetilde{Q}_{i,j,k} \in \mathcal{K}_3$ is on the boundary of Δ_4^* our near the boundary, the spline $s|_{\widetilde{T}^1}$ is influenced by less interpolation points, depending on the distance to the boundary of Δ_4^*.

Restricted to the tetrahedra $\widetilde{T}^l$, $l = 2,3,4$, the spline s can be determined in the same way as on the tetrahedron $\widetilde{T}^1$. Then $s|_{\widetilde{T}^l}$, $l = 2,3,4$, also depends on the values of the interpolation points in $\mathcal{L}_4$ contained in the tetrahedra $\widetilde{T}^1, \ldots, \widetilde{T}^l$ and the B-coefficients of s restricted to the other cubes in lower classes touching $\widetilde{Q}_{i,j,k}$. Thus, if $\widetilde{Q}_{i,j,k}$ is a cube in class $\mathcal{K}_1$, the B-coefficients of $s|_{\widetilde{T}^1 \cup \ldots \cup \widetilde{T}^4}$ depend on the values associated with the interpolation points of $\mathcal{L}_4$ in the four cubes in class $\mathcal{K}_0$ with a distance of one from $\widetilde{Q}_{i,j,k}$. For the case that $\widetilde{Q}_{i,j,k} \in \mathcal{K}_2$ is the interior of $\diamond$, the spline $s|_{\widetilde{T}^1 \cup \ldots \cup \widetilde{T}^4}$ depends on the values corresponding to the interpolation points of $\mathcal{L}_4$ in the 16 cubes in class $\mathcal{K}_0$ and $\mathcal{K}_1$ within a distance of two from $\widetilde{Q}_{i,j,k}$. If $\widetilde{Q}_{i,j,k}$ is a cube in class $\mathcal{K}_3$ in the interior of $\diamond$, the B-coefficients of $s|_{\widetilde{T}^1\cup\ldots\cup\widetilde{T}^4}$ depend on the values of the interpolation points of $\mathcal{L}_4$ in 40 cubes in the classes $\mathcal{K}_0, \mathcal{K}_1$, and $\mathcal{K}_2$ within a distance of three from $\widetilde{Q}_{i,j,k}$. If $\widetilde{Q}_{i,j,k}$ is a cube on or near the boundary of $\diamond$, then the spline $s|_{\widetilde{T}^1\cup\ldots\cup\widetilde{T}^4}$ is influenced by values associated with interpolation points in less cubes.

As the last tetrahedron in $\widetilde{Q}_{i,j,k}$, the spline s is determined on $\widetilde{T}^5$. Following Algorithm 4.8, the tetrahedron $\widetilde{T}^5$ has been subjected to a fourth order partial Worsey-Farin split. Moreover, $\widetilde{T}^5$ shares exactly one triangular face with each of the tetrahedra $\widetilde{T}^1$, $\widetilde{T}^2$, $\widetilde{T}^3$, and $\widetilde{T}^4$. The tetrahedra $\widetilde{T}^3$ and $\widetilde{T}^4$ have already been refined with a partial Worsey-Farin split, such that the common face with $\widetilde{T}^5$ is also subjected to a Clough-Tocher split. In the tetrahedra $\widetilde{T}^1$ and $\widetilde{T}^2$, these faces have not been subjected to a Clough-Tocher split yet. Thus, the spline s restricted to the two tetrahedra $\widetilde{T}^1$ and $\widetilde{T}^2$ has to be subdivided with a partial Worsey-Farin split using the de Casteljau algorithm, such that these faces are refined with a Clough-Tocher split. Then, all B-coefficients of $s|_{\widetilde{T}^5}$ corresponding to the domain points within a distance of one from the faces of $\widetilde{T}^5$ can be uniquely and stably determined using the C^1 smoothness conditions. Thus, also all B-coefficients of the minimal determining set $\mathcal{M}_4^1$ from Theorem 3.3 for this tetrahedron are determined and the remaining B-coefficients of $s|_{\widetilde{T}^5}$ can be uniquely and stably determined as shown in the proof of Theorem 3.3. The computation of these B-coefficients of s is local, since $s|_{\widetilde{T}^5}$ depends on the same values of interpolation points as the other four tetrahedra in the cube $\widetilde{Q}_{i,j,k}$.

Step 3: $s|_{\mathcal{K}_4}$

Finally, s is determined on the cubes in class $\mathcal{K}_4$. Here, it also suffices to consider the triangular faces of the tetrahedra contained in the cubes in class $\mathcal{K}_4$ (cf. statement (v) in Lemma 4.1). Therefore, the spline s can be determined on the cubes in class $\mathcal{K}_4$ in a similar way as on the cubes in the classes considered previously. Let $\hat{Q}_{i,j,k}$ be a cube in class $\mathcal{K}_4$ and let $\hat{T}^l := \hat{T}_{i,j,k}$, $l = 1,\ldots,5$, be the five tetrahedra in $\hat{Q}_{i,j,k}$ (cf. Notation 4.6).

First, the faces of $\hat{T}^1$ are examined in order to determine $s|_{\hat{T}^1}$. In contrast to the way $s|_{\widetilde{T}^1}$ is determined in the cubes in class $\mathcal{K}_1$, $\mathcal{K}_2$, and $\mathcal{K}_3$ in the previous step, only three cases have to be distinguished here for each face $\hat{F}$ of $\hat{T}^1$:

Case 1: The face $\hat{F}$ has exactly one marked edge at the time it is considered in Algorithm 4.8. Then, by the C^1 smoothness conditions, the B-coefficients of $s|_{\hat{F}}$ corresponding to the domain points within a distance of one from the vertices of $\hat{F}$ and within a distance of one from the marked edge of $\hat{F}$ are already uniquely and stably determined. Therefore, $s|_{\hat{F}}$ is determined.

Case 2: The face $\hat{F}$ has exactly two marked edges at the time it is considered in Algorithm 4.8. In this case $\hat{F}$ has been subjected to a Clough-Tocher split, and the B-coefficients of $s|_{\hat{F}}$ corresponding to the domain points within a distance of one from the vertices of $\hat{F}$ and within a distance of one from the two marked edges of $\hat{F}$ are already uniquely and stably determined by C^1 smoothness conditions. Furthermore, the split point $v_{\hat{F}}$ of $\hat{F}$ is contained in the set $\mathcal{L}_4$ of interpolation points. Therefore, by Lemma 2.45, the remaining undetermined B-coefficients of $s|_{\hat{F}}$ can be uniquely and stably determined.

Case 3: The face $\hat{F}$ has exactly three marked edges at the time it is considered in Algorithm 4.8. Then, $\hat{F}$ has been subjected to a Clough-Tocher split. If $\hat{T}^1$ has a neighboring tetrahedron sharing the face $\hat{F}$, on which the spline s is already uniquely determined, the de Casteljau algorithm can be used to subdivide the spline on this tetrahedron with a partial Worsey-Farin split, such that $\hat{F}$ is also refined with a Clough-Tocher split in this tetrahedron. In this case all B-coefficients of $s|_{\hat{F}}$ are already uniquely and stably determined. If there is no such neighboring tetrahedron sharing the face $\hat{F}$ with $\hat{T}^1$, the B-coefficients of $s|_{\hat{F}}$ corresponding to the domain points within a distance of one from the vertices of $\hat{F}$ and within a distance of one from the edges of $\hat{F}$ are already uniquely and stably determined by C^1 smoothness conditions. Thus, the remaining undetermined B-coefficients of $s|_{\hat{F}}$ can be uniquely and stably computed by the C^1 smoothness conditions across the interior edges of $\hat{F}_{CT}$ (cf. Theorem 6.6 and [6, 59]).

Note that for the case that $\hat{Q}_{i,j,k}$ lies in the interior of $\diamond$ the tetrahedron $\hat{T}^1$ shares common faces with exactly three tetrahedra on which the spline s is already uniquely determined. Thus, the spline s can be determined in the way described in case three on all four faces of $\hat{T}^1$.

Since at least one face of $\hat{T}^1$ is refined with a Clough-Tocher split, let m be the number faces of $\hat{T}^1$ that are refined with a Clough-Tocher split. It can easily be seen that $m \geq 1$, see Table 4.3. Then, the tetrahedron $\hat{T}^1$ has been subjected to an m-th order partial Worsey-Farin split and the remaining undetermined B-coefficients of $s|_{\hat{T}^1}$ can be uniquely and stably determined by smoothness conditions in the way described in the proof of Theorem 3.3, since already all B-coefficients corresponding to the minimal determining set $\mathcal{M}_m^1$ from Theorem 3.3 are uniquely determined.

The computation of the B-coefficients of $s|_{\hat{T}^1}$ is local, since the spline $s|_{\hat{T}^1}$ is uniquely determined by the values corresponding to the interpolation points of $\mathcal{L}_4$ contained in $\hat{T}^1$ and the B-coefficients of s restricted to the two cubes in class $\mathcal{K}_0$, sharing a common vertex and a common face with $\hat{T}^1$, respectively, and the at most two cubes in class $\mathcal{K}_l$, $l = 1,2,3$, sharing a common edge or face with $\hat{T}^1$ (cf. Lemma 4.1 and Figures 4.1, 4.2, and 4.3). Note that the spline $s|_{\hat{T}^1}$ does not depend on the B-coefficients of s restricted to other cubes in class $\mathcal{K}_4$ sharing an edge with $\hat{T}^1$, since there are always two other cubes in the classes $\mathcal{K}_0,\ldots,\mathcal{K}_3$ sharing the same edge with $\hat{T}^1$ (cf. statement (iv) in Lemma 4.1 and Figure 4.4).

The spline s on the tetrahedra $\hat{T}^l$, $l = 2,\ldots,5$, can be determined in the same way as on the tetrahedron $\hat{T}^1$. Then, $s|_{\hat{T}^l}$, $l = 2,\ldots,5$, also depends on the values of the interpolation points in $\mathcal{L}_4$ contained in the tetrahedra $\hat{T}^1,\ldots,\hat{T}^l$ and the B-coefficients of s restricted to the cubes in the classes $\mathcal{K}_0,\ldots,\mathcal{K}_3$ touching $\hat{Q}_{i,j,k}$. Note that no interpolation points have to be chosen in $\hat{Q}_{i,j,k}$ in case the cube is situated in the interior of $\diamond$. Now, for the case that the indices i and j of $\hat{Q}_{i,j,k}$ are even and k is odd, the spline s restricted to $\hat{Q}_{i,j,k}$ depends on the values of the interpolation points of $\mathcal{L}_4$ in at most 52 cubes in lower classes contained in the collection of $5 \times 9 \times 3$ cubes, where $\hat{Q}_{i,j,k}$ is in the center of this collection. In case that the indices i and k of $\hat{Q}_{i,j,k}$ are even and j is odd, $s|_{\hat{Q}_{i,j,k}}$ is influenced by the values of the interpolation points of $\mathcal{L}_4$ in at most 62 cubes in lower classes contained in the set of $5 \times 7 \times 5$ cubes, where $\hat{Q}_{i,j,k}$ is the cube in the middle of this set. For the case that the two indices j and k of the cube $\hat{Q}_{i,j,k}$ are even and the index i is odd, the spline $s|_{\hat{Q}_{i,j,k}}$ can be determined from the values of the interpolation points of $\mathcal{L}_4$ in at most 24 cubes in lower classes, which are contained in the cuboid formed by $3 \times 5 \times 5$ cubes, where $\hat{Q}_{i,j,k}$ is in the center of the cuboid. In the last case, where all three indices of the cube $\hat{Q}_{i,j,k}$ are odd, the spline s restricted to $\hat{Q}_{i,j,k}$ depends on the values of the interpolation points of $\mathcal{L}_4$ in at most 56 cubes in lower classes contained in the collection of $7 \times 7 \times 3$ cubes, where $\hat{Q}_{i,j,k}$ is in the center of this collection of cubes. Note that, if $\hat{Q}_{i,j,k}$ is a cube on or near the boundary of $\diamond$ the spline $s|_{\hat{Q}_{i,j,k}}$ is influenced by less cubes in lower classes. □

Theorem 4.10:
Let $\mathcal{L}_4$ and Δ_4^* be the set of points and the refined tetrahedral partition of a type-4 partition based on $\diamond$ obtained by Algorithm 4.8, respectively. Then the set $\mathcal{N}_4 := \{\epsilon_\kappa\}_{\kappa \in \mathcal{L}_4}$ is a local and stable nodal minimal determining set for $\mathcal{S}_3^1(\Delta_4^*)$.

Proof:
The proof of Theorem 4.9 shows that each spline $s \in \mathcal{S}_3^1(\Delta_4^*)$ is uniquely and stably determined by the values $s(\kappa)$, $\kappa \in \mathcal{L}_4$. Since ϵ_κ denotes the point-evaluation at κ, these values are given by $\mathcal{N}_4$. Moreover, following the proof of Theorem 4.9, it can be seen that $\mathcal{N}_4$ is a local nodal minimal determining set. □

Remark 4.11:
Since the set $\mathcal{L}_4$ of interpolation points, defined in section 4.3, only contains domain points, it is easy to see from the proof of Theorem 4.9 that $\mathcal{L}_4$ is also a minimal determining set (cf. subsection 2.3.5).

Remark 4.12:
Note that for cubes in the interior of a cube partition $\diamond$ there are only eight different ways a spline $s \in \mathcal{S}_3^1(\Delta_4^*)$ on this cube is determined, according to the class this cube is in. Moreover, since only smaller linear systems are needed to determine s on each individual tetrahedron in a cube, the computation of a spline $s \in \mathcal{S}_3^1(\Delta_4^*)$ can be done quite fast.

4.5 Bounds on the error of the interpolant

In his section, the error $\|f - \mathcal{I}f\|$ of a smooth function f and the corresponding interpolant $\mathcal{I}f$, constructed in this chapter, is considered, where the error is measured in the maximum norm on Ω. Therefore, let $\mathcal{L}_4$ be the Lagrange interpolation set for the spline space $\mathcal{S}_3^1(\Delta_4^*)$ constructed in section 4.3. Then, for each function $f \in C(\Omega)$ there is a unique spline $\mathcal{I}f \in \mathcal{S}_3^1(\Delta_4^*)$, with

$$\mathcal{I}f(\kappa) = f(\kappa), \ \forall\, \kappa \in \mathcal{L}_4. \tag{4.1}$$

Therefore, a linear projector $\mathcal{I}$ mapping $C(\Omega)$ onto $\mathcal{S}_3^1(\Delta_4^*)$ is defined. Now, for a compact set $B \subseteq \Omega$ and an integer $m \geq 1$, let $W_\infty^m(B)$ be the Sobolev space defined on the set B and let

$$|f|_{m,B} := \sum_{|\alpha|=m} \|D^\alpha f\|_B,$$

where $\|\cdot\|_B$ denotes the maximum norm on B and D^α is defined as $D^\alpha := D_x^{\alpha_1} D_y^{\alpha_1} D_z^{\alpha_3}$ for $\alpha = (\alpha_1, \alpha_2, \alpha_3)$. Moreover, let $|\Delta_4^*|$ be the mesh size of Δ_4^*, i.e. the maximum diameter of the tetrahedra in Δ_4^*.

Theorem 4.13:
For each $f \in W_\infty^{m+1}(\Omega)$, $0 \le m \le 3$, there exists an absolute constant K, such that

$$\|D^\alpha(f - \mathcal{I}f)\|_\Omega \le K |\Delta_4^*|^{m+1-|\alpha|} |f|_{m+1,\Omega} \tag{4.2}$$

holds for all tuples α with $0 \le |\alpha| \le m$.

Proof:
Fix m, and let $f \in W_\infty^{m+1}$ for some $0 \le m \le 3$. Moreover, let T be a tetrahedron in Δ_4^* in a cube Q and let Ω_T be the set of tetrahedra contained in the collection of $7 \times 9 \times 5$ cubes, where Q is in the center of Ω_T. Following Lemma 4.3.8 of [19], a cubic polynomial p exists such that

$$\|D^\beta(f - p)\|_{\Omega_T} \le K_1 |\Omega_T|^{m+1-|\beta|} |f|_{m+1,\Omega_T}, \tag{4.3}$$

for all $0 \le |\beta| \le m$, where $|\Omega_T|$ denotes the diameter of Ω_T and K_1 is an absolute constant. Since $\mathcal{I}p = p$, it follows that

$$\|D^\alpha(f - \mathcal{I}f)\|_T \le \|D^\alpha(f - p)\|_T + \|D^\alpha \mathcal{I}(f - p)\|_T.$$

Following equation (4.3) for $\beta = \alpha$, it suffices to estimate the second term $\|D^\alpha \mathcal{I}(f - p)\|_T$. In the proof of Theorem 4.9 it is shown, that the constructed Lagrange interpolation method is stable in the sense that

$$|c_\xi| \le K_2 \max_{\kappa \in \Omega_T} |z_\kappa|$$

holds for each B-coefficients c_ξ of a spline $s \in \mathcal{S}_3^1(\Delta_4^*)$, with an absolute constant K_2. Thus, also

$$\|\mathcal{I}f\|_T \le K_3 \|f\|_{\Omega_T} \tag{4.4}$$

holds, with an absolute constant K_3. Then, by the Markov inequality, see [102], and (4.4)

$$\|D^\alpha \mathcal{I}(f-p)\|_T \le K_4|T|^{-|\alpha|}\|\mathcal{I}(f-p)\|_T \le K_5|T|^{-|\alpha|}\|f-p\|_{\Omega_T},$$

where $|T|$ is the diameter of T.

Now, due to the geometry of the partition considered here, two absolute constants exist, with $|\Omega_T| \le K_6|T|$ and $|T| \le K_7|\Delta_4^*|$. With these estimations and by using (4.3) with $\beta = 0$ to estimate $\|(f-p)\|_{\Omega_T}$,

$$\|D^\alpha(f-\mathcal{I}f)\|_T \le K_8|\Omega_T|^{m+1-|\alpha|}|f|_{m+1,\Omega_T}$$

follows. Then, taking the maximum over all tetrahedra in Δ_4^* leads to the estimation (4.2). □

Remark 4.14:
Following Theorem 4.13, it can be seen that the Lagrange interpolation method constructed in this chapter yields optimal approximation order of four (cf. subsection 2.4.3).

4.6 Numerical tests and visualizations

In this section, the local Lagrange interpolation method constructed in this chapter is illustrated. Therefore, the Marschner-Lobb test function

$$p(x,y,z) := \frac{1-\sin(\pi\frac{z}{2}) + \alpha(1+\rho_r(\sqrt{x^2+y^2}))}{2(1+\alpha)},$$

with $\rho_r := cos(2\pi f_M cos(\pi\frac{r}{2}))$, where $f_M = 6$ and $\alpha = 0.25$, is used (cf. [65]). The function is interpolated on $\Omega = [[-0.1, 0.9], [-0.1, 0.9], [-0.6, 0.4]]$.

4.6.1 Numerical tests

In this subsection, numerical tests for the Lagrange interpolation method constructed in this chapter are considered. Table 4.4 shows the results of the error for the interpolation on the increasingly refined domain Ω. The error of the interpolation with a spline $s \in \mathcal{S}_3^1(\Delta_4^*)$ is computed at 8 points per edge, 36 points per subface and 56 points per subtetrahedron. Thereby, n denotes the number of cubes, the dimension of the spline space is shown in the column headed by "dim", the error of interpolation on the edges is shown in the column headed by "Error$_E$", the error on the faces is denoted by "Error$_F$", the error in the interior of the tetrahedra with "Error$_T$" and the decay exponent is shown in the column headed by "decay". The tests confirm the results shown in the previous section, that the Lagrange interpolation method yields optimal approximation order.

n	dim	Error$_E$	Error$_F$	Error$_T$	decay
128^3	$1.70 \cdot 10^7$	$4.84 \cdot 10^{-5}$	$2.08 \cdot 10^{-4}$	$2.31 \cdot 10^{-4}$	-
256^3	$1.35 \cdot 10^8$	$3.07 \cdot 10^{-6}$	$1.11 \cdot 10^{-5}$	$1.27 \cdot 10^{-5}$	4.19
512^3	$1.08 \cdot 10^9$	$1.88 \cdot 10^{-7}$	$6.02 \cdot 10^{-7}$	$6.95 \cdot 10^{-7}$	4.19
1024^3	$8.61 \cdot 10^9$	$1.18 \cdot 10^{-8}$	$3.24 \cdot 10^{-8}$	$3.59 \cdot 10^{-8}$	4.27
2048^3	$6.88 \cdot 10^{10}$	$7.36 \cdot 10^{-10}$	$2.01 \cdot 10^{-9}$	$2.25 \cdot 10^{-9}$	3.99
4096^3	$5.50 \cdot 10^{11}$	$4.43 \cdot 10^{-11}$	$1.23 \cdot 10^{-10}$	$1.41 \cdot 10^{-10}$	4.00

Table 4.4: Numerical tests for interpolation with $\mathcal{S}_3^1(\Delta_4^*)$.

Next, the errors of the interpolation obtained by the Lagrange interpolation method constructed in this chapter with exact values is compared to the errors of interpolation with a spline constructed from data of linear splines. Therefore, three different splines have to be considered: A spline $s \in \mathcal{S}_3^1(\Delta_4^*)$ constructed from exact data of the Marschner-Lobb function at all interpolation points; A linear spline $\tilde{s}_{lin} \in \mathcal{S}_1^0(\Delta_4)$ using only data at the vertices of the cubes in $\diamond$; A spline $\tilde{s} \in \mathcal{S}_3^1(\Delta_4^*)$ interpolating the Marschner-Lobb test function only at the vertices of the cubes in $\diamond$, the remaining interpolation points are obtained from the linear spline $\tilde{s}_{lin}$. Thereby, the linear spline is constructed on a cube partition that has been refined into much more cubes, in order to achieve a quite similar error for the two cubic splines in the end. The data from the linear spline are then used to construct a spline $\tilde{s} \in \mathcal{S}_3^1(\Delta_4^*)$ on the same partition as the spline s. In Table 4.5 the number of cubes for the cubic splines is denoted by n and the number of cubes for the linear spline with N.

It can be seen that much more cubes are needed in order to achieve a similar accuracy with linear splines as with the cubic splines used here. Though it can also be seen that the cubic spline constructed with the linear data has an accuracy similar to the one of the cubic spline with exact data values. The results show that by using the cubic C^1 splines constructed

n	Error s	N	Error $\widetilde{s}_{lin}$	n	Error $\widetilde{s}$
256^3	$1{,}27 \cdot 10^{-5}$	5632^3	$2{,}72 \cdot 10^{-6}$	256^3	$3{,}57 \cdot 10^{-5}$
360^3	$3{,}86 \cdot 10^{-6}$	10080^3	$8{,}39 \cdot 10^{-7}$	360^3	$4{,}15 \cdot 10^{-6}$
512^3	$6{,}95 \cdot 10^{-7}$	22528^3	$1{,}67 \cdot 10^{-7}$	512^3	$7{,}98 \cdot 10^{-7}$

Table 4.5: Comparison with linear splines and splines from linear data.

in this chapter compared to spaces of linear splines which yield approximately the same errors, a data reduction up to the factor of 10^3 can be gained. Therefore, see Table 4.6, where again n is the number of cubes for the cubic splines and N the number of cubes for the linear splines, respectively. In this table, also the quotient of the dimension of the linear spline and the cubic spline is listed.

n	dimension cubic spline	N	dimension linear spline	quotient
256^3	135 399 430	1600^3	4 096 000 000	30,25
360^3	375 583 686	4600^3	97 336 000 000	259,16
512^3	1 078 464 518	10800^3	1 259 712 000 000	1 168,06

Table 4.6: Data reduction for linear splines.

4.6.2 Visualizations

In the following, the applications of the Lagrange interpolation method to the Marschner-Lobb test function are illustrated. Therefore, isosurfaces of the Marschner-Lobb function with a value of 0.5 are visualized with a cubic spline on a partition $\diamond$ consisting of $150 \times 150 \times 150$ and $200 \times 200 \times 200$ cubes (see Figures 4.9 (left) and 4.10 (left)). In order to show the smoothness of the interpolant in detail, enlargements of smaller sections of the isosurfaces are illustrated (see Figures 4.9 (right) and 4.10 (right)). In Figures 4.11 and 4.12 the error of the interpolation is shown. Thereby, a correct value is shown in green and the error is marked red, where the intensity of the red color shows the corresponding inaccuracy at this point.

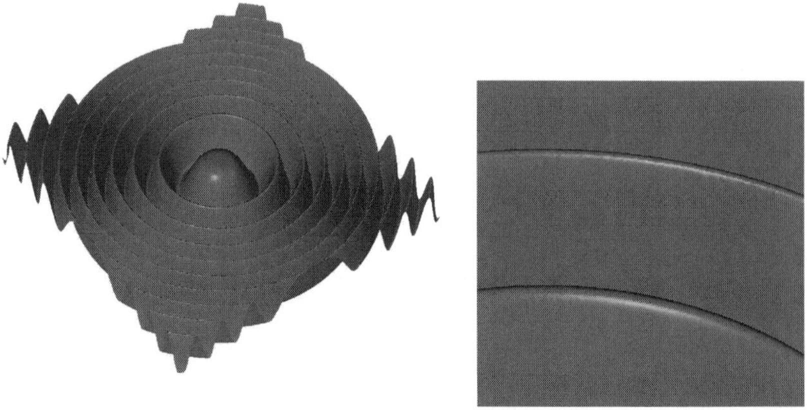

Figure 4.9: Isosurface with value 0.5 of the Marschner-Lobb test function and enlarged section with $150 \times 150 \times 150$ cubes.

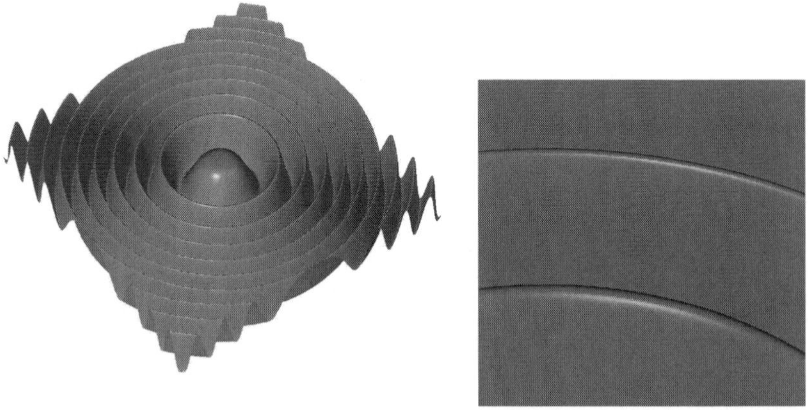

Figure 4.10: Isosurface with value 0.5 of the Marschner-Lobb test function and enlarged section with $200 \times 200 \times 200$ cubes.

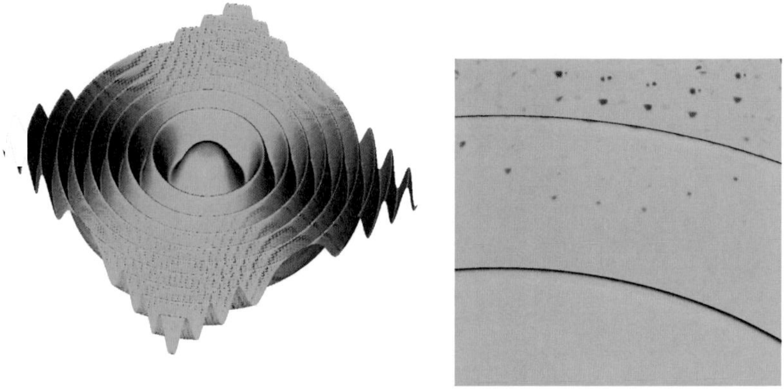

Figure 4.11: Visualization of the error of the interpolant for the isosurface with value 0.5 of the Marschner-Lobb test function and enlarged section with $150 \times 150 \times 150$ cubes.

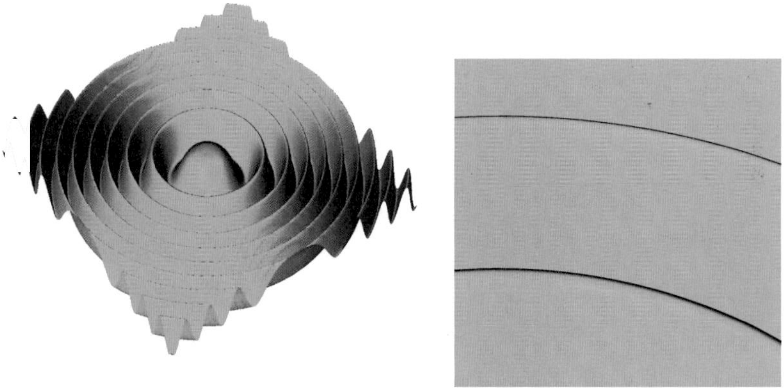

Figure 4.12: Visualization of the error of the interpolant for the isosurface with value 0.5 of the Marschner-Lobb test function and enlarged section with $200 \times 200 \times 200$ cubes.

Acknowledgment: We would like to thank Georg Schneider for performing the numerical tests and constructing the visualizations for the Lagrange interpolation method considered in this chapter.

5 Local Lagrange Interpolation with C^2 splines of degree nine on arbitrary tetrahedral partitions

In this chapter, we construct a local Lagrange interpolation method for trivariate C^2 splines of degree nine on arbitrary tetrahedral partitions. In section 5.1, we consider a decomposition of a tetrahedral partition, which forms a basis for the construction of the Lagrange interpolation method in this chapter. In the next section, we construct a refined tetrahedral partition according to the decomposition obtained in the previous section. Moreover, the spline space used for the Lagrange interpolation in this chapter, which is endowed with several additional smoothness conditions, is defined. In section 5.3, the Lagrange interpolation with the spline space and the refined partition constructed in section 5.2 is investigated. Thereby, we show that the interpolation is 11-local and stable. We also give a nodal minimal determining set for the spline space. Finally, in section 5.4, we prove that the Lagrange interpolation method constructed in this chapter yields optimal approximation order.

5.1 Decomposition of tetrahedral partitions

In this section we describe an algorithm which can be used to decompose an arbitrary tetrahedral partition Δ into classes of tetrahedra. The algorithm is based on considering the vertices, edges, and faces of Δ in order to classify the different tetrahedra. It also imposes an order for the tetrahedra in Δ, which will be used in the following sections, together with the classification, in order to define the Lagrange interpolation method considered in this chapter.

Algorithm 5.1:

1. For $i = 0,\dots 4$, repeat until no longer possible:
   Choose a tetrahedron T which has exactly i common vertices with the tetrahedra chosen earlier, but no edges or faces. Put T in class $\mathcal{A}_i$.

2. For $i = 1,\dots 6$, for $j = 2,\dots,4$, repeat until no longer possible:
   Choose a tetrahedron T which has exactly i common edges and a total number of exactly j common vertices with the tetrahedra chosen earlier, but no faces. Put T in class $\mathcal{B}_{i,j}$.

3. For $i = 1,\dots 4$, for $j = 3,\dots,6$, for $k = 3,4$, repeat until no longer possible:
   Choose a tetrahedron T which has exactly i common faces, a total number of exactly j common edges and exactly k common vertices with the tetrahedra chosen earlier. Put T in class $\mathcal{C}_{i,j,k}$.

This algorithm decomposes the partition Δ into 24 classes. Moreover, it produces an order $T_1,\dots,T_n$ of the tetrahedra in Δ. Note that for a given tetrahedral partition, the algorithm does not produce a unique decomposition, since at each step there are various choices of the first tetrahedron to consider.

Next, we consider the relations between the tetrahedra in the 24 different classes. Therefore, we need the following notation. If a tetrahedron $T_i \in \{T_1,\dots,T_n\}$ shares l common edges with the tetrahedra $T_1,\dots,T_{i-1}$, then we call the edge e_1 the first common edge if T_i shares e_1 with a tetrahedron T_j, such that T_i does not share a common edge with a tetrahedron T_k, with $k < j$. In the same way we denote the second common edge of T_i and so on. The notation is analogue for common faces.

Lemma 5.2:

(i) Any two tetrahedra in class $\mathcal{A}_0$ are disjoint.

(ii) Any two tetrahedra sharing a common face must be in different classes.

(iii) The tetrahedra in class $\mathcal{C}_{i,j,k}$, $i = 1,\dots,4$, $j = 3,\dots,6$, $k = 3,4$, must share the first common face F with a tetrahedron in $\mathcal{A}_0 \cup \dots \cup \mathcal{A}_4 \cup \mathcal{B}_{1,2} \cup \dots \cup \mathcal{B}_{j,k}$.

(iv) The tetrahedra in class $\mathcal{C}_{i,j,4}$, $i = 2,\dots,4$, $j = 5,6$, must share the second common face F with a tetrahedron in $\mathcal{A}_0 \cup \dots \cup \mathcal{A}_4 \cup \mathcal{B}_{1,2} \cup \dots \cup \mathcal{B}_{6,4} \cup \mathcal{C}_{1,3,3} \cup \dots \cup \mathcal{C}_{i-1,j,4}$.

(v) If two tetrahedra in the same class share a common edge e, then e must also be an edge of a tetrahedron in a class considered earlier.

(vi) The tetrahedra in class $\mathcal{B}_{i,j}$, $i = 1,\ldots,6$, $j = 2,3,4$, must share the first common edge e with a tetrahedron in $\mathcal{A}_0 \cup \ldots \cup \mathcal{A}_i$.

(vii) The tetrahedra in class $\mathcal{B}_{i,j}$, $i = 2,\ldots,6$, $j = 3,4$, must share the second common edge e with a tetrahedron in $\mathcal{A}_0 \cup \ldots \cup \mathcal{A}_4 \cup \mathcal{B}_{1,2} \cup \ldots \cup \mathcal{B}_{1,j}$.

(viii) The tetrahedra in class $\mathcal{B}_{i,j}$, $i = 3,\ldots,6$, $j = 3,4$, must share the third common edge e with a tetrahedron in $\mathcal{A}_0 \cup \ldots \cup \mathcal{A}_4 \cup \mathcal{B}_{1,2} \cup \ldots \cup \mathcal{B}_{2,j}$.

(ix) If two tetrahedra in the same class share a common vertex v, then v must also be a vertex of a tetrahedron in a class considered earlier.

Proof:
The first assertion is obvious, since the tetrahedra in class $\mathcal{A}_0$ have no common vertices, edges or faces at the time they are chosen.

Next, we establish (ii). The claim is obvious for the classes $\mathcal{A}_i$, $i = 0,\ldots,4$ and $\mathcal{B}_{i,j}$, $i = 1,\ldots 6, j = 2,\ldots,4$, since the tetrahedra in these classes only have common vertices or edges with those chosen earlier. Thus, the tetrahedra cannot have common faces with a tetrahedron chosen earlier. To check (ii) for the remaining classes, suppose T and $\widetilde{T}$ are two tetrahedra in the same class $\mathcal{C}_{i,j,k}$ sharing a common face, and that T was chosen before $\widetilde{T}$. Then, before T was chosen, $\widetilde{T}$ would have had exactly $i-1$ common faces with the tetrahedra chosen earlier. Thus, for $i = 1$, $\widetilde{T}$ would have been assigned to one of the classes $\mathcal{A}_i$, $i = 0,\ldots,4$, or $\mathcal{B}_{i,j}$, $i = 1,\ldots 6, j = 2,\ldots,4$, according to the number of common edges and vertices of $\widetilde{T}$ with tetrahedra chosen earlier. For $i > 1$, $\widetilde{T}$ would have been assigned to one of the classes $\mathcal{C}_{i-1,j,k}$, $j = 3,\ldots,6$, $k = 3,4$, according to the number of common edges and vertices of $\widetilde{T}$ with tetrahedra chosen earlier.

Subsequently, we prove statement (iii). Let $T \in \mathcal{C}_{i,j,k}$, $i = 1,\ldots,4$, $j = 3,\ldots,6$, $k = 3,4$, and let $\widetilde{T}$ be the first tetrahedron, according to the order imposed by Algorithm 5.1, sharing the first common face F with T. Then let $\widetilde{T}$ be in one of the classes $\widetilde{\mathcal{B}},\ldots,\widetilde{\mathcal{C}}$, where $\widetilde{\mathcal{B}}$ is the class considered immediately after $\mathcal{B}_{j,k}$ and $\widetilde{\mathcal{C}}$ is the class considered immediately before $\mathcal{C}_{i,j,k}$ in Algorithm 5.1. Then, before $\widetilde{T}$ was chosen, the tetrahedron T would have had at most j common edges and k common vertices with the tetrahedra chosen earlier. Thus, T would have been put in one of the classes $\mathcal{A}_0,\ldots,\mathcal{A}_4,\mathcal{B}_{1,2},\ldots,\mathcal{B}_{j,k}$.

Next, we establish statement (iv). Let $T \in \mathcal{C}_{i,j,4}$, $i = 2,\ldots,4$, $j = 5,6$, and let $\widetilde{T}$ be the first tetrahedron, according to the order imposed by Algorithm 5.1, sharing the second common face F with T. Now, let $\widetilde{T}$ be in one of the classes $\widetilde{\mathcal{C}}_1,\ldots,\widetilde{\mathcal{C}}_2$, where $\widetilde{\mathcal{C}}_1$ is the class considered immediately after $\mathcal{C}_{i-1,j,4}$ and $\widetilde{\mathcal{C}}_2$ is the class considered immediately before $\mathcal{C}_{i,j,4}$ in Algorithm 5.1. Then, before $\widetilde{T}$ was chosen, the tetrahedron T would have had exactly one common face, at most j common edges and four common vertices with the tetrahedra chosen earlier. Thus, T would have been put in one of the classes $\mathcal{C}_{1,3,3}\ldots,\mathcal{C}_{i-1,j,4}$.

Now, we establish (v). Therefore, we only have to consider the classes $\mathcal{B}_{1,2},\ldots,\mathcal{C}_{4,6,4}$, since the tetrahedra in the classes $\mathcal{A}_0,\ldots,\mathcal{A}_4$ do not share any edges with the tetrahedra chosen earlier. Suppose two or more tetrahedra in the same class share a common edge e, and let T and $\widetilde{T}$ be the first and second of those tetrahedra that were put in this class. First, let T and $\widetilde{T}$ be in one of the classes $\mathcal{B}_{i,j}$, $i = 1,\ldots,6$, and $j = 2,3,4$. Now, if e is not also an edge of a tetrahedron in one of the classes $\mathcal{A}_0,\ldots,\mathcal{A}_4,\mathcal{B}_{1,2},\ldots,\mathcal{B}_{i-1,j}$, then before T was chosen, $\widetilde{T}$ would have touched other tetrahedra chosen earlier at exactly $i-1$ edges. Thus, for $i = 1$, $\widetilde{T}$ would have been put in one of the classes $\mathcal{A}_{j-2},\mathcal{A}_{j-1}$, or $\mathcal{A}_j$, and for $i > 1$, $\widetilde{T}$ would have been put in one of the classes $\mathcal{B}_{i-1,j-2},\mathcal{B}_{i-1,j-1}$, or $\mathcal{B}_{i-1,j}$, according to the number of common vertices with the other tetrahedra chosen earlier, since $\widetilde{T}$ can also touch other tetrahedra at the two vertices of the edge e. Now, let T and $\widetilde{T}$ be in one of the classes $\mathcal{C}_{i,j,k}$, $i = 1,\ldots,4$, $j = 4,\ldots,6$, and $k = 3,4$. For $j = 3$, the assertion (v) is obvious, since two tetrahedra sharing a common face must be in different classes (cf. (ii)). Then, if e is not also an edge of a tetrahedron in one of the classes $\mathcal{A}_0,\ldots,\mathcal{A}_4,\mathcal{B}_{1,2},\ldots,\mathcal{B}_{6,4},\ldots,\mathcal{C}_{1,3,3},\ldots,\mathcal{C}_{i,j-1,k}$, $\widetilde{T}$ would have touched other tetrahedra chosen earlier at exactly $i-1$ edges, before T was chosen. Thus, $\widetilde{T}$ would have been put in one of the classes $\mathcal{C}_{i,j-1,k-1}$, or $\mathcal{C}_{i,j-1,k}$, according to the number of common vertices with the other tetrahedra chosen earlier.

Subsequently, we prove statement (vi). Let $T \in \mathcal{B}_{i,j}$, $i = 1,\ldots,6$, $j = 2,3,4$, and let $\widetilde{T}$ be the first tetrahedron, according to the order imposed by Algorithm 5.1, sharing the first common edge e with T. For $j = 2$ let $\widetilde{T}$ be in one of the classes $\mathcal{A}_3$ or $\mathcal{A}_4$. Then before $\widetilde{T}$ was chosen, T would have had at most two common vertices with the tetrahedra chosen earlier. Thus, T would have been put in one of the classes $\mathcal{A}_0$, $\mathcal{A}_1$, or $\mathcal{A}_2$. For $j = 3$ let $\widetilde{T}$ be in one of the classes $\mathcal{A}_4,\ldots,\widetilde{\mathcal{B}}$, where $\widetilde{\mathcal{B}}$ is the class considered immediately before $\mathcal{B}_{i,3}$ in Algorithm 5.1. Then before $\widetilde{T}$ was chosen, T would have had at most three common vertices with the tetrahedra chosen earlier. Therefore, T would have been put in one of the classes $\mathcal{A}_0$, $\mathcal{A}_1$, $\mathcal{A}_2$, or $\mathcal{A}_3$. For $j=4$ let $\widetilde{T}$ be in one of the classes $\mathcal{B}_{1,2},\ldots,\widetilde{\mathcal{B}}$, where $\widetilde{\mathcal{B}}$ is the class considered immediately before $\mathcal{B}_{i,4}$ in Algorithm 5.1. Then before $\widetilde{T}$ was chosen, T would have had at most four common vertices with the tetrahedra chosen earlier. Thus, T would have been put in one of the classes $\mathcal{A}_j$, $j=0,\ldots,4$.

Next, we establish statement (vii). Let $T\in\mathcal{B}_{i,j}$, $i=2,\ldots,6$, $j=3,4$, and let $\widetilde{T}$ be the first tetrahedron, according to the order imposed by Algorithm 5.1, sharing the second common edge e with T. For $j=3$ let $\widetilde{T}$ be in $\mathcal{B}_{1,4}\cup\ldots\cup\widetilde{\mathcal{B}}$, where $\widetilde{\mathcal{B}}$ is the class considered immediately before $\mathcal{B}_{i,3}$ in Algorithm 5.1. Then before $\widetilde{T}$ was chosen, T would have had at most one common edge and three common vertices with the tetrahedra chosen earlier. Thus, T would have been put in one of the classes $\mathcal{B}_{1,2}$ or $\mathcal{B}_{1,3}$. For $j=4$ let $\widetilde{T}$ be in class $\mathcal{B}_{2,3}\cup\ldots\cup\widetilde{\mathcal{B}}$, where $\widetilde{\mathcal{B}}$ is the class considered immediately before $\mathcal{B}_{i,4}$ in Algorithm 5.1. Then before $\widetilde{T}$ was put in one of the classes $\mathcal{B}_{2,3},\ldots,\widetilde{\mathcal{B}}$, the tetrahedron T would have had at most one common edge and four common vertices with the tetrahedra chosen earlier. Thus, T would have been put in one of the classes $\mathcal{B}_{1,2}$, $\mathcal{B}_{1,3}$, or $\mathcal{B}_{1,4}$.

Now, we prove statement (viii). Let $T\in\mathcal{B}_{i,j}$, $i=3,\ldots,6$, $j=3,4$, and let $\widetilde{T}$ be the first tetrahedron, according to the order imposed by Algorithm 5.1, sharing the third common edge e with T. For $j=3$ let $\widetilde{T}$ be in $\mathcal{B}_{2,4}\cup\ldots\cup\widetilde{\mathcal{B}}$, where $\widetilde{\mathcal{B}}$ is the class considered immediately before $\mathcal{B}_{i,3}$ in Algorithm 5.1. Then before $\widetilde{T}$ was put in one of the classes $\mathcal{B}_{2,4},\ldots,\widetilde{\mathcal{B}}$, the tetrahedron T would have had at most two common edges and three common vertices with the tetrahedra chosen earlier. Thus, T would have been put in class $\mathcal{B}_{2,3}$. For $j=4$ let $\widetilde{T}$ be in class $\mathcal{B}_{3,3}\cup\ldots\cup\widetilde{\mathcal{B}}$, where $\widetilde{\mathcal{B}}$ is the class considered immediately before $\mathcal{B}_{i,4}$ in Algorithm 5.1. Then before $\widetilde{T}$ was chosen, the tetrahedron T would have had at most two common edges and four common vertices with the tetrahedra chosen earlier. Thus, T would have been put in one of the classes $\mathcal{B}_{2,3}$ or $\mathcal{B}_{2,4}$.

Finally, we prove (ix). Suppose two or more tetrahedra in the same class touch at a vertex v, and let T and $\widetilde{T}$ be the first and second of those tetrahedra according to the order imposed by Algorithm 5.1. First, let T and $\widetilde{T}$ be in one of the classes $\mathcal{A}_i$, $i=1,\ldots,4$, since the tetrahedra in class $\mathcal{A}_0$ are disjoint they need not be considered here. Now, if v is not also a vertex of a tetrahedron in one of the classes $\mathcal{A}_0,\dots,\mathcal{A}_{i-1}$, then before T was chosen, $\widetilde{T}$ would have touched other tetrahedra chosen earlier at exactly $i-1$ vertices. Therefore, $\widetilde{T}$ would have been put in class $\mathcal{A}_{i-1}$. Next, let T and $\widetilde{T}$ be in one of the classes $\mathcal{B}_{i,j}$, $i=1,\dots,6$, and $j=3,4$. For $j=2$, (ix) is obvious, since a tetrahedron $\mathcal{B}_{1,2}$ must share an edge with a tetrahedron in one of the classes $\mathcal{A}_0,\dots,\mathcal{A}_4$ (cf. (v)) Now, if v is not also a vertex of a tetrahedron in one of the classes $\mathcal{A}_0,\dots,\mathcal{A}_4,\mathcal{B}_{1,2},\dots,\mathcal{B}_{i,j-1}$, the tetrahedron $\widetilde{T}$ would have touched other tetrahedra chosen earlier at exactly $j-1$ vertices, before T was chosen. Thus, $\widetilde{T}$ would have been put in class $\mathcal{B}_{i,j-1}$. □

5.2 Construction of Δ^* and $\mathcal{S}_9^2(\Delta^*)$

In this section we describe an algorithm to construct a refined tetrahedral partition Δ^* from an arbitrary partition Δ and we define our spline space $\mathcal{S}_9^2(\Delta^*)$, which is endowed with several additional smoothness conditions. In the algorithm an arbitrary tetrahedral partition Δ is refined by splitting some of the tetrahedra with partial Worsey-Farin splits (see Definition 2.8) according to the number of common edges with tetrahedra considered earlier. For the construction of the spline space $\mathcal{S}_9^2(\Delta^*)$, which is used for the Lagrange interpolation in section 5.3, we enforce some additional smoothness conditions across some of the vertices, edges and faces of the refined partition Δ^*, in order to remove some unnecessary degrees of freedom, and thus lower the dimension of the spline space and facilitate the corresponding Lagrange interpolation set. These additional smoothness conditions were already considered in section 3.2.

For our construction, the split points v_F, used to create the Clough-Tocher splits of the faces of Δ, must be chosen in the following way:

- if F is a boundary face of Δ, choose v_F to be the barycenter of F,
- if F is an interior face of Δ, choose v_F to be the intersection of F with the line connecting the incenters of the two tetrahedra sharing F.

Now, we are ready to describe the refinement algorithm, which proceeds in two steps. In the first step we apply a Clough-Tocher split to some faces of Δ according to the number of edges in common with tetrahedra considered earlier. Therefore, the tetrahedra are regarded in the order $T_1,\dots,T_n$ imposed by Algorithm 5.1. In the second step we apply partial Worsey-Farin splits to the tetrahedra of Δ that have one or more split faces.

Algorithm 5.3:

1. For $i = 1, \ldots, n$, if F is a face of T_i that is not shared with any tetrahedron T_j with $j < i$, apply a Clough-Tocher split to F when either two or three of the edges of F are in common with tetrahedra considered earlier.

2. For each tetrahedron $T \in \Delta$, let m be the number of its faces that have been divided with a Clough-Tocher split. If $m > 0$, apply an m-th order partial Worsey-Farin split to T using its incenter.

The resulting tetrahedral partition is called Δ^*.

The types of splits that may be applied to the tetrahedra in the different classes are listed in table 5.1, where m-WF stands for the m-th order partial Worsey-Farin split. Since the tetrahedra in the classes $\mathcal{A}_0, \ldots, \mathcal{A}_4$ can only have common vertices with the tetrahedra considered earlier, but no edges or faces, and are therefore not split, they are omitted in this table. The symbol "-" indicates that the corresponding split is not applied to the tetrahedra in the indicated class. The cases where a tetrahedron T_i shares an already split face with a tetrahedron T_j, $j < i$, are indicated with the symbol "o". The cases where a tetrahedron has no split face in common with a tetrahedron considered earlier are identified with the symbol "x".

In the following we will define the spline space $\mathcal{S}_9^2(\Delta^*)$. Therefore, we need some more notation. Let $\mathcal{V}$ be the set of vertices and $\mathcal{E}$ the set of edges of the initial tetrahedral partition Δ. Moreover, let $\mathcal{V}_c$ be the set of incenters of the tetrahedra in Δ that have been subjected to a partial Worsey-Farin split in algorithm 5.3 and $\mathcal{E}_c$ the set of edges connecting the incenter in these tetrahedra with the split points on the subdivided faces.
Then let $S_9^{2,3,4,7}(\Delta^*)$ be the space of supersplines defined on Δ^*, with

$$\begin{aligned} S_9^{2,3,4,7}(\Delta^*) := \{s \in S_9^2(\Delta) : & s \in C^3(e) \text{ for all } e \in \mathcal{E}, \\ & s \in C^7(e) \text{ for all } e \in \mathcal{E}_c, \\ & s \in C^4(v) \text{ for all } v \in \mathcal{V}, \\ & s \in C^7(v) \text{ for all } v \in \mathcal{V}_c\}. \end{aligned}$$

	no split	1-WF	2-WF	3-WF	4-WF
$\mathcal{B}_{1,2}$	x	-	-	-	-
$\mathcal{B}_{1,3}$	x	-	-	-	-
$\mathcal{B}_{1,4}$	x	-	-	-	-
$\mathcal{B}_{2,3}$	-	x	-	-	-
$\mathcal{B}_{2,4}$	x	x	-	-	-
$\mathcal{B}_{3,3}$	-	x	-	-	-
$\mathcal{B}_{3,4}$	-	x	x	x	-
$\mathcal{B}_{4,4}$	-	-	-	x	x
$\mathcal{B}_{5,4}$	-	-	-	-	x
$\mathcal{B}_{6,4}$	-	-	-	-	x
$\mathcal{C}_{1,3,3}$	x	o	-	-	-
$\mathcal{C}_{1,3,4}$	x	o	-	-	-
$\mathcal{C}_{1,4,4}$	-	-	x	o	-
$\mathcal{C}_{1,5,4}$	-	-	-	x	o
$\mathcal{C}_{1,6,4}$	-	-	-	x	o
$\mathcal{C}_{2,5,4}$	-	-	x	o	o
$\mathcal{C}_{2,6,4}$	-	-	x	o	o
$\mathcal{C}_{3,6,4}$	-	-	o	o	o
$\mathcal{C}_{4,6,4}$	-	-	o	o	o

Table 5.1: Possible splits for tetrahedra in the different classes.

We also need a special set of additional smoothness conditions:

Definition 5.4:
Let Θ be the set of additional smoothness conditions, such that for each tetrahedron $T \in \Delta$ in class

- $\mathcal{B}_{2,3}$ the corresponding additional smoothness conditions from the spline space $\hat{\mathcal{S}}_9^2(T_{WF}^1)$ in Theorem 3.9 and those in Lemma 3.11
- $\mathcal{B}_{2,4}$, where T is refined with a first order partial Worsey-Farin split, the corresponding additional smoothness conditions from the spline space $\hat{\mathcal{S}}_9^2(T_{WF}^1)$ in Theorem 3.9 and those in Lemma 3.10
- $\mathcal{B}_{3,3}$ the corresponding additional smoothness conditions from the spline space $\widetilde{\mathcal{S}}_9^2(T_{WF}^1)$ in Theorem 3.4 and those in Lemma 3.8

- $\mathcal{B}_{3,4}$, where T is refined with a
  - first order partial Worsey-Farin split, the corresponding additional smoothness conditions from the spline space $\widetilde{\mathcal{S}}_9^2(T_{WF}^1)$ in Theorem 3.4 and those in Lemma 3.7
  - second order partial Worsey-Farin split, the corresponding additional smoothness conditions from the spline space $\widetilde{\mathcal{S}}_9^2(T_{WF}^2)$ in Theorem 3.12 and those in Lemma 3.13
  - third order partial Worsey-Farin split, the corresponding additional smoothness conditions from the spline space $\widetilde{\mathcal{S}}_9^2(T_{WF}^3)$ in Theorem 3.24 and those in Lemma 3.25

- $\mathcal{B}_{4,4}$, where T is refined with a
  - third order partial Worsey-Farin split, the corresponding additional smoothness conditions from the spline space $\widetilde{\mathcal{S}}_9^2(T_{WF}^3)$ in Theorem 3.24 and those in Lemma 3.26
  - fourth order partial Worsey-Farin split, the corresponding additional smoothness conditions from the spline space $\widetilde{\mathcal{S}}_9^2(T_{WF}^4)$ in Theorem 3.37 and those in Lemma 3.38

- $\mathcal{B}_{5,4}$ the corresponding additional smoothness conditions from the spline space $\widetilde{\mathcal{S}}_9^2(T_{WF}^4)$ in Theorem 3.37 and those in Lemma 3.39

- $\mathcal{B}_{6,4}$ the corresponding additional smoothness conditions from the spline space $\widetilde{\mathcal{S}}_9^2(T_{WF}^4)$ in Theorem 3.37 and those in Lemma 3.40

- $\mathcal{C}_{1,3,3}$, where T is refined with a first order partial Worsey-Farin split, the corresponding additional smoothness conditions from the spline space $\widetilde{\mathcal{S}}_9^2(T_{WF}^1)$ in Theorem 3.4 and those in Lemma 3.6

- $\mathcal{C}_{1,3,4}$, where T is refined with a first order partial Worsey-Farin split, the corresponding additional smoothness conditions from the spline space $\widetilde{\mathcal{S}}_9^2(T_{WF}^1)$ in Theorem 3.4 and those in Lemma 3.5

- $\mathcal{C}_{1,4,4}$, where T is refined with a
  - second order partial Worsey-Farin split, the corresponding additional smoothness conditions from the spline space $\widetilde{\mathcal{S}}_9^2(T_{WF}^2)$ in Theorem 3.12 and those in Lemma 3.15
  - third order partial Worsey-Farin split, the corresponding additional smoothness conditions from the spline space $\hat{\mathcal{S}}_9^2(T_{WF}^3)$ in Theorem 3.29 and those in Lemma 3.30
- $\mathcal{C}_{1,5,4}$, where T is refined with a
  - third order partial Worsey-Farin split, the corresponding additional smoothness conditions from the spline space $\widetilde{\mathcal{S}}_9^2(T_{WF}^3)$ in Theorem 3.24 and those in Lemma 3.27
  - fourth order partial Worsey-Farin split, the corresponding additional smoothness conditions from the spline space $\hat{\mathcal{S}}_9^2(T_{WF}^4)$ in Theorem 3.41 and those in Lemma 3.42
- $\mathcal{C}_{1,6,4}$, where T is refined with a
  - third order partial Worsey-Farin split, the corresponding additional smoothness conditions from the spline space $\widetilde{\mathcal{S}}_9^2(T_{WF}^3)$ in Theorem 3.24 and those in Lemma 3.28
  - fourth order partial Worsey-Farin split, the corresponding additional smoothness conditions from the spline space $\hat{\mathcal{S}}_9^2(T_{WF}^4)$ in Theorem 3.41 and those in Lemma 3.43
- $\mathcal{C}_{2,5,4}$, where T is refined with a
  - second order partial Worsey-Farin split, the corresponding additional smoothness conditions from the spline space $\hat{\mathcal{S}}_9^2(T_{WF}^2)$ in Theorem 3.16 and those in Lemma 3.17
  - third order partial Worsey-Farin split, the corresponding additional smoothness conditions from the spline space $\hat{\mathcal{S}}_9^2(T_{WF}^3)$ in Theorem 3.29 and those in Lemma 3.31
  - fourth order partial Worsey-Farin split, the corresponding additional smoothness conditions from the spline space $\check{\mathcal{S}}_9^2(T_{WF}^4)$ in Theorem 3.44 and those in Lemma 3.45
- $\mathcal{C}_{2,6,4}$, where T is refined with a
  - second order partial Worsey-Farin split, the corresponding additional smoothness conditions from the spline space $\bar{\mathcal{S}}_9^2(T_{WF}^2)$ in Theorem 3.18 and those in Lemma 3.19

- third order partial Worsey-Farin split, the corresponding additional smoothness conditions from the spline space $\hat{\mathcal{S}}_9^2(T_{WF}^3)$ in Theorem 3.29 and those in Lemma 3.32
- fourth order partial Worsey-Farin split, the corresponding additional smoothness conditions from the spline space $\check{\mathcal{S}}_9^2(T_{WF}^4)$ in Theorem 3.44 and those in Lemma 3.46

- $C_{3,6,4}$, where T is refined with a
  - second order partial Worsey-Farin split, the corresponding additional smoothness conditions from the spline space $\check{\mathcal{S}}_9^2(T_{WF}^2)$ in Theorem 3.20 and those in Lemma 3.21
  - third order partial Worsey-Farin split, the corresponding additional smoothness conditions from the spline space $\check{\mathcal{S}}_9^2(T_{WF}^3)$ in Theorem 3.33 and those in Lemma 3.34
  - fourth order partial Worsey-Farin split, the corresponding additional smoothness conditions from the spline space $\dot{\mathcal{S}}_9^2(T_{WF}^4)$ in Theorem 3.47 and those in Lemma 3.48

- $C_{4,6,4}$, where T is refined with a
  - second order partial Worsey-Farin split, the corresponding additional smoothness conditions in Lemma 3.23
  - third order partial Worsey-Farin split, the corresponding additional smoothness conditions in Lemma 3.36
  - fourth order partial Worsey-Farin split, the corresponding additional smoothness in Lemma 3.50

are contained.

Now, we can define the spline space $\mathcal{S}_9^2(\Delta^*)$.

Definition 5.5:
Let Θ be as in Definition 5.4. Then let $\mathcal{S}_9^2(\Delta^*)$ be the subspace of splines in $S_9^{2,3,4,7}(\Delta^*)$ that satisfy the additional smoothness conditions in Θ.

5.3 Lagrange interpolation with $\mathcal{S}_9^2(\Delta^*)$

In this section we construct an 11-local and stable Lagrange interpolation set for the spline space $\mathcal{S}_9^2(\Delta^*)$ constructed in the previous section. We go through an arbitrary tetrahedral partition Δ in the order of the tetrahedra imposed by Algorithm 5.1 and chose interpolation points according to the number of common vertices, edges, and faces with the tetrahedra considered earlier. This order and especially the classification of the tetrahedra in Δ is essential in order to get a local Lagrange interpolation set for the spline space $\mathcal{S}_9^2(\Delta^*)$ defined on the refined partition Δ^*.

Algorithm 5.6:
Let $T_1,\ldots,T_n$ be the order of the tetrahedra in Δ obtained from Algorithm 5.1. For $i = 1,\ldots,n$,

- if T_i is not refined with a partial Worsey-Farin split, choose the points

$$\mathcal{D}_{T_i,9} \setminus \left(\bigcup_{v \in V_P} D_4^{T_i}(v) \cup \bigcup_{e \in E_P} E_3^{T_i}(e) \cup \bigcup_{F \in F_P} \bigcup_{j=0}^{2} F_j^{T_i}(F) \right),$$

  where V_P, E_P, and F_P are the sets of common vertices, edges, and faces of T_i with the tetrahedra $T_1,\ldots,T_{i-1}$, respectively;

- if T_i is refined with a partial Worsey-Farin split, choose the incenter v_{T_i} of T_i.

- Choose all remaining vertices of Δ.

The set of points chosen in Algorithm 5.6 is denoted by $\mathcal{L}$.

Now, we show that this set of points is a local and stable Lagrange interpolation set for the space $\mathcal{S}_9^2(\Delta^*)$, as given in Definition 5.5.

Theorem 5.7:
Let Δ be an arbitrary tetrahedral partition and Δ^* the refined partition obtained from Algorithm 5.3. Moreover, let $\mathcal{L}$ be the set of points chosen in Algorithm 5.6. Then $\mathcal{L}$ is a 11-local and stable Lagrange interpolation set for the spline space $\mathcal{S}_9^2(\Delta^*)$, given in Definition 5.5.

Proof:
In order to prove that $\mathcal{L}$ is a 11-local and stable Lagrange interpolation set for the spline space $\mathcal{S}_9^2(\Delta^*)$, we set the values $\{z_\kappa\}_{\kappa\in\mathcal{L}}$ to arbitrary real numbers and show that all B-coefficients of a spline $s \in \mathcal{S}_9^2(\Delta^*)$ are uniquely and stably determined. Moreover, we show that each B-coefficient c_ξ^T, $T \in \Delta$, only depends on values $\{z_\kappa\}_{\kappa\in\Gamma_\xi}$, with $\Gamma_\xi \subset (\mathcal{L} \cap \mathrm{star}^m(T))$, $m \leq 11$. Therefore, we go through the tetrahedra $T_1,\ldots,T_n$ of Δ in the order imposed by Algorithm 5.1 and show how the corresponding B-coefficients of s are determined according to the classes $\mathcal{A}_i, \mathcal{B}_{j,i}$, and $C_{k,j,i}$, $i = 0,\ldots,4$, $j = 1,\ldots,6$, $k = 1,\ldots,4$. Thereby we assume for the tetrahedra in some classes that they share specific vertices, edges, or faces with tetrahedra considered earlier. Though, it is insignificant which special vertex, edge, or face of a tetrahedron is shared with the ones considered previously, as long as the choice results in the same partial Worsey-Farin split. So the proof works analogue for any of these choices of common vertices, edges, or faces.

First we consider the tetrahedron $T := T_1$. Following Algorithm 5.1, this tetrahedron is in class $\mathcal{A}_0$, and thus the set $\mathcal{L}$ contains the points $\mathcal{D}_{T,9}$. In order to compute the B-coefficients of $s|_T$ from the values $\{z_\kappa\}_{\kappa\in\mathcal{D}_{T,9}}$, we have to solve the linear system of equations corresponding to the interpolation at the points $\mathcal{D}_{T,9}$. The corresponding matrix is $M_{\mathcal{A}_0} := \left(B_\xi^{9,T}(\kappa)\right)_{\xi,\kappa\in\mathcal{D}_{T,9}}$, which has a positive determinant and is therefore nonsingular. Moreover, the B-coefficients c_ξ for all domain points $\xi \in T$ only depend on the values $\{z_\kappa\}_{\kappa\in\Gamma_\xi}$, with $\Gamma_\xi \subseteq (\mathcal{L} \cap \mathrm{star}^0(T))$, and (2.12) holds with $C_{\mathcal{A}_0} := \|M_{\mathcal{A}_0}^{-1}\|$. In the same way s can be uniquely and stably determined on the other tetrahedra in class $\mathcal{A}_0$, since by Lemma 5.2, any two tetrahedra in class $\mathcal{A}_0$ are disjoint. Now, using the C^2 smoothness conditions at the faces F, the C^3 supersmoothness conditions at the edges e, and the C^4 supersmoothness conditions at the vertices v of the tetrahedra in class $\mathcal{A}_0$, we can uniquely and stably compute the remaining B-coefficients of s corresponding to the domain points in the balls $D_4(v)$, the tubes $E_3(e)$, and the domain points in $F_i(F)$, $i = 0,1,2$, respectively. Any of these B-coefficients corresponding to a domain point ξ in a tetrahedron $\widetilde{T}$, sharing a vertex, an edge, or a face with T, only depends on values $\{z_\kappa\}_{\kappa\in\Gamma_\xi}$, with $\Gamma_\xi \subset (\mathcal{L} \cap \mathrm{star}^1(\widetilde{T}))$, and (2.12) holds with a constant $\widetilde{C}_{\mathcal{A}_0}$ depending on $C_{\mathcal{A}_0}$ and the smallest solid and faces angles in Δ.

Suppose we have uniquely and stably determined all B-coefficients of s restricted to the tetrahedra $T_1,\ldots,T_{i-1}$, and let $T := T_i \in \mathcal{A}_1$. By the definition of class $\mathcal{A}_1$ in Algorithm 5.1 and following statement (ix) of Lemma 5.2, the tetrahedron T must share exactly one vertex v with a tetrahedron in $\{T_1,\dots,T_{i-1}\}$ contained in class $\mathcal{A}_0$. Thus, the B-coefficients of $s|_T$ associated with the domain points in $D_4^T(v)$ are already uniquely and stably determined. The B-coefficients corresponding to the remaining domain points in T are uniquely determined by interpolation at the points $(\mathcal{D}_{T,9} \setminus D_4^T(v)) \subset \mathcal{L}$. The matrix of the corresponding linear system of equations is $M_{\mathcal{A}_1} := \left(B_\xi^{9,T}(\kappa)\right)_{\xi,\kappa \in \mathcal{D}_{T,9} \setminus D_4^T(v)}$, which has a positive determinant. Then, for all $\xi \in \mathcal{D}_{T,9}$, (2.12) holds with $\Gamma_\xi \subset (\mathcal{L} \cap \mathrm{star}^1(T))$ and $C_{\mathcal{A}_1} := \|M_{\mathcal{A}_1}^{-1}\|(1 + \widetilde{C}_{\mathcal{A}_0}\|M_{\mathcal{A}_0}^{-1}\|)$. In the same way we can uniquely and stably compute the B-coefficients of s restricted to the other tetrahedra in class $\mathcal{A}_1$. Then, for each vertex v, edge e, and face F of these tetrahedra, we can uniquely and stably determine the B-coefficients of s restricted to the domain points in the ball $D_4(v)$, the tube $E_3(e)$, and the domain points in $F_i(F)$, $i = 0,1,2$, using the C^2 smoothness conditions at the faces, the C^3 supersmoothness conditions at the edges, and the C^4 supersmoothness conditions at the vertices. For any of these B-coefficients corresponding to a domain point ξ in a tetrahedron $\widetilde{T}$, sharing a vertex, an edge, or a face with T, (2.12) holds with $\Gamma_\xi \subset (\mathcal{L} \cap \mathrm{star}^2(\widetilde{T}))$ and a constant $\widetilde{C}_{\mathcal{A}_1}$ depending on $C_{\mathcal{A}_1}$ and the smallest solid and faces angles in Δ.

Now, suppose we have uniquely and stably determined all B-coefficients of s restricted to the tetrahedra $T_1,\dots,T_{i-1}$, and let $T := T_i \in \mathcal{A}_2$. By the definition of $\mathcal{A}_2$ the tetrahedron T shares exactly two vertices, u and v, with tetrahedra in $\{T_1,\dots,T_{i-1}\}$. Following statement (ix) of Lemma 5.2, the first two tetrahedra in the order $T_1,\dots,T_{i-1}$ containing the vertices u and v, respectively, have to be in $\mathcal{A}_0 \cup \mathcal{A}_1$. Thus, the B-coefficients of $s|_T$ associated with the domain points in the balls $D_4^T(u)$ and $D_4^T(v)$ are already uniquely and stably determined. Following Algorithm 5.6, the set $\mathcal{L}$ contains the points $\mathcal{D}_{T,9} \setminus \left(D_4^T(u) \cup D_4^T(v)\right)$. Thus, by solving the linear system of equations corresponding to the interpolation at these points, which has the matrix $M_{\mathcal{A}_2} := \left(B_\xi^{9,T}(\kappa)\right)_{\xi,\kappa \in \mathcal{D}_{T,9} \setminus \left(D_4^T(u) \cup D_4^T(v)\right)}$, with positive determinant, we can uniquely compute the remaining B-coefficients of $s|_T$. For the B-coefficients of s associated with the domain points $\xi \in \mathcal{D}_{T,9} \setminus \left(D_4^T(u) \cup D_4^T(v)\right)$, (2.12) holds with $\Gamma_\xi \subset (\mathcal{L} \cap \mathrm{star}^2(T))$ and a constant $C_{\mathcal{A}_2}$, which depends on $\|M_{\mathcal{A}_2}^{-1}\|$ and $\widetilde{C}_{\mathcal{A}_1}$. In the same way the B-coefficients of s restricted to the remaining tetrahedra in class $\mathcal{A}_2$ can be uniquely and stably computed. Subsequently, using the C^2 smoothness conditions at the faces F, the C^3 supersmoothness conditions at the edges e, and the C^4 supersmoothness conditions at the vertices v of the tetrahedra in $\mathcal{A}_2$, the B-coefficients corresponding to the domain points in the balls $D_4(v)$, the tubes $E_3(e)$, and the domain points in $F_i(F)$, $i = 0,1,2$, can be uniquely and stably determined. The B-coefficients corresponding to these domain points ξ in a tetrahedron $\widetilde{T}$ only depend on the values $\{z_\kappa\}_{\kappa\in\Gamma_\xi}$, with $\Gamma_\xi \subset (\mathcal{L} \cap \operatorname{star}^3(\widetilde{T}))$, and (2.12) holds with a constant $\widetilde{C}_{\mathcal{A}_2}$ depending on $C_{\mathcal{A}_2}$ and the smallest solid and faces angles of Δ.

Next, suppose we have uniquely and stably determined all B-coefficients of s restricted to the tetrahedra $T_1,\ldots,T_{i-1}$, and let $T := T_i \in \mathcal{A}_3$. By the definition of class $\mathcal{A}_3$ the tetrahedron T shares exactly three vertices u,v, and w with tetrahedra in $\{T_1,\ldots,T_{i-1}\}$. Following statement (ix) of Lemma 5.2, the first tetrahedra containing these three vertices must be in $\mathcal{A}_0 \cup \mathcal{A}_1 \cup \mathcal{A}_2$. Thus, the B-coefficients of s corresponding to the domain points in the balls $D_4^T(u), D_4^T(v)$, and $D_4^T(w)$ are already uniquely and stably determined. Following Algorithm 5.6, the points $\mathcal{D}_{T,9} \setminus \big(D_4^T(u) \cup D_4^T(v) \cup D_4^T(w)\big)$ are contained in the set $\mathcal{L}$ of interpolation points. Thus, by solving a linear system of equations with matrix $M_{\mathcal{A}_3} := \Big(B_\xi^{9,T}(\kappa)\Big)_{\xi,\kappa\in\mathcal{D}_{T,9}\setminus\left(D_4^T(u)\cup D_4^T(v)\cup D_4^T(w)\right)}$, which has a positive determinant, we can uniquely determine the remaining B-coefficients of the polynomial $s|_T$. Thereby, the B-coefficients corresponding to the domain points $\xi \in \mathcal{D}_{T,9} \setminus \big(D_4^T(u) \cup D_4^T(v) \cup D_4^T(w)\big)$ only depend on the values $\{z_\kappa\}_{\kappa\in\Gamma_\xi}$, with $\Gamma_\xi \subset (\mathcal{L} \cap \operatorname{star}^3(T))$, and (2.12) holds with a constant $C_{\mathcal{A}_3}$ that depends on $\|M_{\mathcal{A}_3}^{-1}\|$ and $\widetilde{C}_{\mathcal{A}_2}$. In the same way we can uniquely and stably determine the B-coefficients of s restricted to the remaining tetrahedra in class $\mathcal{A}_3$. Now, using the C^2 smoothness conditions at the faces F, the C^3 supersmoothness conditions at the edges e, and the C^4 supersmoothness conditions at the vertices v of the tetrahedra in class $\mathcal{A}_3$, the B-coefficients corresponding to the domain points in the balls $D_4(v)$, the tubes $E_3(e)$, and the domain points in $F_i(F)$, $i = 0,1,2$, can be uniquely and stably computed. Each of these B-coefficients corresponding to a domain point ξ in a tetrahedron $\widetilde{T}$, sharing a vertex, an edge, or a face with T, depends on values $\{z_\kappa\}_{\kappa\in\Gamma_\xi}$, with $\Gamma_\xi \subset (\mathcal{L} \cap \operatorname{star}^4(\widetilde{T}))$, and (2.12) holds with a constant $\widetilde{C}_{\mathcal{A}_3}$ that depends on $C_{\mathcal{A}_3}$ and the smallest solid and face angles of Δ.

Suppose we have uniquely and stably determined all B-coefficients of s restricted to the tetrahedra $T_1,\ldots,T_{i-1}$, and let $T := T_i \in \mathcal{A}_4$. Following the definition of $\mathcal{A}_4$, the tetrahedron T shares all four vertices v_i, $i = 1,\dots,4$, with tetrahedra in $\{T_1,\dots,T_{i-1}\}$. By statement (ix) of Lemma 5.2 it is obvious that the first tetrahedra in $\{T_1,\dots,T_{i-1}\}$ that contain the four vertices of T are in $\mathcal{A}_0 \cup \dots \cup \mathcal{A}_3$. Thus, the B-coefficients of s associated with the domain points in $D_4^T(v_i)$, $i = 1,\dots,4$, are already uniquely and stably determined. The remaining B-coefficients of $s|_T$ can be uniquely determined by interpolation at the points $\mathcal{D}_{T,9} \setminus \bigcup_{i=1}^4 D_4^T(v_i)$, which are contained in $\mathcal{L}$. Therefore, we have to solve a linear system with matrix $M_{\mathcal{A}_4} := \left(B_\xi^{9,T}(\kappa)\right)_{\xi,\kappa \in \mathcal{D}_{T,9} \setminus \bigcup_{i=1}^4 D_4^T(v_i)}$, which has a positive determinant. Moreover, for the computation of the B-coefficients of s corresponding to the domain points ξ in T, (2.12) holds with $\Gamma_\xi \subset (\mathcal{L} \cap \operatorname{star}^4(T))$ and a constant $C_{\mathcal{A}_4}$ that depends on $\|M_{\mathcal{A}_4}^{-1}\|$ and $\widetilde{C}_{\mathcal{A}_3}$. Now, in the same way we can uniquely and stably determine the B-coefficients of s restricted to all other tetrahedra in class $\mathcal{A}_4$. Then, we can uniquely and stably determine the B-coefficients of s restricted to the domain points in the tubes $E_3(e)$ and the domain points in $F_i(F)$, $i = 0,1,2$, using the C^2 smoothness conditions at the faces F and the C^3 supersmoothness conditions at the edges e of the tetrahedra in $\mathcal{A}_4$. For any of these domain points ξ in a tetrahedron $\widetilde{T}$ the corresponding B-coefficient of s only depends on values $\{z_\kappa\}_{\kappa \in \Gamma_\xi}$, with $\Gamma_\xi \subset (\mathcal{L} \cap \operatorname{star}^5(\widetilde{T}))$. Moreover, (2.12) holds with a constant $\widetilde{C}_{\mathcal{A}_4}$ that depends on $C_{\mathcal{A}_4}$ and the smallest solid and face angles of Δ.

So far, we have uniquely and stably determined s restricted to the tetrahedra in the classes $\mathcal{A}_0$, $\mathcal{A}_1$, $\mathcal{A}_2$, $\mathcal{A}_3$, and $\mathcal{A}_4$.

Now, suppose we have uniquely and stably determined all B-coefficients of s restricted to the tetrahedra $T_1,\dots,T_{i-1}$, and let $T := T_i \in \mathcal{B}_{1,2}$. Following Algorithm 5.1, the tetrahedron T shares exactly one edge, $e := \langle u,v \rangle$, with the tetrahedra $T_1,\dots,T_{i-1}$. By statement (vi) of Lemma 5.2, the first tetrahedron of $T_1,\dots,T_{i-1}$ sharing an edge with T must be in $\mathcal{A}_0 \cup \mathcal{A}_1 \cup \mathcal{A}_2$. Thus, the B-coefficients of $s|_T$ associated with the domain points in $E_3^T(e) \cup D_4^T(u) \cup D_4^T(v)$ are already uniquely and stably determined. Since $\mathcal{L} \cap T$ contains the points $\mathcal{D}_{T,9} \setminus (E_3^T(e) \cup D_4^T(u) \cup D_4^T(v))$, the remaining B-coefficients of $s|_T$ can be uniquely determined by interpolation at these points. This involves solving a linear system of equations, where the corresponding matrix $M_{\mathcal{B}_{1,2}} := \left(B_\xi^{9,T}(\kappa)\right)_{\xi,\kappa \in \mathcal{D}_{T,9} \setminus (E_3^T(e) \cup D_4^T(u) \cup D_4^T(v))}$ has a positive determinant. Then, for all $\xi \in \mathcal{D}_{T,9}$, (2.12) holds with $\Gamma_\xi \subset (\mathcal{L} \cap \operatorname{star}^3(T))$ and a constant $C_{\mathcal{B}_{1,2}}$ which depends on $\|M_{\mathcal{B}_{1,2}}^{-1}\|$ and $\widetilde{C}_{\mathcal{A}_2}$. In the same way s

restricted to the other tetrahedra in class $\mathcal{B}_{1,2}$ can be uniquely and stably determined. Now, for each vertex v, edge e, and face F of the tetrahedra in $\mathcal{B}_{1,2}$, we can uniquely and stably determine the B-coefficients of s restricted to the domain points in the ball $D_4(v)$, the tube $E_3(e)$, and the domain points in $F_i(F)$, $i=0,1,2$, using the C^2 smoothness conditions at the faces, the C^3 supersmoothness conditions at the edges, and the C^4 supersmoothness conditions at the vertices. For any of these B-coefficients corresponding to a domain point ξ in a tetrahedron $\widetilde{T}$, sharing a vertex, an edge, or a face with T, (2.12) holds with $\Gamma_\xi \subset (\mathcal{L} \cap \text{star}^4(\widetilde{T}))$ and a constant $\widetilde{C}_{\mathcal{B}_{1,2}}$ depending on $C_{\mathcal{B}_{1,2}}$ and the smallest solid and faces angles in Δ.

Next, suppose we have uniquely and stably determined all B-coefficients of s restricted to the tetrahedra $T_1,\dots,T_{i-1}$, and let $T := T_i \in \mathcal{B}_{1,3}$. By the definition of class $\mathcal{B}_{1,3}$ the tetrahedron T shares exactly one edge $e := \langle u,v\rangle$ and one other vertex w with tetrahedra in $\{T_1,\dots,T_{i-1}\}$. Following statement (vi) of Lemma 5.2, the first tetrahedron of $T_1,\dots,T_{i-1}$ containing the edge e must be in $\mathcal{A}_0 \cup \ldots \cup \mathcal{A}_3$. The first tetrahedron of $T_1,\dots,T_{i-1}$ containing the vertex w must be in $\mathcal{A}_0 \cup \ldots \cup \mathcal{A}_3 \cup \mathcal{B}_{1,2}$. It can not be in class $\mathcal{A}_4$, since all vertices of the tetrahedra in class $\mathcal{A}_4$ are common vertices with tetrahedra in $\mathcal{A}_0 \cup \ldots \cup \mathcal{A}_3$. Then, the B-coefficients of s corresponding to the domain points $E_3^T(e) \cup D_4^T(u) \cup D_4^T(v) \cup D_4^T(w)$ are already uniquely and stably determined. By Algorithm 5.6, the points $\mathcal{D}_{T,9} \setminus \big(E_3^T(e) \cup D_4^T(u) \cup D_4^T(v) \cup D_4^T(w)\big)$ are contained in the set $\mathcal{L}$ of interpolation points. Then, by solving a linear system of equations with matrix $M_{\mathcal{B}_{1,3}} := \Big(B_\xi^{9,T}(\kappa)\Big)_{\xi,\kappa \in \mathcal{D}_{T,9} \setminus \big(E_3^T(e) \cup D_4^T(u) \cup D_4^T(v) \cup D_4^T(w)\big)}$, which has a positive determinant, we can uniquely determine the remaining B-coefficients of the polynomial $s|_T$. The B-coefficients corresponding to the domain points $\xi \in \mathcal{D}_{T,9} \setminus \big(E_3^T(e) \cup D_4^T(u) \cup D_4^T(v) \cup D_4^T(w)\big)$ only depend on the values $\{z_\kappa\}_{\kappa \in \Gamma_\xi}$, with $\Gamma_\xi \subset (\mathcal{L} \cap \text{star}^4(T))$, and (2.12) holds with a constant $C_{\mathcal{B}_{1,3}}$ that depends on $\|M_{\mathcal{B}_{1,3}}^{-1}\|$, $\widetilde{C}_{\mathcal{B}_{1,2}}$ and $\widetilde{C}_{\mathcal{A}_3}$. In the same way we can uniquely and stably determine the B-coefficients of s restricted to the remaining tetrahedra in class $\mathcal{B}_{1,3}$. Then, using the C^2 smoothness conditions at the faces F, the C^3 supersmoothness conditions at the edges e, and the C^4 supersmoothness conditions at the vertices v of the tetrahedra in $\mathcal{B}_{1,3}$, the B-coefficients corresponding to the domain points in the balls $D_4(v)$, the tubes $E_3(e)$, and the domain points in $F_i(F)$, $i=0,1,2$, can be uniquely and stably determined. For any of these B-coefficients corresponding to a domain point ξ in a tetrahedron $\widetilde{T}$, sharing a vertex, an edge, or a face with

T, (2.12) holds with $\Gamma_\xi \subset (\mathcal{L} \cap \mathrm{star}^5(\widetilde{T}))$ and a constant $\widetilde{C}_{\mathcal{B}_{1,3}}$ that depends on $C_{\mathcal{B}_{1,3}}$ and the smallest solid and face angles of Δ.

Suppose we have uniquely and stably determined all B-coefficients of s restricted to the tetrahedra $T_1,\ldots,T_{i-1}$, and let $T := T_i \in \mathcal{B}_{1,4}$. Following Algorithm 5.1, the tetrahedron $T := \langle v_1,v_2,v_3,v_4\rangle$ shares exactly one edge, $e := \langle v_1,v_2\rangle$, and all four vertices with the tetrahedra $T_1,\ldots,T_{i-1}$. By statement (vi) of Lemma 5.2, the first tetrahedron of $T_1,\ldots,T_{i-1}$ containing the edge e must be in $\mathcal{A}_0 \cup \ldots \cup \mathcal{A}_4$. Following statement (ix) of Lemma 5.2, the first tetrahedra of $T_1,\ldots,T_{i-1}$ containing the vertices v_3 and v_4 must be in $\mathcal{A}_0 \cup \ldots \cup \mathcal{A}_3 \cup \mathcal{B}_{1,2} \cup \mathcal{B}_{1,3}$. Thus, the B-coefficients of s corresponding to the domain points $E_3^T(e) \cup \bigcup_{i=1}^4 D_4^T(v_i)$ are already uniquely and stably determined. Since the set $\mathcal{L}$ contains the points $\mathcal{D}_{T,9} \setminus \left(E_3^T(e) \cup \bigcup_{i=1}^4 D_4^T(v_i)\right)$, the remaining B-coefficients of $s|_T$ can be uniquely determined by interpolation at these points. To this end, a linear system of equations with matrix $M_{\mathcal{B}_{1,4}} := \left(B_\xi^{9,T}(\kappa)\right)_{\xi,\kappa \in \mathcal{D}_{T,9} \setminus \left(E_3^T(e) \cup \bigcup_{i=1}^4 D_4^T(v_i)\right)}$, which has a positive determinant, has to be solved. For the B-coefficients of s corresponding to the domain points $\xi \in \mathcal{D}_{T,9} \setminus \left(E_3^T(e) \cup \bigcup_{i=1}^4 D_4^T(v_i)\right)$, (2.12) holds with $\Gamma_\xi \subset (\mathcal{L} \cap \mathrm{star}^5(T))$ and a constant $C_{\mathcal{B}_{1,4}}$, that depends on $\|M_{\mathcal{B}_{1,4}}^{-1}\|$, $\widetilde{C}_{\mathcal{B}_{1,3}}$, and $\widetilde{C}_{\mathcal{A}_4}$. In the same way we can uniquely and stably compute the B-coefficients of s restricted to the remaining tetrahedra in $\mathcal{B}_{1,4}$. Then, for each edge e and face F of the tetrahedra in class $\mathcal{B}_{1,4}$, we can uniquely and stably determine the B-coefficients of s restricted to the domain points in the tube $E_3(e)$ and the domain points in $F_i(F)$, $i = 0,1,2$, using the C^2 smoothness conditions at the faces and the C^3 supersmoothness conditions at the edges of the tetrahedra in $\mathcal{B}_{1,4}$. For any of these B-coefficients corresponding to a domain point ξ in a tetrahedron $\widetilde{T}$, sharing an edge or a face with T, (2.12) holds with $\Gamma_\xi \subset (\mathcal{L} \cap \mathrm{star}^6(\widetilde{T}))$ and a constant $\widetilde{C}_{\mathcal{B}_{1,4}}$ that depends on $C_{\mathcal{B}_{1,4}}$ and the smallest solid and face angles of Δ.

Next, suppose we have uniquely and stably determined all B-coefficients of s restricted to the tetrahedra $T_1,\ldots,T_{i-1}$, and let $T := T_i \in \mathcal{B}_{2,3}$. Following Algorithm 5.1, the tetrahedron T shares exactly two edges, $e_1 := \langle u,v\rangle$ and $e_2 := \langle v,w\rangle$, with the tetrahedra $T_1,\ldots,T_{i-1}$. Following Algorithm 5.3, the tetrahedron T is refined with a first order partial Worsey-Farin split. Moreover, by statement (vi) of Lemma 5.2, the first common edge of T is shared with a tetrahedron in $\mathcal{A}_0 \cup \mathcal{A}_1 \cup \mathcal{A}_3$. Following (vii) of the same lemma, the second common edge is shared with a tetrahedron in

$\mathcal{A}_0 \cup \ldots \cup \mathcal{A}_4 \cup \mathcal{B}_{1,2} \cup \mathcal{B}_{1,3}$. Now, since the set $\mathcal{L}$ contains the incenter and the fourth vertex of T, it follows from Lemma 3.11, that $s|_T$ is uniquely and stably determined by interpolation at these two points and the additional smoothness conditions of s. For the B-coefficients associated with a domain point ξ in T_{WF}^1, (2.12) holds with $\Gamma_\xi \subset (\mathcal{L} \cap \text{star}^5(T))$ and a constant $C_{\mathcal{B}_{2,3}}$ depending on $\widetilde{C}_{\mathcal{B}_{1,3}}$, $\widetilde{C}_{\mathcal{A}_4}$, and the smallest solid and face angles of Δ. In the same way we can uniquely and stably determine the B-coefficients of s restricted to all other tetrahedra in class $\mathcal{B}_{2,3}$. Then, using the C^2 smoothness conditions at the faces F, the C^3 supersmoothness conditions at the edges e, and the C^4 supersmoothness conditions at the vertices v of the tetrahedra in $\mathcal{B}_{2,3}$, the B-coefficients corresponding to the domain points in the balls $D_4(v)$, the tubes $E_3(e)$, and the domain points in $F_i(F)$, $i = 0,1,2$, can be uniquely and stably determined. For any of these B-coefficients corresponding to a domain point ξ in a tetrahedron $\widetilde{T}$, sharing a vertex, an edge, or a face with T, (2.12) holds with $\Gamma_\xi \subset (\mathcal{L} \cap \text{star}^6(\widetilde{T}))$ and a constant $\widetilde{C}_{\mathcal{B}_{2,3}}$ that depends on $C_{\mathcal{B}_{2,3}}$ and the smallest solid and face angles of Δ.

Now, suppose we have uniquely and stably determined all B-coefficients of s restricted to the tetrahedra $T_1, \ldots, T_{i-1}$, and let $T := T_i \in \mathcal{B}_{2,4}$. By Algorithm 5.1, the tetrahedron $T := \langle v_1, v_2, v_3, v_4 \rangle$ shares all four vertices and exactly two edges with the tetrahedra $T_1, \ldots, T_{i-1}$. We have to distinguish between two cases (see Table 5.1):

- T is not refined with a partial Worsey-Farin split. Then, following statement (vi) of Lemma 5.2, the first common edge $e_1 := \langle v_1, v_2 \rangle$ of T is shared with a tetrahedron in $\mathcal{A}_0 \cup \ldots \cup \mathcal{A}_4$. By statement (vii) of this lemma, the second common edge $e_2 := \langle v_3, v_4 \rangle$ of T is shared with a tetrahedron in $\mathcal{A}_0 \cup \ldots \cup \mathcal{A}_4 \cup \mathcal{B}_{1,2} \cup \mathcal{B}_{1,3} \cup \mathcal{B}_{1,4}$. Thus, the B-coefficients of s corresponding to the domain points $P_{\mathcal{B}}^{2,4} := E_3^T(e_1) \cup E_3^T(e_2) \cup \bigcup_{i=1}^4 D_4^T(v_i)$ are already uniquely and stably determined. Since the set $\mathcal{L}$ contains the points $\mathcal{D}_{T,9} \setminus P_{\mathcal{B}}^{2,4}$, the remaining B-coefficients of $s|_T$ can be uniquely determined by interpolation at these points. The matrix of the corresponding linear system is $M_{\mathcal{B}_{2,4}} := \left(B_\xi^{9,T}(\kappa)\right)_{\xi,\kappa \in \mathcal{D}_{T,9} \setminus P_{\mathcal{B}}^{2,4}}$, which has a positive determinant. Then, for the B-coefficients of s associated with the domain points $\xi \in \mathcal{D}_{T,9} \setminus P_{\mathcal{B}}^{2,4}$, (2.12) holds with $\Gamma_\xi \subset (\mathcal{L} \cap \text{star}^6(T))$ and a constant $C_{\mathcal{B}_{2,4}}$, that depends on $\|M_{\mathcal{B}_{2,4}}^{-1}\|$ and $\widetilde{C}_{\mathcal{B}_{1,4}}$. In the same way we can uniquely and stably compute the B-coefficients of s restricted to all other non-split tetrahedra in $\mathcal{B}_{2,4}$.

- T is refined with a first order partial Worsey-Farin split. Then, by statement (vi) of Lemma 5.2, the first common edge $e_1 := \langle v_1, v_2 \rangle$ of T is shared with a tetrahedron in $\mathcal{A}_0 \cup \ldots \cup \mathcal{A}_4$. Following statement (vii) of the same lemma, the second common edge $e_2 := \langle v_2, v_3 \rangle$ of T is shared with a tetrahedron in $\mathcal{A}_0 \cup \ldots \cup \mathcal{A}_4 \cup \mathcal{B}_{1,2} \cup \mathcal{B}_{1,3} \cup \mathcal{B}_{1,4}$. Moreover, following statement (ix) of Lemma 5.2, T shares the vertex v_4 with a tetrahedron in $\mathcal{A}_0 \cup \ldots \cup \mathcal{A}_3 \cup \mathcal{B}_{1,2} \cup \mathcal{B}_{1,3} \cup \mathcal{B}_{2,3}$. Then, since $\mathcal{L}$ contains the incenter of T, it follows from Lemma 3.10, that $s|_T$ is uniquely and stably determined by interpolation at this point and the additional smoothness conditions of s. Then for the B-coefficients of s associated with a domain point ξ in T_{WF}^1, (2.12) holds with $\Gamma_\xi \subset (\mathcal{L} \cap \text{star}^6(T))$ and a constant $\hat{C}_{\mathcal{B}_{2,4}}$ depending on $\widetilde{C}_{\mathcal{B}_{1,4}}$ and the smallest solid and face angles of Δ. In the same way s can be uniquely and stably determined restricted to all other tetrahedra in $\mathcal{B}_{2,4}$ that are refined with a first order partial Worsey-Farin split.

Then, for each edge e and face F of the tetrahedra in class $\mathcal{B}_{2,4}$, we can uniquely and stably determine the B-coefficients of s restricted to the domain points in the tube $E_3(e)$ and the domain points in $F_i(F)$, $i = 0,1,2$, using the C^2 smoothness conditions at the faces and the C^3 supersmoothness conditions at the edges of the tetrahedra in $\mathcal{B}_{2,4}$. For any of these B-coefficients corresponding to a domain point ξ in a tetrahedron $\widetilde{T}$, sharing an edge or a face with T, (2.12) holds with $\Gamma_\xi \subset (\mathcal{L} \cap \text{star}^7(\widetilde{T}))$ and a constant $\widetilde{C}_{\mathcal{B}_{2,4}}$ that depends on $C_{\mathcal{B}_{2,4}}$, $\hat{C}_{\mathcal{B}_{2,4}}$ and the smallest solid and face angles of Δ.

Next, suppose we have uniquely and stably determined all B-coefficients of s restricted to the tetrahedra $T_1, \ldots, T_{i-1}$, and let $T := T_i \in \mathcal{B}_{3,3}$. Following Algorithm 5.1, the tetrahedron $T := \langle v_1, v_2, v_3, v_4 \rangle$ shares exactly three edges, $e_1 := \langle v_1, v_2 \rangle$, $e_2 := \langle v_2, v_3 \rangle$, and $e_3 := \langle v_3, v_1 \rangle$ with the tetrahedra $T_1, \ldots, T_{i-1}$. Following Algorithm 5.3, the tetrahedron T is refined with a first order partial Worsey-Farin split. Moreover, by statement (vi) of Lemma 5.2, the first common edge of T is shared with a tetrahedron in $\mathcal{A}_0 \cup \ldots \cup \mathcal{A}_3$. Following statement (vii) of the same lemma, the second common edge of T is shared with a tetrahedron in $\mathcal{A}_0 \cup \ldots \cup \mathcal{A}_4 \cup \mathcal{B}_{1,2} \cup \mathcal{B}_{1,3}$. By statement (viii) of Lemma 5.2, the third common edge of T is shared with a tetrahedron in $\mathcal{A}_0 \cup \ldots \cup \mathcal{A}_4 \cup \mathcal{B}_{1,2} \cup \ldots \cup \mathcal{B}_{2,3}$. According to Algorithm 5.6, the set $\mathcal{L}$ contains the incenter and the vertex v_4 of T. Thus, following Lemma 3.8, we can uniquely and stably determined all B-coefficients of $s|_T$ by interpolation at these two points and the additional smoothness conditions of s. For the B-coefficients of s corresponding to a domain point ξ in T_{WF}^1, (2.12) holds with $\Gamma_\xi \subset (\mathcal{L} \cap \text{star}^6(T))$ and a constant $C_{\mathcal{B}_{3,3}}$ depending on $\widetilde{C}_{\mathcal{B}_{2,3}}$, $\widetilde{C}_{\mathcal{B}_{1,4}}$, and the smallest solid and face angles of Δ. In the same way we can uniquely and stably determine s restricted to the remaining tetrahedra in class $\mathcal{B}_{3,3}$. Then, using the C^2 smoothness conditions at the faces F, the C^3 supersmoothness conditions at the edges e, and the C^4 supersmoothness conditions at the vertices v of the tetrahedra in $\mathcal{B}_{3,3}$, the B-coefficients corresponding to the domain points in the balls $D_4(v)$, the tubes $E_3(e)$, and the domain points in $F_i(F)$, $i = 0,1,2$, can be uniquely and stably determined. For each of these B-coefficients corresponding to a domain point ξ in a tetrahedron $\widetilde{T}$, sharing a vertex or an edge with T, (2.12) holds with $\Gamma_\xi \subset (\mathcal{L} \cap \text{star}^7(\widetilde{T}))$ and a constant $\widetilde{C}_{\mathcal{B}_{3,3}}$ that depends on $C_{\mathcal{B}_{3,3}}$ and the smallest solid and face angles of Δ. For the B-coefficients corresponding to a domain point ξ in a tetrahedron $\widetilde{T}$, sharing a common face with T, (2.12) holds with $\Gamma_\xi \subset (\mathcal{L} \cap \text{star}^6(\widetilde{T}))$ and a constant $\widetilde{C}_{\mathcal{B}_{3,3},F}$ that depends on $C_{\mathcal{B}_{3,3}}$ and the smallest solid and face angles of Δ.

Subsequently, suppose we have uniquely and stably determined all B-coefficients of s restricted to the tetrahedra $T_1,\ldots,T_{i-1}$, and let $T := T_i \in \mathcal{B}_{3,4}$. By Algorithm 5.1, the tetrahedron $T := \langle v_1,v_2,v_3,v_4 \rangle$ shares all four vertices and exactly three edges with the tetrahedra $T_1,\ldots,T_{i-1}$. Following statement (vi) of Lemma 5.2, the first common edge of T is shared with a tetrahedron in $\mathcal{A}_0 \cup \ldots \cup \mathcal{A}_4$. By statement (vii) of this lemma, the second common edge of T is shared with a tetrahedron in $\mathcal{A}_0 \cup \ldots \cup \mathcal{A}_4 \cup \mathcal{B}_{1,2} \cup \mathcal{B}_{1,3} \cup \mathcal{B}_{1,4}$. Following statement (viii) of Lemma 5.2, the third common edge of T is shared with a tetrahedron in $\mathcal{A}_0 \cup \ldots \cup \mathcal{A}_4 \cup \mathcal{B}_{1,2} \cup \ldots \cup \mathcal{B}_{2,4}$. Then, we have to distinguish between three cases (see Table 5.1):

- T is refined with a first order partial Worsey-Farin split. Then the first common edge is $e_1 := \langle v_1,v_2 \rangle$, the second is $e_2 := \langle v_2,v_3 \rangle$, and the third one is $e_3 := \langle v_3,v_1 \rangle$. Moreover, by statement (ix) of Lemma 5.2, T shares the vertex v_4 with a tetrahedron in $\mathcal{A}_0 \cup \ldots \cup \mathcal{A}_3 \cup \mathcal{B}_{1,2} \cup \ldots \cup \mathcal{B}_{3,3}$. Then, since $\mathcal{L}$ contains the incenter of T, it follows from Lemma 3.7, that $s|_T$ is uniquely and stably determined by interpolation at this point and the additional smoothness conditions of s. Then, for any B-coefficient of s associated with a domain point ξ in T_{WF}^1, (2.12) holds with $\Gamma_\xi \subset (\mathcal{L} \cap \text{star}^7(T))$ and a constant $C_{\mathcal{B}_{3,4}}$ depending on $\widetilde{C}_{\mathcal{B}_{3,3}}$, $\widetilde{C}_{\mathcal{B}_{1,4}}$, and the smallest solid and face angles of Δ. Then, for any B-coefficient of s associated with a domain point ξ in T_{WF}^1, (2.12) holds with $\Gamma_\xi \subset (\mathcal{L} \cap \text{star}^7(T))$ and a constant $C_{\mathcal{B}_{3,4}}$ depending on $\widetilde{C}_{\mathcal{B}_{3,3}}$, $\widetilde{C}_{\mathcal{B}_{1,4}}$, and the smallest solid and face angles of Δ. In the same way we can uniquely and stably determine the B-coefficients of s restricted to all other tetrahedra in $\mathcal{B}_{3,4}$ that are refined with a first order partial Worsey-Farin split.

- T is refined with a second order partial Worsey-Farin split. Then the first common edge is $e_1 := \langle v_1, v_2 \rangle$, the second one is $e_2 := \langle v_2, v_3 \rangle$, and the third common edge of T is $e_3 := \langle v_3, v_4 \rangle$. Now, since $\mathcal{L}$ contains the incenter of T, following Lemma 3.13, we can uniquely and stably determine $s|_T$ by interpolation at this point and the additional smoothness conditions of s. Then for each B-coefficient of s associated with a domain point ξ in T_{WF}^2, (2.12) holds with $\Gamma_\xi \subset (\mathcal{L} \cap \text{star}^7(T))$ and a constant $\hat{C}_{\mathcal{B}_{3,4}}$ depending on $\widetilde{C}_{\mathcal{B}_{2,4}}$ and the smallest solid and face angles of Δ. In the same way s restricted to all other tetrahedra in $\mathcal{B}_{3,4}$ that are refined with a second order partial Worsey-Farin split can be uniquely and stably determined.
- T is refined with a third order partial Worsey-Farin split. In this case the first common edge is $e_1 := \langle v_1, v_3 \rangle$, the second one is $e_2 := \langle v_2, v_3 \rangle$, and the third common edge of T is $e_3 := \langle v_4, v_3 \rangle$. Following Lemma 3.25, we can uniquely and stably determine $s|_T$ by interpolation at the incenter of T, which is contained in $\mathcal{L}$, and the additional smoothness conditions of s. For the B-coefficients of s associated with a domain point ξ in T_{WF}^3, (2.12) holds with $\Gamma_\xi \subset (\mathcal{L} \cap \text{star}^7(T))$ and a constant $\check{C}_{\mathcal{B}_{3,4}}$ depending on $\widetilde{C}_{\mathcal{B}_{2,4}}$ and the smallest solid and face angles of Δ. In the same way we can uniquely and stably compute the B-coefficients of s restricted to the remaining tetrahedra in $\mathcal{B}_{3,4}$ that are refined with a third order partial Worsey-Farin split.

Then, using the C^2 smoothness conditions at the faces F and the C^3 supersmoothness conditions at the edges e of the tetrahedra in $\mathcal{B}_{3,4}$, the B-coefficients corresponding to the domain points in the tubes $E_3(e)$ and the domain points in $F_i(F)$, $i = 0,1,2$, can be uniquely and stably determined. For any of these B-coefficients corresponding to a domain point ξ in a tetrahedron $\widetilde{T}$, sharing an edge or a face with T, (2.12) holds with $\Gamma_\xi \subset (\mathcal{L} \cap \text{star}^8(\widetilde{T}))$ and a constant $\widetilde{C}_{\mathcal{B}_{3,4}}$ that depends on $C_{\mathcal{B}_{3,4}}$ and the smallest solid and face angles of Δ.

Suppose we have uniquely and stably determined all B-coefficients of s restricted to the tetrahedra $T_1, \ldots, T_{i-1}$, and let $T := T_i \in \mathcal{B}_{4,4}$. Then, following Algorithm 5.1, the tetrahedron $T := \langle v_1, v_2, v_3, v_4 \rangle$ shares all four vertices and exactly four edges with the tetrahedra $T_1, \ldots, T_{i-1}$. By statement (v) of Lemma 5.2, T shares the four common edges with tetrahedra in $\mathcal{A}_0 \cup \ldots \cup \mathcal{A}_4 \cup \mathcal{B}_{1,2} \cup \ldots \cup \mathcal{B}_{3,4}$. Then, we have to distinguish between two cases (see Table 5.1):

- T is refined with a third order partial Worsey-Farin split. In this case, the common edges of T with the tetrahedra considered earlier are $e_1 := \langle v_1, v_3 \rangle$, $e_2 := \langle v_2, v_3 \rangle$, $e_3 := \langle v_4, v_3 \rangle$, and $e_4 := \langle v_1, v_2 \rangle$. Since by Algorithm 5.6 the set $\mathcal{L}$ contains the incenter of T, we can uniquely and stably compute the B-coefficients of $s|_T$ by interpolation at this point and the additional smoothness conditions of s, see Lemma 3.26. For each of the B-coefficients of s corresponding to a domain point ξ in T_{WF}^3, we have $\Gamma_\xi \subset (\mathcal{L} \cap \text{star}^8(T))$, and (2.12) holds with a constant $C_{\mathcal{B}_{4,4}}$ that depends on $\widetilde{C}_{\mathcal{B}_{3,4}}$ and the smallest solid and face angles of Δ. In the same way we can uniquely and stably compute the B-coefficients of s restricted to the remaining tetrahedra in $\mathcal{B}_{4,4}$ that are refined with a third order partial Worsey-Farin split.

- T is refined with a fourth order partial Worsey-Farin split. Then, the common edges of T with the tetrahedra considered earlier are $e_1 := \langle v_1, v_3 \rangle$, $e_2 := \langle v_1, v_4 \rangle$, $e_3 := \langle v_2, v_3 \rangle$, and $e_4 := \langle v_2, v_4 \rangle$. Then, following Lemma 3.38, we can uniquely and stably determine the B-coefficients of s restricted to T by interpolation at the incenter of T, which is contained in $\mathcal{L}$, and the additional smoothness conditions of s. For each of these B-coefficients, which are associated with a domain point ξ in T_{WF}^4, (2.12) holds with $\Gamma_\xi \subset (\mathcal{L} \cap \text{star}^8(T))$ and a constant $\hat{C}_{\mathcal{B}_{4,4}}$ that depends on $\widetilde{C}_{\mathcal{B}_{3,4}}$ and the smallest solid and face angles of Δ. In the same way the B-coefficients of s restricted to all other tetrahedra in $\mathcal{B}_{4,4}$ that are refined with a fourth order partial Worsey-Farin split can be uniquely and stably determined.

Then, for each edge e and face F of the tetrahedra in class $\mathcal{B}_{4,4}$, we can uniquely and stably determine the B-coefficients of s restricted to the domain points in the tube $E_3(e)$ and the domain points in $F_i(F)$, $i = 0,1,2$, using the C^2 smoothness conditions at the faces and the C^3 supersmoothness conditions at the edges of the tetrahedra in $\mathcal{B}_{4,4}$. For each of these B-coefficients corresponding to a domain point ξ in a tetrahedron $\widetilde{T}$, sharing an edge or a face with T, (2.12) holds with $\Gamma_\xi \subset (\mathcal{L} \cap \text{star}^9(\widetilde{T}))$ and a constant $\widetilde{C}_{\mathcal{B}_{4,4}}$ that depends on $C_{\mathcal{B}_{4,4}}$, $\hat{C}_{\mathcal{B}_{4,4}}$ and the smallest solid and face angles of Δ.

Next, suppose we have uniquely and stably determined all B-coefficients of s restricted to the tetrahedra $T_1,\dots,T_{i-1}$, and let $T := T_i \in \mathcal{B}_{5,4}$. Following Algorithm 5.1, the tetrahedron $T := \langle v_1,v_2,v_3,v_4\rangle$ shares all four vertices and exactly five edges with the tetrahedra $T_1,\dots,T_{i-1}$ and by statement (v) of Lemma 5.2, T shares these five common edges with tetrahedra in $\mathcal{A}_0 \cup \dots \cup \mathcal{A}_4 \cup \mathcal{B}_{1,2} \cup \dots \cup \mathcal{B}_{4,4}$. Then, the common edges of T with the tetrahedra considered earlier are $e_1 := \langle v_1,v_3\rangle$, $e_2 := \langle v_1,v_4\rangle$, $e_3 := \langle v_2,v_3\rangle$, $e_4 := \langle v_2,v_4\rangle$, and $e_5 := \langle v_1,v_2\rangle$. Since the incenter of T is contained in $\mathcal{L}$, following Lemma 3.39, we can uniquely and stably determine all B-coefficients of s restricted to T by interpolation at this point and the additional smoothness conditions of s. Then for each of the B-coefficients of s associated with a domain point ξ in T^4_{WF}, (2.12) holds with $\Gamma_\xi \subset (\mathcal{L} \cap \mathrm{star}^9(T))$ and a constant $\hat{C}_{\mathcal{B}_{5,4}}$ that depends on $\widetilde{C}_{\mathcal{B}_{4,4}}$ and the smallest solid and face angles of Δ. In the same way we can uniquely and stably determine the B-coefficients of s restricted to the remaining tetrahedra in $\mathcal{B}_{5,4}$. Then, using the C^2 smoothness conditions at the faces F and the C^3 supersmoothness conditions at the edges e of the tetrahedra in $\mathcal{B}_{5,4}$, the B-coefficients corresponding to the domain points in the tubes $E_3(e)$ and the domain points in $F_i(F)$, $i = 0,1,2$, can be uniquely and stably determined. For each of these B-coefficients corresponding to a domain point ξ in a tetrahedron $\widetilde{T}$, sharing an edge or a face with T, (2.12) holds with $\Gamma_\xi \subset (\mathcal{L} \cap \mathrm{star}^{10}(\widetilde{T}))$ and a constant $\widetilde{C}_{\mathcal{B}_{5,4}}$ that depends on $C_{\mathcal{B}_{4,4}}$ and the smallest solid and face angles of Δ.

Now, suppose we have uniquely and stably determined all B-coefficients of s restricted to the tetrahedra $T_1,\dots,T_{i-1}$, and let $T := T_i \in \mathcal{B}_{6,4}$. Then by Algorithm 5.1, the tetrahedron T shares all four vertices and six edges with the tetrahedra $T_1,\dots,T_{i-1}$. Following statement (v) of Lemma 5.2, T shares all edges with tetrahedra in $\mathcal{A}_0 \cup \dots \cup \mathcal{A}_4 \cup \mathcal{B}_{1,2} \cup \dots \cup \mathcal{B}_{5,4}$. By Algorithm 5.6, the set $\mathcal{L}$ contains the incenter of T. Then, following Lemma 3.40, by interpolation at the incenter and the additional smoothness conditions of s, we can uniquely and stably compute the B-coefficients of s restricted to T. For each of these B-coefficients, associated with a domain point ξ in T^4_{WF}, we have $\Gamma_\xi \subset (\mathcal{L} \cap \mathrm{star}^{10}(T))$ and (2.12) holds with a constant $C_{\mathcal{B}_{6,4}}$ that depends on $\widetilde{C}_{\mathcal{B}_{5,4}}$ and the smallest solid and face angles of Δ. In the same way the B-coefficients of s restricted to all other tetrahedra in $\mathcal{B}_{6,4}$ can be uniquely and stably determined. Now, using the C^2 smoothness conditions at the faces F of the tetrahedra in $\mathcal{B}_{6,4}$, the B-coefficients corresponding to the domain points in $F_i(F)$, $i = 0,1,2$, can be uniquely and stably determined. For each of these B-coefficients corresponding to a domain point ξ in a tetrahedron $\widetilde{T}$, sharing a face with T, (2.12) holds with $\Gamma_\xi \subset (\mathcal{L} \cap \text{star}^{10}(\widetilde{T}))$ and a constant $\widetilde{C}_{\mathcal{B}_{6,4}}$ that depends on $C_{\mathcal{B}_{6,4}}$ and the smallest solid and face angles of Δ.

At this point, we have uniquely and stably determined all B-coefficients of s restricted to the tetrahedra in the classes $\mathcal{A}_0,\ldots,\mathcal{A}_4,\mathcal{B}_{1,2},\ldots,\mathcal{B}_{6,4}$.

Suppose we have uniquely and stably determined all B-coefficients of s restricted to the tetrahedra $T_1,\ldots,T_{i-1}$, and let $T := T_i \in \mathcal{C}_{1,3,3}$. Following Algorithm 5.1, the tetrahedron T shares exactly one face $F := \langle u,v,w \rangle$ with the tetrahedra $T_1,\ldots,T_{i-1}$. By statement (iii) of Lemma 5.2, T shares the common face F with a tetrahedron in $\mathcal{A}_0 \cup \ldots \cup \mathcal{A}_4 \cup \mathcal{B}_{1,2} \cup \ldots \cup \mathcal{B}_{3,3}$. Then, we have to distinguish between two cases (see Table 5.1):

- T is not refined with a partial Worsey-Farin split. Due to the common face with the tetrahedra considered earlier, the B-coefficients of s corresponding to the domain points $P_{\mathcal{C}}^{1,3,3} := \bigcup_{i=0}^{2} F_i^T(F) \cup E_3^T(\langle u,v \rangle) \cup E_3^T(\langle v,w \rangle) \cup E_3^T(\langle w,u \rangle) \cup D_4^T(u) \cup \cup D_4^T(v) \cup D_4^T(w)$ are already uniquely and stably determined. Then, since the set $\mathcal{L}$ contains the points $\mathcal{D}_{T,9} \setminus P_{\mathcal{C}}^{1,3,3}$, the remaining B-coefficients of $s|_T$ can be uniquely determined by interpolation. The matrix of the corresponding linear system is $M_{\mathcal{C}_{1,3,3}} := \left(B_\xi^{9,T}(\kappa)\right)_{\xi,\kappa \in \mathcal{D}_{T,9} \setminus P_{\mathcal{C}}^{1,3,3}}$, which has a positive determinant. Moreover, for the B-coefficients of s associated with the domain points $\xi \in \mathcal{D}_{T,9} \setminus P_{\mathcal{C}}^{1,3,3}$, (2.12) holds with $\Gamma_\xi \subset (\mathcal{L} \cap \text{star}^6(T))$ and a constant $C_{\mathcal{C}_{1,3,3}}$, that depends on $\|M_{\mathcal{C}_{1,3,3}}^{-1}\|$ and $\widetilde{C}_{\mathcal{B}_{3,3},F}$. In the same way we can uniquely and stably compute the B-coefficients of s restricted to the remaining non-split tetrahedra in $\mathcal{C}_{1,3,3}$.
- T is refined with a first order partial Worsey-Farin split. Following Algorithm 5.6, the set $\mathcal{L}$ contains the incenter and the fourth vertex of T, the one not shared with a tetrahedron in $\{T_1,\ldots,T_{i-1}\}$. Now, by Lemma 3.6, we can uniquely and stably determine all B-coefficients of $s|_T$ by interpolation at these two points and the additional smoothness conditions of s. For each B-coefficient of s associated with a domain point ξ in T_{WF}^1, (2.12) holds with $\Gamma_\xi \subset (\mathcal{L} \cap \text{star}^6(T))$ and a constant $\hat{C}_{\mathcal{C}_{1,3,3}}$ depending on $\widetilde{C}_{\mathcal{B}_{3,3},F}$ and the smallest solid and face angles of Δ. In the same way s restricted to the remaining split tetrahedra in class $\mathcal{C}_{1,3,3}$ can be uniquely and stably determined.

Then, using the C^2 smoothness conditions at the faces F, the C^3 supersmoothness conditions at the edges e, and the C^4 supersmoothness conditions at the vertices v of the tetrahedra in $\mathcal{C}_{1,3,3}$, the B-coefficients corresponding to the domain points in the balls $D_4(v)$, the tubes $E_3(e)$, and the domain points in $F_i(F)$, $i = 0,1,2$, can be uniquely and stably determined. For each of these B-coefficients corresponding to a domain point ξ in a tetrahedron $\widetilde{T}$, sharing a vertex or an edge with T, (2.12) holds with $\Gamma_\xi \subset (\mathcal{L} \cap \text{star}^7(\widetilde{T}))$ and a constant $\widetilde{C}_{\mathcal{C}_{1,3,3}}$ that depends on $C_{\mathcal{C}_{1,3,3}}$, $\hat{C}_{\mathcal{C}_{1,3,3}}$, and the smallest solid and face angles of Δ. For the B-coefficients corresponding to a domain point ξ in a tetrahedron $\widetilde{T}$, sharing a common face with T, (2.12) holds with $\Gamma_\xi \subset (\mathcal{L} \cap \text{star}^6(\widetilde{T}))$ and a constant $\widetilde{C}_{\mathcal{C}_{1,3,3},F}$ that depends on $C_{\mathcal{C}_{1,3,3}}$, $\hat{C}_{\mathcal{C}_{1,3,3}}$, and the smallest solid and face angles of Δ.

Now, suppose we have uniquely and stably determined all B-coefficients of s restricted to the tetrahedra $T_1, \ldots, T_{i-1}$, and let $T := T_i \in \mathcal{C}_{1,3,4}$. By the definition of $\mathcal{C}_{1,3,4}$, the tetrahedron $T := \langle v_1, v_2, v_3, v_4 \rangle$ shares exactly one face $F := \langle v_1, v_2, v_3 \rangle$ and the opposite vertex v_4 with the tetrahedra $T_1, \ldots, T_{i-1}$. Following statement (iii) of Lemma 5.2, T shares the common face F with a tetrahedron in $\mathcal{A}_0 \cup \ldots \cup \mathcal{A}_4 \cup \mathcal{B}_{1,2} \cup \ldots \cup \mathcal{B}_{3,4}$. By statement (ix) of the same lemma, we obtain that T shares the vertex v_4 with a tetrahedron in $\mathcal{A}_0 \cup \ldots \cup \mathcal{A}_3 \cup \mathcal{B}_{1,2} \cup \mathcal{B}_{1,3} \cup \mathcal{B}_{2,3} \cup \mathcal{B}_{3,3} \cup \mathcal{C}_{1,3,3}$. Again, we have to distinguish between two cases (see Table 5.1):

- T is not refined with a partial Worsey-Farin split. The B-coefficients of s corresponding to the domain points in the set $P_{\mathcal{C}}^{1,3,4} := \bigcup_{i=0}^{2} F_i^T(F) \cup E_3^T(\langle v_1, v_2 \rangle) \cup E_3^T(\langle v_2, v_3 \rangle) \cup E_3^T(\langle v_3, v_1 \rangle) \cup \bigcup_{i=1}^{4} D_4^T(v_i)$ are uniquely and stably determined. Then, since the set $\mathcal{L}$ contains the points $\mathcal{D}_{T,9} \setminus P_{\mathcal{C}}^{1,3,4}$, the remaining B-coefficients of $s|_T$ can be uniquely determined by interpolation at these points. The matrix of the corresponding linear system is $M_{\mathcal{C}_{1,3,4}} := \left(B_\xi^{9,T}(\kappa)\right)_{\xi,\kappa \in \mathcal{D}_{T,9} \setminus P_{\mathcal{C}}^{1,3,4}}$, which has a positive determinant. For the B-coefficients of s associated with the domain points $\xi \in P_{\mathcal{C}}^{1,3,4}$, (2.12) holds with $\Gamma_\xi \subset (\mathcal{L} \cap \text{star}^7(T))$ and a constant $C_{\mathcal{C}_{1,3,4}}$, that depends on $\|M_{\mathcal{C}_{1,3,4}}^{-1}\|$ and $\widetilde{C}_{\mathcal{C}_{1,3,3}}$. In the same way the B-coefficients of s restricted to the remaining non-split tetrahedra in $\mathcal{C}_{1,3,4}$ can be uniquely and stably determined.

- T is refined with a first order partial Worsey-Farin split. In this case the set $\mathcal{L}$ contains the incenter of T. Now, following Lemma 3.5, we can uniquely and stably compute the B-coefficients of $s|_T$ by interpolation at the incenter and the additional smoothness conditions of s. Then, for all B-coefficient of s associated with a domain point ξ in T_{WF}^1, (2.12) holds with $\Gamma_\xi \subset (\mathcal{L} \cap \mathrm{star}^7(T))$ and a constant $\hat{C}_{\mathcal{C}_{1,3,4}}$ depending on $\widetilde{C}_{\mathcal{C}_{1,3,3}}$ and the smallest solid and face angles of Δ. In the same way we can uniquely and stably determine the B-coefficients of s restricted to the remaining split tetrahedra in class $\mathcal{C}_{1,3,4}$.

Then, for each edge e and face F of the tetrahedra in class $\mathcal{C}_{1,3,4}$, we can uniquely and stably determine the B-coefficients of s restricted to the domain points in the tube $E_3(e)$ and the domain points in $F_i(F)$, $i = 0,1,2$, using the C^2 smoothness conditions at the faces and the C^3 supersmoothness conditions at the edges of the tetrahedra in $\mathcal{C}_{1,3,4}$. For each of these B-coefficients corresponding to a domain point ξ in a tetrahedron $\widetilde{T}$, sharing an edge or a face with T, (2.12) holds with $\Gamma_\xi \subset (\mathcal{L} \cap \mathrm{star}^7(\widetilde{T}))$ and a constant $\widetilde{C}_{\mathcal{C}_{1,3,4}}$ that depends on $C_{\mathcal{C}_{1,3,4}}$, $\hat{C}_{\mathcal{C}_{1,3,4}}$ and the smallest solid and face angles of Δ.

Next, suppose we have uniquely and stably determined all B-coefficients of s restricted to the tetrahedra $T_1, \ldots, T_{i-1}$, and let $T := T_i \in \mathcal{C}_{1,4,4}$. By Algorithm 5.1, the tetrahedron $T := \langle v_1, v_2, v_3, v_4 \rangle$ shares exactly one face $F := \langle v_1, v_2, v_4 \rangle$ and the edge $e := \langle v_2, v_3 \rangle$ with the tetrahedra $T_1, \ldots, T_{i-1}$. Following statement (iii) of Lemma 5.2, T shares the common face F with a tetrahedron in $\mathcal{A}_0 \cup \ldots \cup \mathcal{A}_4 \cup \mathcal{B}_{1,2} \cup \ldots \cup \mathcal{B}_{4,4}$. Moreover, by statement (v) of this lemma, we conclude that T shares the edge e with a tetrahedron in $\mathcal{A}_0 \cup \ldots \cup \mathcal{A}_4 \cup \mathcal{B}_{1,2} \cup \ldots \cup \mathcal{B}_{5,4} \cup \mathcal{C}_{1,3,3} \cup \mathcal{C}_{1,3,4}$. Now, we have to distinguish between two cases (see Table 5.1):

- T is refined with a second order partial Worsey-Farin split. Then, by Algorithm 5.6, the incenter of T is contained in the set $\mathcal{L}$. Thus, following Lemma 3.15, we can uniquely and stably determine the B-coefficients of $s|_T$ by interpolation at the incenter and the additional smoothness conditions of s. For all B-coefficient of s corresponding to a domain point ξ in T_{WF}^2, (2.12) holds with $\Gamma_\xi \subset (\mathcal{L} \cap \mathrm{star}^{10}(T))$ and a constant $C_{\mathcal{C}_{1,4,4}}$ depending on $\widetilde{C}_{\mathcal{C}_{1,3,4}}$ and the smallest solid and face angles of Δ. Then, we can uniquely and stably determine the B-coefficients of s restricted to all other tetrahedra in class $\mathcal{C}_{1,4,4}$ that are refined with a second order partial Worsey-Farin split in the same way.

- T is refined with a third order partial Worsey-Farin split. In this case $\mathcal{L}$ also contains the incenter of T. Therefore, following Lemma 3.30, the B-coefficients of $s|_T$ can be uniquely and stably determined by interpolation at the incenter and the additional smoothness conditions of s. Then, for the B-coefficient of s corresponding to a domain point ξ in T_{WF}^3, (2.12) holds with $\Gamma_\xi \subset (\mathcal{L} \cap \mathrm{star}^{10}(T))$ and a constant $\hat{C}_{\mathcal{C}_{1,4,4}}$ that depends on $\widetilde{C}_{\mathcal{C}_{1,3,4}}$ and the smallest solid and face angles of Δ. In the same way we can uniquely and stably compute the B-coefficients of s restricted to the remaining tetrahedra in class $\mathcal{C}_{1,4,4}$ that are refined with a third order partial Worsey-Farin split.

Then, using the C^2 smoothness conditions at the faces F and the C^3 supersmoothness conditions at the edges e of the tetrahedra in $\mathcal{C}_{1,4,4}$, the B-coefficients corresponding to the domain points in the tubes $E_3(e)$ and the domain points in $F_i(F)$, $i = 0,1,2$, can be uniquely and stably determined. For each of these B-coefficients corresponding to a domain point ξ in a tetrahedron $\widetilde{T}$, sharing an edge or a face with T, (2.12) holds with $\Gamma_\xi \subset (\mathcal{L} \cap \mathrm{star}^{10}(\widetilde{T}))$ and a constant $\widetilde{C}_{\mathcal{C}_{1,4,4}}$ that depends on $C_{\mathcal{C}_{1,4,4}}$, $\hat{C}_{\mathcal{C}_{1,4,4}}$, and the smallest solid and face angles of Δ.

Now, suppose we have uniquely and stably determined all B-coefficients of s restricted to the tetrahedra $T_1,\dots,T_{i-1}$, and let $T := T_i \in \mathcal{C}_{1,5,4}$. Following Algorithm 5.1, the tetrahedron $T := \langle v_1,v_2,v_3,v_4\rangle$ shares exactly one face $F := \langle v_1,v_2,v_4\rangle$ and the two edges $e_1 := \langle v_1,v_3\rangle$ and $e_2 := \langle v_2,v_3\rangle$ with the tetrahedra $T_1,\dots,T_{i-1}$. By statement (iii) of Lemma 5.2, T shares the common face F with a tetrahedron in $\mathcal{A}_0 \cup \dots \cup \mathcal{A}_4 \cup \mathcal{B}_{1,2} \cup \dots \cup \mathcal{B}_{5,4}$. Following statement (v) of the same lemma, we see that T shares the edges e_1 and e_2 with tetrahedra in $\mathcal{A}_0 \cup \dots \cup \mathcal{A}_4 \cup \mathcal{B}_{1,2} \cup \dots \cup \mathcal{B}_{5,4} \cup \mathcal{C}_{1,3,3} \cup \mathcal{C}_{1,3,4} \cup \mathcal{C}_{1,4,4}$. Now, since T is refined with a partial Worsey-Farin split, the set $\mathcal{L}$ contains the incenter of T. Then, we have to distinguish between two cases (see Table 5.1):

- T is refined with a third order partial Worsey-Farin split. Following Lemma 3.27, we can uniquely and stably compute the B-coefficients of $s|_T$ by interpolation at the incenter of T and the additional smoothness conditions of s. For these B-coefficients, associated with a domain point ξ in T_{WF}^3, (2.12) holds with $\Gamma_\xi \subset (\mathcal{L} \cap \mathrm{star}^{10}(T))$ and a constant $C_{\mathcal{C}_{1,5,4}}$ that depends on $\widetilde{C}_{\mathcal{C}_{1,4,4}}$ and the smallest solid and face angles of Δ. In the same way we can uniquely and stably compute the B-coefficients of s restricted to all other tetrahedra in class $\mathcal{C}_{1,5,4}$ that are refined with a third order partial Worsey-Farin split.

- T is refined with a fourth order partial Worsey-Farin split. Then, following Lemma 3.42, the B-coefficients of $s|_T$ can be uniquely and stably computed by interpolation at the incenter of T and the additional smoothness conditions of s. For these B-coefficients of s corresponding to a domain point ξ in T_{WF}^4, we have $\Gamma_\xi \subset (\mathcal{L} \cap \text{star}^{10}(T))$ and (2.12) holds with a constant $\hat{C}_{\mathcal{C}_{1,5,4}}$ depending on $\widetilde{C}_{\mathcal{C}_{1,4,4}}$ and the smallest solid and face angles of Δ. Then, the B-coefficients of s restricted to the remaining tetrahedra in class $\mathcal{C}_{1,5,4}$ that are refined with a fourth order partial Worsey-Farin split can be uniquely and stably determined in the same way.

Subsequently, using the C^2 smoothness conditions at the faces F and the C^3 supersmoothness conditions at the edges e of the tetrahedra in $\mathcal{C}_{1,5,4}$, we can uniquely and stably determine the B-coefficients corresponding to the domain points in the tubes $E_3(e)$ and the domain points in $F_i(F)$, $i = 0,1,2$. For any of these B-coefficients corresponding to a domain point ξ in a tetrahedron $\widetilde{T}$, sharing an edge or a face with T, (2.12) holds with $\Gamma_\xi \subset (\mathcal{L} \cap \text{star}^{10}(\widetilde{T}))$ and a constant $\widetilde{C}_{\mathcal{C}_{1,5,4}}$ that depends on $C_{\mathcal{C}_{1,5,4}}$, $\hat{C}_{\mathcal{C}_{1,5,4}}$, and the smallest solid and face angles of Δ.

Suppose we have uniquely and stably determined all B-coefficients of s restricted to the tetrahedra $T_1, \ldots, T_{i-1}$, and let $T := T_i \in \mathcal{C}_{1,6,4}$. By the definition of $\mathcal{C}_{1,6,4}$, the tetrahedron $T := \langle v_1, v_2, v_3, v_4 \rangle$ shares exactly one face and all six edges with the tetrahedra $T_1, \ldots, T_{i-1}$. By statement (iii) of Lemma 5.2, T shares the common face F with a tetrahedron in $\mathcal{A}_0 \cup \ldots \cup \mathcal{A}_4 \cup \mathcal{B}_{1,2} \cup \ldots \cup \mathcal{B}_{6,4}$. Moreover, by statement (v) of the this lemma, we obtain that T shares the other three edges with tetrahedra in $\mathcal{A}_0 \cup \ldots \cup \mathcal{A}_4 \cup \mathcal{B}_{1,2} \cup \ldots \cup \mathcal{B}_{5,4} \cup \mathcal{C}_{1,3,3} \cup \ldots \cup \mathcal{C}_{1,5,4}$. Then, since T is refined with a partial Worsey-Farin split, the incenter of T is contained in $\mathcal{L}$ and we have to distinguish between two cases (see Table 5.1):

- T is refined with a third order partial Worsey-Farin split. By Lemma 3.28, we can uniquely and stably determine the B-coefficients of $s|_T$ by interpolation at the incenter of T and the additional smoothness conditions of s. For each of the B-coefficients corresponding to a domain point ξ in T_{WF}^3, we have $\Gamma_\xi \subset (\mathcal{L} \cap \text{star}^{10}(T))$ and (2.12) holds with a constant $C_{\mathcal{C}_{1,6,4}}$ depending on $\widetilde{C}_{\mathcal{C}_{1,5,4}}$ and the smallest solid and face angles of Δ. In the same way the B-coefficients of s restricted to all other tetrahedra in $\mathcal{C}_{1,6,4}$ that are refined with a third order partial Worsey-Farin split can be uniquely and stably determined.

- T is refined with a fourth order partial Worsey-Farin split. In this case, we can follow Lemma 3.43 to uniquely and stably determine the B-coefficients of $s|_T$ by interpolation at the incenter of T and the additional smoothness conditions of s. Then, for the B-coefficients, that correspond to a domain points ξ in T_{WF}^4, (2.12) holds with $\Gamma_\xi \subset (\mathcal{L} \cap \text{star}^{10}(T))$ and a constant $\hat{C}_{\mathcal{C}_{1,6,4}}$ which depends on $\widetilde{C}_{\mathcal{C}_{1,5,4}}$ and the smallest solid and face angles of Δ. In the same way we can uniquely and stably determine the B-coefficients of s restricted to the other tetrahedra in $\mathcal{C}_{1,6,4}$ that are refined with a fourth order partial Worsey-Farin split.

Now, for each face F of the tetrahedra in class $\mathcal{C}_{1,6,4}$, we can uniquely and stably determine the B-coefficients of s restricted to the domain points in $F_i(F)$, $i = 0,1,2$, using the C^2 smoothness conditions at the faces of the tetrahedra in $\mathcal{C}_{1,6,4}$. For all of these B-coefficients corresponding to a domain point ξ in a tetrahedron $\widetilde{T}$, sharing a face with T, (2.12) holds with $\Gamma_\xi \subset (\mathcal{L} \cap \text{star}^{11}(\widetilde{T}))$ and a constant $\widetilde{C}_{\mathcal{C}_{1,6,4}}$ that depends on $C_{\mathcal{C}_{1,6,4}}$, $\hat{C}_{\mathcal{C}_{1,6,4}}$ and the smallest solid and face angles of Δ.

Now, suppose we have uniquely and stably determined all B-coefficients of s restricted to the tetrahedra $T_1, \ldots, T_{i-1}$, and let $T := T_i \in \mathcal{C}_{2,5,4}$. By the definition of $\mathcal{C}_{2,5,4}$ in Algorithm 5.1, the tetrahedron T shares exactly two faces F_1 and F_2 with the tetrahedra $T_1, \ldots, T_{i-1}$. Following statement (iii) of Lemma 5.2, T shares the first common face F_1 with a tetrahedron in $\mathcal{A}_0 \cup \ldots \cup \mathcal{A}_4 \cup \mathcal{B}_{1,2} \cup \ldots \cup \mathcal{B}_{5,4}$. Moreover, by statement (iii) of the same lemma, we see that T shares the second common face F_2 with a tetrahedron in $\mathcal{A}_0 \cup \ldots \cup \mathcal{A}_4 \cup \mathcal{B}_{1,2} \cup \ldots \cup \mathcal{B}_{6,4} \cup \mathcal{C}_{1,3,3} \cup \ldots \cup \mathcal{C}_{1,5,4}$. Due to the fact that T is refined with a partial Worsey-Farin split, the set $\mathcal{L}$ contains the incenter of T and we have to distinguish between three cases (see Table 5.1):

- T is refined with a second order partial Worsey-Farin split. Following Lemma 3.17, we can uniquely and stably determine the B-coefficients of $s|_T$ by interpolation at the incenter of T and the additional smoothness conditions of s. For each of the B-coefficients corresponding to a domain point ξ in T_{WF}^2, (2.12) holds with $\Gamma_\xi \subset (\mathcal{L} \cap \text{star}^{10}(T))$ and a constant $C_{\mathcal{C}_{2,5,4}}$ depending on $\widetilde{C}_{\mathcal{C}_{1,5,4}}$ and the smallest solid and face angles of Δ. In the same way we can uniquely and stably determine the B-coefficients of s restricted to all other tetrahedra in $\mathcal{C}_{2,5,4}$ that are refined with a second order partial Worsey-Farin split.

- T is refined with a third order partial Worsey-Farin split. Then, by Lemma 3.31, the B-coefficients of $s|_T$ can be uniquely and stably determined by interpolation at the incenter of T and the additional smoothness conditions of s. For all B-coefficients of s corresponding to a domain point ξ in T_{WF}^3, we have $\Gamma_\xi \subset (\mathcal{L} \cap \text{star}^{10}(T))$ and (2.12) holds with a constant $\hat{C}_{\mathcal{C}_{2,5,4}}$ that depends on $\widetilde{C}_{\mathcal{C}_{1,5,4}}$ and the smallest solid and face angles of Δ. In the same way the B-coefficients of s restricted to the remaining tetrahedra in $\mathcal{C}_{2,5,4}$ that are refined with a third order partial Worsey-Farin split can be uniquely and stably determined.

- T is refined with a fourth order partial Worsey-Farin split. Following Lemma 3.45, we can uniquely and stably compute the B-coefficients of $s|_T$ by interpolation at the incenter of T and the additional smoothness conditions of s. Then, for each B-coefficient of s associated with a domain point ξ in T_{WF}^4, (2.12) holds with $\Gamma_\xi \subset (\mathcal{L} \cap \text{star}^{10}(T))$ and a constant $\check{C}_{\mathcal{C}_{2,5,4}}$ that depends on $\widetilde{C}_{\mathcal{C}_{1,5,4}}$ and the smallest solid and face angles of Δ. In the same way we can uniquely and stably compute the B-coefficients of s restricted to all other tetrahedra in $\mathcal{C}_{2,5,4}$ that are refined with a fourth order partial Worsey-Farin split.

Then, using the C^2 smoothness conditions at the faces F and the C^3 supersmoothness conditions at the edges e of the tetrahedra in $\mathcal{C}_{2,5,4}$, the B-coefficients corresponding to the domain points in the tubes $E_3(e)$ and the domain points in $F_i(F)$, $i = 0,1,2$, can be uniquely and stably determined. For each of these B-coefficients corresponding to a domain point ξ in a tetrahedron $\widetilde{T}$, sharing an edge or a face with T, (2.12) holds with $\Gamma_\xi \subset (\mathcal{L} \cap \text{star}^{10}(\widetilde{T}))$ and a constant $\widetilde{C}_{\mathcal{C}_{2,5,4}}$ that depends on $C_{\mathcal{C}_{2,5,4}}$, $\hat{C}_{\mathcal{C}_{2,5,4}}$, $\check{C}_{\mathcal{C}_{2,5,4}}$, and the smallest solid and face angles of Δ.

Next, suppose we have uniquely and stably determined all B-coefficients of s restricted to the tetrahedra $T_1, \ldots, T_{i-1}$, and let $T := T_i \in \mathcal{C}_{2,6,4}$. Following Algorithm 5.1, the tetrahedron T shares exactly two faces F_1 and F_2, and also the last edge e, which is not contained in F_1 or F_2, with the tetrahedra $T_1, \ldots, T_{i-1}$. By statements (iii) and (iv) of Lemma 5.2, T shares the first common face F_1 with a tetrahedron in $\mathcal{A}_0 \cup \ldots \cup \mathcal{A}_4 \cup \mathcal{B}_{1,2} \cup \ldots \cup \mathcal{B}_{6,4}$ and the second common face F_2 with a tetrahedron in $\mathcal{A}_0 \cup \ldots \cup \mathcal{A}_4 \cup \mathcal{B}_{1,2} \cup \ldots \cup \mathcal{B}_{6,4} \cup \mathcal{C}_{1,3,3} \cup \ldots \cup \mathcal{C}_{1,6,4}$. Moreover, following statement (v) of the same lemma, we can see that T shares the edge e with a tetrahedron in $\mathcal{A}_0 \cup \ldots \cup \mathcal{A}_4 \cup \mathcal{B}_{1,2} \cup \ldots \cup \mathcal{B}_{5,4} \cup \mathcal{C}_{1,3,3} \cup \ldots \cup \mathcal{C}_{1,5,4} \cup \mathcal{C}_{2,5,4}$. Now, since T is refined with a partial Worsey-Farin split, the set $\mathcal{L}$ contains the incenter of T. Then, we have to distinguish between three cases (see Table 5.1):

- T is refined with a second order partial Worsey-Farin split. Following Lemma 3.19, we can uniquely and stably determine the B-coefficients of $s|_T$ by interpolation at the incenter of T and the additional smoothness conditions of s. Then, for each B-coefficient of s corresponding to a domain point ξ in T_{WF}^2, (2.12) holds with $\Gamma_\xi \subset (\mathcal{L} \cap \text{star}^{11}(T))$ and a constant $C_{\mathcal{C}_{2,6,4}}$ that depends $\widetilde{C}_{\mathcal{C}_{2,5,4}}$ and the smallest solid and face angles of Δ. In the same way the B-coefficients of s restricted to all other tetrahedra in $\mathcal{C}_{2,6,4}$ that are refined with a second order partial Worsey-Farin split can be uniquely and stably determined.

- T is refined with a third order partial Worsey-Farin split. Then, by Lemma 3.32, we can uniquely and stably determine the B-coefficients of $s|_T$ by interpolation at the incenter of T and the additional smoothness conditions of s. For all B-coefficients of s corresponding to a domain point ξ in T_{WF}^3, we have $\Gamma_\xi \subset (\mathcal{L} \cap \text{star}^{11}(T))$ and (2.12) holds with a constant $\hat{C}_{\mathcal{C}_{2,6,4}}$ that depends on $\widetilde{C}_{\mathcal{C}_{2,5,4}}$ and the smallest solid and face angles of Δ. In the same way the B-coefficients of s restricted to the remaining tetrahedra in $\mathcal{C}_{2,5,4}$ that are refined with a third order partial Worsey-Farin split can be uniquely and stably determined.

- T is refined with a fourth order partial Worsey-Farin split. In this case, it follows from Lemma 3.46, that we can uniquely and stably compute the B-coefficients of $s|_T$ by interpolation at the incenter of T and the additional smoothness conditions of s. Then, for the B-coefficients of s associated with a domain point ξ in T_{WF}^4, (2.12) holds with $\Gamma_\xi \subset (\mathcal{L} \cap \text{star}^{11}(T))$ and a constant $\check{C}_{\mathcal{C}_{2,6,4}}$ that depends on $\widetilde{C}_{\mathcal{C}_{2,5,4}}$ and the smallest solid and face angles of Δ. In the same way we can uniquely and stably compute the B-coefficients of s restricted to all other tetrahedra in class $\mathcal{C}_{2,6,4}$ that are refined with a fourth order partial Worsey-Farin split.

Subsequently, for each face F of the tetrahedra in class $\mathcal{C}_{2,6,4}$, we can uniquely and stably determine the B-coefficients of s restricted to the domain points in $F_i(F)$, $i = 0,1,2$, using the C^2 smoothness conditions at the faces of the tetrahedra in $\mathcal{C}_{2,6,4}$. For each of these B-coefficients corresponding to a domain point ξ in a tetrahedron $\widetilde{T}$, sharing a face with T, (2.12) holds with $\Gamma_\xi \subset (\mathcal{L} \cap \text{star}^{11}(\widetilde{T}))$ and a constant $\widetilde{C}_{\mathcal{C}_{2,6,4}}$ depending on $C_{\mathcal{C}_{2,6,4}}$, $\hat{C}_{\mathcal{C}_{2,6,4}}$, $\check{C}_{\mathcal{C}_{2,6,4}}$ and the smallest solid and face angles of Δ.

Suppose we have uniquely and stably determined all B-coefficients of s restricted to the tetrahedra $T_1, \ldots, T_{i-1}$, and let $T := T_i \in \mathcal{C}_{3,6,4}$. By the definition of $\mathcal{C}_{3,6,4}$, the tetrahedron T shares exactly three faces F_1, F_2, and F_3 with the tetrahedra $T_1, \ldots, T_{i-1}$. Following statement (iii) of Lemma 5.2, T shares the first common face F_1 with a tetrahedron in $\mathcal{A}_0 \cup \ldots \cup \mathcal{A}_4 \cup \mathcal{B}_{1,2} \cup \ldots \cup \mathcal{B}_{6,4}$. By statements (ii) and (iv) of the same lemma, we obtain that T shares the faces F_2 and F_3 with tetrahedra in $\mathcal{A}_0 \cup \ldots \cup \mathcal{A}_4 \cup \mathcal{B}_{1,2} \cup \ldots \cup \mathcal{B}_{6,4} \cup \mathcal{C}_{1,3,3} \cup \ldots \cup \mathcal{C}_{2,6,4}$. Then, since T is refined with a partial Worsey-Farin split, the set $\mathcal{L}$ contains the incenter of T, and we have to distinguish between three cases (see Table 5.1):

- T is refined with a second order partial Worsey-Farin split. Following Lemma 3.21, the B-coefficients of $s|_T$ are uniquely and stably determined by interpolation at the incenter of T and the additional smoothness conditions of s. Then, for each B-coefficient of s associated with a domain point ξ in T_{WF}^2, (2.12) holds with $\Gamma_\xi \subset (\mathcal{L} \cap \text{star}^{11}(T))$ and a constant $C_{\mathcal{C}_{3,6,4}}$ depending on $\widetilde{C}_{\mathcal{C}_{2,6,4}}$ and the smallest solid and face angles of Δ. We can uniquely and stably determine the B-coefficients of s restricted to the remaining tetrahedra in $\mathcal{C}_{3,6,4}$ that are refined with a second order partial Worsey-Farin split in the same way.

- T is refined with a third order partial Worsey-Farin split. It follows from Lemma 3.34 that the B-coefficients of $s|_T$ are uniquely and stably determined by interpolation at the incenter of T and the additional smoothness conditions of s. For each of the B-coefficients of s corresponding to a domain point ξ in T_{WF}^3, we have $\Gamma_\xi \subset (\mathcal{L} \cap \text{star}^{11}(T))$ and (2.12) holds with a constant $\hat{C}_{\mathcal{C}_{3,6,4}}$ that depends on $\widetilde{C}_{\mathcal{C}_{2,6,4}}$ and the smallest solid and face angles of Δ. In the same way we can uniquely and stably compute the B-coefficients of s restricted to all other tetrahedra in $\mathcal{C}_{3,6,4}$ that are refined with a third order partial Worsey-Farin.

- T is refined with a fourth order partial Worsey-Farin split. In this case, it follows from Lemma 3.48, that we can uniquely and stably determine the B-coefficients of $s|_T$ by interpolation at the incenter of T and the additional smoothness conditions of s. Then, for each B-coefficient of s corresponding to a domain point ξ in T_{WF}^4, (2.12) holds with $\Gamma_\xi \subset (\mathcal{L} \cap \text{star}^{11}(T))$ and a constant $\check{C}_{\mathcal{C}_{3,6,4}}$ depending on $\widetilde{C}_{\mathcal{C}_{2,6,4}}$

and the smallest solid and face angles of Δ. In the same way the B-coefficients of s corresponding to the domain points in the remaining tetrahedra in $\mathcal{C}_{3,6,4}$ that are refined with a fourth order partial Worsey-Farin split can be uniquely and stably determined.

Then, using the C^2 smoothness conditions at the only face F of T, that is not shared with a tetrahedron considered earlier, the B-coefficients corresponding to the domain points in $F_i(F)$, $i = 0,1,2$, can be uniquely and stably determined. For each of these B-coefficients associated with a domain point ξ in a tetrahedron $\widetilde{T}$, sharing the face F with T, (2.12) holds with $\Gamma_\xi \subset (\mathcal{L} \cap \text{star}^{11}(\widetilde{T}))$ and a constant $\widetilde{C}_{\mathcal{C}_{3,6,4}}$ that depends on $C_{\mathcal{C}_{3,6,4}}$, $\hat{C}_{\mathcal{C}_{3,6,4}}$, $\check{C}_{\mathcal{C}_{3,6,4}}$, and the smallest solid and face angles of Δ.

Finally, we consider the last class $\mathcal{C}_{4,6,4}$. To this end, suppose we have uniquely and stably determined all B-coefficients of s restricted to the tetrahedra $T_1, \ldots, T_{i-1}$, and let $T := T_i \in \mathcal{C}_{4,6,4}$. Then following the definition of $\mathcal{C}_{4,6,4}$ in Algorithm 5.1, the tetrahedron T shares all four faces with the tetrahedra $T_1, \ldots, T_{i-1}$. By statement (ii) of Lemma 5.2, we can see that T shares its faces with the tetrahedra in $\mathcal{A}_0 \cup \ldots \cup \mathcal{A}_4 \cup \mathcal{B}_{1,2} \cup \ldots \cup \mathcal{B}_{6,4} \cup \mathcal{C}_{1,3,3} \cup \ldots \cup \mathcal{C}_{3,6,4}$. Then, following Algorithm 5.6, the set $\mathcal{L}$ contains the incenter of T, since T is refined with a partial Worsey-Farin split, and we have to distinguish between three cases (see Table 5.1):

- T is refined with a second order partial Worsey-Farin split. It follows from Lemma 3.23, that the B-coefficients of $s|_T$ are uniquely and stably determined by interpolation at the incenter of T and the additional smoothness conditions of s. For each B-coefficient of s corresponding to a domain point ξ in T_{WF}^2, (2.12) holds with $\Gamma_\xi \subset (\mathcal{L} \cap \text{star}^{11}(T))$ and a constant $C_{\mathcal{C}_{4,6,4}}$ depending on $\widetilde{C}_{\mathcal{C}_{3,6,4}}$ and the smallest solid and face angles of Δ. In the same way we can uniquely and stably determine the B-coefficients of s restricted to the remaining tetrahedra in $\mathcal{C}_{4,6,4}$ that are refined with a second order partial Worsey-Farin split.

- T is refined with a third order partial Worsey-Farin split. Following Lemma 3.36, we can uniquely and stably determined the B-coefficients of $s|_T$ by interpolation at the incenter of T and the additional smoothness conditions of s. Then, for each B-coefficient of s associated with a domain point ξ in T_{WF}^3, (2.12) holds with $\Gamma_\xi \subset (\mathcal{L} \cap \text{star}^{11}(T))$ and a constant $\hat{C}_{\mathcal{C}_{4,6,4}}$ that depends on $\widetilde{C}_{\mathcal{C}_{3,6,4}}$ and the smallest solid and face angles of Δ. In the same way the B-coefficients of s restricted to all other tetrahedra in $\mathcal{C}_{4,6,4}$ that are refined with a third order partial Worsey-Farin can be uniquely and stably computed.

- T is refined with a fourth order partial Worsey-Farin split. Then, by Lemma 3.50, we can uniquely and stably determine the B-coefficients of $s|_T$ by interpolation at the incenter of T and the additional smoothness conditions of s. For all B-coefficients of s corresponding to a domain point ξ in T^4_{WF}, we have $\Gamma_\xi \subset (\mathcal{L} \cap \text{star}^{11}(T))$ and (2.12) holds with a constant $\check{C}_{\mathcal{C}_{4,6,4}}$ that depends on $\widetilde{C}_{\mathcal{C}_{3,6,4}}$ and the smallest solid and face angles of Δ. In the same way the B-coefficients of s restricted to the remaining tetrahedra in $\mathcal{C}_{4,6,4}$ that are refined with a fourth order partial Worsey-Farin split can be uniquely and stably determined. □

Theorem 5.8:
Let Δ be an arbitrary tetrahedral partition and Δ^* the refined partition obtained from Algorithm 5.3. Moreover, let $\mathcal{L}$ be the set of points chosen in Algorithm 5.6. Then the set $\mathcal{N} := \{\epsilon_\kappa\}_{\kappa \in \mathcal{L}}$ is an 11-local and stable nodal minimal determining set for $\mathcal{S}^2_9(\Delta^*)$.

Proof:
Following the proof of Theorem 5.7, each spline $s \in \mathcal{S}^2_9(\Delta^*)$ is uniquely and stably determined by the values of $s(\kappa)$, $\kappa \in \mathcal{L}$. Thus, a spline $s \in \mathcal{S}^2_9(\Delta^*)$ is uniquely and stably determined by the nodal data in $\mathcal{N}$. Moreover, by the proof of Theorem 5.7, the nodal minimal determining set is 11-local. □

Remark 5.9:
Since the local and stable Lagrange interpolation set $\mathcal{L}$ only contains domain points in Δ^*, it can easily be seen that this set is also a local and stable minimal determining set for $\mathcal{S}^2_9(\Delta^*)$.

5.4 Bounds on the error of the interpolant

In this section, we consider the error $\|f - \mathcal{I}f\|$ of a smooth function f and the corresponding interpolant $\mathcal{I}f$, define on an initial tetrahedral partition Δ of a domain $\Omega \subset \mathbb{R}^3$, which is constructed by the method described in this chapter. The error is measured in the maximum norm on Ω. To this end, let Δ^* be the refined partition obtained from Δ by Algorithm 5.3, and let $\mathcal{L}$ be the corresponding Lagrange interpolation set for the spline space $\mathcal{S}_9^2(\Delta^*)$ constructed in Algorithm 5.6. Then, for each function $f \in C(\Omega)$ there is a unique spline $\mathcal{I}f \in \mathcal{S}_9^2(\Delta^*)$, with

$$\mathcal{I}f(\kappa) = f(\kappa),\ \forall\, \kappa \in \mathcal{L}. \tag{5.1}$$

Thus, a linear projector $\mathcal{I}$ mapping $C(\Omega)$ onto $\mathcal{S}_9^2(\Delta^*)$ is defined. Moreover, for a compact set $B \subseteq \Omega$ and an integer $m \geq 1$, let $W_\infty^m(B)$ be the Sobolev space defined on the set B and let

$$|f|_{m,B} := \sum_{|\alpha|=m} \|D^\alpha f\|_B,$$

where $\|\cdot\|_B$ denotes the maximum norm on B and D^α is defined as $D^\alpha := D_x^{\alpha_1} D_y^{\alpha_1} D_z^{\alpha_3}$ for $\alpha = (\alpha_1, \alpha_2, \alpha_3)$. Moreover, let $|\Delta|$ be the mesh size of Δ.

Theorem 5.10:
For each $f \in W_\infty^{m+1}(\Omega)$, $0 \leq m \leq 9$, there exists an absolute constant K, such that

$$\|D^\alpha(f - \mathcal{I}f)\|_\Omega \leq K|\Delta|^{m+1-|\alpha|}|f|_{m+1,\Omega} \tag{5.2}$$

holds for all tuples α with $0 \leq |\alpha| \leq m$.

Proof:
Fix $0 \leq m \leq 9$, and let $f \in W_\infty^{m+1}$. Let T be a tetrahedron in Δ^* and let $\Omega_T := \mathrm{star}^{11}(\widetilde{T})$, where $\widetilde{T} \in \Delta$ with $T \subseteq \widetilde{T}$. In case $\widetilde{T}$ is a non-split tetrahedron $T = \widetilde{T}$ holds. Now, following Lemma 4.3.8 of [19], there exists a polynomial p of degree nine such that

$$\|D^\beta(f - p)\|_{\Omega_T} \leq K_1|\Omega_T|^{m+1-|\beta|}|f|_{m+1,\Omega_T}, \tag{5.3}$$

for all $0 \leq |\beta| \leq m$, where $|\Omega_T|$ denotes the diameter of Ω_T and K_1 is an absolute constant. Then, from

$$|\Omega_T| \leq 23|\Delta|,$$

we get

$$\|D^\beta(f - p)\|_{\Omega_T} \leq K_2|\Delta|^{m+1-|\beta|}|f|_{m+1,\Omega_T}, \tag{5.4}$$

with $K_2 := 23^{m+1}K_1$. Since $\mathcal{I}p = p$, it follows that

$$\|D^\alpha(f-\mathcal{I}f)\|_T \leq \|D^\alpha(f-p)\|_T + \|D^\alpha\mathcal{I}(f-p)\|_T.$$

Following equation (5.4) for $\beta = \alpha$, it suffices to estimate the second term $\|D^\alpha\mathcal{I}(f-p)\|_T$. In the proof of Theorem 5.7 it is shown, that the Lagrange interpolation set $\mathcal{L}$ is stable in the sense that

$$|c_\xi| \leq K_3 \max_{z_\kappa \in \Omega_T} |f(\kappa)|$$

holds for each B-coefficients c_ξ of a spline $s \in \mathcal{S}_9^2(\Delta^*)$, with an absolute constant K_3. Thus, also

$$\|\mathcal{I}f\|_T \leq K_4\|f\|_{\Omega_T} \tag{5.5}$$

holds, with an absolute constant K_4. Now, by the Markov inequality, see [102], and (5.5)

$$\|D^\alpha\mathcal{I}(f-p)\|_T \leq K_5|\Delta|^{-|\alpha|}\|\mathcal{I}(f-p)\|_T \leq K_6|\Delta|^{-|\alpha|}\|f-p\|_{\Omega_T}.$$

Then, by using (5.3) with $\beta = 0$ to estimate $\|(f-p)\|_{\Omega_T}$,

$$\|D^\alpha(f-\mathcal{I}f)\|_T \leq K_7|\Delta|^{m+1-|\alpha|}|f|_{m+1,\Omega_T}$$

holds. Then, taking the maximum over all tetrahedra in Δ^* leads to the estimation (5.2). □

Remark 5.11:
Following Theorem 4.13, we can see that the Lagrange interpolation at the points in $\mathcal{L}$ with splines in $\mathcal{S}_9^2(\Delta^*)$ yields optimal approximation order of ten (cf. subsection 2.4.3).

6 Macro-elements of arbitrary smoothness over the Clough-Tocher split of a triangle

In this chapter, the bivariate C^r macro-elements over the Clough-Tocher split of a triangle from Matt [66] are considered. They are more general than the bivariate C^r macro-elements based on the Clough-Tocher split constructed in [59], since they can be extended to splines of higher degree. Thus, the bivariate macro-elements considered here form a basis for the construction of the trivariate macro-elements of arbitrary smoothness over the Worsey-Farin split of a tetrahedron (see chapter 8).

In section 6.1, the minimal conditions of the degree of polynomials and the supersmoothness at the vertices of a triangle for bivariate macro-elements are investigated. These results are also used in chapter 7 and 8 in order to examine the minimal conditions for trivariate macro-elements. Thus, more general results are studied here, not only those for macro-elements based on the Clough-Tocher split. In the following section 6.2, minimal determining sets for the macro-elements examined in this chapter are considered. In the last section of this chapter, some examples for the bivariate macro-elements shown here are given.

6.1 Minimal degrees of supersmoothness and polynomials for bivariate C^r macro-elements

In this section the minimal conditions for the degree of polynomials and the supersmoothness at the vertices of a triangle, which are needed to construct C^r macro-elements based on non-split triangles, Clough-Tocher split triangles, and Powell-Sabin split triangles, are considered. For non-split triangles, a polynomial degree of $4r + 1$ and C^{2r} supersmoothness at the vertices is needed in order to construct C^r macro-elements (cf. [106]). This result is also proved here. However, since also triangles refined by the Clough-Tocher split and the Powell-Sabin split are considered here, a lower degree of polynomials and supersmoothness at the vertices of the triangles can be used. These lower degrees are investigated here.

For bivariate splines the following general result is known for the degree of supersmoothness at the vertices of a triangle.

Lemma 6.1 ([61]):
Suppose that $F := \langle v_1, v_2, v_3 \rangle$ is a triangle, and that Δ_F is a refinement of F such that there are $n \geq 0$ interior edges connected to the vertex v_1. Let s be a spline of degree d and smoothness r defined on Δ_F. Then the cross derivatives of s up to order r on the edges $e_1 := \langle v_1, v_2 \rangle$ and $e_2 := \langle v_1, v_3 \rangle$ can be specified independently only if we require $s \in C^\rho(v_1)$, with

$$\rho \geq \left\lceil \frac{(n+2)r - n}{n+1} \right\rceil.$$

Thus, the following result concerning the degree of polynomials and supersmoothness for bivariate C^r macro-elements over non-split triangles can be deduced (see also [105], cf. [68] for C^1).

Corollary 6.2:
For $r \geq 0$, a bivariate C^r macro-element over a non-split triangle F can be constructed if and only if the degree of polynomials d and the degree of supersmoothness ρ at the vertices of F are at least as in Table 6.1.

ρ	$2r$
d	$4r+1$

Table 6.1: Minimal degree of polynomials d and minimal supersmoothness ρ for C^r macro-elements over non-split triangles.

Proof:
Let s be a spline with smoothness r defined on a triangulation Δ consisting of non-split triangles. Each non-split triangle has zero interior edges. Thus, following Lemma 6.1, the supersmoothness ρ at the vertices of F must be at least equal to $2r$ such that s is r-times continuously differentiable.

For each edge $e := \langle u, v \rangle \in \Delta$, the B-coefficients of s associated with the domain points in $D_\rho(u)$ and $D_\rho(v)$ are uniquely determined by the partial derivatives up to order ρ at the vertices u and v. From each of the sets

$D_\rho(u)$ and $D_\rho(v)$ a total number of $\rho+1$ domain points lie on e, respectively. In order that the sets do not overlap and the derivatives at the vertices can be chosen independently, $s|_e$ has to be at least of degree $d=4r+1$. Thus, the polynomial degree of s must be at least $d=4r+1$. □

For C^r macro-elements based on the Clough-Tocher split of a triangle the following result for the polynomial degree and the degree of supersmoothness can be stated (cf. [59]).

Corollary 6.3:
For $r \geq 0$, a bivariate C^r macro-element over a Clough-Tocher split triangle F_{CT} can be constructed if and only if the degree of polynomials d and the degree of supersmoothness ρ at the vertices of F are at least as in Table 6.2, for $m \geq 0$.

r	$2m$	$2m+1$
ρ	$3m$	$3m+1$
d	$6m+1$	$6m+3$

Table 6.2: Minimal degree of polynomials d and minimal supersmoothness ρ for C^r macro-elements over Clough-Tocher split triangles.

Proof:
Let Δ be a triangulation and Δ_{CT} the triangulation obtained by splitting each triangle in Δ with a Clough-Tocher split. Moreover, let s be a spline of smoothness r defined on Δ_{CT}. Since each triangle in Δ is refined with a Clough-Tocher split it has exactly one interior edge at each vertex. Thus, following Lemma 6.1, the supersmoothness ρ of s at the vertices of Δ must satisfy

$$\rho \geq \left\lceil \frac{3r-1}{2} \right\rceil,$$

such that s is r-times continuously differentiable at the edges of Δ. Therefore, for $r=2m$ and $m \geq 0$

$$\rho \geq \left\lceil \frac{6m-1}{2} \right\rceil = 3m$$

and for $r=2m+1$

$$\rho \geq \left\lceil \frac{6m+2}{2} \right\rceil = 3m+1.$$

Now, let $e := \langle u,v \rangle$ be an edge in Δ. Then the B-coefficients of s associated with the domain points in $D_\rho(u)$ and $D_\rho(v)$ are uniquely determined by the partial derivatives of s up to order ρ at the vertices u and v. Then for each of the sets $D_\rho(u)$ and $D_\rho(v)$ $\rho+1$ domain points lie on e, respectively. In order that these sets do not overlap and the partial derivatives of s at the vertices u and v can be set independently, the polynomial degree of $s|_e$ has to be at least $d = 2\rho + 1$. Thus, the degree of polynomials of s must be at least $d = 6m + 1$ for $r = 2m$ and $d = 6m + 3$ for $r = 2m + 1$. □

Since it is desirable that the supersmoothness η at the split point v_F of a triangle F, which has been refined with a Clough-Tocher split, is as high as possible such that the dimension of the corresponding spline space is as low as possible, the degree of supersmoothness η is considered separately in the following Lemma (cf. [59]).

Lemma 6.4:
For $r \geq 0$, a bivariate C^r macro-element over a Clough-Tocher split triangle F_{CT} with C^ρ supersmoothness at the vertices of F and polynomial degree d can be constructed if and only if the degree of supersmoothness η at the split point v_F, at which F is refined, is at most $d + \rho - 2r$.

Proof:
Let s be a triangle defined on the Clough-Tocher split triangle F_{CT}. The number of domain points in $R_{\rho+1}(v_1)$ which are also contained in the set $D_\eta(v_F)$ is equal to $2(\rho + 1 - d + \eta) + 1$. From the cross derivatives of s up to order r at the edges $\langle v_1,v_2 \rangle$ and $\langle v_1,v_3 \rangle$, the B-coefficients of s corresponding to $r - d + \eta + 1$ of these domain points are uniquely determined. Thus, only $2(\rho - r) + 1$ B-coefficients of s associated with domain points in $R_{\rho+1}(v_1) \cap D_\eta(v_F)$ are still undetermined. These B-coefficients must satisfy the $\rho + 1 - d + \eta$ smoothness conditions at the edge $\langle v_1,v_F \rangle$. Unless $2(\rho - r) + 1 \geq \rho + 1 - d + \eta$ holds, this is impossible. Therefore, the supersmoothness η of s at v_F is bounded by $d + \rho - 2r$. □

Considering C^r macro-elements based on the Powell-Sabin split of a triangle the following result for the polynomial degree and the degree of supersmoothness can be obtained (cf. [61]).

Corollary 6.5:
For $r \geq 0$, a bivariate C^r macro-element over a Powell-Sabin split triangle F_{PS} can be constructed if and only if the degree of polynomials d and the degree of supersmoothness ρ at the vertices of F are at least as in Table 6.3, for $m \geq 0$.

r	$4m$	$4m+1$	$4m+2$	$4m+3$
ρ	$6m$	$6m+1$	$6m+3$	$6m+4$
d	$9m+1$	$9m+2$	$9m+5$	$9m+7$

Table 6.3: Minimal degree of polynomials d and minimal supersmoothness ρ for C^r macro-elements over Powell-Sabin split triangles.

Proof:
Let Δ be a triangulation and Δ_{PS} the triangulation obtained by splitting each triangle in Δ with a Powell-Sabin split at the incenter. Then, let s be a spline of smoothness r defined on Δ_{PS}. Since each triangle in Δ is subjected to a Powell-Sabin split it has exactly on interior edge at each vertex. Therefore, following Lemma 6.1,

$$\rho \geq \left\lceil \frac{3r-1}{2} \right\rceil = \begin{cases} 3k, & \text{for } r = 2k, \\ 3k+1, & \text{for } r = 2k+1, \end{cases}$$

has to hold for the degree of supersmoothness ρ at the vertices of Δ, such that s is r-times continuously differentiable at the edges of Δ.

Now, in order to examine the degree of polynomials, first the supersmoothness at the split point has to be considered. To this end, let F be a single triangle and F_{PS} the Powell-Sabin split of F at the split point v_F. Then, following Lemma 3.2 of [45], the dimension of the space $\mathcal{S}_d^{r,\eta}(F_{PS}) := \{s|_{F_i} \in \mathcal{P}_d^2 \text{ for all } F_i \in F_{PS} : s \in C^\eta(v_F)\}$ is equal to

$$\binom{\eta+2}{2} + 6\left(\binom{d-r+1}{2} - \binom{\eta-r+1}{2}\right) + \sum_{j=\eta-r+1}^{d-r} (r+j+1-je)_+,$$

where $3 \leq e \leq 6$ is the number of interior edges of F_{PS} with different slopes. Thus, to obtain a stable dimension for all configurations of F_{PS}, η has to be chosen such that the sum

$$\sum_{j=\eta-r+1}^{d-r} (r+j+1-je)_+$$

is equal to zero for all $3 \leq e \leq 6$. This leads to $\eta + 2 - 3(\eta - r + 1) \leq 0$, which is equivalent to

$$\eta \geq \left\lceil \frac{3r-1}{2} \right\rceil = \begin{cases} 3k, & \text{for } r = 2k, \\ 3k+1, & \text{for } r = 2k+1. \end{cases}$$

Now, it is possible to investigate the minimal degree of polynomials needed to construct C^r macro-elements based on the Powell-Sabin split. Therefore, let e be an interior edge of F_{PS}, which belongs to the two subtriangles $F_{e,1}$ and $F_{e,2}$ in F_{PS}. Then, the B-coefficients of a spline s defined on F_{PS} with C^ρ supersmoothness at the vertices of F and C^η supersmoothness at the split point of F_{PS} associated with the domain points in the rings $R_\eta^{F_{e,1}}(v_F)$ and $R_\eta^{F_{e,2}}(v_F)$ are considered. The set $R_\eta^{F_{e,1}}(v_F) \cup R_\eta^{F_{e,2}}(v_F)$ contains a total number of $2\eta+1$ distinct domain points. Setting the B-coefficients of s corresponding to the domain points in the disks $D_\rho(u)$ and $D_\rho(v)$ for the two vertices u and v of F contained in $F_{e,1}$ and $F_{e,2}$ leaves exactly $2\eta+1-2(\rho+\eta-d+1)$ B-coefficients associated with the domain points in $R_\eta^{F_{e,1}}(v_F) \cup R_\eta^{F_{e,2}}(v_F)$ undetermined. Since these B-coefficients have to satisfy η smoothness conditions across the edge e, such that s has C^η supersmoothness at the split point of F_{PS}, there have to be at least η undetermined B-coefficients left. Thus,

$$d \geq \left\lceil \frac{\eta + 2\rho + 1}{2} \right\rceil$$

has to hold in order to construct a C^r macro-element based on the Powell-Sabin split. Then, inserting the minimal values for ρ and η leads to

$$d \geq \begin{cases} \left\lceil \frac{9k+1}{2} \right\rceil, & \text{for } r = 2k, \\ \left\lceil \frac{9k+4}{2} \right\rceil, & \text{for } r = 2k+1. \end{cases}$$

Thus, considering the two cases for an even and an odd k, it follows that

$$d \geq \begin{cases} 9m+1, & \text{for } r = 4m, \\ 9m+2, & \text{for } r = 4m+1, \\ 9m+5, & \text{for } r = 4m+2, \\ 9m+7, & \text{for } r = 4m+3. \end{cases}$$

For these four cases, it follows that

$$\rho = \begin{cases} 6m, & \text{for } r = 4m, \\ 6m+1, & \text{for } r = 4m+1, \\ 6m+3, & \text{for } r = 4m+2, \\ 6m+4, & \text{for } r = 4m+3. \end{cases}$$

□

6.2 Minimal determining sets for C^r macro-elements over the Clough-Tocher split of a triangle

In this section minimal determining sets for macro-elements over the Clough-Tocher split of a triangle that can be extended to higher polynomial degrees are constructed. Therefore, let F be a triangle and F_{CT} the corresponding Clough-Tocher split. For $r \geq 0$ and $m = \lfloor \frac{r}{2} \rfloor$ it follows from Corollary 6.3 and Corollary 6.4, that it is possible to construct C^r macro-elements based on the Clough-Tocher split of a triangle, if

$$(\rho,\eta,d) = \begin{cases} (3m+q+1, 5m+3q+p+2, 6m+2q+p+3), & r = 2m+1, \\ (3m+q, 5m+3q+p+1, 6m+2q+p+1), & r = 2m, \end{cases} \tag{6.1}$$

with $0 \leq p$ and $0 \leq q \leq \lceil \frac{r}{2} \rceil$, holds for the supersmoothness ρ at the vertices of F, the supersmoothness η at the split point of F and the degree of polynomials d.

The macro-elements considered are in the superspline space

$$\mathcal{S}_d^{r,\rho,\eta}(F_{CT}) := \{s \in \mathcal{S}_d^r(F_{CT}) : s \in C^\rho(v_i),\ i = 1,2,3,\ s \in C^\eta(v_F)\}.$$

Theorem 6.6:
Let $r \in \mathbb{N}_0$, and let ρ, η, and d be as in (6.1) and $F := \langle v_1, v_2, v_3 \rangle$ a triangle. Moreover, let F_{CT} be the refinement of F by a Clough-Tocher split with subtriangles $F_i := \langle v_i, v_{i+1}, v_F \rangle$, $i = 1,2,3$, where $v_4 := v_1$ and v_F is the split point strictly inside F. Let $\mathcal{M}_{F_{CT}}$ be the union of the following sets of domain points:

($M1_{F_{CT}}$) $D_{\rho}^{F_i}(v_i)$, $i = 1,2,3$
($M2_{F_{CT}}$) $\{\xi_{i,j,k}^{F_l}\}$, $l = 1,2,3$ with $i,j < \rho + p + 1$, $k \leq r$ and $i + j + k = d$
($M3_{F_{CT}}$) $D_{r+p-2}^{F_1}(v_F)$
with $0 \leq p$ and $0 \leq q \leq \lceil \frac{\mu}{2} \rceil$.
Then $\mathcal{M}_{F_{CT}}$ is a minimal determining set for $\mathcal{S}_d^{r,\rho,\eta}(F_{CT})$ and

$$\dim \mathcal{S}_d^{r,\rho,\eta}(F_{CT}) = \begin{cases} \frac{43m^2+(16p+18q+65)m+p^2+13p+3q^2+15q+24}{2}, & \text{for } r = 2m+1, \\ \frac{43m^2+(16p+18q+31)m+p^2+5p+3q^2+9q+6}{2}, & \text{for } r = 2m. \end{cases} \tag{6.2}$$

Proof:
First, it is proved that $\mathcal{M}_{F_{CT}}$ is a minimal determining set. Therefore, the B-coefficients c_ξ for a bivariate spline $s \in \mathcal{S}_d^{r,\rho,\eta}(F_{CT})$ are set to prescribed values for all domain points ξ contained in the set $\mathcal{M}_{F_{CT}}$. Then it is shown, that all remaining B-coefficients of s corresponding to domain points in F_{CT} are uniquely and stably determined.

The undetermined B-coefficients of s associated with domain points in the disks $D_\rho(v_i)$, $i = 1,2,3$, can be uniquely and stably determined from the B-coefficients corresponding to the domain points in the set ($M1_{F_{CT}}$) by using the C^ρ supersmoothness at the three vertices of F.

At this time, together with the B-coefficients associated with the domain points in the set ($M2_{F_{CT}}$), all B-coefficients corresponding to the domain points within a distance of r from the three edges of F are uniquely and stably determined.

Now, the last undetermined B-coefficients of s can be determined. These are all associated with domain points in the disk $D_\eta(v_F)$. Since the spline s has C^η supersmoothness at the split point v_F, the B-coefficients of s corresponding to the domain points in $D_\eta(v_F)$ can be considered as those of a bivariate polynomial g of degree η defined over the triangle $\widetilde{F}$, which is bounded by the domain points in the ring $R_\eta(v_F)$. For the spline s some B-coefficients associated with domain points in the disk $D_\eta(v_F)$ are already uniquely determined. Thus, considering the polynomial g, the B-coefficients corresponding to the domain points within a distance of $2(m+q)$ from the vertices and a distance of $m+q$ from the edges of $\widetilde{F}$ are already known. Now, the B-coefficients associated with the domain points in the set ($M3_{F_{CT}}$) uniquely and stably determine the partial derivatives at the point v_F up to order $r+p-2$. Therefore, following Lemma 2.42 for the case

$n = 3$, the remaining undetermined B-coefficients of the polynomial g are uniquely and stably determined. Thus, all B-coefficients of s are uniquely and stably computed.

Since the set $\mathcal{M}_{F_{CT}}$ is a minimal determining set for the superspline space $\mathcal{S}_d^{r,\rho,\eta}(F_{CT})$, it suffices to compute the cardinality of $\mathcal{M}_{F_{CT}}$ to determine the dimension of $\mathcal{S}_d^{r,\rho,\eta}(F_{CT})$. It can easily be seen that the set (M1$_{F_{CT}}$) contains

$$3\binom{\rho+2}{2}$$

domain points. The set (M2$_{F_{CT}}$) consists of

$$3\left(\binom{r+1}{2} + p(r+1)\right)$$

domain points, and

$$\binom{r+p}{2}$$

domain points are contained in the last set (M3$_{F_{CT}}$). Thus

$$\#\mathcal{M}_{F_{CT}} = 3\left(\binom{\rho+2}{2} + \binom{r+1}{2} + p(r+1)\right) + \binom{r+p}{2},$$

which reduces to the number in (6.2). □

6.3 Examples for macro-elements over the Clough-Tocher split of a triangle

In this section examples for the minimal determining sets for macro-elements based on the Clough-Tocher split, constructed in section 6.2, are given. As in [66], different cases of C^r macro-elements, $r = 0,\dots,9$, are considered on one triangle $F := \langle v_1, v_2, v_3 \rangle$. These examples are also used in section 8.3.

6.3.1 C^0 macro-elements over the Clough-Tocher split

One example for minimal determining sets for C^0 macro-elements based on the Clough-Tocher split of a triangle is shown in this subsection.

Example 6.7:
Let $r = 0$, $p = 1$, and $q = 0$. Therefore, $m = 0$, and $\rho = 0$, $\eta = 2$, and $d = 2$. Then the macro-elements are in the superspline space $\mathcal{S}_2^{0,0,2}(F_{CT})$, which reduces to a space of polynomials of degree two defined on F_{CT}. By Theorem 6.6, the dimension of $\mathcal{S}_2^{0,0,2}(F_{CT})$ is equal to 6 and a corresponding minimal determining set $\mathcal{M}_{F_{CT}}$ is given by the union of the following sets of domain points:
($C^0_{1,0}$:M1$_{F_{CT}}$) $D_0^{F_i}(v_i)$, $i = 1,2,3$
($C^0_{1,0}$:M2$_{F_{CT}}$) $\{\xi^{F_l}_{i,j,k}\}$, $l = 1,2,3$, with $i,j < 2$, $k = 0$, and $i + j + k = 2$
This minimal determining set is illustrated in Figure 6.1.

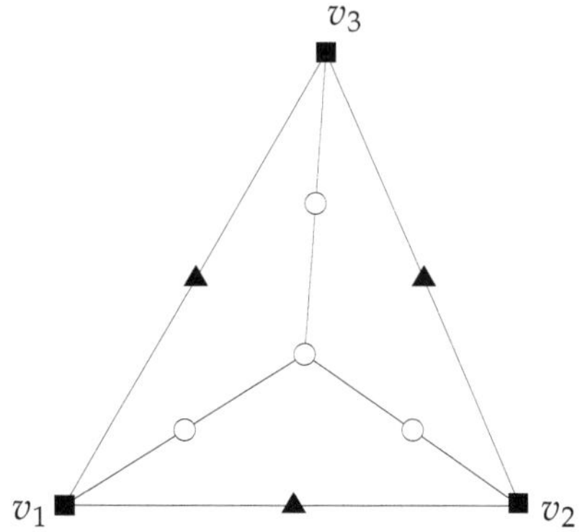

Figure 6.1: Minimal determining set of a C^0 macro-element in $\mathcal{S}_2^{0,0,2}(F_{CT})$. Domain points contained in ($C^0_{1,0}$:M1$_{F_{CT}}$) are marked with ■, and those in ($C^0_{1,0}$:M2$_{F_{CT}}$) with ▲.

6.3.2 C^1 macro-elements over the Clough-Tocher split

Two examples for minimal determining sets for C^1 macro-elements based on the Clough-Tocher split of a triangle are presented in this subsection.

Example 6.8:
Let $r = 1$, $p = 0$, and $q = 0$. In this case $m = 0$, and thus $\rho = 1$, $\eta = 2$, and $d = 3$. Therefore, the macro-elements are in the superspline space $\mathcal{S}_3^{1,1,2}(F_{CT})$. Following Theorem 6.6, the dimension of $\mathcal{S}_3^{1,1,2}(F_{CT})$ is equal to 12 and a

corresponding minimal determining set $\mathcal{M}_{F_{CT}}$ is given by the union of the following sets of domain points:
($C^1_{0,0}$:M1$_{F_{CT}}$) $D_1^{F_i}(v_i)$, $i = 1,2,3$
($C^1_{0,0}$:M2$_{F_{CT}}$) $\{\xi^{F_l}_{i,j,k}\}$, $l = 1,2,3$, with $i,j < 2$, $k \leq 1$, and $i+j+k = 3$
This minimal determining set is illustrated in Figure 6.2.

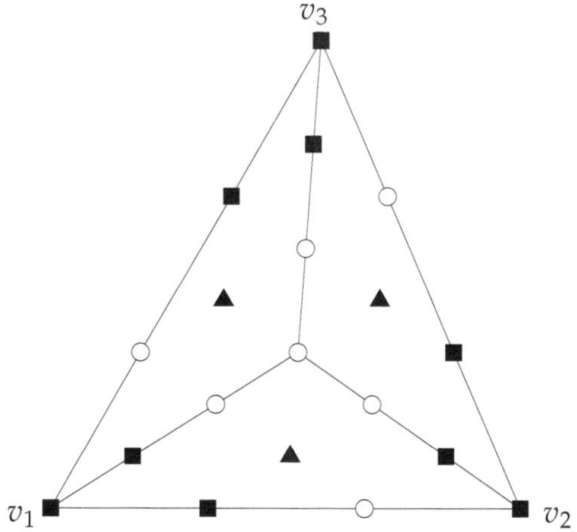

Figure 6.2: Minimal determining set of a C^1 macro-element in $\mathcal{S}_3^{1,1,2}(F_{CT})$. Domain points contained in ($C^1_{0,0}$:M1$_{F_{CT}}$) are marked with ■, and those in ($C^1_{0,0}$:M2$_{F_{CT}}$) with ▲.

Example 6.9:
Let $r = 1$, $p = 2$, and $q = 1$. Therefore, $m = 0$, and thus $\rho = 2$, $\eta = 7$, and $d = 7$. Then the macro-elements are in the superspline space $\mathcal{S}_7^{1,2,7}(F_{CT})$, which reduces to a space of polynomials of degree seven defined on F_{CT}. By Theorem 6.6, the dimension of $\mathcal{S}_7^{1,2,7}(F_{CT})$ is equal to 36 and a corresponding minimal determining set $\mathcal{M}_{F_{CT}}$ is given by the union of the following sets of domain points:
($C^1_{2,1}$:M1$_{F_{CT}}$) $D_2^{F_i}(v_i)$, $i = 1,2,3$
($C^1_{2,1}$:M2$_{F_{CT}}$) $\{\xi^{F_l}_{i,j,k}\}$, $l = 1,2,3$, with $i,j < 5$, $k \leq 1$, and $i+j+k = 7$
($C^1_{2,1}$:M3$_{F_{CT}}$) $D_1^{F_1}(v_F)$
This minimal determining set is illustrated in Figure 6.3.

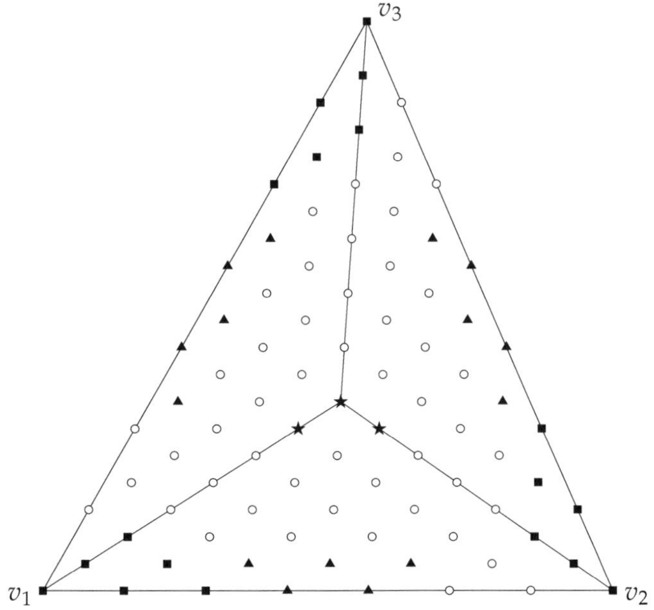

Figure 6.3: Minimal determining set of a C^1 macro-element in $\mathcal{S}_7^{1,2,7}(F_{CT})$. Domain points contained in ($C^1_{2,1}$:M1$_{F_{CT}}$) are marked with ■, those in ($C^1_{2,1}$:M2$_{F_{CT}}$) with ▲, and those in ($C^1_{2,1}$:M3$_{F_{CT}}$) with ★.

6.3.3 C^2 macro-elements over the Clough-Tocher split

In this subsection, two examples for minimal determining sets of C^2 macro-elements over the Clough-Tocher split of a triangle are considered.

Example 6.10:
Let $r = 2$, $p = 1$, and $q = 0$. Then $m = 1$, and therefore $\rho = 3$, $\eta = 7$, and $d = 8$. Thus, the macro-elements are in the superspline space $\mathcal{S}_8^{2,3,7}(F_{CT})$. Following Theorem 6.6, the dimension of $\mathcal{S}_8^{2,3,7}(F_{CT})$ is equal to 51 and a corresponding minimal determining set $\mathcal{M}_{F_{CT}}$ is given by the union of the following sets of domain points:
($C^2_{1,0}$:M1$_{F_{CT}}$) $D_3^{F_i}(v_i)$, $i = 1,2,3$
($C^2_{1,0}$:M2$_{F_{CT}}$) $\{\xi^{F_l}_{i,j,k}\}$, $l = 1,2,3$, with $i,j < 5$, $k \leq 2$, and $i+j+k = 8$
($C^2_{1,0}$:M3$_{F_{CT}}$) $D_1^{F_1}(v_F)$

This minimal determining set is illustrated in Figure 6.4.

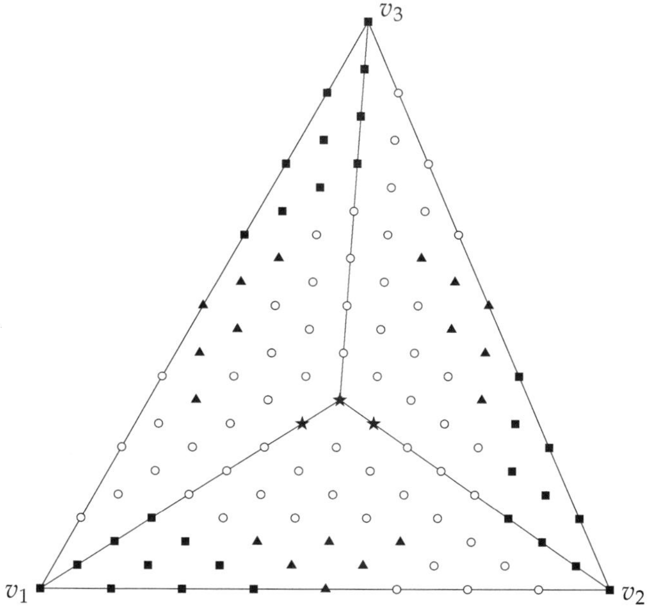

Figure 6.4: Minimal determining set of a C^2 macro-element in $\mathcal{S}_8^{2,3,7}(F_{CT})$. Domain points contained in ($C^2_{1,0}$:M1$_{F_{CT}}$) are marked with ■, those in ($C^2_{1,0}$:M2$_{F_{CT}}$) with ▲, and those in ($C^2_{1,0}$:M3$_{F_{CT}}$) with ★.

Example 6.11:
Let $r = 2$, $p = 2$, and $q = 1$. Then $m = 1$, and thus $\rho = 4$, $\eta = 11$, and $d = 11$. In this case, the macro-elements are in the superspline space $\mathcal{S}_{11}^{2,4,11}(F_{CT})$, which is in fact just a space of polynomials defined on F_{CT}, since the supersmoothness η at the split point v_F is equal to the degree of polynomials d. Following Theorem 6.6, the dimension of $\mathcal{S}_{11}^{2,4,11}(F_{CT})$ is equal to 78 and a corresponding minimal determining set $\mathcal{M}_{F_{CT}}$ consists of the union of the following sets of domain points:

($C^2_{2,1}$:M1$_{F_{CT}}$) $D_4^{F_i}(v_i)$, $i = 1,2,3$

($C^2_{2,1}$:M2$_{F_{CT}}$) $\{\xi^{F_l}_{i,j,k}\}$, $l = 1,2,3$, with $i,j < 7$, $k \leq 2$, and $i+j+k = 11$

($C^2_{2,1}$:M3$_{F_{CT}}$) $D_2^{F_1}(v_F)$

This minimal determining set is illustrated in Figure 6.5.

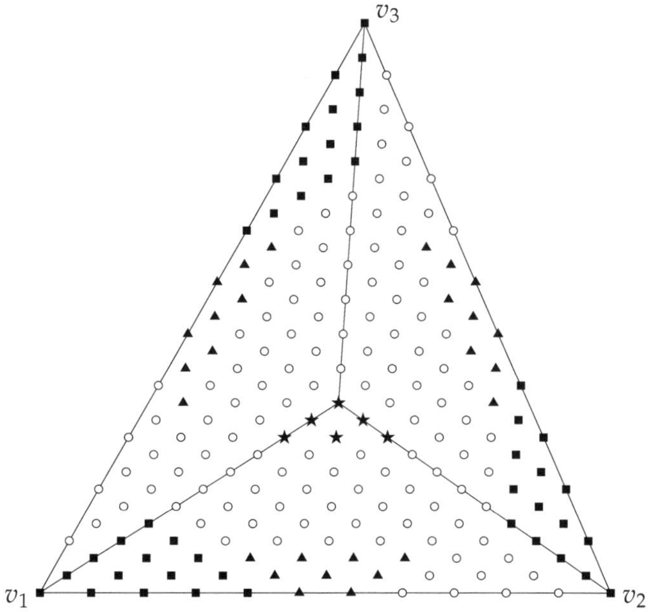

Figure 6.5: Minimal determining set of a C^2 macro-element in $\mathcal{S}_{11}^{2,4,11}(F_{CT})$. Domain points contained in $(C^2_{2,1}\text{:M1}_{F_{CT}})$ are marked with ■, those in $(C^2_{2,1}\text{:M2}_{F_{CT}})$ with ▲, and those in $(C^2_{2,1}\text{:M3}_{F_{CT}})$ with ★.

6.3.4 C^3 macro-elements over the Clough-Tocher split

In this subsection, four examples for minimal determining sets of C^3 macro-elements based on the Clough-Tocher split of a triangle are presented.

Example 6.12:
Let $r = 3$, $p = 0$, and $q = 0$. Therefore, $m = 1$, and thus $\rho = 4$, $\eta = 7$, and $d = 9$. Then the macro-elements are in the superspline space $\mathcal{S}_9^{3,4,7}(F_{CT})$, which has also already been studied by Lai and Schumaker [59] since p and q are equal to zero. By Theorem 6.6, the dimension of $\mathcal{S}_9^{3,4,7}(F_{CT})$ is equal to 66 and a corresponding minimal determining set $\mathcal{M}_{F_{CT}}$ is given by the union of the following sets of domain points:

($C^3_{0,0}$:M1$_{F_{CT}}$) $D_4^{F_i}(v_i)$, $i = 1,2,3$
($C^3_{0,0}$:M2$_{F_{CT}}$) $\{\xi^{F_l}_{i,j,k}\}$, $l = 1,2,3$, with $i,j < 5$, $k \leq 3$, and $i+j+k = 9$
($C^3_{0,0}$:M3$_{F_{CT}}$) $D_1^{F_1}(v_F)$
This minimal determining set is illustrated in Figure 6.6.

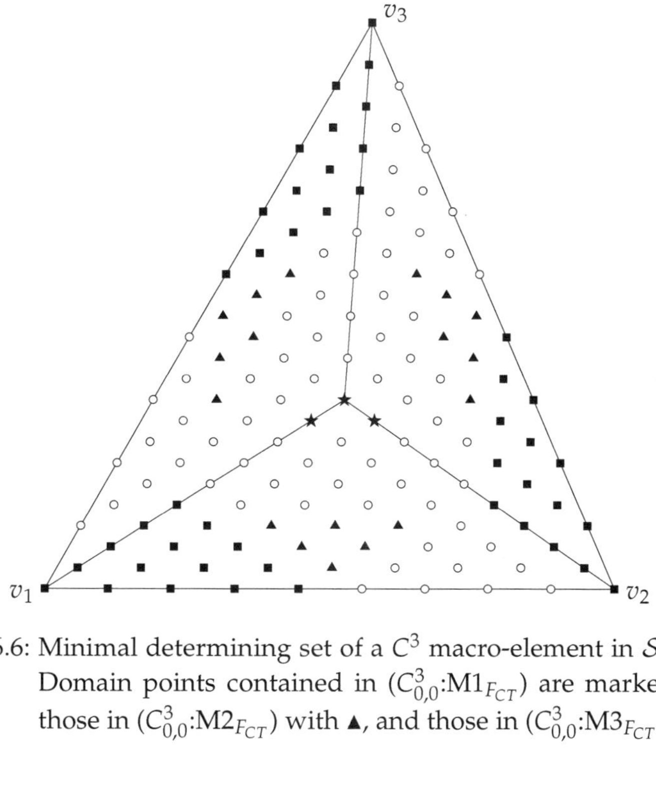

Figure 6.6: Minimal determining set of a C^3 macro-element in $\mathcal{S}_9^{3,4,7}(F_{CT})$. Domain points contained in ($C^3_{0,0}$:M1$_{F_{CT}}$) are marked with ■, those in ($C^3_{0,0}$:M2$_{F_{CT}}$) with ▲, and those in ($C^3_{0,0}$:M3$_{F_{CT}}$) with ★.

Example 6.13:
Let $r = 3$, $p = 1$, and $q = 1$. Therefore, $m = 1$, and thus $\rho = 5$, $\eta = 11$, and $d = 12$. Then the macro-elements are in the superspline space $\mathcal{S}_{12}^{3,5,11}(F_{CT})$. Following Theorem 6.6, the dimension of $\mathcal{S}_{12}^{3,5,11}(F_{CT})$ is equal to 99 and a corresponding minimal determining set $\mathcal{M}_{F_{CT}}$ is given by the union of the following sets of domain points:

($C^3_{1,1}$:M1$_{F_{CT}}$) $D_5^{F_i}(v_i)$, $i = 1,2,3$
($C^3_{1,1}$:M2$_{F_{CT}}$) $\{\xi^{F_l}_{i,j,k}\}$, $l = 1,2,3$, with $i,j < 7$, $k \le 3$, and $i+j+k = 12$
($C^3_{1,1}$:M3$_{F_{CT}}$) $D_2^{F_1}(v_F)$
This minimal determining set is illustrated in Figure 6.7.

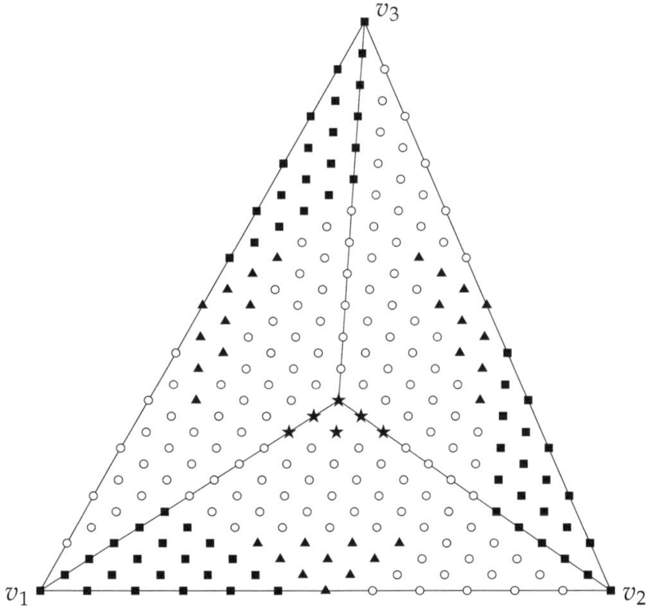

Figure 6.7: Minimal determining set of a C^3 macro-element in $\mathcal{S}_{12}^{3,5,11}(F_{CT})$. Domain points contained in ($C^3_{1,1}$:M1$_{F_{CT}}$) are marked with ■, those in ($C^3_{1,1}$:M2$_{F_{CT}}$) with ▲, and those in ($C^3_{1,1}$:M3$_{F_{CT}}$) with ★.

Example 6.14:
Let $r = 3$, $p = 3$, and $q = 2$. Then $m = 1$, and thus $\rho = 6$, $\eta = 16$, and $d = 16$. In this case, the macro-elements are in the superspline space $\mathcal{S}_{16}^{3,6,16}(F_{CT})$, which is in fact just a space of polynomials defined on F_{CT}, since the supersmoothness η is equal to the degree of polynomials d. Following Theorem 6.6, the dimension of $\mathcal{S}_{16}^{3,6,16}(F_{CT})$ is equal to 153 and a corresponding minimal determining set $\mathcal{M}_{F_{CT}}$ consists of the union of the following sets of domain points:

($C^3_{3,2}$:M1$_{F_{CT}}$) $D_6^{F_i}(v_i)$, $i = 1,2,3$
($C^3_{3,2}$:M2$_{F_{CT}}$) $\{\xi^{F_l}_{i,j,k}\}$, $l = 1,2,3$, with $i,j < 10$, $k \leq 3$, and $i+j+k = 16$
($C^3_{3,2}$:M3$_{F_{CT}}$) $D_4^{F_1}(v_F)$
This minimal determining set is illustrated in Figure 6.8.

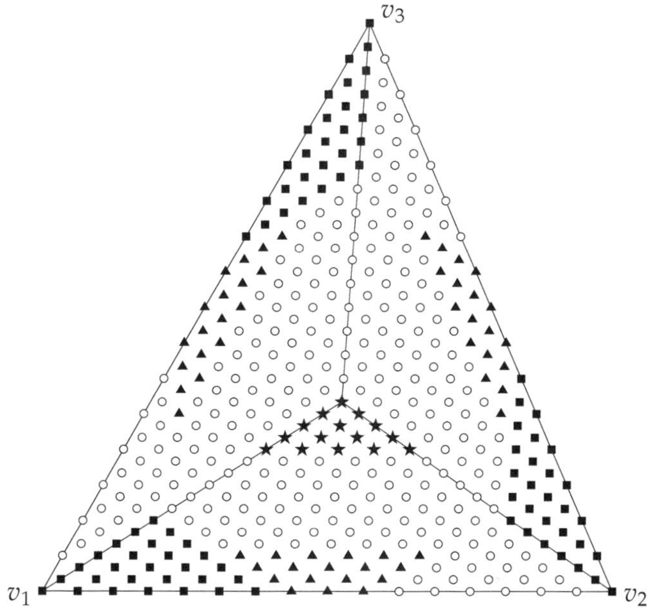

Figure 6.8: Minimal determining set of a C^3 macro-element in $\mathcal{S}_{16}^{3,6,16}(F_{CT})$. Domain points contained in ($C^3_{3,2}$:M1$_{F_{CT}}$) are marked with ■, those in ($C^3_{3,2}$:M2$_{F_{CT}}$) with ▲, and those in ($C^3_{3,2}$:M3$_{F_{CT}}$) with ★.

Example 6.15:
Let $r = 3$, $p = 4$, and $q = 2$. Therefore, $m = 1$, and thus $\rho = 6$, $\eta = 17$, and $d = 17$. Then the macro-elements are in the superspline space $\mathcal{S}_{17}^{3,6,17}(F_{CT})$, which again reduces to a space of polynomials, as in Example 6.14. By Theorem 6.6, the dimension of $\mathcal{S}_{17}^{3,6,17}(F_{CT})$ is equal to 171 and a corresponding minimal determining set $\mathcal{M}_{F_{CT}}$ is given by the union of the following sets of domain points:

($C^3_{4,2}$:M1$_{F_{CT}}$) $D_6^{F_i}(v_i)$, $i = 1,2,3$
($C^3_{4,2}$:M2$_{F_{CT}}$) $\{\xi^{F_l}_{i,j,k}\}$, $l = 1,2,3$, with $i,j < 11$, $k \leq 3$, and $i+j+k = 17$
($C^3_{4,2}$:M3$_{F_{CT}}$) $D_5^{F_1}(v_F)$
This minimal determining set is illustrated in Figure 6.9.

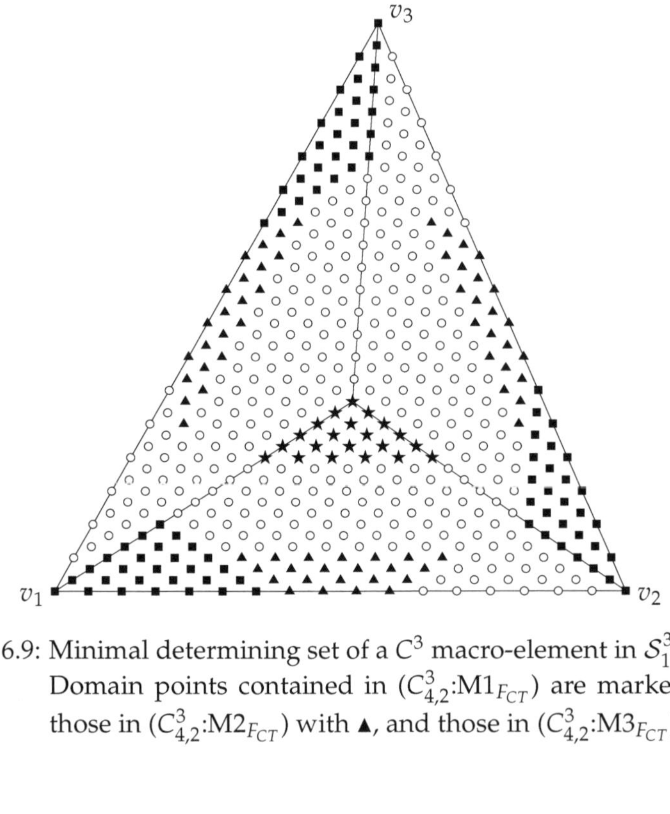

Figure 6.9: Minimal determining set of a C^3 macro-element in $\mathcal{S}_{17}^{3,6,17}(F_{CT})$. Domain points contained in ($C^3_{4,2}$:M1$_{F_{CT}}$) are marked with ■, those in ($C^3_{4,2}$:M2$_{F_{CT}}$) with ▲, and those in ($C^3_{4,2}$:M3$_{F_{CT}}$) with ★.

6.3.5 C^4 macro-elements over the Clough-Tocher split

Four examples for minimal determining sets of C^4 macro-elements over the Clough-Tocher split of a triangle are shown in this subsection.

Example 6.16:
Let $r = 4$, $p = 0$, and $q = 0$. Then $m = 2$, and thus $\rho = 6$, $\eta = 11$, and $d = 13$. In this case the macro-elements are in the space $\mathcal{S}_{13}^{4,6,11}(F_{CT})$. These macro-elements have already been studied by Lai and Schumaker [59] since the parameters q and p are equal to zero. Following Theorem 6.6, the dimension of $\mathcal{S}_{13}^{4,6,11}(F_{CT})$ is equal to 120 and a corresponding minimal determining set $\mathcal{M}_{F_{CT}}$ consists of the following sets of domain points:
($C_{0,0}^4$:M1$_{F_{CT}}$) $D_6^{F_i}(v_i)$, $i = 1,2,3$
($C_{0,0}^4$:M2$_{F_{CT}}$) $\{\xi_{i,j,k}^{F_l}\}$, $l = 1,2,3$, with $i,j < 7$, $k \leq 4$, and $i+j+k = 13$
($C_{0,0}^4$:M3$_{F_{CT}}$) $D_2^{F_1}(v_F)$
This minimal determining set is illustrated in Figure 6.10.

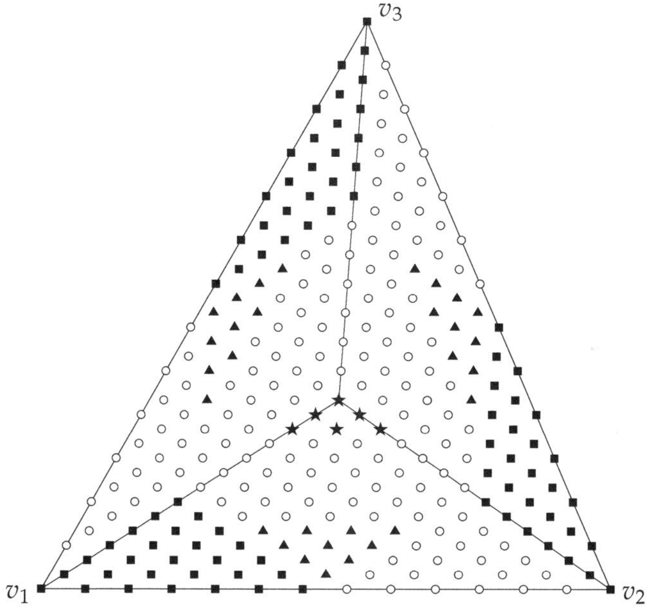

Figure 6.10: Minimal determining set of a C^4 macro-element in $\mathcal{S}_{13}^{4,6,11}(F_{CT})$. Domain points contained in ($C_{0,0}^4$:M1$_{F_{CT}}$) are marked with ■, those in ($C_{0,0}^4$:M2$_{F_{CT}}$) with ▲, and those in ($C_{0,0}^4$:M3$_{F_{CT}}$) with ★.

Example 6.17:
Let $r = 4$, $p = 2$, and $q = 1$. In this case $m = 2$, and thus $\rho = 7$, $\eta = 16$, and $d = 17$. Therefore, the macro-elements are in the superspline space $\mathcal{S}_{17}^{4,7,16}(F_{CT})$. Following Theorem 6.6, the dimension of $\mathcal{S}_{17}^{4,7,16}(F_{CT})$ is equal to 183 and a corresponding minimal determining set $\mathcal{M}_{F_{CT}}$ consists of the following sets of domain points:
($C^4_{2,1}$:M1$_{F_{CT}}$) $D_7^{F_i}(v_i)$, $i = 1,2,3$
($C^4_{2,1}$:M2$_{F_{CT}}$) $\{\xi^{F_l}_{i,j,k}\}$, $l = 1,2,3$, with $i,j < 10$, $k \leq 4$, and $i + j + k = 17$
($C^4_{2,1}$:M3$_{F_{CT}}$) $D_4^{F_1}(v_F)$
This minimal determining set is illustrated in Figure 6.11.

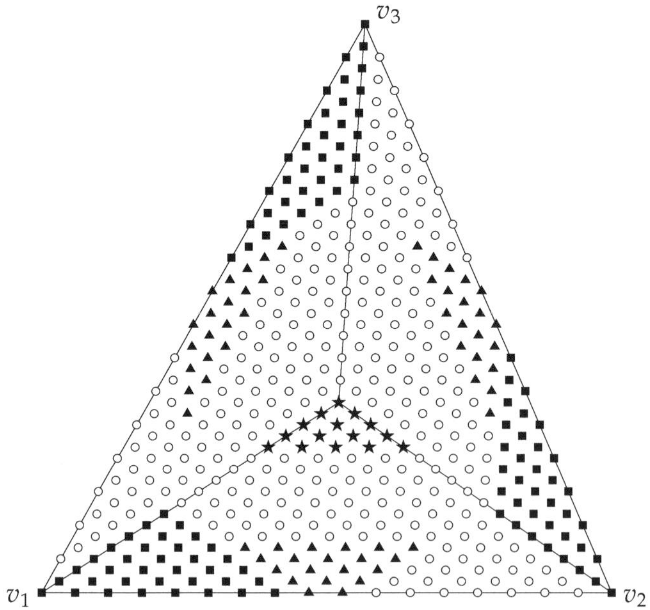

Figure 6.11: Minimal determining set of a C^4 macro-element in $\mathcal{S}_{17}^{4,7,16}(F_{CT})$. Domain points contained in ($C^4_{2,1}$:M1$_{F_{CT}}$) are marked with ■, those in ($C^4_{2,1}$:M2$_{F_{CT}}$) with ▲, and those in ($C^4_{2,1}$:M3$_{F_{CT}}$) with ★.

Example 6.18:
Let $r = 4$, $p = 3$, and $q = 1$. Therefore, $m = 2$, and thus $\rho = 7$, $\eta = 17$, and $d = 18$. Then the macro-elements are in the superspline space $\mathcal{S}_{18}^{4,7,17}(F_{CT})$. By Theorem 6.6, the dimension of $\mathcal{S}_{18}^{4,7,17}(F_{CT})$ is equal to 204 and a corresponding minimal determining set $\mathcal{M}_{F_{CT}}$ is given by the union of the following sets of domain points:
($C^4_{3,1}$:M1$_{F_{CT}}$) $D_7^{F_i}(v_i)$, $i = 1,2,3$
($C^4_{3,1}$:M2$_{F_{CT}}$) $\{\xi_{i,j,k}^{F_l}\}$, $l = 1,2,3$, with $i,j < 11$, $k \leq 4$, and $i + j + k = 18$
($C^4_{3,1}$:M3$_{F_{CT}}$) $D_5^{F_1}(v_F)$
This minimal determining set is illustrated in Figure 6.12.

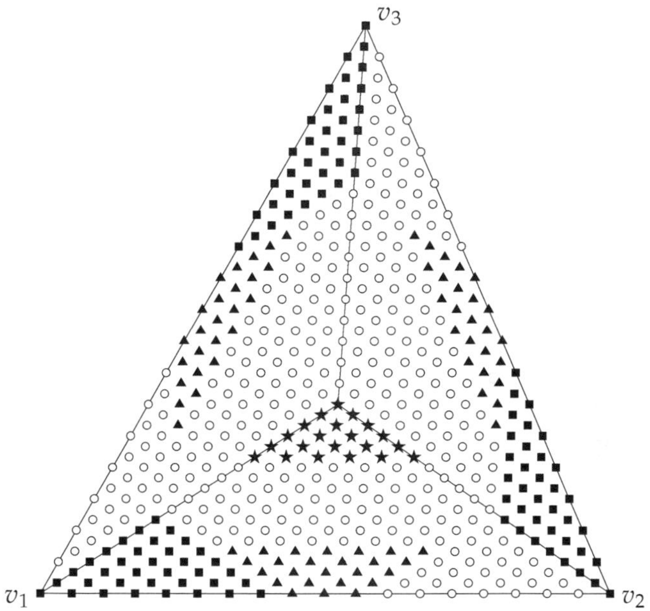

Figure 6.12: Minimal determining set of a C^4 macro-element in $\mathcal{S}_{18}^{4,7,17}(F_{CT})$. Domain points contained in ($C^4_{3,1}$:M1$_{F_{CT}}$) are marked with ■, those in ($C^4_{3,1}$:M2$_{F_{CT}}$) with ▲, and those in ($C^4_{3,1}$:M3$_{F_{CT}}$) with ★.

Example 6.19:
Let $r = 4$, $p = 5$, and $q = 2$. Then $m = 2$, and therefore $\rho = 8$, $\eta = 22$, and $d = 22$. Thus, the macro-elements are in the superspline space $\mathcal{S}_{22}^{4,8,22}(F_{CT})$, which is in fact just the space of polynomials of degree 22 defined on F_{CT}, since the supersmoothness η at the split point v_F is equal to the degree of polynomials d. Following Theorem 6.6, the dimension of $\mathcal{S}_{22}^{4,8,22}(F_{CT})$ is equal to 276 and a corresponding minimal determining set $\mathcal{M}_{F_{CT}}$ is given by the union of the following sets of domain points:

($C^4_{5,2}$:M1$_{F_{CT}}$) $D_8^{F_i}(v_i)$, $i = 1,2,3$

($C^4_{5,2}$:M2$_{F_{CT}}$) $\{\xi_{i,j,k}^{F_l}\}$, $l = 1,2,3$, with $i,j < 14$, $k \leq 4$, and $i + j + k = 22$

($C^4_{5,2}$:M3$_{F_{CT}}$) $D_7^{F_1}(v_F)$

This minimal determining set is illustrated in Figure 6.13.

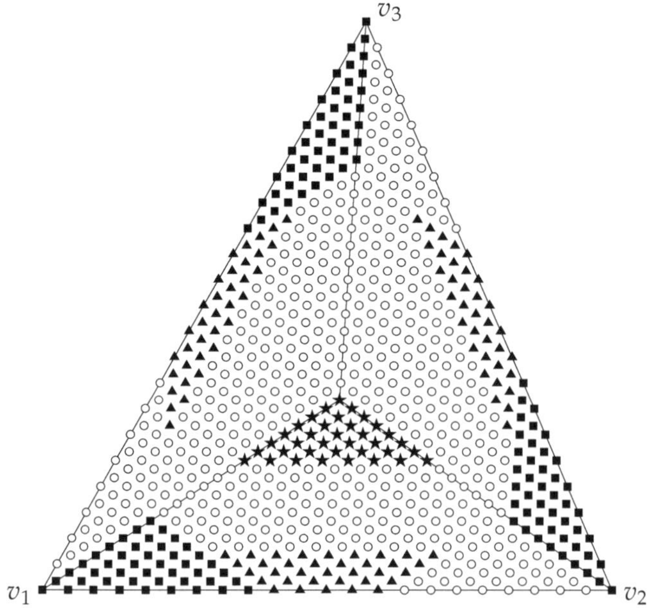

Figure 6.13: Minimal determining set of a C^4 macro-element in $\mathcal{S}_{22}^{4,8,22}(F_{CT})$. Domain points contained in ($C^4_{5,2}$:M1$_{F_{CT}}$) are marked with ■, those in ($C^4_{5,2}$:M2$_{F_{CT}}$) with ▲, and those in ($C^4_{5,2}$:M3$_{F_{CT}}$) with ★.

6.3.6 C^5 macro-elements over the Clough-Tocher split

In this subsection, three examples for minimal determining sets of C^5 macro-elements based on the Clough-Tocher split of a triangle are shown.

Example 6.20:
Let $r = 5$, $\rho = 1$, and $q = 1$. In this case $m = 2$, and therefore $\rho = 8$, $\eta = 16$, and $d = 18$. Thus, the macro-elements are in the superspline space $\mathcal{S}_{18}^{5,8,16}(F_{CT})$. By Theorem 6.6, the dimension of $\mathcal{S}_{18}^{5,8,16}(F_{CT})$ is equal to 213 and a corresponding minimal determining set $\mathcal{M}_{F_{CT}}$ consists of the following sets of domain points:
($C_{1,1}^5$:M1$_{F_{CT}}$) $D_8^{F_i}(v_i)$, $i = 1,2,3$
($C_{1,1}^5$:M2$_{F_{CT}}$) $\{\xi_{i,j,k}^{F_l}\}$, $l = 1,2,3$, with $i,j < 10$, $k \leq 5$, and $i + j + k = 18$
($C_{1,1}^5$:M3$_{F_{CT}}$) $D_4^{F_1}(v_F)$
This minimal determining set is illustrated in Figure 6.14.

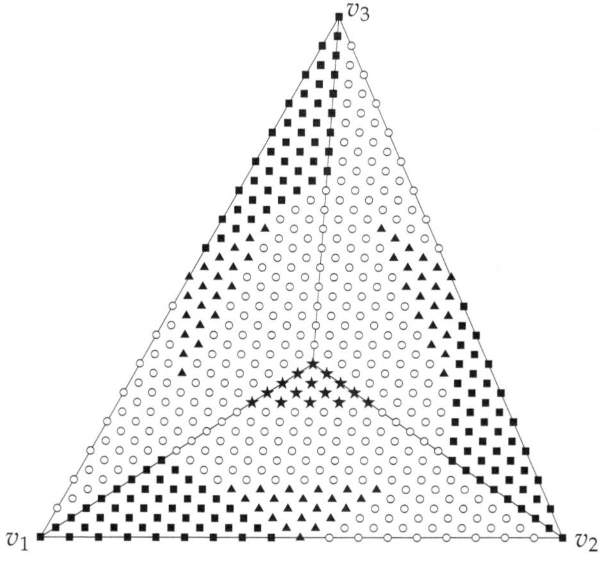

Figure 6.14: Minimal determining set of a C^5 macro-element in $\mathcal{S}_{18}^{5,8,16}(F_{CT})$. Domain points contained in ($C_{1,1}^5$:M1$_{F_{CT}}$) are marked with ■, those in ($C_{1,1}^5$:M2$_{F_{CT}}$) with ▲, and those in ($C_{1,1}^5$:M3$_{F_{CT}}$) with ★.

Example 6.21:
Let $r = 5$, $p = 2$, and $q = 1$. Therefore, $m = 2$, and thus $\rho = 8$, $\eta = 17$, and $d = 19$. Then the macro-elements are in the superspline space $\mathcal{S}_{19}^{5,8,17}(F_{CT})$. Following Theorem 6.6, the dimension of $\mathcal{S}_{19}^{5,8,17}(F_{CT})$ is equal to 237 and a corresponding minimal determining set $\mathcal{M}_{F_{CT}}$ is given by the union of the following sets of domain points:
($C^5_{2,1}$:M1$_{F_{CT}}$) $D_8^{F_i}(v_i)$, $i = 1,2,3$
($C^5_{2,1}$:M2$_{F_{CT}}$) $\{\xi^{F_l}_{i,j,k}\}$, $l = 1,2,3$, with $i,j < 11$, $k \leq 5$, and $i + j + k = 19$
($C^5_{2,1}$:M3$_{F_{CT}}$) $D_5^{F_1}(v_F)$
This minimal determining set is illustrated in Figure 6.15.

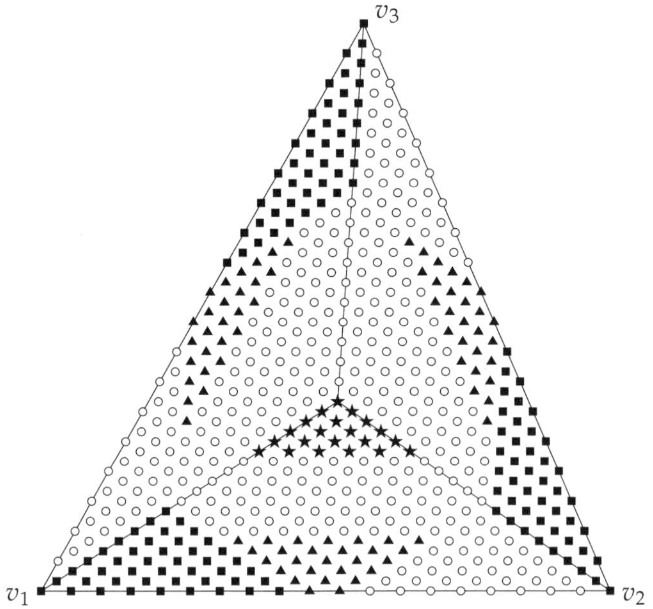

Figure 6.15: Minimal determining set of a C^5 macro-element in $\mathcal{S}_{19}^{5,8,17}(F_{CT})$. Domain points contained in ($C^5_{2,1}$:M1$_{F_{CT}}$) are marked with ■, those in ($C^5_{2,1}$:M2$_{F_{CT}}$) with ▲, and those in ($C^5_{2,1}$:M3$_{F_{CT}}$) with ★.

Example 6.22:
Let $r=5$, $p=4$, and $q=2$. Then $m=2$, and thus $\rho=9$, $\eta=22$, and $d=23$. In this case the macro-elements are in the superspline space $\mathcal{S}_{23}^{5,9,22}(F_{CT})$. By Theorem 6.6, the dimension of $\mathcal{S}_{23}^{5,9,22}(F_{CT})$ is equal to 318 and a corresponding minimal determining set $\mathcal{M}_{F_{CT}}$ consists of the following sets of domain points:

($C_{4,2}^3$:M1$_{F_{CT}}$) $D_9^{F_i}(v_i)$, $i=1,2,3$

($C_{4,2}^3$:M2$_{F_{CT}}$) $\{\xi_{i,j,k}^{F_l}\}$, $l=1,2,3$, with $i,j<14$, $k\leq 5$, and $i+j+k=23$

($C_{4,2}^3$:M3$_{F_{CT}}$) $D_7^{F_1}(v_F)$

This minimal determining set is illustrated in Figure 6.16.

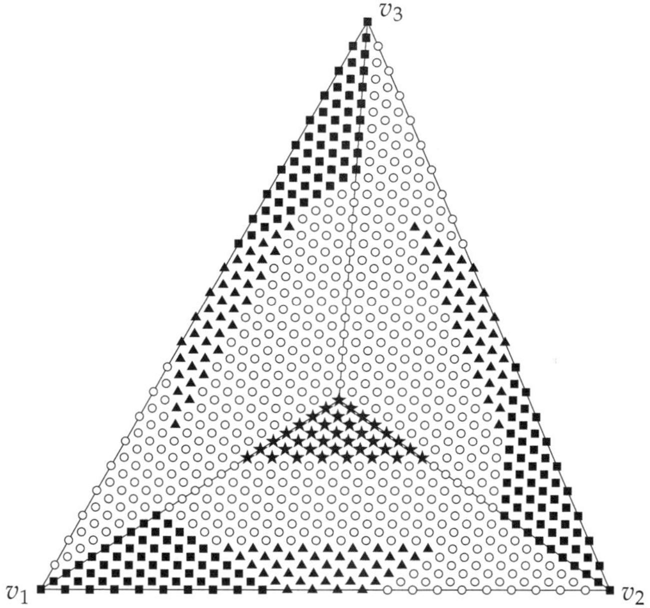

Figure 6.16: Minimal determining set of a C^5 macro-element in $\mathcal{S}_{23}^{5,9,22}(F_{CT})$. Domain points contained in ($C_{4,2}^3$:M1$_{F_{CT}}$) are marked with ■, those in ($C_{4,2}^3$:M2$_{F_{CT}}$) with ▲, and those in ($C_{4,2}^3$:M3$_{F_{CT}}$) with ★.

6.3.7 C^6 macro-elements over the Clough-Tocher split

Three examples for minimal determining sets of C^6 macro-elements over the Clough-Tocher split are presented in this subsection.

Example 6.23:
Let $r = 6$, $p = 0$, and $q = 0$. Thus, $m = 3$, and therefore $\rho = 9$, $\eta = 16$, and $d = 19$. In this case the macro-elements are in the superspline space $\mathcal{S}_{19}^{6,9,16}(F_{CT})$. Since p and q are equal to zero, these macro-elements have already been considered by Lai and Schumaker [59]. Following Theorem 6.6, the dimension of $\mathcal{S}_{19}^{6,9,16}(F_{CT})$ is equal to 243 and a corresponding minimal determining set $\mathcal{M}_{F_{CT}}$ consists of the following sets of domain points:

($C_{0,0}^6$:M1$_{F_{CT}}$) $D_9^{F_i}(v_i)$, $i = 1,2,3$

($C_{0,0}^6$:M2$_{F_{CT}}$) $\{\xi_{i,j,k}^{F_l}\}$, $l = 1,2,3$, with $i,j < 10$, $k \leq 6$, and $i + j + k = 19$

($C_{0,0}^6$:M3$_{F_{CT}}$) $D_4^{F_1}(v_F)$

This minimal determining set is illustrated in Figure 6.17.

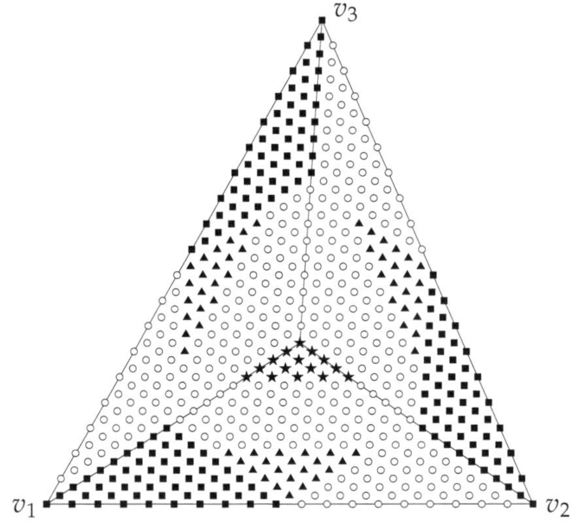

Figure 6.17: Minimal determining set of a C^6 macro-element in $\mathcal{S}_{19}^{6,9,16}(F_{CT})$. Domain points contained in ($C_{0,0}^6$:M1$_{F_{CT}}$) are marked with ■, those in ($C_{0,0}^6$:M2$_{F_{CT}}$) with ▲, and those in ($C_{0,0}^6$:M3$_{F_{CT}}$) with ★.

Example 6.24:
Let $r = 6$, $p = 1$, and $q = 0$. Therefore, $m = 3$, and thus $\rho = 9$, $\eta = 17$, and $d = 20$. Then the macro-elements are in the superspline space $\mathcal{S}_{20}^{6,9,17}(F_{CT})$. By Theorem 6.6, the dimension of $\mathcal{S}_{20}^{6,9,17}(F_{CT})$ is equal to 270 and a corresponding minimal determining set $\mathcal{M}_{F_{CT}}$ is given by the union of the following sets of domain points:
($C^6_{1,0}$:M1$_{F_{CT}}$) $D_9^{F_i}(v_i)$, $i = 1,2,3$
($C^6_{1,0}$:M2$_{F_{CT}}$) $\{\xi^{F_l}_{i,j,k}\}$, $l = 1,2,3$, with $i,j < 11$, $k \leq 6$, and $i+j+k = 20$
($C^6_{1,0}$:M3$_{F_{CT}}$) $D_5^{F_1}(v_F)$
This minimal determining set is illustrated in Figure 6.18.

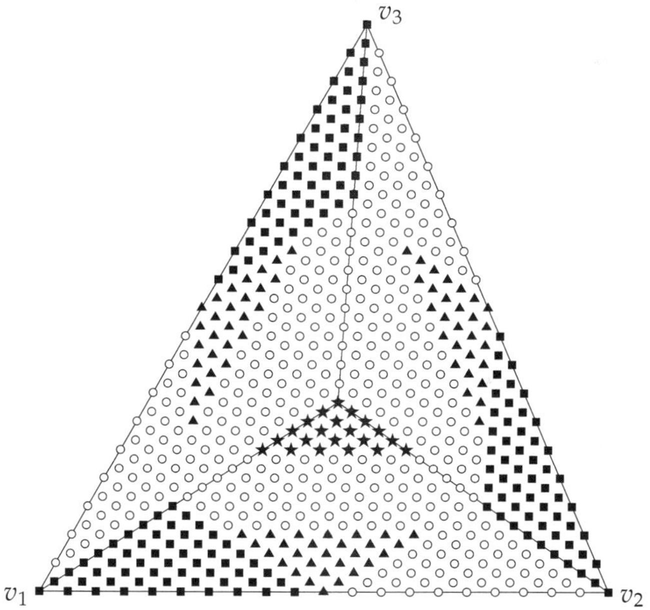

Figure 6.18: Minimal determining set of a C^6 macro-element in $\mathcal{S}_{20}^{6,9,17}(F_{CT})$. Domain points contained in ($C^6_{1,0}$:M1$_{F_{CT}}$) are marked with ■, those in ($C^6_{1,0}$:M2$_{F_{CT}}$) with ▲, and those in ($C^6_{1,0}$:M3$_{F_{CT}}$) with ★.

Example 6.25:
Let $r = 6$, $p = 3$, and $q = 1$. Then $m = 3$, and therefore $\rho = 10$, $\eta = 22$, and $d = 24$. Thus, the macro-elements are in the superspline space $\mathcal{S}_{24}^{6,10,22}(F_{CT})$. Following Theorem 6.6, the dimension of $\mathcal{S}_{24}^{6,10,22}(F_{CT})$ is equal to 360 and a corresponding minimal determining set $\mathcal{M}_{F_{CT}}$ is given by the union of the following sets of domain points:
($C^6_{3,1}$:M1$_{F_{CT}}$) $D_{10}^{F_i}(v_i)$, $i = 1,2,3$
($C^6_{3,1}$:M2$_{F_{CT}}$) $\{\xi_{i,j,k}^{F_l}\}$, $l = 1,2,3$, with $i,j < 14$, $k \leq 6$, and $i+j+k = 24$
($C^6_{3,1}$:M3$_{F_{CT}}$) $D_7^{F_1}(v_F)$
This minimal determining set is illustrated in Figure 6.19.

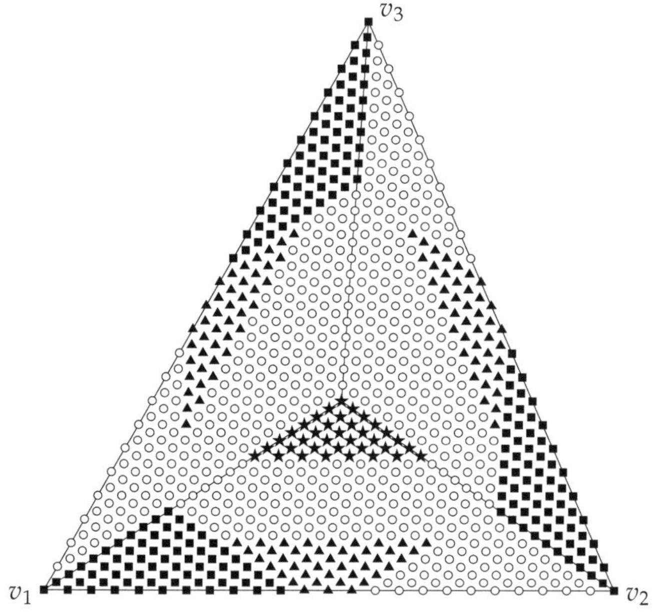

Figure 6.19: Minimal determining set of a C^6 macro-element in $\mathcal{S}_{24}^{6,10,22}(F_{CT})$. Domain points contained in ($C^6_{3,1}$:M1$_{F_{CT}}$) are marked with ■, those in ($C^6_{3,1}$:M2$_{F_{CT}}$) with ▲, and those in ($C^6_{3,1}$:M3$_{F_{CT}}$) with ★.

6.3.8 C^7 macro-elements over the Clough-Tocher split

In this subsection, two examples for minimal determining sets of C^7 macro-elements over the Clough-Tocher split of a triangle are considered.

Example 6.26:
Let $r = 7$, $p = 0$, and $q = 0$. Therefore, $m = 3$, and thus $\rho = 10$, $\eta = 17$, and $d = 21$. Then the macro-elements are in the superspline space $\mathcal{S}_{21}^{7,10,17}(F_{CT})$. Since $p = q = 0$, the macro-elements considered here reduce to those from Lai and Schumaker [59]. By Theorem 6.6, the dimension of $\mathcal{S}_{21}^{7,10,17}(F_{CT})$ is equal to 303 and a corresponding minimal determining set $\mathcal{M}_{F_{CT}}$ is given by the union of the following sets of domain points:
($C^7_{0,0}$:M1$_{F_{CT}}$) $D_{10}^{F_i}(v_i)$, $i = 1,2,3$
($C^7_{0,0}$:M2$_{F_{CT}}$) $\{\xi_{i,j,k}^{F_l}\}$, $l = 1,2,3$, with $i,j < 11$, $k \leq 7$, and $i + j + k = 21$
($C^7_{0,0}$:M3$_{F_{CT}}$) $D_5^{F_1}(v_F)$
This minimal determining set is illustrated in Figure 6.20.

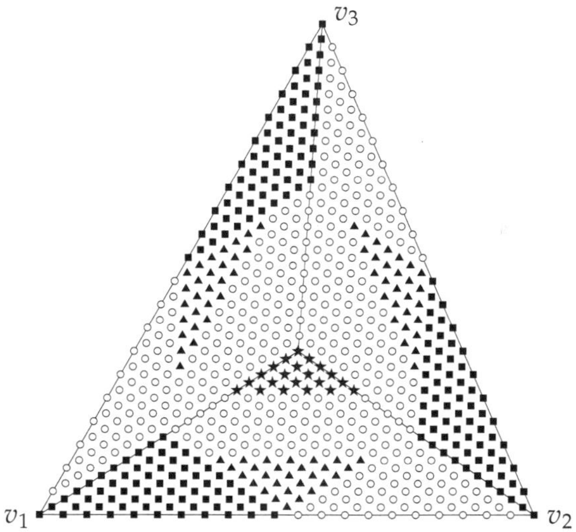

Figure 6.20: Minimal determining set of a C^7 macro-element in $\mathcal{S}_{21}^{7,10,17}(F_{CT})$. Domain points contained in ($C^7_{0,0}$:M1$_{F_{CT}}$) are marked with ■, those in ($C^7_{0,0}$:M2$_{F_{CT}}$) with ▲, and those in ($C^7_{0,0}$:M3$_{F_{CT}}$) with ★.

Example 6.27:
Let $r = 7$, $p = 2$, and $q = 1$. Then $m = 3$, and thus $\rho = 11$, $\eta = 22$, and $d = 25$. In this case, the macro-elements are in the superspline space $\mathcal{S}_{25}^{7,11,22}(F_{CT})$. Following Theorem 6.6, the dimension of $\mathcal{S}_{25}^{7,11,22}(F_{CT})$ is equal to 402 and a corresponding minimal determining set $\mathcal{M}_{F_{CT}}$ consists of the union of the following sets of domain points:
($C_{2,1}^7$:M1$_{F_{CT}}$) $D_{11}^{F_i}(v_i)$, $i = 1,2,3$
($C_{2,1}^7$:M2$_{F_{CT}}$) $\{\xi_{i,j,k}^{F_l}\}$, $l = 1,2,3$, with $i,j < 14$, $k \leq 7$, and $i + j + k = 25$
($C_{2,1}^7$:M3$_{F_{CT}}$) $D_7^{F_1}(v_F)$
This minimal determining set is illustrated in Figure 6.21.

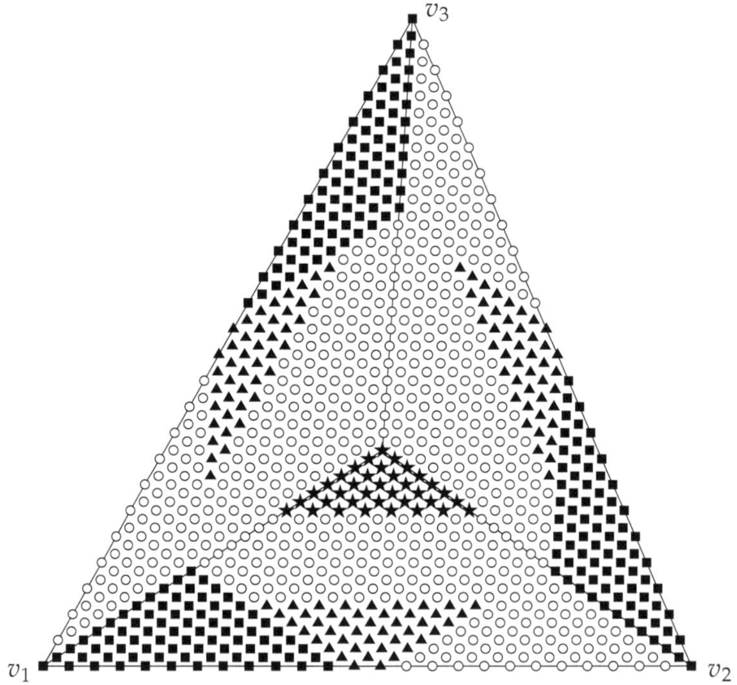

Figure 6.21: Minimal determining set of a C^7 macro-element in $\mathcal{S}_{25}^{7,11,22}(F_{CT})$. Domain points contained in ($C_{2,1}^7$:M1$_{F_{CT}}$) are marked with ■, those in ($C_{2,1}^7$:M2$_{F_{CT}}$) with ▲, and those in ($C_{2,1}^7$:M3$_{F_{CT}}$) with ★.

6.3.9 C^8 macro-elements over the Clough-Tocher split

In this subsection, one example for a minimal determining set of a C^8 macro-element over the Clough-Tocher split of a triangle is shown.

Example 6.28:
Let $r = 8$, $p = 1$, and $q = 0$. Therefore, $m = 4$, and thus $\rho = 12$, $\eta = 22$, and $d = 26$. Then the macro-elements are in the superspline space $\mathcal{S}_{26}^{8,12,22}(F_{CT})$. By Theorem 6.6, the dimension of $\mathcal{S}_{26}^{8,12,22}(F_{CT})$ is equal to 444 and a corresponding minimal determining set $\mathcal{M}_{F_{CT}}$ is given by the union of the following sets of domain points:
($C^8_{1,0}$:M1$_{F_{CT}}$) $D_{12}^{F_i}(v_i)$, $i = 1,2,3$
($C^8_{1,0}$:M2$_{F_{CT}}$) $\{\xi_{i,j,k}^{F_l}\}$, $l = 1,2,3$, with $i,j < 14$, $k \leq 8$, and $i + j + k = 26$
($C^8_{1,0}$:M3$_{F_{CT}}$) $D_7^{F_1}(v_F)$
This minimal determining set is illustrated in Figure 6.22.

6.3.10 C^9 macro-elements over the Clough-Tocher split

One example for a minimal determining set of a C^9 macro-element based on the Clough-Tocher split of a triangle is presented in this subsection.

Example 6.29:
Let $r = 9$, $p = 0$, and $q = 0$. Then $m = 4$, and therefore $\rho = 13$, $\eta = 22$, and $d = 27$. Thus, the macro-elements are in the superspline space $\mathcal{S}_{27}^{9,13,22}(F_{CT})$. These macro-elements have already been studied by Lai and Schumaker [59] since q and p are equal to zero. Following Theorem 6.6, the dimension of $\mathcal{S}_{27}^{9,13,22}(F_{CT})$ is equal to 486 and a corresponding minimal determining set $\mathcal{M}_{F_{CT}}$ is given by the union of the following sets of domain points:
($C^9_{0,0}$:M1$_{F_{CT}}$) $D_{13}^{F_i}(v_i)$, $i = 1,2,3$
($C^9_{0,0}$:M2$_{F_{CT}}$) $\{\xi_{i,j,k}^{F_l}\}$, $l = 1,2,3$, with $i,j < 14$, $k \leq 4$, and $i + j + k = 27$
($C^9_{0,0}$:M3$_{F_{CT}}$) $D_7^{F_1}(v_F)$
This minimal determining set is illustrated in Figure 6.23.

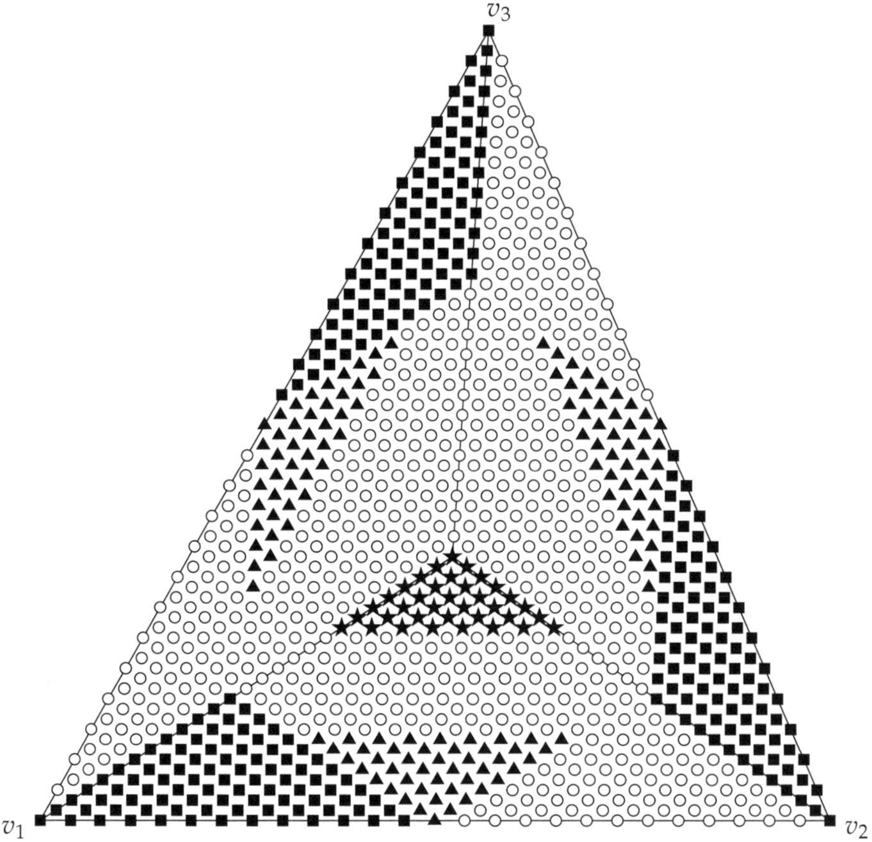

Figure 6.22: Minimal determining set of a C^8 macro-element in $\mathcal{S}_{26}^{8,12,22}(F_{CT})$. Domain points contained in $(C_{1,0}^{8}$:M1$_{F_{CT}})$ are marked with ■, those in $(C_{1,0}^{8}$:M2$_{F_{CT}})$ with ▲, and those in $(C_{1,0}^{8}$:M3$_{F_{CT}})$ with ★.

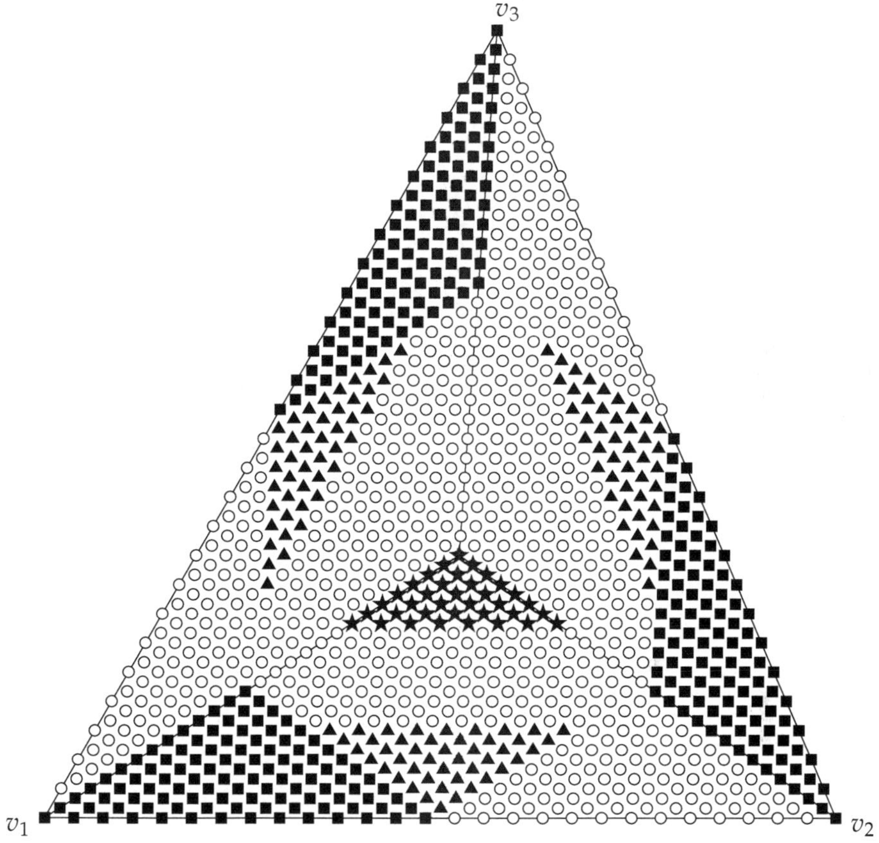

Figure 6.23: Minimal determining set of a C^9 macro-element in $\mathcal{S}_{27}^{9,13,22}(F_{CT})$. Domain points contained in $(C_{0,0}^9\text{:M1}_{F_{CT}})$ are marked with ■, those in $(C_{0,0}^9\text{:M2}_{F_{CT}})$ with ▲, and those in $(C_{0,0}^9\text{:M3}_{F_{CT}})$ with ★.

7 Macro-elements of arbitrary smoothness over the Alfeld split of a tetrahedron

In this chapter the trivariate C^r macro-elements of Lai and Matt [54] based on the Alfeld split of a tetrahedron (see Definition 2.6) are considered. In section 7.1, we investigate the minimal conditions for the polynomial degree and the degree supersmoothness for splines based on the Alfeld split. In the next section, we consider minimal determining sets for C^r macro-elements over the Alfeld split. In the following section 7.3, in order to ease the understanding of the constructed minimal determining sets, we give some examples for these. In section 7.4, we illustrate nodal minimal determining sets for the macro-elements. Finally, in section 7.5, we construct a Hermite interpolant based on the Alfeld split, which yields optimal approximation order.

7.1 Minimal degrees of supersmoothness and polynomials for macro-elements over the Alfeld split of a tetrahedron

In this section the constraints on the degree of polynomials and the degree of supersmoothness at the vertices and edges of a tetrahedron, which are needed to construct C^r macro-elements based on the Alfeld split, are investigated. In general it is desirable to use splines with a low degree of polynomials. However, there exist some restrictions due to the geometry of the considered partition, as well as desired locality of the applied splines. Normally a polynomial degree of at least $8r + 1$ is needed to construct C^r macro-elements over an arbitrary tetrahedral partition (cf. [63] and [105]). Thus, to be able to construct macro-elements with a lower degree of polynomials, the tetrahedra of a partition have to be subdivided. In this chapter, the Alfeld split is used to this end. In order to analyze the restrictions on the degree of supersmoothness and the polynomial degree, some bivariate layers in the split tetrahedra have to be considered more closely. Since a trivariate spline defined on a tetrahedron is just a bivariate spline, when restricted to a triangle, for the trivariate spline the same restrictions have to be fulfilled as for the bivariate spline.

With the results from section 6.1 it is possible to determine the minimal degree of polynomials and the minimal and maximal degree of supersmoothnesses for C^r macro-elements over the Alfeld split of a tetrahedron.

Theorem 7.1:
Let $T := \langle v_1, v_2, v_3, v_4 \rangle$ be a tetrahedron and T_A the Alfeld split of T at the point v_T strictly in the interior of T. For $r \geq 0$, a trivariate C^r macro-element based on T_A can be constructed if and only if the degree of polynomials d, the degree of supersmoothness ρ at the vertices of T, and the degree of supersmoothness μ at the edges of T are at least as in Table 7.1 and the degree of supersmoothness η at v_T is at most as in Table 7.1, for $m \geq 0$.

r	μ	ρ	η	d
$2m$	$3m$	$6m$	$11m+1$	$12m+1$
$2m+1$	$3m+1$	$6m+2$	$11m+4$	$12m+5$

Table 7.1: Minimal degree of polynomials d, minimal supersmoothnesses ρ and μ, and maximal supersmoothness η for C^r macro-elements over Alfeld split tetrahedra.

Proof:
Let F be a plane intersecting T at the three points v_1, v_2, and v_T. The proof works analog for any other combination of vertices of T. Since T was subdivided with the Alfeld split, a Clough-Tocher split triangle on F is induced. This triangle is marked with light gray in Figure 7.1. In Figure 7.1, right, only this triangle is shown, where $\langle v_1, v_2 \rangle$ is an edge of the tetrahedron T, v_e is the intersection of the edge $\langle v_3, v_4 \rangle$ with the plane F, the two edges $\langle v_1, v_e \rangle$ and $\langle v_2, v_e \rangle$ are the intersection of the plane F with the triangular faces $\langle v_1, v_3, v_4 \rangle$ and $\langle v_2, v_3, v_4 \rangle$ of T, $\langle v_1, v_T \rangle$ and $\langle v_2, v_T \rangle$ are interior edges of T_A, and the edge $\langle v_T, v_e \rangle$ is the intersection of the plane F with the interior triangular face $\langle v_3, v_4, v_T \rangle$ of T_A.

Since each trivariate spline s defined on T_A restricted to any triangle in T_A results in a bivariate spline with the same degree of polynomials and supersmoothness, the spline s has to fulfill the same conditions as a bivariate

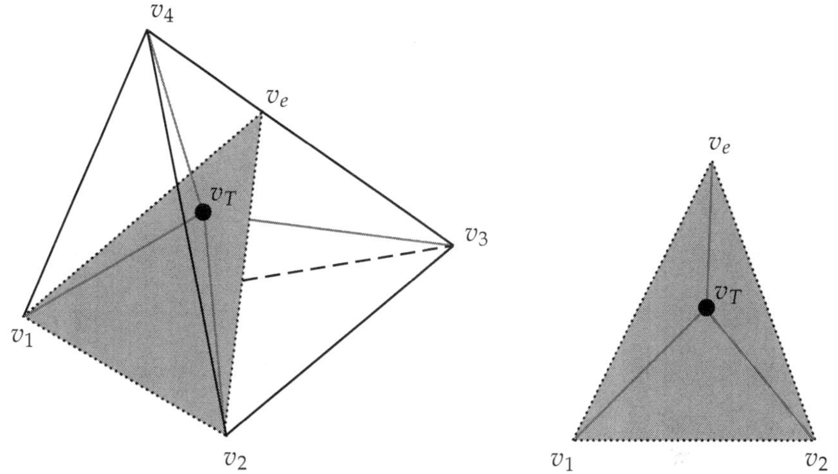

Figure 7.1: Bivariate layer with a Clough-Tocher split triangle in a tetrahedron $T := \langle v_1, v_2, v_3, v_4 \rangle$ subdivided with the Alfeld split at the point v_T in the interior of T.

spline in Corollary 6.2 and 6.3. Therefore, the degree of supersmoothness of s at the three vertices v_1, v_2, and v_e of the Clough-Tocher split triangle in F has to be at least $3m$ or $3m + 1$, such that $s \in C^{2m}$ or $s \in C^{2m+1}$, respectively. Since the point v_e, where the spline s has C^{3m} or C^{3m+1} supersmoothness, as the case may be, is a point in the interior of the edge $\langle v_3, v_4 \rangle$ of T, the degree of supersmoothness of s at the edges of T has to be at least $3m$ or $3m + 1$, respectively.

Next, the triangular face $\langle v_1, v_2, v_3 \rangle$ of T is considered. The proof is analogue for any other face of T. This triangle is not subdivided and, following the above considerations, the spline $s|_{\langle v_1, v_2, v_3 \rangle}$ is a bivariate spline defined on this triangle with C^{3m} or C^{3m+1} supersmoothness at the edges of $\langle v_1, v_2, v_3 \rangle$. Thus, following Corollary 6.2, the degree of supersmoothness of s at the vertices of $\langle v_1, v_2, v_3 \rangle$ has to be at least C^{6m} or C^{6m+2}, in order to construct a C^{2m} or C^{2m+1} macro-element based on the Alfeld split of T with C^{3m} or C^{3m+1} supersmoothness at the edges of T, respectively. In addition, it follows that the degree of polynomials of s has to be at least $12m + 1$ or $12m + 5$ for a C^{2m} or C^{2m+1} macro-element, as the case may be.

Finally, the supersmoothness at the point v_T is considered. To this end the Clough-Tocher split triangle in the plane F from above is analyzed again. Restricted to this triangle the spline s has C^{6m} or C^{6m+2} supersmoothness at v_1 and v_2, and C^{3m} or C^{3m+1} supersmoothness at v_e, respectively. Considering the two vertices v_1 and v_2, following Lemma 6.4, the degree of supersmoothness of s at v_T can be at most $14m+1$ or $14m+5$, respectively. But considering the vertex v_e the supersmoothness at v_T can be at most of degree $11m+1$ or $11m+4$, as the case may be. □

7.2 Minimal determining sets for C^r macro-elements over the Alfeld split

In this section minimal determining sets for C^r macro-elements based on the Alfeld split of a tetrahedron and a tetrahedral partition are constructed. First, some notations are established to define the considered space of supersplines. Subsequently, minimal determining sets for the C^r macro-elements over the Alfeld split of a tetrahedron are constructed and proved. Afterwards, more notations are introduced and the space of supersplines for C^r macro-elements over the Alfeld split of a tetrahedral partition is defined. Then, minimal determining sets for theses macro-elements are constructed. We also prove this result.

Notation 7.2:
Let $T := \langle v_1, v_2, v_3, v_4 \rangle$ be a tetrahedron with the four triangular faces $F_i := \langle v_i, v_{i+1}, v_{i+2} \rangle$, $i = 1, \ldots, 4$. Then, let T_A be the Alfeld split of T at the point v_T in the interior of T and $T_i := \langle v_i, v_{i+1}, v_{i+2}, v_T \rangle$, $i = 1, \ldots, 4$, the four subtetrahedra of T_A, where $v_5 := v_1$ and $v_6 := v_2$. Moreover, let $\mathcal{V}_T, \mathcal{E}_T$, and $\mathcal{F}_T$ be the sets of vertices, edges, and faces of T, respectively, and let $\mathcal{F}_{I,T}$ be the set of six interior triangular faces of T_A of the form $\langle u, v, v_T \rangle$, where $\langle u, v \rangle \in \mathcal{E}_T$. Furthermore, for each face $F := \langle u, v, v_T \rangle \in \mathcal{F}_{I,T}$, let $T_F := \langle u, v, v_T, w \rangle \in T_A$ be a subtetrahedron in T_A containing F.

The here considered macro-elements over the Alfeld split of a tetrahedron T are in the superspline space

$$\begin{aligned}\mathcal{S}_d^{r,\rho,\mu,\eta}(T_A) := \{s \in C^r(\widetilde{T}) : \; & s|_{\widetilde{T}} \in \mathcal{P}_d^3 \text{ for all } \widetilde{T} \in T_A,\\ & s \in C^{\mu}(e) \text{ for all } e \in \mathcal{E}_T,\\ & s \in C^{\rho}(v) \text{ for all } v \in \mathcal{V}_T,\\ & s \in C^{\eta}(v_T)\},\end{aligned}$$

where r, ρ, μ, η, and d are as in Table 7.1.

Theorem 7.3:
Let $\mathcal{M}_{T_A}$ be the union of the following sets of domain points:

(M1$_{T_A}$) $D_{\rho}^{T_i}(v_i)$, $i = 1, \ldots, 4$
(M2$_{T_A}$) $\bigcup_{e:=\langle u,v\rangle \in \mathcal{E}_T} E_{\mu}^{T_e}(e) \setminus (D_{\rho}(u) \cup D_{\rho}(v))$, where $e \in T_e$
(M3$_{T_A}$) $\bigcup_{j=0}^{r} F_j^{T_i}(F_i) \setminus (\bigcup_{v \in F_i} D_{\rho}(v) \cup \bigcup_{e \in F_i} E_{\mu}(e))$, $i = 1, \ldots, 4$
(M4$_{T_A}$) $\bigcup_{F \in \mathcal{F}_{I,T}} \{\xi^{T_F}_{\mu-2i+j, d-\rho-j, \mu+2i, 0} : i = 1, \ldots, \lfloor \frac{r}{3} \rfloor, \; j = 1, \ldots, \mu + 2i\}$
(M5$_{T_A}$) $\bigcup_{j=1}^{\lfloor \frac{r}{3} \rfloor} \left(F_{r+j}^{T_i}(F_i) \setminus (\bigcup_{v \in F_i} D_{\rho}(v) \cup \bigcup_{e \in F_i} E_{\mu+2j}(e)) \right)$, $i = 1, \ldots, 4$
(M6$_{T_A}$) $D^{T_1}_{\eta - 4(m + \lfloor \frac{r}{3} \rfloor + 1)}(v_T)$,

where m is as in Table 7.1. Then $\mathcal{M}_{T_A}$ is a stable minimal determining set for the space $\mathcal{S}_d^{r,\rho,\mu,\eta}(T_A)$ and the dimension of the superspline space is given as

$$\dim \mathcal{S}_d^{r,\rho,\mu,\eta}(T_A) = \begin{cases} \frac{2035m^3 + 1317m^2 + 314m + 24}{6}, & \text{for } r = 2m, \\ \frac{2035m^3 + 3582m^2 + 2057m + 390}{6}, & \text{for } r = 2m + 1. \end{cases} \tag{7.1}$$

Proof:
First, it is proved that $\mathcal{M}_{T_A}$ is a stable minimal determining set for the superspline space $\mathcal{S}_d^{r,\rho,\mu,\eta}(T_A)$. Therefore, the B-coefficients c_{ξ} of a spline $s \in \mathcal{S}_d^{r,\rho,\mu,\eta}(T_A)$ are set to arbitrary values for each domain point $\xi \in \mathcal{M}_{T_A}$ and it is shown that all the remaining B-coefficients of s are uniquely and stably determined.

Initially, the B-coefficients of s associated with the domain points in $D_\rho(v_i)$, $i = 1,\ldots,4$, are determined. Using the C^ρ supersmoothness conditions of s at each vertex of T, these B-coefficients are uniquely determined from B-coefficients corresponding to the domain points in the set $(M1_{T_A})$. The computation of these B-coefficients is stable, since they are directly determined from smoothness conditions.

Next, the remaining undetermined B-coefficients associated with domain points in the tubes $E_\mu(e)$, $e \in \mathcal{E}_T$, are considered. Using the C^μ supersmoothness conditions at the edges of T, these can be uniquely determined from the B-coefficients of s associated with the domain points in the set $(M2_{T_A})$. Their computation is also stable, since they are directly computed from other B-coefficients using smoothness conditions.

Subsequently, the B-coefficients of s associated with the domain points in the shells $R_d(v_T),\ldots,R_{d-r}(v_T)$ are considered. All B-coefficients of s corresponding to the domain points in the balls $D_\rho(v)$, $v \in \mathcal{V}_T$, and the tubes $E_\mu(e)$, $e \in \mathcal{E}_T$, are already uniquely and stably determined. The remaining undetermined B-coefficients of s corresponding to domain points in the shells $R_d(v_T),\ldots,R_{d-r}(v_T)$ are those associated with the domain points in the set $(M3_{T_A})$. Thus, all B-coefficients of s associated with the domain points in the shells $R_d(v_T),\ldots,R_{d-r}(v_T)$ are already uniquely and stably determined.

In order to compute the remaining undetermined B-coefficients of s, those associated with the domain points in the ball $D_\eta(v_T)$ are examined more thoroughly. Since s has C^η supersmoothness at the point v_T, the B-coefficients of s corresponding to the domain points in $D_\eta(v_T)$ can be regarded as those of a polynomial p of degree η, where p is defined on the tetrahedron $\widetilde{T}$, which is bounded by the domain points in the shell $R_\eta(v_T)$. The tetrahedron $\widetilde{T}$ is also refined with the Alfeld split. Since some of the B-coefficients of s are already determined, also some of the B-coefficients of p are determined. In particular, the B-coefficients of p associated with the domain points within a distance of m from the faces of $\widetilde{T}$, within a distance of $2m$ from the edges of $\widetilde{T}$, and within a distance of $\rho - d + \eta$ from the vertices of $\widetilde{T}$ are already determined.

Now, the remaining undetermined B-coefficients of p are determined. In case $\lfloor \frac{r}{3} \rfloor > 0$, the B-coefficients of p corresponding to the domain points in the shells $R_{d-r-i}(v_T)$, $i = 1,\ldots,\lfloor \frac{r}{3} \rfloor$, are considered first. They are determined by repeating the following two steps for each i.

First, following Lemma 2.33 for $j = 0$, the remaining undetermined B-coefficients of p associated with domain points within a distance of

$2(m+i)-1$ from the edges of $\widetilde{T}$ are uniquely and stably determined by the $C^{2(m+i)-1}$ smoothness conditions at these edges. Next, together with the B-coefficients associated with the domain points in the set $(\mathrm{M4}_{T_A})$ with a distance of $2(m+i)$ from the edges of $\widetilde{T}$, the undetermined B-coefficients of p corresponding to the domain points within a distance of $2(m+i)$ from these edges can be uniquely and stably determined by the $C^{2(m+i)}$ smoothness conditions at the edges of $\widetilde{T}$, following Lemma 2.33 for the case $j=1$. Thus, with the B-coefficients associated with the domain points in the set $(\mathrm{M5}_{T_A})$ with a distance of $m+i$ from the faces of $\widetilde{T}$, all B-coefficients of p corresponding to the domain points in the shell $R_{d-r-i}(v_T)$ are uniquely and stably determined.

Consequently, all B-coefficients of p corresponding to domain points within a distance of $m+\lfloor\frac{r}{3}\rfloor$ from the faces of $\widetilde{T}$ are uniquely and stably determined. Now, the partial derivatives at v_T up to order $\eta-4(m+\lfloor\frac{r}{3}\rfloor+1)$ can be uniquely and stably determined from the B-coefficients corresponding to the domain points in the set $(\mathrm{M6}_{T_A})$. Then, following Lemma 2.42 for the case $n=4$, all remaining undetermined B-coefficients of p are uniquely and stably determined. Thus, by subdivision using the de Casteljau algorithm, all B-coefficients of s are uniquely and stably determined.

Finally, the dimension of the space $\mathcal{S}_d^{r,\rho,\mu,\eta}(T_A)$ can be determined. Since $\mathcal{M}_{T_A}$ is a minimal determining set for $\mathcal{S}_d^{r,\rho,\mu,\eta}(T_A)$,

$$\dim \mathcal{S}_d^{r,\rho,\mu,\eta}(T_A) = \#\mathcal{M}_{T_A}$$

holds, where

$$\begin{aligned}
\#\mathcal{M}_{T_A} =& 4\binom{\rho+3}{3} + 12\binom{\mu+2}{3} \\
&+ 4\sum_{i=0}^{r}\left(\binom{d-i+2}{2} - 3\binom{\rho-i+2}{2} - 3\binom{\mu-i+1}{2}\right. \\
&\left. - 3i(\mu-i+1)\right) + 4\sum_{i=1}^{\lfloor\frac{r}{3}\rfloor}\left(\binom{d-r-i+2}{2} - 3\binom{\rho-r-i+2}{2}\right. \\
&\left. - 3\binom{\mu-r+i+1}{2} - 3(r+i)(\mu-r+i+1)\right) + 6\sum_{i=1}^{\lfloor\frac{r}{3}\rfloor}(\mu+2i) \\
&+ \binom{\eta-4(r+\lfloor\frac{r}{3}\rfloor+\eta-d)-1}{3},
\end{aligned}$$

which reduces to the number in (7.1). □

For a spline defined on a tetrahedron T, which has been subjected to the Alfeld split, all B-coefficients associated with the domain points within a distance of r from the faces, μ from the edges, and ρ from the vertices of T can be uniquely determined from the B-coefficients corresponding to the domain points in the sets ($M1_{T_A}$) - ($M4_{T_A}$). Thus, all smoothness conditions connecting T_A with a neighboring tetrahedron $\widetilde{T}_A$, which has also been subdivided with the Alfeld split, are fulfilled and the macro-elements considered in Theorem 7.3 can be extended to a tetrahedral partition.

Notation 7.4:
Let Δ be an arbitrary tetrahedral partition and Δ_A the partition obtained by subdividing each tetrahedron in Δ with the Alfeld split. Moreover, let $\mathcal{V}, \mathcal{E}$, and $\mathcal{F}$ be the sets of vertices, edges, and faces of Δ, respectively, and $\#\mathcal{V}, \#\mathcal{E}, \#\mathcal{F}$, and $\#\mathcal{N}$ the number of vertices, edges, faces, and tetrahedra of Δ. Furthermore, let $\mathcal{V}_I$ be the set of split points in the interior of Δ, where the tetrahedra of Δ have been subdivided with the Alfeld split, and let $\mathcal{F}_I$ be the set of triangular faces in Δ_A of the form $\langle u,v,w \rangle$, with $u,v \in \mathcal{V}$ and $w \in \mathcal{V}_I$.

The macro-elements based on the tetrahedral partition Δ_A are in the superspline space

$$\begin{aligned} \mathcal{S}_d^{r,\rho,\mu,\eta}(\Delta_A) := \{ s \in C^r(\Omega) :\ & s|_{\widetilde{T}} \in \mathcal{P}_d^3 \text{ for all } \widetilde{T} \in \Delta_A, \\ & s \in C^{\mu}(e) \text{ for all } e \in \mathcal{E}, \\ & s \in C^{\rho}(v) \text{ for all } v \in \mathcal{V}, \\ & s \in C^{\eta}(v_T) \text{ for all } v_T \in \mathcal{V}_I \}. \end{aligned}$$

Theorem 7.5:
Let $\mathcal{M}_{\Delta_A}$ be the union of the following sets of domain points:

($M1_{\Delta_A}$) $\bigcup_{v \in \mathcal{V}} D_{\rho}^{T_v}(v)$, with $v \in T_v$

($M2_{\Delta_A}$) $\bigcup_{e:=\langle u,v \rangle \in \mathcal{E}} E_{\mu}^{T_e}(e) \setminus (D_{\rho}(u) \cup D_{\rho}(v))$, with $e \in T_e$

($M3_{\Delta_A}$) $\bigcup_{F \in \mathcal{F}} \bigcup_{j=0}^{r} F_j^{T_F}(F) \setminus (\bigcup_{v \in F} D_{\rho}(v) \cup \bigcup_{e \in F} E_{\mu}(e))$, with $F \in T_F$

($M4_{\Delta_A}$) $\bigcup_{F \in \mathcal{F}_I} \{\xi^{T_F}_{\mu-2i+j,d-\rho-j,\mu+2i,0} : \quad i = 1,\dots,\lfloor \frac{r}{3} \rfloor, \quad j = 1,\dots,\mu+2i\}$, with $F \in T_F$

(M5$_{\Delta_A}$) $\bigcup_{T\in\Delta}\bigcup_{F\in T}\bigcup_{j=r+1}^{\lfloor\frac{r}{3}\rfloor}\left(F_j^T(F)\setminus\big(\bigcup_{v\in F}D_\rho(v)\cup\bigcup_{e\in F}E_{\mu+2j}(e)\big)\right)$

(M6$_{\Delta_A}$) $\bigcup_{v\in\mathcal{V}_I} D^{T_v}_{\eta-4(m+\lfloor\frac{r}{3}\rfloor+1)}(v)$, with $v\in T_v$,

where m is as in Table 7.1. Then $\mathcal{M}_{\Delta_A}$ is a 1-local and stable minimal determining set for the space $\mathcal{S}_d^{r,\rho,\mu,\eta}(\Delta_A)$ and the dimension of the superspline space is given as

$$\dim\mathcal{S}_d^{r,\rho,\mu,\eta}(\Delta_A)=\begin{cases}(36m^3+36m^2+11m+1)\#\mathcal{V}+(9m^3+9m^2+2m)\#\mathcal{E}\\+(\frac{134m^3+57m^2-5m}{6})\#\mathcal{F}+(\frac{311m^3-99m^2-2m}{6})\#\mathcal{N},\\ \text{for } r=2m,\\ \\ (36m^3+72m^2+47m+10)\#\mathcal{V}+(9m^3+18m^2\\+11m+2)\#\mathcal{E}+(\frac{134m^3+228m^2+112m+18}{6})\#\mathcal{F}\\+(\frac{311m^3+294m^2+85m+6}{6})\#\mathcal{N},\ \text{for } r=2m+1.\end{cases}\tag{7.2}$$

Proof:
First, we show that $\mathcal{M}_\Delta$ is a minimal determining set for $\mathcal{S}_d^{r,\rho,\mu,\eta}(\Delta_A)$. Therefore, we set the B-coefficients c_ξ of a spline $s\in\mathcal{S}_d^{r,\rho,\mu,\eta}(\Delta_A)$ to arbitrary values for all $\xi\in\mathcal{M}_\Delta$. Then, all remaining B-coefficients of s associated with domain points in Δ_A are uniquely and stably determined.

Since the balls $D_\rho(v)$, $v\in\mathcal{V}$, do not overlap, we can uniquely and stably compute the undetermined B-coefficients of $D_\rho(v)$, $v\in\mathcal{V}$, from those associated with the domain points in (M1$_{\Delta_A}$) using the C^ρ supersmoothness at the vertices in $\mathcal{V}$. These computations are 1-local and only depend on B-coefficients corresponding to domain points in $\Gamma_\xi\subset\mathcal{M}_\Delta\cap\operatorname{star}(T_v)$, for all $\xi\in D_\rho(v)$ and $v\in T_v$.

Considering the tubes $E_\mu(e)$, $e\in\mathcal{E}$, we see that the set (M2$_{\Delta_A}$) does not overlap the balls $D_\rho(v)$, $v\in\mathcal{V}$. Thus, we can uniquely and stably compute the undetermined B-coefficients of s associated with domain points in $E_\mu(e)$, $e\in\mathcal{E}$, from those corresponding to the domain points in (M2$_{\Delta_A}$) using the C^μ supersmoothness at the edges in $\mathcal{E}$. For these B-coefficients (2.11) holds with $\Gamma_\xi\subset\mathcal{M}_\Delta\cap\operatorname{star}(T_e)$, for all $\xi\in E_\mu(e)$ and $e\in T_e$.

Since the set (M3$_{\Delta_A}$) does not overlap the balls $D_\rho(v)$, $v\in\mathcal{V}$, or the tubes $E_\mu(e)$, $e\in\mathcal{E}$, we can use the C^r smoothness conditions at the triangular faces of Δ to uniquely and stably determine the B-coefficients of s associated with domain points in $F_i^{T_F}(F)$, $i = 0,\ldots,r$, for all $F \in \mathcal{F}$ and $F \in T_F$, from those corresponding to the domain points in $(\mathrm{M3}_{\Delta_A})$. Moreover, the involved computations are 1-local and (2.11) holds with $\Gamma_\xi \subset \mathcal{M}_\Delta \cap \mathrm{star}(T_F)$, for all $\xi \in F_r^{T_F}(F)$.

Now, we have determined all B-coefficients of s, such that two splines $s|_{T_A}$ and $s|_{\widetilde{T}_A}$, with $T, \widetilde{T} \in \Delta$, join with C^ρ, C^μ, and C^r continuity at the vertices, edges, and faces of T and $\widetilde{T}$. Since the sets $(\mathrm{M4}_{\Delta_A})$, $(\mathrm{M5}_{\Delta_A})$, and $(\mathrm{M6}_{\Delta_A})$ contain the sets $(\mathrm{M4}_{T_A})$, $(\mathrm{M5}_{T_A})$, and $(\mathrm{M6}_{T_A})$ from Theorem 7.3 for each tetrahedron $T \in \Delta$, the remaining undetermined B-coefficients of s can be uniquely and stably determined in the way described in the proof of Theorem 7.3. For these B-coefficients of s, (2.11) holds with $\Gamma_\xi \subset \mathcal{M}_\Delta \cap \mathrm{star}(T)$, for all $\xi \in T$ and $T \in \Delta$.

To compute the dimension of $\mathcal{S}_d^{r,\rho,\mu,\eta}(\Delta_A)$, we have to consider the cardinalities of the sets $(\mathrm{M1}_{\Delta_A})$ - $(\mathrm{M6}_{\Delta_A})$.

For each vertex $v \in \mathcal{V}$, the set $(\mathrm{M1}_{\Delta_A})$ contains

$$\binom{\rho+3}{3}$$

domain points. For each edge $e \in \mathcal{E}$, the set $(\mathrm{M2}_{\Delta_A})$ contains

$$2\binom{\mu+2}{3}$$

domain points. For each triangular face $F \in \mathcal{F}$, the set $(\mathrm{M3}_{\Delta_A})$ contains

$$\sum_{i=0}^{r}\left(\binom{d-i+2}{2} - 3\binom{\rho-i+2}{2} - 3\binom{\mu-i+1}{2} - 3i(\mu-i+1)\right)$$

domain points. For each tetrahedron $T \in \Delta$, the sets $(\mathrm{M4}_{\Delta_A})$, $(\mathrm{M5}_{\Delta_A})$, and $(\mathrm{M6}_{\Delta_A})$ contain

$$6\sum_{i=1}^{\lfloor \frac{r}{3} \rfloor}(\mu+2i) + 4\sum_{i=1}^{\lfloor \frac{r}{3} \rfloor}\left(\binom{d-r-i+2}{2} - 3\binom{\rho-r-i+2}{2}\right.$$
$$\left. - 3\binom{\mu-r+i+1}{2} - 3(r+i)(\mu-r+i+1)\right)$$
$$+ \binom{\eta - 4(r+\lfloor \frac{r}{3} \rfloor + \eta - d) - 1}{3}$$

domain points. This reduces to the number in (7.2). □

Remark 7.6:
For $r = 1$ and $r = 2$ the macro-elements constructed in this section are consistent with the macro-elements on the Alfeld split investigated in [62]. Here, the superspline space $\mathcal{S}_5^{1,2,1,4}(\Delta_A)$ is used in order to construct C^1 macro-elements over the Alfeld split of a tetrahedral partition. In [62] the same space of supersplines is used to construct C^1 macro-elements based on the Alfeld split. In order to construct C^2 macro-elements over the Alfeld split of a tetrahedral partition the superspline space $\mathcal{S}_{13}^{2,6,3,12}(T_A)$ is used here. Again, this is the same superspline space as the one considered in [62] to construct C^2 macro-elements on a tetrahedral partition refined with the Alfeld split. Furthermore, the same space of supersplines is also used in [10] to construct C^2 macro-elements based on a tetrahedral partition that has been refined with the Alfeld split. However, in [10] no minimal determining set was constructed.

7.3 Examples for macro-elements over the Alfeld split of a tetrahedron

In this section some examples for the macro-elements constructed in section 7.2 are given. Since the minimal determining sets for the macro-elements are quite complex, the examples in this section are supposed to ease the understanding of the macro-elements. Therefore, as in [54], the macro-elements on one tetrahedron $T := \langle v_1, v_2, v_3, v_4 \rangle$ for the cases $r = 1, \ldots, 4$ are considered more closely, as well as the cases $r = 5$ and $r = 6$. Thereby it is also shown that the macro-elements reduce to those already known from Lai and Schumaker [62] for $r = 1$ and $r = 2$.

7.3.1 C^1 macro-elements over the Alfeld split

The macro-elements are in the superspline space $\mathcal{S}_d^{r,\rho,\mu,\eta}(T_A)$. For the case $r = 1$, m is equal to zero and, following Table 7.1 (third row), the superspline space reduces to the space $\mathcal{S}_5^{1,2,1,4}(T_A)$. As shown in Theorem 7.3, the dimension of this space is equal to 65 and the corresponding minimal determining set $\mathcal{M}_{T_A,1}$ is the union of the following sets of domain points:

(C^1:M1$_{T_A}$) $D_2^{T_i}(v_i)$, $i = 1, \ldots, 4$

(C^1:M2$_{T_A}$) $\bigcup\limits_{e := \langle u,v \rangle \in \mathcal{E}_T} E_1^{T_e}(e) \setminus (D_2(u) \cup D_2(v))$, with $e \in T_e$

(C^1:M3$_{T_A}$) $F_1^{T_i}(F_i) \setminus (\bigcup_{v \in F_i} D_\rho(v) \cup \bigcup_{e \in F_i} E_\mu(e))$, $i = 1,\dots,4$

(C^1:M4$_{T_A}$) v_T

Proof:
First, it is shown that $\mathcal{M}_{T_A,1}$ is a stable minimal determining set for the superspline space $\mathcal{S}_5^{1,2,1,4}(T_A)$. Therefore, the B-coefficients c_ξ of a spline $s \in \mathcal{S}_5^{1,2,1,4}(T_A)$ are set to arbitrary values for each domain point $\xi \in \mathcal{M}_{T_A,1}$ and it is shown that all the remaining B-coefficients of s are uniquely and stably determined.

The B-coefficients of s associated with the domain points in the balls $D_2(v_i)$, $i = 1,\dots,4$, are uniquely determined from those corresponding to the domain points in (C^1:M1$_{T_A}$) by the C^2 supersmoothness at the vertices of T. Since the B-coefficients are directly determined from smoothness conditions, their computation is stable

The remaining undetermined B-coefficients associated with domain points in the tubes $E_\mu(e)$, $e \in \mathcal{E}_T$, can be determined from the B-coefficients of s corresponding to the domain points in (C^1:M2$_{T_A}$) by the C^1 smoothness conditions at the edges of T. The corresponding computations are also stable.

Next, the B-coefficients of s associated with the domain points in the two shells $R_5(v_T)$ and $R_4(v_T)$ are considered. Since those corresponding to the domain points in the balls $D_\rho(v)$, $v \in \mathcal{V}_T$, and the tubes $E_\mu(e)$, $e \in \mathcal{E}_T$, are already uniquely determined, all B-coefficients of s associated with domain points on the faces of T, in the shell $R_5(v_T)$, are already determined. The remaining undetermined B-coefficients of s corresponding to domain points in the shell $R_4(v_T)$ are contained in the set (C^1:M3$_{T_A}$). Thus, all B-coefficients of s corresponding to the domain points in the shells $R_5(v_T)$ and $R_4(v_T)$ are uniquely and stably determined (cf. Figure 7.2).

To determine the remaining B-coefficients of s, those corresponding to the domain points in the ball $D_4(v_T)$ are examined more closely. The spline s has C^4 supersmoothness at the point v_T. Thus, the B-coefficients of s corresponding to the domain points in $D_4(v_T)$ can be regarded as those of a quartic polynomial p defined on the tetrahedron $\widetilde{T}$, which is bounded by the domain points in the shell $R_4(v_T)$ and also refined with the Alfeld split.

The B-coefficients of s associated with domain points in $R_4(v_T)$ are already determined. Thus, these B-coefficients are also known for the polynomial p. Since the point v_T is contained in the minimal determining set $\mathcal{M}_{T_A,1}$, in the subset (C^1:M4$_{T_A}$), the corresponding B-coefficient has been set to an arbitrary value. Now, since this value is also known for p, following Lemma 2.42 for $n = 4$, the remaining undetermined B-coefficients of p are uniquely and stably determined. Thus, all B-coefficients of s are uniquely and stably determined.

Then, the dimension of the space $\mathcal{S}_5^{1,2,1,4}(T_A)$ is equal to the cardinality of the minimal determining set $\mathcal{M}_{T_A,1}$. The set (C^1:M1$_{T_A}$) contains

$$4\binom{2+3}{3} = 40$$

domain points. The cardinality of the set (C^1:M2$_{T_A}$) is equal to

$$12\binom{1+2}{3} = 12.$$

The set (C^1:M3$_{T_A}$) consists of

$$4\left(\binom{4+2}{2} - 3\binom{1+2}{2} - 3\right) = 12$$

domain points. Finally, the set (C^1:M4$_{T_A}$) contains only one single domain point, v_T. Thus, the dimension of $\mathcal{S}_5^{1,2,1,4}(T_A)$ is equal to 65. □

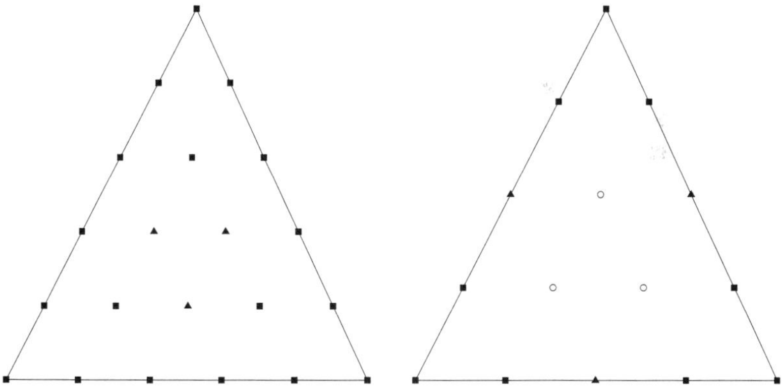

Figure 7.2: The shells $R_5^{\hat{T}}(v_T)$ (left) and $R_4^{\hat{T}}(v_T)$ (right) for a tetrahedron $\hat{T}$ in T_A. Domain points whose B-coefficients are determined from (C^1:M1$_{T_A}$) are marked with ■, those from (C^1:M2$_{T_A}$) with ▲, and those from (C^1:M3$_{T_A}$) with ○.

Remark 7.7:
The macro-elements considered in this example are exactly the same as the ones constructed by Lai and Schumaker [62] in section 18.3.

7.3.2 C^2 macro-elements over the Alfeld split

For the case of C^2 macro-elements the superspline space $\mathcal{S}_d^{r,\rho,\mu,\eta}(T_A)$ reduces to the space $\mathcal{S}_{13}^{2,6,3,12}(T_A)$, by the second row of Table 7.1 and $m = 1$. Following (7.1), the dimension of the superspline space is equal to 615 and the minimal determining set $\mathcal{M}_{T_A,2}$ is the union of the following sets of domain points:

(C^2:M1$_{T_A}$) $D_6^{T_i}(v_i)$, $i = 1,\dots,4$

(C^2:M2$_{T_A}$) $\bigcup\limits_{e:=\langle u,v\rangle \in \mathcal{E}_T} E_3^{T_e}(e) \setminus (D_6(u) \cup D_6(v))$, with $e \in T_e$

(C^2:M3$_{T_A}$) $\bigcup\limits_{j=0}^{2} F_j^{T_i}(F_i) \setminus (\bigcup\limits_{v\in F_i} D_6(v) \cup \bigcup\limits_{e\in F_i} E_3(e))$, $i = 1,\dots,4$

(C^2:M4$_{T_A}$) $D_4^{T_1}(v_T)$

Proof:
In order to show that $\mathcal{M}_{T_A,2}$ is a minimal determining set for the superspline space $\mathcal{S}_{13}^{2,6,3,12}(T_A)$, the B-coefficients of a spline $s \in \mathcal{S}_{13}^{2,6,3,12}(T_A)$ associated with the domain points in $\mathcal{M}_{T_A,2}$ are set to arbitrary values and it is shown that the remaining B-coefficients of s can be uniquely and stably determined from these.

Using the C^6 supersmoothness at the vertices of T, the B-coefficients of s corresponding to the domain points in the balls $D_6(v_i)$, $i = 1,\dots,4$, can be uniquely and stably computed from the B-coefficients associated with the domain points in the set (C^2:M1$_{T_A}$).

The remaining undetermined B-coefficients of s corresponding to domain points in the tubes $E_3(e)$, $e \in \mathcal{E}_T$, can be uniquely and stably determined by the C^3 supersmoothness conditions at the edges of T and the B-coefficients associated with the domain points in the set (C^2:M2$_{T_A}$).

Now, the remaining B-coefficients of s corresponding to the domain points in the shells $R_{13}(v_T)$, $R_{12}(v_T)$, and $R_{11}(v_T)$ are investigated. The B-coefficients of s associated with the domain points in the balls $D_6(v)$, $v \in \mathcal{V}_T$, and the tubes $E_3(e)$, $e \in \mathcal{E}_T$, are already determined. Considering the shell $R_{13}(v_T)$ containing the domain points in the faces of T, there are still twelve undetermined B-coefficients of s corresponding to three domain points on each face of T. Since these twelve domain points are contained in the set (C^2:M3$_{T_A}$), all B-coefficients of s corresponding to domain points on the faces of T are uniquely and stably determined (cf. Figure 7.3 (left)). Since the domain points corresponding to the remaining $4 \cdot 10 + 4 \cdot 18$ undetermined B-coefficients of s restricted to the shells $R_{12}(v_T)$ and $R_{11}(v_T)$ are also contained in the set (C^2:M3$_{T_A}$), s is uniquely and stably determined on these shells (cf. Figure 7.3 (right), and Figure 7.4).

Finally, the last undetermined B-coefficients of s associated with domain points in the ball $D_{12}(v_T)$ are determined. The spline s has C^{12} supersmoothness at the split point v_T in the interior of T. Therefore, the B-coefficients of s corresponding to the domain points in $D_{12}(v_T)$ can be considered as those of a polynomial p of degree twelve defined on the tetrahedron $\widetilde{T}$, which is formed by the domain points in $D_{12}(v_T)$. Note that the tetrahedron $\widetilde{T}$ is also subdivided with the Alfeld split at the point v_T. Since some B-coefficients of s corresponding to the domain point in the considered ball are already determined, the corresponding B-coefficients of p are also already known. These are the B-coefficients associated with the domain points within a distance of one from the faces of $\widetilde{T}$, within a distance of two from the edges of $\widetilde{T}$, and within a distance of five from the vertices of $\widetilde{T}$. Now, from the B-coefficients corresponding to the domain points in the set (C^2:M4$_{T_A}$), the partial derivatives at v_T up to order four are uniquely and stably determined. Thus, following Lemma 2.42, with $n = 4$, the remaining undetermined B-coefficients of p are uniquely and stably determined form these derivatives. Then, all B-coefficients of s are uniquely and stably determined.

Now, the dimension of the superspline space $\mathcal{S}_{13}^{2,6,3,12}(T_A)$ can be determined. Since the dimension is equal to the cardinality of the minimal determining set $\mathcal{M}_{T_A,2}$, it suffices to count the number of domain points contained in the set. The set (C^2:M1$_{T_A}$) consists of

$$4\binom{6+3}{3} = 336$$

domain points. The set (C^2:M2$_{T_A}$) contains

$$12\binom{3+2}{3} = 120$$

domain points.

In the set (C^2:M3$_{T_A}$) a total number of

$$4\sum_{i=0}^{2}\left(\binom{15-i}{2}-3\binom{8-i}{2}-3\binom{4-i}{2}-3i(4-i)\right)=124$$

domain points are contained. The cardinality of the set (C^2:M4$_{T_A}$) is equal to

$$\binom{4+3}{3}=35.$$

Therefore, the cardinality of $\mathcal{M}_{T_A,2}$, and thus also the dimension of the spline space $\mathcal{S}_{13}^{2,6,3,12}(T_A)$, is equal to 615. □

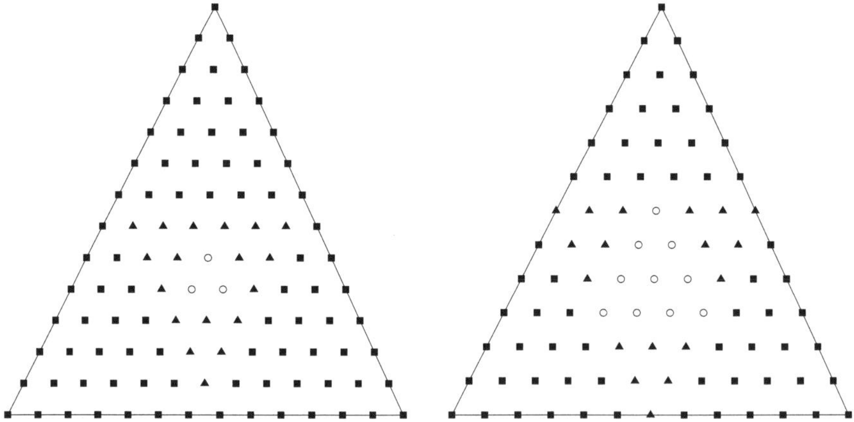

Figure 7.3: The shells $R_{13}^{\hat{T}}(v_T)$ (left) and $R_{12}^{\hat{T}}(v_T)$ (right) for a tetrahedron $\hat{T}$ in T_A. Domain points whose B-coefficients are determined from (C^2:M1$_{T_A}$) are marked with ■, those from (C^2:M2$_{T_A}$) with ▲, and those from (C^2:M3$_{T_A}$) with ○.

Remark 7.8:
The C^2 macro-elements investigated in this example are exactly the same as the C^2 macro-elements considered in Lai and Schumaker [62] in section 18.7. The same superspline space is also used in Alfeld and Schumaker [9] in order to construct C^2 macro-elements over the Alfeld split of a tetrahedron.

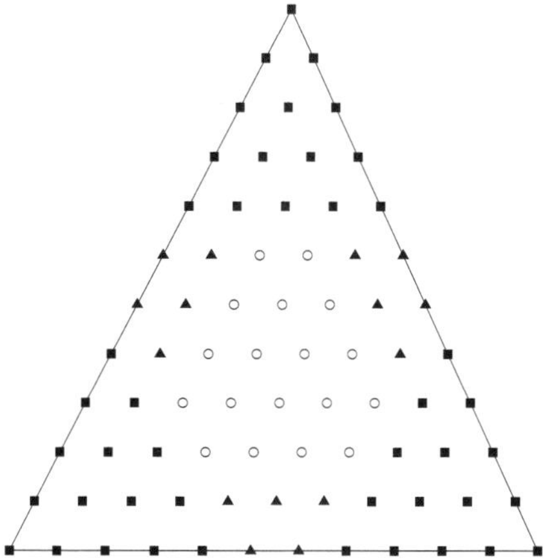

Figure 7.4: The shell $R_{11}^{\hat{T}}(v_T)$ for a tetrahedron $\hat{T}$ in T_A. Domain points whose B-coefficients are determined from (C^2:M1$_{T_A}$) are marked with ■, those from (C^2:M2$_{T_A}$) with ▲, and those from (C^2:M3$_{T_A}$) with ○.

7.3.3 C^3 macro-elements over the Alfeld split

The C^3 macro-elements are in the superspline space $\mathcal{S}_{17}^{3,8,4,15}(T_A)$, which is derived from the space $\mathcal{S}_d^{r,\rho,\mu,\eta}(T_A)$ with the values from the third row of Table 7.1 and $m = 1$. Following Theorem 7.3, the dimension of the superspline space is equal to 1344 and the corresponding minimal determining set $\mathcal{M}_{T_A,3}$ is the union of the following sets of domain points:

(C^3:M1$_{T_A}$) $D_8^{T_i}(v_i)$, $i = 1, \ldots, 4$

(C^3:M2$_{T_A}$) $\bigcup\limits_{e:=\langle u,v\rangle \in \mathcal{E}_T} E_4^{T_e}(e) \setminus (D_8(u) \cup D_8(v))$, with $e \in T_e$

(C^3:M3$_{T_A}$) $\bigcup\limits_{j=0}^{3} F_j^{T_i}(F_i) \setminus (\bigcup\limits_{v \in F_i} D_8(v) \cup \bigcup\limits_{e \in F_i} E_4(e))$, $i = 1, \ldots, 4$

(C^3:M4$_{T_A}$) $\bigcup\limits_{F \in \mathcal{F}_{I,T}} \{\xi_{2+j,9-j,6,0}^{T_F} : j = 1, \ldots, 6\}$

(C^3:M5$_{T_A}$) $F_4^{T_i}(F_i) \setminus (\bigcup\limits_{v \in F_i} D_8(v) \cup \bigcup\limits_{e \in F_i} E_6(e))$, $i = 1, \ldots, 4$

(C^3:M6$_{T_A}$) $D_3^{T_1}(v_T)$

Proof:
To show that $\mathcal{M}_{T_A,3}$ is a stable minimal determining set for the superspline space $\mathcal{S}_{17}^{3,8,4,15}(T_A)$, the B-coefficients c_ξ of a spline $s \in \mathcal{S}_{17}^{3,8,4,15}(T_A)$ are set to arbitrary values for each domain point $\xi \in \mathcal{M}_{T_A,3}$. Then, it is shown that all the remaining B-coefficients of s are uniquely and stably determined.

The B-coefficients of s corresponding to the domain points in the balls $D_8(v_i)$, $i = 1,\dots,4$, are uniquely and stably determined from the B-coefficients associated with the domain points in (C^3:M1$_{T_A}$) by the C^8 supersmoothness at the vertices of T.

The remaining B-coefficients of s corresponding to domain points in the tubes $E_4(e)$, $e \in \mathcal{E}_T$, are uniquely determined by the C^4 supersmoothness conditions at the edges of T from the B-coefficients associated with the domain points in the set (C^3:M2$_{T_A}$). These computations are stable, since they only involve smoothness conditions.

Next, the remaining B-coefficients associated with domain points in the shells $R_{17}(v_T)$, $R_{16}(v_T)$, $R_{15}(v_T)$, and $R_{14}(v_T)$ are considered. Since the B-coefficients of s corresponding to the domain points in the balls $D_8(v_i)$, $i = 1,\dots,4$, and the tubes $E_4(e)$, $e \in \mathcal{E}_T$, are already determined, the only undetermined B-coefficients of s associated with domain points in the investigated shells correspond to the domain points in the set (C^3:M3$_{T_A}$). Since the B-coefficients of s corresponding to the domain points in the balls $D_8(v_i)$, $i = 1,\dots,4$, and the tubes $E_4(e)$, $e \in \mathcal{E}_T$, are already determined, the only undetermined B-coefficients of s associated with domain points in the investigated shells correspond to the domain points in the set (C^3:M3$_{T_A}$). Thus, all B-coefficients of s associated with the domain points in the sets $R_{17}(v_T)$, $R_{16}(v_T)$, $R_{15}(v_T)$, and $R_{14}(v_T)$ are uniquely and stably determined (cf. Figure 7.5 and Figure 7.6).

Now, the remaining undetermined B-coefficients of s associated with domain points in the ball $D_{15}(v_T)$ are determined. Since the spline s has C^{15} supersmoothness at the vertex v_T in the interior of T, the B-coefficients of s associated with the domain points in the set $D_{15}(v_T)$ can be regarded as B-coefficients of a polynomial p of degree fifteen defined on the tetrahedron $\widetilde{T}$, which is bounded by the domain points in $R_{15}(v_T)$. The tetrahedron $\widetilde{T}$ is also refined with the Alfeld split at the point v_T. Since some of the B-coefficients of s associated with domain points in $D_{15}(v_T)$ are already determined, also some of the B-coefficients of p are determined. Thus, the B-coefficients of p corresponding to the domain points within a distance of one from the faces, to those within a distance of two from the edges, and to those within a distance of six from the vertices of $\widetilde{T}$ are determined.

Now, the B-coefficients of p associated with the domain points within a distance of two from the faces of $\widetilde{T}$ are considered more closely. Therefore, first the remaining B-coefficients corresponding to the domain points with a distance of three from the edges of $\widetilde{T}$ are computed, and subsequently, the undetermined B-coefficients associated with domain with a distance of four from the edges. Then, it is shown how the remaining B-coefficients of p corresponding to the domain points with a distance of two from the faces of $\widetilde{T}$ are determined. See Figure 7.7 for the domain points in this layer determined in the different steps. The domain points whose B-coefficients are already determined are marked with ■ and ▲. Those, whose B-coefficients are still undetermined so far, are highlighted with ✚, ✖ and ●.

First, the B-coefficients of p associated with the domain points with a distance of three from the edges of $\widetilde{T}$ are considered. Note that these domain points have a distance of five from the edges of T. The B-coefficients corresponding to the domain points in shells $R_{14}(v_T)$ and $R_{15}(v_T)$ are already determined. Thus, the only undetermined B-coefficients associated with domain points with a distance of three from the edges of $\widetilde{T}$ are those corresponding to domain points with a distance of one from the interior faces of $\widetilde{T}_A$ in $R_{13}(v_T)$ and those on the interior faces in $R_{12}(v_T)$. The corresponding domain points in $R_{13}(v_T)$ are shown in Figure 7.7, they are marked with ✚. Now, considering the C^1, C^2, and C^3 smoothness conditions across the interior faces of $\widetilde{T}_A$, it can be seen that each triple of these conditions connects three of the undetermined B-coefficients. One of the B-coefficients is associated with a domain point on an interior face of $\widetilde{T}_A$ in $R_{12}(v_T)$ and the other two are associated with the two corresponding domain points with a distance of one from this interior face in $R_{13}(v_T)$. In Figure 7.7, each of the domain points marked with ✚ corresponds to one of these triples of smoothness conditions. Therefore, for each triple of undetermined B-coefficients there are three equations from the C^1, C^2, and C^3 conditions containing no other undetermined B-coefficients. Then, following Lemma 2.33 for the case $j = 0$, the corresponding system of equations has a unique solution and these B-coefficients are uniquely and stably determined. Thus, all B-coefficients corresponding to domain points with a distance of three from the edges of $\widetilde{T}$ are determined.

Next, the B-coefficients of p associated with the domain points with a distance of four from the edges of $\widetilde{T}$ can be considered. The only undetermined B-coefficients associated with these domain points correspond to the domain points with a distance of two from the interior faces of $\widetilde{T}_A$ in

$R_{13}(v_T)$, those with a distance of one from the interior faces in $R_{12}(v_T)$ and those on the interior faces in $R_{11}(v_T)$. The corresponding domain points in $R_{13}(v_T)$ are marked with ✖ in Figure 7.7. Since the set (C^3:M4$_{T_A}$) contains the domain points on the interior faces in $R_{11}(v_T)$, the associated B-coefficients are determined. Now, each quadruple of associated C^1, C^2, C^3, and C^4 smoothness conditions across the interior faces of $\widetilde{T}_A$ connects a total number of four undetermined B-coefficients corresponding to domain points with a distance of four from the edges of $\widetilde{T}$. Of these domain points, two have a distance of one from an interior face of $\widetilde{T}_A$ in $R_{12}(v_T)$ and the other two have a distance of two from the same interior face and lie in $R_{13}(v_T)$ (labeled by ✖ in Figure 7.7). So, each domain point in Figure 7.7 marked with ✖ corresponds to one quadruple of smoothness conditions connecting the B-coefficient of the marked domain point with three other undetermined B-coefficients. Then, following Lemma 2.33 for the case $j = 1$, the four equations from the smoothness conditions uniquely and stably determine the four undetermined B-coefficients. This way, all B-coefficients corresponding to domain points with a distance of four from the edges of $\widetilde{T}$ are determined.

Now, the last undetermined B-coefficients of p corresponding to domain points in the shell $R_{13}(v_T)$ are considered. These B-coefficients are just associated with the domain points contained in the set (C^3:M5$_{T_A}$). In Figure 7.7 these domain points are marked with ●. Thus, all B-coefficients of the polynomial p corresponding to the domain points in the set $R_{13}(v_T)$, those with a distance of two from the faces of $\widetilde{T}$, are uniquely and stably determined.

Finally, the last undetermined B-coefficients of p can be determined. Up to now, all B-coefficients of p corresponding to the domain points within a distance of two from the faces of $\widetilde{T}$ are determined. Then, from the B-coefficients of p associated with the domain points in the set (C^3:M6$_{T_A}$), contained in $\mathcal{M}_{T_A,3}$, the partial derivatives of p at the point v_T up to order three can be directly computed. Now, following Lemma 2.42 for $n = 4$, the remaining undetermined B-coefficients of the polynomial p can be uniquely and stably determined. Thus, all B-coefficients of p and therefore also all B-coefficients of s are uniquely and stably determined.

Now, the dimension of the superspline space $\mathcal{S}_{17}^{3,8,4,15}(T_A)$ is considered. The dimension is equal to the cardinality of the minimal determining set $\mathcal{M}_{T_A,3}$. Thus, it suffices to count the number of domain points contained in each of the sets (C^3:M1$_{T_A}$) - (C^3:M6$_{T_A}$). The set (C^3:M1$_{T_A}$) contains

$$4\binom{8+3}{3} = 660$$

domain points. The set (C^3:M2$_{T_A}$) consists of

$$12\binom{4+2}{3} = 240$$

domain points. A total number of

$$4\sum_{i=0}^{3}\left(\binom{19-i}{2} - 3\binom{10-i}{2} - 3\binom{5-i}{2} - 3i(5-i)\right) = 328$$

domain points are contained in the set (C^3:M3$_{T_A}$). The set (C^3:M4$_{T_A}$) contains

$$6(4+2) = 36$$

domain points. The set (C^3:M5$_{T_A}$) holds a total number of

$$4\left(\binom{15}{2} - 3\binom{6}{2} - 3\binom{3}{2} - 36\right) = 60$$

domain points. The cardinality of the last set (C^3:M6$_{T_A}$) contained in the minimal determining set $\mathcal{M}_{T_A,3}$ is equal to

$$\binom{6}{3} = 20.$$

Thus, the cardinality of $\mathcal{M}_{T_A,3}$, and therefore also the dimension of the superspline space$\mathcal{S}_{17}^{3,8,4,15}(T_A)$, is equal to 1344. □

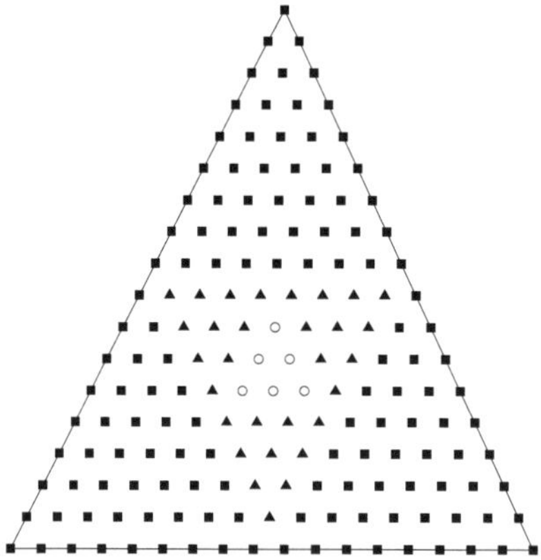

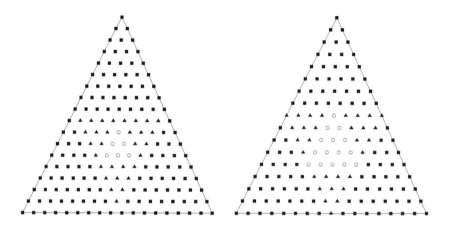

Figure 7.5: The shells $R^{\hat{T}}_{17}(v_T)$ (left) and $R^{\hat{T}}_{16}(v_T)$ (right) for a tetrahedron $\hat{T}$ in T_A. Domain points whose B-coefficients are determined from (C^3:M1$_{T_A}$) are marked with ■, those from (C^3:M2$_{T_A}$) with ▲, and those from (C^3:M3$_{T_A}$) with ○.

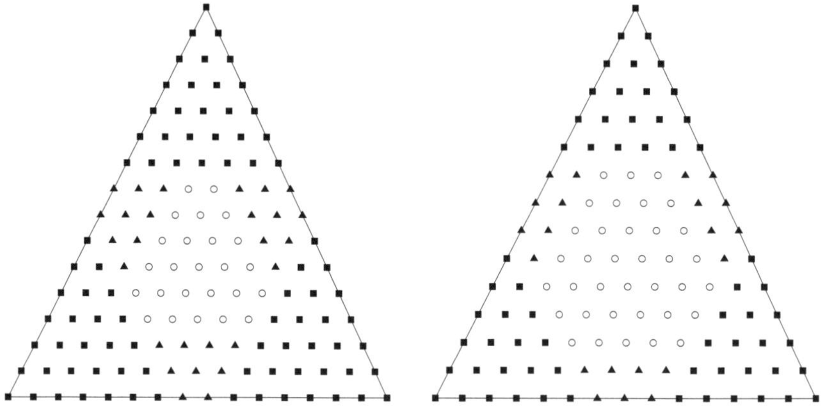

Figure 7.6: The shells $R^{\hat{T}}_{15}(v_T)$ (left) and $R^{\hat{T}}_{14}(v_T)$ (right) for a tetrahedron $\hat{T}$ in T_A. Domain points whose B-coefficients are determined from (C^3:M1$_{T_A}$) are marked with ■, those from (C^3:M2$_{T_A}$) with ▲, and those from (C^3:M3$_{T_A}$) with ○.

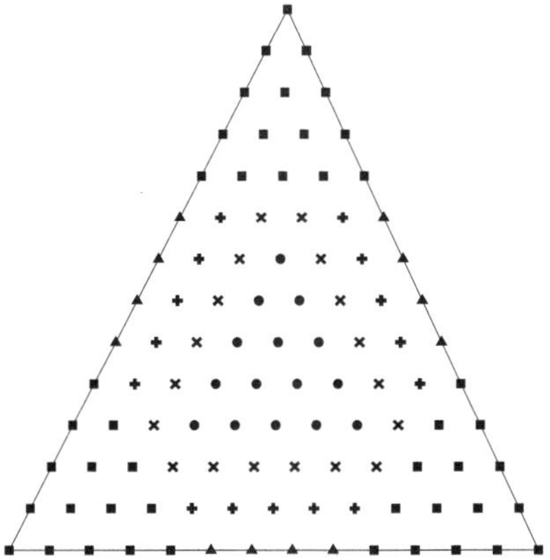

Figure 7.7: The shell $R_{13}^{\hat{T}}(v_T)$ for a tetrahedron $\hat{T}$ in T_A. Domain points whose B-coefficients are determined from (C^3:M1$_{T_A}$) are marked with ■, those from (C^3:M2$_{T_A}$) with ▲, those from smoothness conditions using Lemma 2.33 ($j = 0$) with ✚, those from (C^3:M4$_{T_A}$) and smoothness conditions using Lemma 2.33 ($j = 1$) with ✖, and those from (C^3:M5$_{T_A}$) with ●.

7.3.4 C^4 macro-elements over the Alfeld split

For the C^4 macro-elements the superspline space $\mathcal{S}_d^{r,\rho,\mu,\eta}(T_A)$ reduces to the space $\mathcal{S}_{25}^{4,12,6,23}(T_A)$, by the second row of Table 7.1 and $m = 2$. As shown in Theorem 7.3, the dimension of the space is equal to 3700 and the corresponding minimal determining set $\mathcal{M}_{T_A,4}$ is the union of the following sets of domain points:

(C^4:M1$_{T_A}$) $D_{12}^{T_i}(v_i)$, $i = 1,\dots,4$

(C^4:M2$_{T_A}$) $\bigcup\limits_{e:=\langle u,v\rangle \in \mathcal{E}_T} E_6^{T_e}(e) \setminus (D_{12}(u) \cup D_{12}(v))$, with $e \in T_e$

(C^4:M3$_{T_A}$) $\bigcup\limits_{j=0}^{4} F_j^{T_i}(F_i) \setminus (\bigcup\limits_{v\in F_i} D_{12}(v) \cup \bigcup\limits_{e\in F_i} E_6(e))$, $i = 1,\dots,4$

(C^4:M4$_{T_A}$) $\bigcup\limits_{F\in \mathcal{F}_{I,T}} \{\xi_{4+j,13-j,8,0}^{T_F} : j = 1,\dots,8\}$

(C^4:M5$_{T_A}$) $F_5^{T_i}(F_i) \setminus (\bigcup\limits_{v\in F_i} D_{12}(v) \cup \bigcup\limits_{e\in F_i} E_8(e))$, $i = 1,\dots,4$

(C^4:M6$_{T_A}$) $D_7^{T_1}(v_T)$

Proof:
To show that $\mathcal{M}_{T_A,4}$ is a minimal determining set for the superspline space $\mathcal{S}_{25}^{4,12,6,23}(T_A)$, the B-coefficients of a spline $s \in \mathcal{S}_{25}^{4,12,6,23}(T_A)$ corresponding to the domain points in $\mathcal{M}_{T_A,4}$ are set to arbitrary values and it is shown that the remaining B-coefficients of s can be uniquely and stably determined from these.

The B-coefficients of s corresponding to the domain points in $D_{12}(v_i)$, $i = 1,\ldots,4$, are uniquely and stably determined from those associated with the domain points in (C^4:M1$_{T_A}$) by the C^{12} supersmoothness at the vertices of T.

The remaining undetermined B-coefficients of s associated with the domain points in the tubes $E_6(e)$, $e \in \mathcal{E}_T$, can be uniquely and stably computed from the C^6 supersmoothness conditions at the edges of T and the B-coefficients corresponding to the domain points in the set (C^4:M2$_{T_A}$).

Then, the last undetermined B-coefficients of s corresponding to domain points in the shells $R_{25}(v_T)$, $R_{24}(v_T)$, $R_{23}(v_T)$, $R_{22}(v_T)$, and $R_{21}(v_T)$ are considered. The B-coefficients of s corresponding to the domain points in the balls $D_{12}(v_i)$, $i = 1,\ldots,4$, and the tubes $E_6(e)$, $e \in \mathcal{E}_T$, are already determined. Thus, the only undetermined B-coefficients corresponding to domain points in the considered shells are associated with the domain points contained in the set (C^4:M3$_{T_A}$). Therefore, s is uniquely and stably determined restricted to the shells $R_{25}(v_T)$, $R_{24}(v_T)$, $R_{23}(v_T)$, $R_{22}(v_T)$, and $R_{21}(v_T)$ (cf. Figure 7.8, Figure 7.9, and Figure 7.10 left).

Next, it is shown how the remaining undetermined B-coefficients of s corresponding to domain points in the ball $D_{23}(v_T)$ are determined. The spline s has C^{23} supersmoothness at the point v_T. Thus, the B-coefficients of s associated with the domain points in $D_{23}(v_T)$ can be regarded as those of a polynomial p of degree twenty-three on the tetrahedron $\widetilde{T}$, which is bounded by the domain points in $R_{23}(v_T)$. Since some of the B-coefficients of s corresponding to the domain points in $D_{23}(v_T)$ are already determined, some of the B-coefficients of p are also already determined. Therefore, the B-coefficients of p associated with the domain points within a distance of two from the faces of $\widetilde{T}$, those within a distance of four from the edges of $\widetilde{T}$, and those within a distance of ten from the vertices of $\widetilde{T}$ are already known.

Now, the remaining undetermined B-coefficients of p corresponding to the domain points with a distance of three from the faces of $\widetilde{T}$, in the shell $R_{20}(v_T)$, are determined. This involves the following three steps. First, the undetermined B-coefficients associated with the domain points with a distance of five from the edges of $\widetilde{T}$ are computed. Then, the undetermined B-coefficients corresponding to domain with a distance of six from the edges of $\widetilde{T}$ are determined. In the third step, the remaining undetermined B-coefficients of p associated with domain points with a distance of three from the faces of $\widetilde{T}$ are determined. In Figure 7.10 (right) the domain points in the shell $R_{20}(v_T)$ of one subtetrahedron of $\widetilde{T}_A$ are shown. Those, whose B-coefficients are already determined, are marked with ■ and ▲. The domain points whose B-coefficients are still undetermined are marked with ✚, ✖ and ●.

First, the B-coefficients of p corresponding to the domain points with a distance of five from the edges of $\widetilde{T}$ are investigated. These domain points have a distance of seven from the edges of T. Note that the B-coefficients associated with the domain points in the shells $R_{23}(v_T)$, $R_{22}(v_T)$, and $R_{21}(v_T)$ are already determined. Therefore, the only undetermined B-coefficients corresponding to domain points with a distance of five from the edges of $\widetilde{T}$ are associated with the domain points in the shell $R_{20}(v_T)$ with a distance of two from the interior faces of $\widetilde{T}_A$, with the domain points in the shell $R_{19}(v_T)$ with a distance of one from the interior faces, and the domain points in the shell $R_{18}(v_T)$ on the interior faces of $\widetilde{T}_A$. In Figure 7.10 (right), the domain points of the undetermined B-coefficients considered here are marked with ✚. Now, considering each of the C^5 smoothness condition across the interior faces of $\widetilde{T}_A$ and the corresponding $C^1,\ldots,C^4$ conditions, it can be seen that a total number of five undetermined B-coefficients are involved. Two of the B-coefficients correspond to domain points in $R_{20}(v_T)$, two others are associated with domain points in $R_{19}(v_T)$, and the last one corresponds to a domain point in $R_{18}(v_T)$. In fact, for each domain point in the shell $R_{18}(v_T)$ considered here, there are two undetermined B-coefficients. But from the C^0 smoothness at the interior faces of $\widetilde{T}_A$, it can be obtained that these two B-coefficients are equal. Thus, there are five smoothness conditions for exactly five undetermined B-coefficients. Following Lemma 2.33 for the case $j = 0$, the matrix of the corresponding linear system to compute these five B-coefficients is nonsingular. Therefore, all remaining B-coefficients of p corresponding to domain points with a distance of five from the edges of $\widetilde{T}$ are uniquely and stably determined. So far, in Figure 7.10 (right), the only undetermined B-coefficients are marked with ✖ and ●.

In the second step, the B-coefficients of p corresponding to the domain points with a distance of six from the edges of $\widetilde{T}$ are considered. These domain points have a distance of eight from the edges of T. Since the B-coefficients associated with the domain points in the shells $R_{23}(v_T)$, $R_{22}(v_T)$, and $R_{21}(v_T)$ are already determined, the only undetermined B-coefficients of the considered ones correspond to the domain points in the shell $R_{20}(v_T)$ with a distance of three from the interior faces of $\widetilde{T}_A$, to the domain points in the shell $R_{19}(v_T)$ with a distance of two from the interior faces, to the domain points in the shell $R_{18}(v_T)$ with a distance of one from the interior faces, and to those in the shell $R_{17}(v_T)$ lying on the interior faces of $\widetilde{T}_A$. The domain points in one subtetrahedron of $\widetilde{T}_A$ in the shell $R_{20}(v_T)$ whose B-coefficients are determined in this step are marked with ✖ in Figure 7.10 (right). Now, each individual C^6 smoothness condition across an interior face of $\widetilde{T}_A$ and the corresponding $C^1,\ldots,C^5$ conditions are considered. Each of these sets of conditions involves a total number of seven undetermined B-coefficients. Since the set (C^4:M4$_{T_A}$) contains the domain points in $R_{17}(v_T)$ corresponding to the undetermined B-coefficients considered here, there are only six undetermined B-coefficients left for each set of C^6 and the corresponding $C^1,\ldots,C^5$ smoothness conditions. Then, following Lemma 2.33 for the case $j = 1$, the remaining undetermined B-coefficients of p associated with domain points with a distance of six from the edges of $\widetilde{T}$ can be uniquely and stably determined by the linear system of the $C^1,\ldots,C^6$ smoothness conditions across the interior faces of $\widetilde{T}_A$. In Figure 7.10 (right), the domain points in $R_{20}(v_T)$ corresponding to the determined B-coefficients in this step are marked with ✖. Thus, the domain points of the only undetermined B-coefficients associated with domain points within a distance of three from the faces of $\widetilde{T}$ and marked with ● in Figure 7.10 (right).

Now, the domain points associated with the remaining undetermined B-coefficients of p restricted to the shell $R_{20}(v_T)$ are those contained in the set (C^4:M5$_{T_A}$). In Figure 7.10 (right) these domain points are marked with ●. Therefore, all B-coefficients of p corresponding to the domain points within a distance of three from the faces of $\widetilde{T}$ are uniquely and stably determined.

Finally, the last undetermined B-coefficients of p are considered. All B-coefficients of p associated with the domain points within a distance of three from the faces of $\widetilde{T}$ are already determined. Now, from the B-coefficients corresponding to the domain points in the set (C^4:M6$_{T_A}$), the derivatives of p at v_T up to order seven can be uniquely and stably computed. Thus, following Lemma 2.42, with $n = 4$, the remaining undetermined B-coefficients of p can be uniquely and stably determined. Then, all B-coefficients of p are uniquely and stably determined and therefore, all B-coefficients of s.

Now, the dimension of the superspline space $\mathcal{S}_{25}^{4,12,6,23}(T_A)$ can be considered. The cardinality of the minimal determining set $\mathcal{M}_{T_A,4}$ is equal to the dimension, thus it suffices to count the domain points contained in $\mathcal{M}_{T_A,4}$. The set (C^4:M1$_{T_A}$) consists of

$$4\binom{12+3}{3} = 1820$$

domain points. The set (C^4:M2$_{T_A}$) contains

$$12\binom{6+2}{3} = 672$$

domain points. In the set (C^4:M3$_{T_A}$) a total number of

$$4\sum_{i=0}^{4}\left(\binom{27-i}{2} - 3\binom{14-i}{2} - 3\binom{7-i}{2} - 3i(7-i)\right) = 860$$

domain points are contained. The set (C^4:M4$_{T_A}$) consists of

$$6(6+2) = 48$$

domain points. The set (C^4:M5$_{T_A}$) contains

$$4\left(\binom{22}{2} - 3\binom{9}{2} - 3\binom{4}{2} - 60\right) = 180$$

domain points. At last, the cardinality of the set (C^4:M6$_{T_A}$) is equal to

$$\binom{7+3}{3} = 120.$$

Thus, the cardinality of $\mathcal{M}_{T_A,4}$, and therefore also the dimension of the spline space $\mathcal{S}_{25}^{4,12,6,23}(T_A)$, is equal to 3700. □

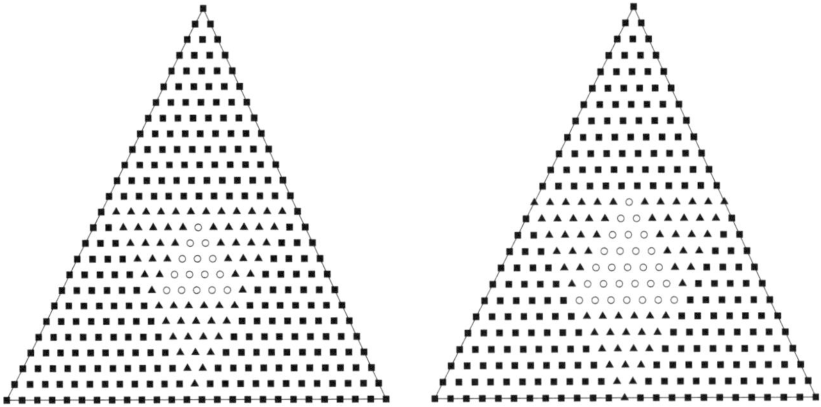

Figure 7.8: The shells $R_{25}^{\hat{T}}(v_T)$ (left) and $R_{24}^{\hat{T}}(v_T)$ (right) for a tetrahedron $\hat{T}$ in T_A. Domain points whose B-coefficients are determined from (C^4:M1$_{T_A}$) are marked with ■, those from (C^4:M2$_{T_A}$) with ▲, and those from (C^4:M3$_{T_A}$) with ○.

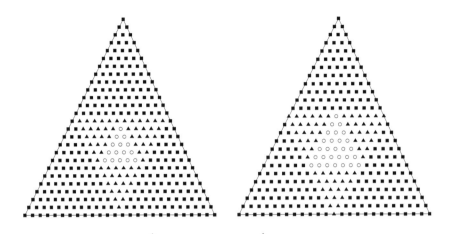

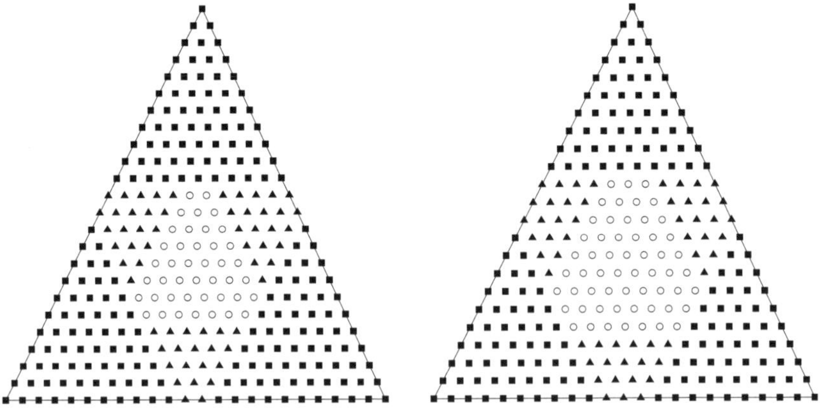

Figure 7.9: The shells $R_{23}^{\hat{T}}(v_T)$ (left) and $R_{22}^{\hat{T}}(v_T)$ (right) for a tetrahedron $\hat{T}$ in T_A. Domain points whose B-coefficients are determined from (C^4:M1$_{T_A}$) are marked with ■, those from (C^4:M2$_{T_A}$) with ▲, and those from (C^4:M3$_{T_A}$) with ○.

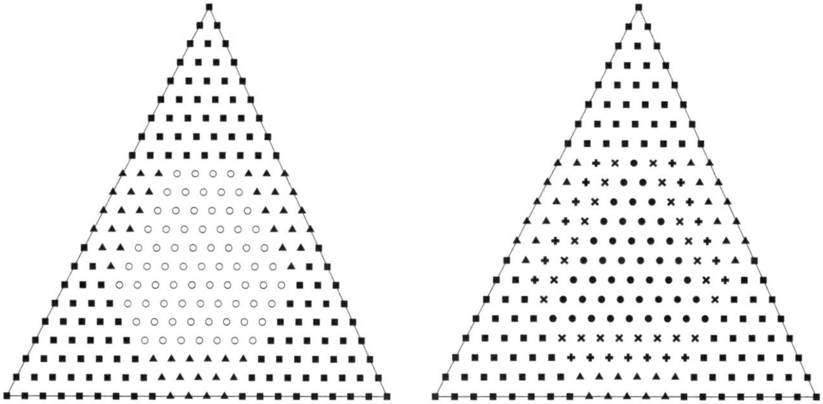

Figure 7.10: The shells $R_{21}^{\hat{T}}(v_T)$ (left) and $R_{20}^{\hat{T}}(v_T)$ (right) for a tetrahedron $\hat{T}$ in T_A. Domain points whose B-coefficients are determined from (C^4:M1$_{T_A}$) are marked with ■, those from (C^4:M2$_{T_A}$) with ▲, those from (C^4:M3$_{T_A}$) with ○, those from smoothness conditions using Lemma 2.33 ($j = 0$) with ✚, those from (C^4:M4$_{T_A}$) and smoothness conditions using Lemma 2.33 ($j = 1$) with ✖, and those from (C^4:M5$_{T_A}$) with ●.

7.3.5 C^5 macro-elements over the Alfeld split

The C^5 macro-elements over the Alfeld split of a tetrahedron T are in the superspline space $\mathcal{S}_{29}^{5,14,7,26}(T_A)$, which is derived from the space $\mathcal{S}_d^{r,\rho,\mu,\eta}(T_A)$ with the values from the third row of Table 7.1 and $m = 2$. Following Theorem 7.3, the dimension of the superspline space is equal to 5852 and the corresponding minimal determining set $\mathcal{M}_{T_A,5}$ is the union of the following sets of domain points:

(C^5:M1$_{T_A}$) $D_{14}^{T_i}(v_i)$, $i = 1,\dots,4$

(C^5:M2$_{T_A}$) $\bigcup\limits_{e:=\langle u,v\rangle \in \mathcal{E}_T} E_7^{T_e}(e) \setminus (D_{14}(u) \cup D_{14}(v))$, where $e \in T_e$

(C^5:M3$_{T_A}$) $\bigcup\limits_{j=0}^{5} F_j^{T_i}(F_i) \setminus (\bigcup\limits_{v \in F_i} D_{14}(v) \cup \bigcup\limits_{e \in F_i} E_7(e))$, $i = 1,\dots,4$

(C^5:M4$_{T_A}$) $\bigcup\limits_{F \in \mathcal{F}_{I,T}} \{\xi_{5+j,15-j,9,0}^{T_F} : j = 1,\dots,9\}$

(C^5:M5$_{T_A}$) $F_6^{T_i}(F_i) \setminus (\bigcup_{v \in F_i} D_{14}(v) \cup \bigcup_{e \in F_i} E_9(e))$, $i = 1,\ldots,4$

(C^5:M6$_{T_A}$) $D_{10}^{T_1}(v_T)$,

Proof:
In order to show that $\mathcal{M}_{T_A,5}$ is a stable minimal determining set for the superspline space $\mathcal{S}_{29}^{5,14,7,26}(T_A)$, we set the B-coefficients c_ξ of a spline $s \in \mathcal{S}_{29}^{5,14,7,26}(T_A)$ to arbitrary values for each domain point $\xi \in \mathcal{M}_{T_A,5}$. Then we show that all the remaining B-coefficients of s are uniquely and stably determined.

First, the B-coefficients of s associated with the domain points in $D_{14}(v_i)$, $i = 1,\ldots,4$, are uniquely and stably determined from the B-coefficients corresponding to the domain points in (C^5:M1$_{T_A}$) by the C^{14} supersmoothness at the vertices of T.

Next, the remaining B-coefficients of s associated with the domain points in the tubes $E_7(e)$, $e \in \mathcal{E}_T$, are determined. They can be uniquely and stably computed from the B-coefficients corresponding to the domain points in the set (C^5:M2$_{T_A}$) and the already determined ones by the C^7 smoothness conditions at the edges of T.

Following, we investigate the B-coefficients of s associated with the domain points in the shells $R_{29}(v_T),\ldots,R_{24}(v_T)$. Note that all B-coefficients of s corresponding to the domain points in the balls $D_{14}(v)$, $v \in \mathcal{V}_T$, and the tubes $E_7(e)$, $e \in \mathcal{E}_T$, are already determined. Now, considering the first shell $R_{29}(v_T)$, we can see that the only undetermined B-coefficients of s associated with domain points in this shell correspond to the domain points in the set (C^5:M3$_{T_A}$). Thus, all B-coefficients of s restricted to the faces of T are uniquely and stably determined. The other domain points corresponding to the undetermined B-coefficients of s restricted to the shells $R_{28}(v_T),\ldots,R_{24}(v_T)$ are also contained in the set (C^5:M3$_{T_A}$). Therefore, all B-coefficients of s corresponding to the domain points in the shells $R_{29}(v_T),\ldots,R_{24}(v_T)$ are uniquely and stably determined (cf. Figure 7.11, Figure 7.12, and Figure 7.13).

Now, we consider the remaining undetermined B-coefficients of s associated with domain points in the ball $D_{26}(v_T)$. The spline s has C^{26} supersmoothness at the split point v_T. Thus, the B-coefficients of s corresponding to the domain points in the ball $D_{26}(v_T)$ can be regarded as those of a polynomial p of degree twenty-six defined on the tetrahedron $\widetilde{T}$. The tetrahedron $\widetilde{T}$ is formed by the convex hull of the domain points in the shell $R_{26}(v_T)$. Note that the tetrahedron $\widetilde{T}$ is also refined with the Alfeld split. Above, we have already determined some of the B-coefficients of s associated with domain points in the ball $D_{26}(v_T)$. Thus, we already know the B-coefficients of p corresponding to the domain points within a distance of two from the faces of $\widetilde{T}$, within a distance of four from the edges of $\widetilde{T}$, and within a distance of eleven from the vertices of $\widetilde{T}$.

In the following, we show how the remaining B-coefficients of the polynomial p are determined. Therefore, we consider the B-coefficients of p associated with the domain points in the shell $R_{23}(v_T)$. These are determined in the following steps. First, the undetermined B-coefficients corresponding to the domain points with a distance of five from the edges of $\widetilde{T}$ are computed. Subsequently, the undetermined B-coefficients corresponding to domain with a distance of six from the edges of $\widetilde{T}$ are determined. At last, the remaining undetermined B-coefficients of p associated with domain points with a distance of three from the faces of $\widetilde{T}$ are determined. The domain points in the shell $R_{23}(v_T)$ of one subtetrahedron of $\widetilde{T}$ are shown in Figure 7.14. Those, whose B-coefficients are already determined, are marked with ■ and ▲. The domain points whose B-coefficients are still undetermined are marked with ✚, ✖ and ●.

First, we compute the remaining undetermined B-coefficients of p corresponding to the domain points with a distance of five from the edges of $\widetilde{T}$. We already know the B-coefficients of p associated with the domain points in the shells $R_{26}(v_T)$, $R_{25}(v_T)$, and $R_{24}(v_T)$. Thus, the only undetermined B-coefficients corresponding to domain points with a distance of five from the edges of $\widetilde{T}$ are associated with the domain points in the shell $R_{23}(v_T)$ with a distance of two from the interior faces of $\widetilde{T}_A$, with the domain points in the shell $R_{22}(v_T)$ with a distance of one from the interior faces, and the domain points in the shell $R_{21}(v_T)$ on the interior faces of $\widetilde{T}_A$. In Figure 7.14, these domain points in the shell $R_{23}(v_T)$ are marked with ✚. Now, we investigate each of the C^5 smoothness conditions across the interior faces of $\widetilde{T}_A$ including one of the undetermined B-coefficients of p. In each set of equations formed by a C^5 smoothness condition and the corresponding $C^1,\ldots,C^4$ conditions, a total number of five undetermined B-coefficients are involved. Following Lemma 2.33 for the case $j = 0$, the matrix of the corresponding linear system is nonsingular. Hence, we can uniquely and stably determine all B-coefficients of p corresponding to domain points with a distance of five from the edges of $\widetilde{T}$.

Now, we show how the remaining undetermined B-coefficients of p corresponding to the domain points with a distance of six from the edges of $\widetilde{T}$

are determined. Here, the only undetermined B-coefficients are associated with the domain points in the shell $R_{23}(v_T)$ with a distance of three from the interior faces of $\widetilde{T}_A$, with the domain points in the shell $R_{22}(v_T)$ with a distance of two from the interior faces, with the domain points in the shell $R_{21}(v_T)$ with a distance of one from the interior faces, and with those in the shell $R_{20}(v_T)$ lying on the interior faces of $\widetilde{T}_A$. In Figure 7.14 the domain points in $R_{23}(v_T)$ whose B-coefficients are considered here are marked with ✖. The B-coefficients of p corresponding to the domain points in the shells $R_{26}(v_T)$, $R_{25}(v_T)$, and $R_{24}(v_T)$ are already determined. Considering the C^6 smoothness conditions across the interior faces of $\widetilde{T}_A$, we see that exactly seven undetermined B-coefficients are covered by one set consisting of a C^6 smoothness condition and the corresponding $C^1,\ldots,C^5$ conditions. Now, the set (C^5:M4$_{T_A}$) contains the domain points in the shell $R_{20}(v_T)$ corresponding to the undetermined B-coefficients considered here. Thus, there are only six undetermined B-coefficients left for one set of $C^1,\ldots,C^5$, and C^6 smoothness conditions. Following Lemma 2.33 for the case $j = 1$, the system of linear equations corresponding to the conditions has a unique solution and thus, the B-coefficients of p associated with the domain points with a distance of six from the edges of $\widetilde{T}$ are uniquely and stably determined. Then, the only undetermined B-coefficients associated with domain points within a distance of three from the faces of $\widetilde{T}$ are in the shell $R_{23}(v_T)$ and the corresponding domain points are marked with ● in Figure 7.14.

The remaining undetermined B-coefficients of p corresponding to domain points in the shell $R_{23}(v_T)$ are associated with the domain points contained in the set (C^5:M5$_{T_A}$). Thus, all B-coefficients of p corresponding to the domain points within a distance of three from the faces of $\widetilde{T}$ are uniquely and stably determined.

In the end, we consider the last undetermined B-coefficients of the polynomial p. So far we have determined the B-coefficients of p corresponding to the domain points within a distance of three from the faces of $\widetilde{T}$. Then, the derivatives of p at v_T up to order ten can be uniquely and stably computed from the B-coefficients associated with the domain points in the set (C^5:M6$_{T_A}$). Now, following Lemma 2.42, with $n = 4$, we can uniquely and stably determine the remaining undetermined B-coefficients of p from these derivatives. Therefore, we have uniquely and stably determined all B-coefficients of p. Then, by subdivision using the de Casteljau algorithm, all B-coefficients of s are also uniquely and stably determined.

Now, the dimension of the superspline space $\mathcal{S}_{29}^{5,14,7,26}(T_A)$ is considered. The dimension is equal to the cardinality of the minimal determining set $\mathcal{M}_{T_A,5}$. Therefore, it suffices to count the number of domain points contained in each of the sets (C^5:M1$_{T_A}$) - (C^5:M6$_{T_A}$). The set (C^5:M1$_{T_A}$) contains

$$4\binom{14+3}{3} = 2720$$

domain points. The set (C^5:M2$_{T_A}$) consists of

$$12\binom{7+2}{3} = 1008$$

domain points. A total number of

$$4\sum_{i=0}^{5}\left(\binom{31-i}{2} - 3\binom{16-i}{2} - 3\binom{8-i}{2} - 3i(8-i)\right) = 1484$$

domain points are contained in the set (C^5:M3$_{T_A}$). The set (C^5:M4$_{T_A}$) contains

$$6(7+2) = 54$$

domain points. The set (C^5:M5$_{T_A}$) consists of a total number of

$$4\left(\binom{25}{2} - 3\binom{10}{2} - 3\binom{4}{2} - 72\right) = 300$$

domain points. The last set (C^5:M6$_{T_A}$) contained in the minimal determining set has a cardinality of

$$\binom{10+3}{3} = 286.$$

Thus, the dimension of $\mathcal{S}_{29}^{5,14,7,26}(T_A)$, is equal to 5852. □

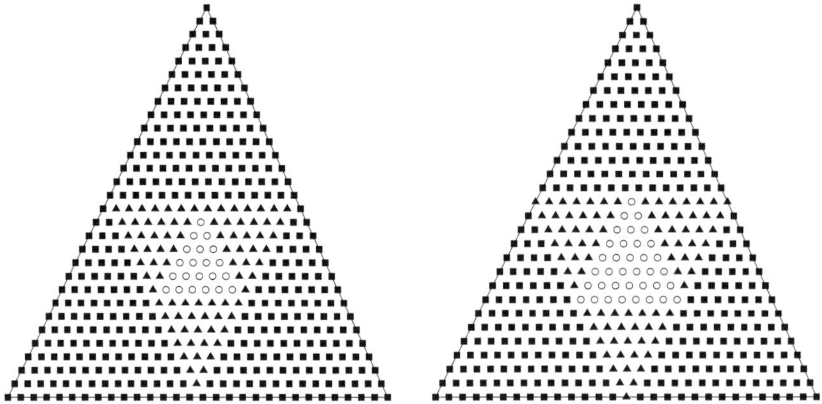

Figure 7.11: The shells $R_{29}^{\hat{T}}(v_T)$ (left) and $R_{28}^{\hat{T}}(v_T)$ (right) for a tetrahedron $\hat{T}$ in T_A. Domain points whose B-coefficients are determined from (C^5:M1$_{T_A}$) are marked with ■, those from (C^5:M2$_{T_A}$) with ▲, and those from (C^5:M3$_{T_A}$) with ○.

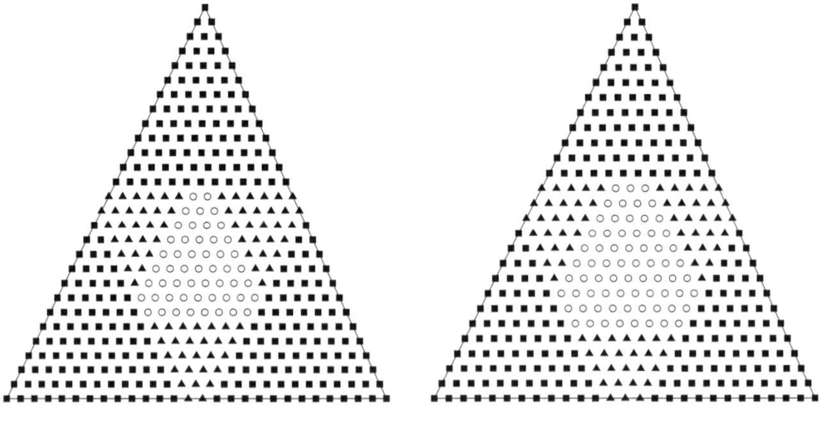

Figure 7.12: The shells $R_{27}^{\hat{T}}(v_T)$ (left) and $R_{26}^{\hat{T}}(v_T)$ (right) for a tetrahedron $\hat{T}$ in T_A. Domain points whose B-coefficients are determined from (C^5:M1$_{T_A}$) are marked with ■, those from (C^5:M2$_{T_A}$) with ▲, and those from (C^5:M3$_{T_A}$) with ○.

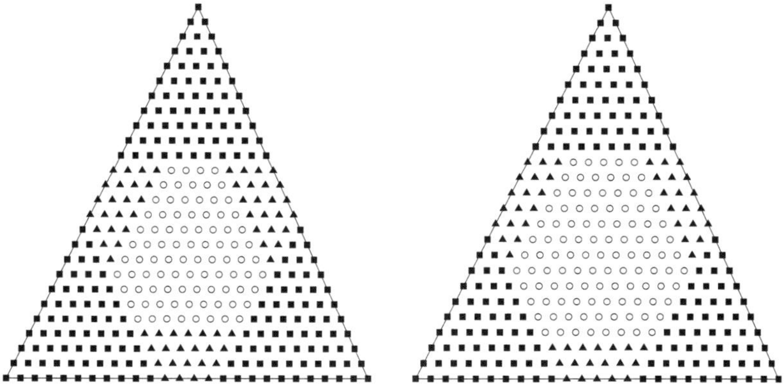

Figure 7.13: The shells $R_{25}^{\hat{T}}(v_T)$ (left) and $R_{24}^{\hat{T}}(v_T)$ (right) for a tetrahedron $\hat{T}$ in T_A. Domain points whose B-coefficients are determined from (C^5:M1$_{T_A}$) are marked with ■, those from (C^5:M2$_{T_A}$) with ▲, and those from (C^5:M3$_{T_A}$) with ○.

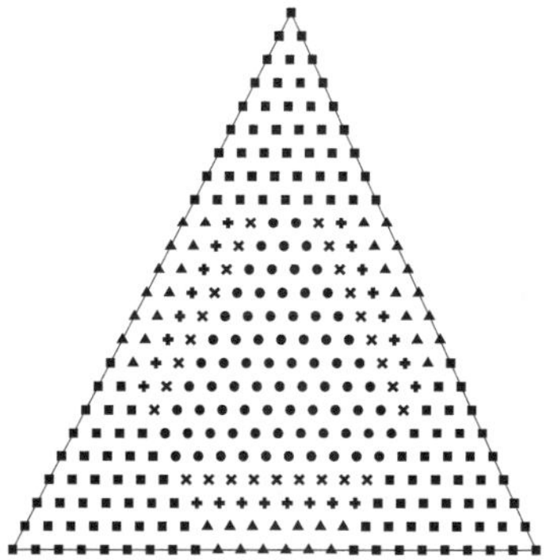

Figure 7.14: The shell $R_{23}^{\hat{T}}(v_T)$ for a tetrahedron $\hat{T}$ in T_A. Domain points whose B-coefficients are determined from (C^5:M1$_{T_A}$) are marked with ■, those from (C^5:M2$_{T_A}$) with ▲, those from smoothness conditions using Lemma 2.33 ($j = 0$) with ✚, those from (C^5:M4$_{T_A}$) and smoothness conditions using Lemma 2.33 ($j = 1$) with ✖, and those from (C^5:M5$_{T_A}$) with ●.

7.3.6 C^6 macro-elements over the Alfeld split

For the case of C^6 macro-elements, the superspline space $\mathcal{S}_d^{r,\rho,\mu,\eta}(T_A)$ reduces to the space $\mathcal{S}_{37}^{6,18,9,34}(T_A)$, by the second row of Table 7.1 and $m=3$. From Theorem 7.3 we know that the dimension of the spline space is equal to 11294 and the corresponding minimal determining set $\mathcal{M}_{T_A,6}$ is the union of the following sets of domain points:

(C^6:M1$_{T_A}$) $D_{18}^{T_i}(v_i)$, $i=1,\dots,4$

(C^6:M2$_{T_A}$) $\bigcup\limits_{e:=\langle u,v\rangle\in\mathcal{E}_T} E_9^{T_e}(e)\setminus(D_{18}(u)\cup D_{18}(v))$, where $e\in T_e$

(C^6:M3$_{T_A}$) $\bigcup\limits_{j=0}^{6} F_j^{T_i}(F_i)\setminus(\bigcup\limits_{v\in F_i} D_{18}(v)\cup\bigcup\limits_{e\in F_i} E_9(e))$, $i=1,\dots,4$

(C^6:M4$_{T_A}$) $\bigcup\limits_{F\in\mathcal{F}_{I,T}} \{\xi_{9-2i+j,19-j,9+2i,0}^{T_F} : i=1,2,\ j=1,\dots,9+2i\}$

(C^6:M5$_{T_A}$) $\bigcup\limits_{j=1}^{2}\left(F_{6+j}^{T_i}(F_i)\setminus(\bigcup\limits_{v\in F_i} D_{18}(v)\cup\bigcup\limits_{e\in F_i} E_{9+2j}(e))\right)$, $i=1,\dots,4$

(C^6:M6$_{T_A}$) $D_{10}^{T_1}(v_T)$,

Proof:
In order to show that $\mathcal{M}_{T_A,6}$ is a stable minimal determining set for the space $\mathcal{S}_{37}^{6,18,9,34}(T_A)$, we set the B-coefficients c_ξ of a spline $s\in\mathcal{S}_{37}^{6,18,9,34}(T_A)$ to arbitrary values for each domain point $\xi\in\mathcal{M}_{T_A,6}$ and show that the remaining B-coefficients of s can be uniquely and stably determined by smoothness conditions.

We can uniquely determine the B-coefficients of s associated with the domain points in $D_{18}(v_i)$, $i=1,\dots,4$, from the B-coefficients corresponding to the domain points in the set (C^6:M1$_{T_A}$). Moreover, since the B-coefficients are directly computed by smoothness conditions, the computation is stable.

Next, we determine the remaining B-coefficients of s associated with the domain points in the tubes $E_9(e)$, $e\in\mathcal{E}_T$. Using the C^9 supersmoothness conditions at the edges of T, these can be uniquely and stably determined from the already known B-coefficients and those corresponding to the domain points in the set (C^6:M2$_{T_A}$).

Following, we consider the B-coefficients of s corresponding to the domain points in the shells $R_{37}(v_T),\dots,R_{31}(v_T)$. We have already determined all B-coefficients of s associated with the domain points in the balls $D_{18}(v)$, $v\in\mathcal{V}_T$, and the tubes $E_9(e)$, $e\in\mathcal{E}_T$. Now, the remaining undetermined

B-coefficients of s restricted to these shells are associated with the domain points in the set $(C^6{:}M3_{T_A})$ and are thus already known (cf. Figure 7.15, Figure 7.16, Figure 7.17, and Figure 7.18 (left)). Therefore, all B-coefficients of s corresponding to the domain points in the shells $R_{37}(v_T),\ldots,R_{31}(v_T)$ are uniquely and stably determined.

Subsequently, we consider the remaining undetermined B-coefficients of s. These are all associated with domain points contained in the ball $D_{34}(v_T)$. Thus, we investigate s restricted to $D_{34}(v_T)$ in detail. The spline s has C^{34} supersmoothness at the split point v_T in T. Due to this reason, we can regard the B-coefficients of s corresponding to the domain points in $D_{34}(v_T)$ as those of a trivariate polynomial p of degree 34, which is defined on the tetrahedron $\widetilde{T}$ that is formed by the convex hull of the domain points in the shell $R_{34}(v_T)$. Note that the tetrahedron $\widetilde{T}$ is also subjected to the Alfeld split. Since we have already determined some of the B-coefficients of s associated with domain points in the ball $D_{34}(v_T)$, these B-coefficients are also already determined for the polynomial p. To be precise, the B-coefficients of p corresponding to the domain points within a distance of three from the faces of $\widetilde{T}$, within a distance of six from the edges of $\widetilde{T}$, and within a distance of fifteen from the vertices of $\widetilde{T}$ are already determined.

Now, we show how the remaining undetermined B-coefficients of p are computed. Therefore, we first consider the B-coefficients corresponding to the domain points with a distance of four from the faces of $\widetilde{T}$ (those in $R_{30}(v_T)$) and then the B-coefficients associated with the domain points with a distance of five from the faces (those in $R_{29}(v_T)$). In order to determine these B-coefficients, we have to regard the B-coefficients corresponding to the domain points with a distance of seven up to ten from the edges of $\widetilde{T}$.

We first consider the B-coefficients of s corresponding to the domain points within a distance of seven from the edges of $\widetilde{T}$. The B-coefficients associated with the domain points in the shells $R_{34}(v_T),\ldots,R_{31}(v_T)$ are already determined. Then, the only undetermined B-coefficients of p corresponding to domain points with a distance of seven from the edges of $\widetilde{T}$ are those associated with the domain points with a distance of $3-i$, $i=0,\ldots,3$, from the interior faces of $\widetilde{T}_A$ in the shell $R_{30-i}(v_T)$. The domain points in the shells $R_{30}(v_T)$ and $R_{29}(v_T)$ of the considered B-coefficients are marked with ✚ in Figure 7.18 (right) and Figure 7.19. Now, each tuple of $C^1,\ldots,C^7$ smoothness conditions across the interior faces of $\widetilde{T}_A$ connects a total number of seven undetermined B-coefficients examined in this step. Note that no other undetermined B-coefficients are involved. Thus, there are seven equations from the $C^1,\dots,C^7$ conditions for exactly seven undetermined B-coefficients. Following Lemma 2.33 for the case $j=0$, the system of equations has a unique solution and all B-coefficients of p associated with the domain points within a distance of seven from the edges of $\widetilde{T}$ are uniquely and stably determined.

Next, we examine the B-coefficients of p corresponding to the domain points with a distance of eight from the edges of $\widetilde{T}$. Here, the only undetermined B-coefficients are associated with the domain points with a distance of $4-i$, $i=0,\dots,4$, from the interior faces of $\widetilde{T}_A$ in the shell $R_{30-i}(v_T)$, since those corresponding to the domain points in the shells $R_{34}(v_T),\dots,R_{31}(v_T)$ are already known. The domain points in the shells $R_{30}(v_T)$ and $R_{29}(v_T)$ of the examined B-coefficients are marked with ✖ in Figure 7.18 (right) and Figure 7.19. Then, each C^8 smoothness condition across an interior face of $\widetilde{T}_A$ and the corresponding $C^1,\dots,C^7$ conditions connect exactly nine of the undetermined B-coefficients. Now, the set (C^6:M4$_{\widetilde{T}_A}$) contains the domain points in $R_{26}(v_T)$ of the undetermined B-coefficients considered here. Thus, there are only eight undetermined B-coefficients left for each tuple of $C^1,\dots,C^8$ smoothness conditions across the interior faces of $\widetilde{T}_A$. Following Lemma 2.33 for the case $j=1$, these undetermined B-coefficients of p can be uniquely and stably computed by the linear system of $C^1,\dots,C^8$ smoothness conditions.

Now, the only undetermined B-coefficients of p in the shell $R_{30}(v_T)$ are those associated with the domain points contained in the set (C^6:M5$_{\widetilde{T}_A}$). These are marked with ● in Figure 7.18 (right). Thus, all B-coefficients of p corresponding to the domain points in the shell $R_{30}(v_T)$ are uniquely and stably determined.

Next, we consider the B-coefficients of p corresponding to the domain points with a distance of nine from the edges of $\widetilde{T}$. Since the B-coefficients associated with the domain points in the shells $R_{34}(v_T),\dots,R_{30}(v_T)$ are already uniquely determined, the only undetermined B-coefficients p corresponding to domain points with a distance of nine from the edges of $\widetilde{T}$ are those associated with the domain points with a distance of $4-i$, $i=0,\dots,4$, from the interior faces of $\widetilde{T}_A$ in the shell $R_{31-i}(v_T)$. In Figure 7.19, the domain points in the shell $R_{29}(v_T)$ of the examined B-coefficients are marked with ✸. Considering the $C^1,\dots,C^9$ smoothness conditions across the interior faces of $\widetilde{T}_A$, we see that each tuple of these connects a total number of nine undetermined B-coefficients. Therefore, for each tuple, we have nine undetermined B-coefficients and nine linear independent equations, from which we can uniquely and stably determine these B-coefficients following Lemma 2.33 for the case $j = 0$. Then, all B-coefficients of p associated with domain points with a distance of nine from the edges of $\widetilde{T}$ are uniquely and stably determined.

Subsequently, we investigate the B-coefficients of p corresponding to the domain points with a distance of ten from the edges of $\widetilde{T}$. Here, the only undetermined B-coefficients are associated with the domain points with a distance of $5 - i$, $i = 0,\dots,5$, from the interior faces of $\widetilde{T}_A$ in the shell $R_{31-i}(v_T)$. The domain points in the shell $R_{29}(v_T)$ corresponding to the B-coefficients examined here are marked with ★ in Figure 7.19. Now, we consider the $C^1,\dots,C^{10}$ smoothness conditions across the interior faces of $\widetilde{T}_A$ covering the undetermined B-coefficients. Then, each tuple, consisting of one C^{10} smoothness condition and the corresponding $C^1,\dots,C^9$ conditions, connects exactly eleven undetermined B-coefficients. Now, for each tuple connecting eleven undetermined B-coefficients, the subset (C^6:M4$_{T_A}$) of the minimal determining set $\mathcal{M}_{T_A,6}$ contains exactly one domain point for one of these eleven undetermined B-coefficients. Thus, we are left with ten equations from smoothness conditions and exactly ten undetermined B-coefficients. Following Lemma 2.33 for the case $j = 1$, the corresponding system has a unique solution and all B-coefficients of p corresponding to the domain points within a distance of ten from the edges of $\widetilde{T}$ can be uniquely and stably computed.

Considering the shell $R_{29}(v_T)$, we see that the only undetermined B-coefficients of p associated with the domain points in this shell are contained in the set (C^6:M5$_{T_A}$). These domain points are marked with ● in Figure 7.19. Now, all B-coefficients of p associated with the domain points in the shell $R_{29}(v_T)$ are uniquely and stably determined.

Finally, we consider the last undetermined B-coefficients of p. The B-coefficients of p corresponding to the domain points within a distance of five from the faces of $\widetilde{T}$ are already determined. From the B-coefficients associated with the domain points in the set (C^6:M6$_{T_A}$), the derivatives of p at v_T up to order ten can be uniquely and stably computed. Then, following Lemma 2.42 for $n = 4$, the remaining undetermined B-coefficients of p can be uniquely and stably determined. Thus, all B-coefficients of p are uniquely determined and therefore, all B-coefficients of s.

Now, we can determine the dimension of the space $\mathcal{S}_{37}^{6,18,9,34}(T_A)$ of supersplines, since the dimension is equal to the cardinality of the minimal determining set $\mathcal{M}_{T_A,6}$. Thus, it suffices to count the number of domain points contained in each of the sets (C^6:M1$_{T_A}$) - (C^6:M6$_{T_A}$). The set (C^6:M1$_{T_A}$) consists of

$$4\binom{18+3}{3} = 5320$$

domain points. The set (C^6:M2$_{T_A}$) contains

$$12\binom{9+2}{3} = 1980$$

domain points. A total number of

$$4\sum_{i=0}^{6}\left(\binom{39-i}{2} - 3\binom{20-i}{2} - 3\binom{10-i}{2} - 3i(10-i)\right) = 2744$$

domain points are contained in the set (C^6:M3$_{T_A}$). The set (C^6:M4$_{T_A}$) contains

$$6\sum_{i=1}^{2}(9+2i) = 144$$

domain points. The set (C^6:M5$_{T_A}$) consists of a total number of

$$4\sum_{i=1}^{2}\left(\binom{33-i}{2} - 3\binom{14-i}{2} - 3\binom{4+i}{2} - (18+3i)(4+i)\right) = 820$$

domain points. The last set (C^6:M6$_{T_A}$) contained in the minimal determining set has a cardinality of

$$\binom{10+3}{3} = 286.$$

Therefore, the dimension of $\mathcal{S}_{37}^{6,18,9,34}(T_A)$, is equal to 11294. □

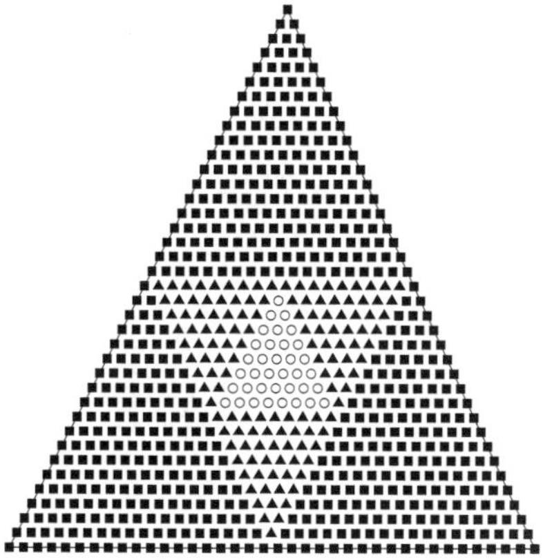

Figure 7.15: The shells $R_{37}^{\hat{T}}(v_T)$ (left) and $R_{36}^{\hat{T}}(v_T)$ (right) for a tetrahedron $\hat{T}$ in T_A. Domain points whose B-coefficients are determined from (C^6:M1$_{T_A}$) are marked with ■, those from (C^6:M2$_{T_A}$) with ▲, and those from (C^6:M3$_{T_A}$) with ○.

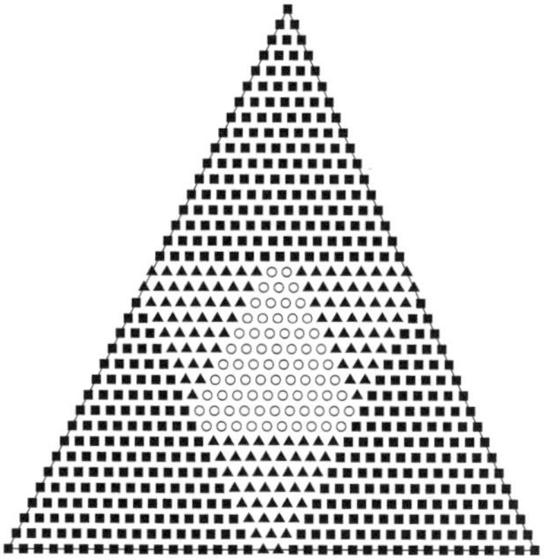

Figure 7.16: The shells $R_{35}^{\hat{T}}(v_T)$ (left) and $R_{34}^{\hat{T}}(v_T)$ (right) for a tetrahedron $\hat{T}$ in T_A. Domain points whose B-coefficients are determined from (C^6:M1$_{T_A}$) are marked with ■, those from (C^6:M2$_{T_A}$) with ▲, and those from (C^6:M3$_{T_A}$) with ○.

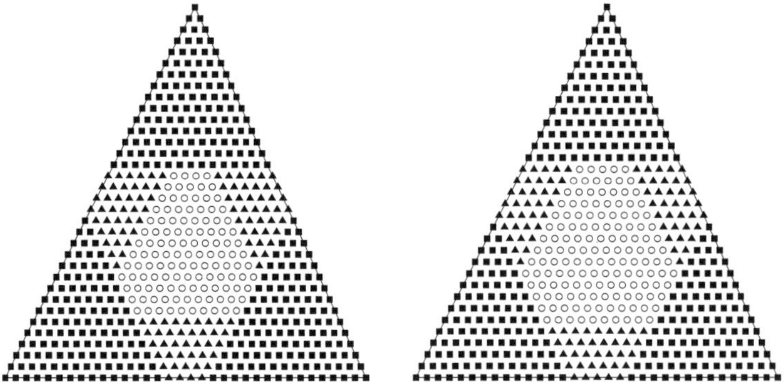

Figure 7.17: The shells $R^{\hat{T}}_{33}(v_T)$ (left) and $R^{\hat{T}}_{32}(v_T)$ (right) for a tetrahedron $\hat{T}$ in T_A. Domain points whose B-coefficients are determined from (C^6:M1$_{T_A}$) are marked with ■, those from (C^6:M2$_{T_A}$) with ▲, and those from (C^6:M3$_{T_A}$) with ○.

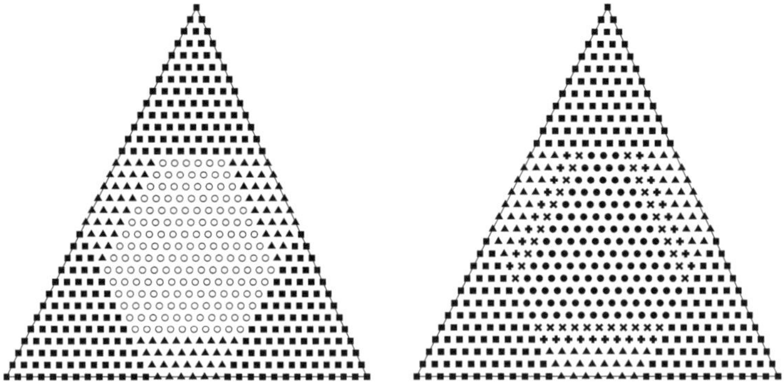

Figure 7.18: The shells $R^{\hat{T}}_{31}(v_T)$ (left) and $R^{\hat{T}}_{30}(v_T)$ (right) for a tetrahedron $\hat{T}$ in T_A. Domain points whose B-coefficients are determined from (C^6:M1$_{T_A}$) are marked with ■, those from (C^6:M2$_{T_A}$) with ▲, those from (C^6:M3$_{T_A}$) with ○, those from smoothness conditions using Lemma 2.33 ($j = 0$) with ✚, those from (C^6:M4$_{T_A}$) and smoothness conditions using Lemma 2.33 ($j = 1$) with ✖, and those from (C^6:M5$_{T_A}$) with ●.

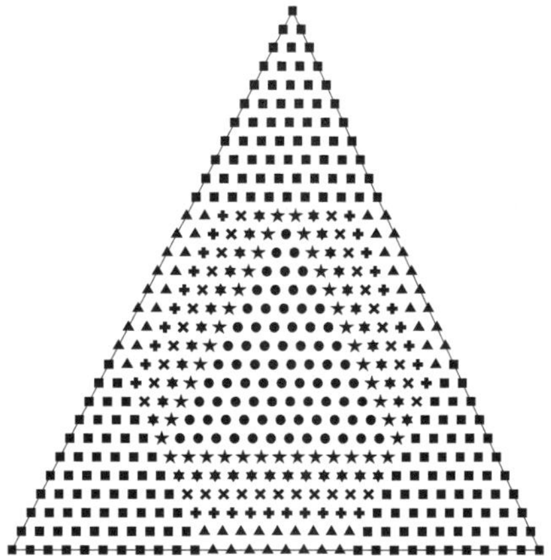

Figure 7.19: The shell $R_{29}^{\hat{T}}(v_T)$ for a tetrahedron $\hat{T}$ in T_A. Domain points whose B-coefficients are determined from (C^6:M1$_{T_A}$) are marked with ■, those from (C^6:M2$_{T_A}$) with ▲, those from smoothness conditions using Lemma 2.33 ($j = 0$) with ✚ and ✱, those from (C^6:M4$_{T_A}$) and smoothness conditions using Lemma 2.33 ($j = 1$) with ✖, those from (C^6:M4$_{T_A}$) and smoothness conditions using Lemma 2.33 ($j = 1$) with ★, and those from (C^6:M5$_{T_A}$) with ●.

7.4 Nodal minimal determining sets for C^r macro-elements over the Alfeld split

In this section nodal minimal determining sets for the C^r macro-elements constructed in section 7.2 are investigated. Therefore, the same notations as in the previous section are used and some definitions for certain derivatives are introduced. Subsequently, we consider and prove nodal minimal determining sets for C^r macro-elements over the Alfeld split of a single tetrahedron. Finally, nodal minimal determining sets for the C^r macro-elements over a tetrahedral partition refined with the Alfeld split are constructed and proved.

First, some derivatives are denoted, which are needed to describe the nodal minimal determining sets.

Notation 7.9:
For each multi-index $\alpha := (\alpha_1, \alpha_2, \alpha_3)$ let D^α be the mixed partial derivative $D_x^{\alpha_1} D_y^{\alpha_2} D_z^{\alpha_3}$. For each edge $e := \langle u, v \rangle$ of a tetrahedron T let X_e be the plane perpendicular to e at the point u, endowed with Cartesian coordinate axes with origin at u. Then, let $D_{e\perp}^{\beta}$, with $\beta := (\beta_1, \beta_2)$, be the partial derivative of order $|\beta| := \beta_1 + \beta_2$ in direction of the two coordinate axes in X_e. Moreover, for an oriented triangular face $F := \langle u, v, w \rangle$ let D_F be the unit normal derivative corresponding to F. For each edge e of a tetrahedron T let $D_{e^{\hat{F}}}$ be the directional derivative corresponding to a unit vector perpendicular to e that lies in the interior face $\hat{F}$ containing e of the Alfeld split tetrahedron T_A. Furthermore, for each face F of a tetrahedron T and each edge e in F let $D_{e\perp F}$ be the directional derivative associated with a unit vector perpendicular to e that lies in F.

In order to describe nodal minimal determining sets in this section also certain points on the edges of a tetrahedron and sets of points on the faces of a tetrahedron are needed.

Definition 7.10:
Let r, μ, and d be as in Table 7.1. For $e := \langle u, v \rangle$, and for $i > 0$ let

$$\varsigma_{e,j}^{i} := \frac{(i - j + 1)u + jv}{i + 1}, \qquad j = 1, \dots i,$$

be equally spaced points in the interior of the edge e.

For a triangle F and $\alpha \geq 0$ let

$$A_\alpha^F := \{\xi_{i,j,k}^F \in \mathcal{D}_{d-\alpha,F} : i,j,k \geq \mu + 1 - \alpha + \lfloor \frac{\alpha}{2} \rfloor)$$

be a set of points in the interior of F. Furthermore, if $\lfloor \frac{r}{3} \rfloor > 0$, for $\beta \geq 1$ let

$$B_\beta^F := \mathcal{D}_{d-r-\beta,F} \setminus \bigcup_{e \in F} E_{\mu-2\beta-1}^F(e)$$

also be a set of points in the interior of F.

Finally, for each point $u \in \mathbb{R}^3$ let ϵ_u be the point evaluation functional defined by

$$\epsilon_u f := f(u).$$

Now, we are able to describe a nodal minimal determining set for the C^r macro-elements based on the Alfeld split of a tetrahedron constructed in section 7.2.

Theorem 7.11:
Let $\mathcal{N}_{T_A}$ be the union of the following sets of nodal data:

(N1$_{T_A}$) $\bigcup\limits_{v \in \mathcal{V}_T} \{\epsilon_v D^\alpha\}_{|\alpha| \leq \rho}$

(N2$_{T_A}$) $\bigcup\limits_{e \in \mathcal{E}_T} \bigcup\limits_{i=1}^{\mu} \bigcup\limits_{j=1}^{i} \{\epsilon_{\zeta^i_{e,j}} D^\beta_{e^\perp}\}_{|\beta|=i}$

(N3$_{T_A}$) $\bigcup\limits_{F \in \mathcal{F}_T} \{\epsilon_\xi D^\alpha_F\}_{\xi \in A^F_\alpha, \alpha=0,\ldots,r}$

(N4$_{T_A}$) $\bigcup\limits_{F \in \mathcal{F}_T} \bigcup\limits_{e \in F} \bigcup\limits_{i=2}^{r} \bigcup\limits_{j=1}^{\lfloor \frac{i}{2} \rfloor} \{\epsilon_{\zeta^{\mu+j}_{e,k}} D^i_{e^F} D^{\mu-i+j}_{e\perp F}\}_{k=1}^{\mu+j}$

(N5$_{T_A}$) $\bigcup\limits_{e \in \mathcal{E}_T} \bigcup\limits_{i=1}^{\lfloor \frac{r}{3} \rfloor} \bigcup\limits_{j=1}^{\mu+2i} \{\epsilon_{\zeta^i_{e,j}} D^\beta_{e^F}\}_{|\beta|=\mu+2i}$

(N6$_{T_A}$) $\bigcup\limits_{F \in \mathcal{F}_T} \bigcup\limits_{e \in F} \bigcup\limits_{i=1}^{\lfloor \frac{r}{3} \rfloor} \bigcup\limits_{j=1}^{r-3i-1} \{\epsilon_{\zeta^{\mu+2i+j}_{e,k}} D^{r+i}_{e^F} D^{\mu-r+i+j}_{e\perp F}\}_{k=1}^{\mu+2i+j}$

(N7$_{T_A}$) $\bigcup\limits_{F \in \mathcal{F}_T} \{\epsilon_\xi D^{r+\beta}_F\}_{\xi \in B^F_\beta}, \beta = 1, \ldots, \lfloor \frac{r}{3} \rfloor\}$

(N8$_{T_A}$) $\{\epsilon_{v_T} D^\alpha\}_{|\alpha| \leq \eta - 4(m + \lfloor \frac{r}{3} \rfloor + 1)}$

Then $\mathcal{N}_{T_A}$ is a stable nodal minimal determining set for the superspline space $\mathcal{S}^{r,\rho,\mu,\eta}_d(T_A)$.

Proof:
It is easy to check, that the cardinality of $\mathcal{N}_{T_A}$ is equal to the dimension of the spline space $\mathcal{S}^{r,\rho,\mu,\eta}_d(T_A)$, as given in Theorem 7.3 in section 7.2. Thus, it suffices to show that setting $\{\lambda s\}_{\lambda \in \mathcal{N}_{T_A}}$ for a spline $s \in \mathcal{S}^{r,\rho,\mu,\eta}_d(T_A)$ determines all B-coefficients of s.

Using the C^ρ supersmoothness of s at all vertices $v \in \mathcal{V}_T$, the B-coefficients associated with the domain points in $D_\rho(v)$ are stably determined form the values of the derivatives of s associated with (N1$_{T_A}$). The remaining undetermined B-coefficients associated with domain points within a distance of μ from the edges $e \in \mathcal{E}_T$ are stably determined from the values of the derivatives of s associated with (N2$_{T_A}$), since s has supersmoothness C^μ at e.

Next, we determine the B-coefficients of s associated with the domain points in the shell $R_d(v_T)$, i.e. the domain points on the faces of T.

For each face F of T_A, we have already determined the B-coefficients associated with the domain points within a distance of μ from the edges of F. Thus, the only undetermined B-coefficients of $s|_F$ are those associated with the domain points in the set A_0^F. Since (N3$_{T_A}$) contains the values of s at these points, the remaining undetermined B-coefficients of s corresponding to the domain points in $R_d(v_T)$ can be computed by solving a linear system with matrix $M_0 := \left[B_\xi^d(\kappa)\right]_{\xi,\kappa \in A_0^F}$. The entries of the matrix only depend on barycentric coordinates, thus M_0 is independent of the size and shape of F. Moreover, by lemma 2.44 the matrix is nonsingular.

Subsequently, we compute the B-coefficients of s associated with the domain points in the shell $R_{d-1}(v_T)$. For a face F of T_A, let F_1 be the layer of domain points with a distance of one from F. We have already determined the B-coefficients of $s|_{F_1}$ associated with the domain points in $D_{\rho-1}^{F_1}(v)$ for each vertex v of F_1, and the B-coefficients corresponding to the domain points in $E_{\mu-1}^{F_1}(e)$ for each edge e of F_1. Thus, the only undetermined B-coefficients of $s|_{F_1}$ are associated with the domain points in the set $A_1^{F_1}$. These B-coefficients can be computed from the values of the first order derivatives in (N3$_{T_A}$) corresponding to F. The computation involves solving a linear system with matrix $M_1 := \left[B_\xi^{d-1}(\kappa)\right]_{\xi,\kappa \in A_1^{F_1}}$, which is independent of the size and shape of F and nonsingular by lemma 2.44.

Next, we consider the B-coefficients of s associated with the domain points in the shells $R_{d-i}(v_T)$, for $i = 2,\ldots,r$. Let F_i be the layer of domain points with a distance of i from a face F of T. Then, we have already determined the B-coefficients of s associated with the domain points within a distance of $\rho - i$ from the vertices of F_i and the B-coefficients of s corresponding to the domain points within a distance of $\mu - i$ from the edges of F_i. Now, we use the values of the derivatives in (N4$_{T_A}$) corresponding to the layer F_i of domain points to stably determine the B-coefficients associated with the domain points with a distance of $\mu - i + j$ from the edges of F_i, for $j = 1,\ldots \lfloor \frac{i}{2} \rfloor$. Then, we use the values of the i-th order derivatives in (N3$_{T_A}$) corresponding to F_i to compute the remaining B-coefficients of $s|_{F_i}$. Therefore, we have to solve a linear system with matrix $M_i := \left[B_\xi^{d-i}(\kappa)\right]_{\xi,\kappa \in A_i^{F_i}}$. By lemma 2.44, this matrix is nonsingular. Moreover, M_i is independent of the size an shape of F. Thus, we have determined all B-coefficients of s associated with the domain points in the shells $R_d(v_T) \cup \ldots \cup R_{d-r}(v_T)$.

Now, we consider the remaining undetermined B-coefficients of s associated with domain points in the ball $D_\eta(v_T)$. By the C^η smoothness at v_T, these B-coefficients of s can be regarded as those of a polynomial p of degree η on the tetrahedron $\widetilde{T}$ which is bounded by the domain points in the shell $R_\eta(v_T)$. Note that the tetrahedron $\widetilde{T}$ is also subdivided with the Alfeld split at the point v_T.

Since some of the B-coefficients of s associated with domain points in the ball $D_\eta(v_T)$ are already determined, also some of the B-coefficients of p are determined. Thus, considering p, we have already determined the B-coefficients associated with the domain points within a distance of m from the faces of $\widetilde{T}$, those within a distance of $2m$ from the edges of $\widetilde{T}$ and those within a distance of $\rho - d + \eta$ from the vertices of $\widetilde{T}$.

Now, for $\lfloor \frac{r}{3} \rfloor > 0$, we consider the B-coefficients of p associated with the domain points in the shells $R_{d-r-i}(v_T)$, $i = 1, \ldots, \lfloor \frac{r}{3} \rfloor$. In order to determine these B-coefficients of p, for $i = 1, \ldots, \lfloor \frac{r}{3} \rfloor$, we repeat the following:

We use Lemma 2.33 for $j = 0$ to stably determine the B-coefficients associated with the domain points within a distance of $2(m+i) - 1$ from the edges of $\widetilde{T}$. In relation to s, these are the B-coefficients associated with the domain points within a distance of $\mu + 2i - 1$ from the edges of T.

Now, we use the values of the derivatives of order $\mu + 2i$ in $(N5_{T_A})$ to stably determine the B-coefficients of s associated with the domain points on the interior faces of T_A with a distance of $\mu + 2i$ from the corresponding edge of T. These domain points have a distance of $2(m+i)$ from the corresponding edge of $\widetilde{T}$. Thus, the B-coefficients of p corresponding to the domain points with a distance of $2(m+i)$ from the edges of $\widetilde{T}$ on the interior triangular faces of $\widetilde{T}_A$ are determined. Following Lemma 2.33 for the case $j = 1$, the remaining B-coefficients of p associated with domain points within a distance of $2(m+i)$ from the edges of $\widetilde{T}$ can be stably determined.

Now, let $\widetilde{F}_i$ be the layer of domain points of p with a distance of $m+i$ from a face $\widetilde{F}$ of $\widetilde{T}$. Using the values of the derivatives in $(N6_{T_A})$ corresponding to $\widetilde{F}_i$, the B-coefficients of p associated with the domain points with a distance of $\mu - r + i + j$, $j = 1, \ldots, r - 3i - 1$, from the edges of $\widetilde{F}_i$ can be stably determined. Subsequently, we use the values of the derivatives of order $r + i$ in $(N7_{T_A})$ corresponding to $\widetilde{F}_i$ to compute the remaining B-coefficients of $p|_{\widetilde{F}_i}$. This involves solving a linear system with matrix $\widetilde{M}_i := \left[B_\xi^{d-r-i}(\kappa) \right]_{\xi, \kappa \in B_i^{\widetilde{F}_i}}$, which is nonsingular by lemma 2.44 and also independent of the size an shape of $\widetilde{F}$.

We have determined all B-coefficients of p within a distance of $m + \lfloor \frac{r}{3} \rfloor$ from the faces of $\widetilde{T}$. Thus, using Lemma 2.42 with $n = 4$, the remaining undetermined B-coefficients of p can be stably determined from the values of the derivatives in ($\mathrm{N8}_{T_A}$). Therefore, p is uniquely determined, which implies that the remaining undetermined B-coefficients of s are uniquely and stably determined by applying subdivision. □

Next, let Δ be a tetrahedral partition and Δ_A the partition obtained by refining each tetrahedron in Δ with the Alfeld split. Now, nodal minimal determining sets for the C^r macro-elements constructed in section 7.2, based on the partition Δ_A, are described.

Theorem 7.12:
Let $\mathcal{N}_{\Delta_A}$ be the union of the following sets of nodal data:

($\mathrm{N1}_{\Delta_A}$) $\bigcup_{v \in \mathcal{V}} \{\epsilon_v D^\alpha\}_{|\alpha| \le \rho}$

($\mathrm{N2}_{\Delta_A}$) $\bigcup_{e \in \mathcal{E}} \bigcup_{i=1}^{\mu} \bigcup_{j=1}^{i} \{\epsilon_{\zeta^i_{e,j}} D^\beta_{e^\perp}\}_{|\beta|=i}$

($\mathrm{N3}_{\Delta_A}$) $\bigcup_{F \in \mathcal{F}} \{\epsilon_\xi D^\alpha_F\}_{\xi \in A^F_\alpha, \alpha=0,\dots,r}$

($\mathrm{N4}_{\Delta_A}$) $\bigcup_{F \in \mathcal{F}} \bigcup_{e \in F} \bigcup_{i=2}^{r} \bigcup_{j=1}^{\lfloor \frac{i}{2} \rfloor} \{\epsilon_{\zeta^{\mu+j}_{e,k}} D^i_{e^F} D^{\mu-i+j}_{e \perp F}\}_{k=1}^{\mu+j}$

($\mathrm{N5}_{\Delta_A}$) $\bigcup_{T \in \Delta} \bigcup_{e \in T} \bigcup_{i=1}^{\lfloor \frac{r}{3} \rfloor} \bigcup_{j=1}^{\mu+2i} \{\epsilon_{\zeta^i_{e,j}} D^\beta_{e^F}\}_{|\beta|=\mu+2i}$

($\mathrm{N6}_{\Delta_A}$) $\bigcup_{T \in \Delta} \bigcup_{F \in T} \bigcup_{e \in F} \bigcup_{i=1}^{\lfloor \frac{r}{3} \rfloor} \bigcup_{j=1}^{r-3i-1} \{\epsilon_{\zeta^{\mu+2i+j}_{e,k}} D^{r+i}_{e^F} D^{\mu-r+i+j}_{e \perp F}\}_{k=1}^{\mu+2i+j}$

($\mathrm{N7}_{\Delta_A}$) $\bigcup_{T \in \Delta} \bigcup_{F \in T} \{\epsilon_\xi D^{r+\beta}_F\}_{\xi \in B^F_\beta}, \beta = 1,\dots,\lfloor \frac{r}{3} \rfloor\}$

($\mathrm{N8}_{\Delta_A}$) $\bigcup_{v_T \in \mathcal{V}_I} \{\epsilon_{v_T} D^\alpha\}_{|\alpha| \le \eta - 4(m + \lfloor \frac{r}{3} \rfloor + 1)}$

Then $\mathcal{N}_{\Delta_A}$ is a 1-local and stable nodal minimal determining set for the superspline space $\mathcal{S}^{r,\rho,\mu,\eta}_d(\Delta_A)$.

Proof:
The cardinality of the set $\mathcal{N}_{\Delta_A}$ is equal to the dimension of the superspline space $\mathcal{S}^{r,\rho,\mu,\eta}_d(\Delta_A)$, as given in (7.2) in Theorem 7.5. Thus, it suffices to show that setting $\{\lambda s\}_{\lambda \in \mathcal{N}_{\Delta_A}}$ stably and locally determines all B-coefficients of a spline $s \in \mathcal{S}^{r,\rho,\mu,\eta}_d(\Delta_A)$.

The B-coefficients of s associated with the domain points in the balls $D_\rho^{T_v}(v)$, for each vertex $v \in \mathcal{V}$ and a tetrahedron T_v containing v, can be uniquely and stably computed from the values of the derivatives in the set $(\mathrm{N1}_{\Delta_A})$. Then, the remaining undetermined B-coefficients of s corresponding to domain points in $D_\rho(v)$ can be uniquely and stably determined using the C^ρ supersmoothness conditions at the vertices in $\mathcal{V}$. These computations are 1-local, since each B-coefficient $c_{i,j,k,l}^T$ of s determined so far only depends on the nodal data from the set $(\mathrm{N1}_{\Delta_A})$ contained in $\mathrm{star}(T)$.

For each edge $e \in \mathcal{E}$, the undetermined B-coefficients of s corresponding to domain points in the tube $E_\mu^{T_e}(e)$ can be uniquely and stably determined from the values of the derivatives in $(\mathrm{N2}_{\Delta_A})$, for a tetrahedron T_e containing e. All other undetermined B-coefficients of s associated with domain points in the tube $E_\mu(e)$ can be uniquely and stably determined from the C^μ supersmoothness conditions at the edges in $\mathcal{E}$. Moreover, the computation of the B-coefficients of s corresponding to the domain points in $E_\mu(e)$ is 1-local, since each B-coefficient $c_{i,j,k,l}^T$ of s associated with a domain point $\xi_{i,j,k,l}^T \in E_\mu(e)$ only depends on the nodal data from the sets $(\mathrm{N1}_{\Delta_A})$ and $(\mathrm{N2}_{\Delta_A})$ contained in $\mathrm{star}(T)$.

For each face $F \in \mathcal{F}$, the sets $(\mathrm{N3}_{\Delta_A})$ and $(\mathrm{N4}_{\Delta_A})$ contain nodal data for exactly one tetrahedron T_F containing F. Since these are the same nodal data as those in the sets $(\mathrm{N3}_{T_A})$ and $(\mathrm{N4}_{T_A})$ from Theorem 7.11, the B-coefficients of $s|_{T_F}$ associated with the domain points in T_F within a distance of r from the face F can be stably determined in the same way as in the proof of Theorem 7.11. Then, if T_F has a neighboring tetrahedron sharing F, the B-coefficients of s corresponding to the domain points within a distance of r from F in this tetrahedron can be stably determined by the C^r smoothness conditions across F. The computation of these B-coefficients is 1-local, since each of these B-coefficient $c_{i,j,k,l}^T$ only depends on the nodal data from the sets $(\mathrm{N1}_{\Delta_A})$ - $(\mathrm{N4}_{\Delta_A})$ contained in $\mathrm{star}(T)$.

Now, for each tetrahedron $T \in \Delta$, all B-coefficients of s, such that the smoothness conditions at the vertices, edges and faces of Δ are satisfied, are determined.

Then, for each tetrahedron $T \in \Delta$ sets $(\mathrm{N5}_{\Delta_A})$ - $(\mathrm{N8}_{\Delta_A})$ contain the same nodal data as the sets $(\mathrm{N5}_{T_A})$ - $(\mathrm{N8}_{T_A})$ from Theorem 7.11. Thus, the remaining undetermined B-coefficients of s can be stably determined in the same way as in the proof of Theorem 7.11. Therefore, all B-coefficients of s are determined and their computation is 1-local, in the sense that they only depend on nodal data contained in $\mathrm{star}(T)$, respectively. □

Remark 7.13:
For $r = 2$, the macro-elements considered in this chapter differ slightly from those constructed by Alfeld and Schumaker [11]. The difference lies in the way the B-coefficients of a spline corresponding to the domain points in the shell $R_{11}(v_T)$ are determined. For the macro-elements investigated in this chapter, some of these B-coefficients are determined from the nodal data in (N3$_{T_A}$) and (N4$_{T_A}$), and accordingly (N3$_{\Delta_A}$) and (N4$_{\Delta_A}$). In [9] the nodal data in (N4$_{T_A}$) and (N4$_{\Delta_A}$) can be omitted, since more derivatives at the faces of the tetrahedra are used. Unfortunately, in order to extend the construction of macro-elements over the Alfeld split to arbitrary smoothness, this more complicated nodal minimal determining set has to be used here. In case that conjecture 2.43 holds, more derivatives at the faces of the tetrahedra could be used also for $r > 2$ and the nodal data in (N4$_{T_A}$) and (N4$_{\Delta_A}$) could be omitted. Unfortunately, since the conjecture has not been proved yet, it is not possible to use the same kind of nodal data as in [11] to compute the B-coefficients on the shells around v_T for macro-elements with higher r.

7.5 Hermite interpolation with C^r macro-elements based on the Alfeld split

In this section we investigate the Hermite interpolation method corresponding to the macro-elements considered in this chapter. Moreover, we also give an error bounds for the interpolation method.

Following Theorem 7.12 in section 7.4, for every function $f \in C^{\rho}(\Omega)$ there is a unique spline $s \in \mathcal{S}_d^{r,\rho,\mu,\eta}(\Delta_A)$, solving the Hermite interpolation problem

$$\lambda s = \lambda f, \quad \text{for all } \lambda \in \mathcal{N}_{\Delta_A},$$

or equivalently,

(H1$_{\Delta_A}$) $D^{\alpha} s(v) = D^{\alpha} f(v)$, for all $|\alpha| \leq \rho$, and all vertices $v \in \mathcal{V}$

(H2$_{\Delta_A}$) $D^{\beta}_{e^{\perp}} s(\varsigma^i_{e,j}) = D^{\beta}_{e^{\perp}} f(\varsigma^i_{e,j})$, for all $|\beta| = i$, $1 \leq j \leq i$, $1 \leq i \leq \mu$, and all edges $e \in \mathcal{E}$

(H3$_{\Delta_A}$) $D^{\alpha}_{F} s(\xi) = D^{\alpha}_{F} f(\xi)$, for all $\xi \in A^F_{\alpha}$, $1 \leq \alpha \leq r$, and all faces $F \in \mathcal{F}$

(H4$_{\Delta_A}$) $D^i_{e^F} D^{\mu-i+j}_{e \perp F} s(\varsigma^{\mu+j}_{e,k}) = D^i_{e^F} D^{\mu-i+j}_{e \perp F} f(\varsigma^{\mu+j}_{e,k})$, for $1 \leq k \leq \mu + j$, $1 \leq j \leq \lfloor \frac{i}{2} \rfloor$, $2 \leq i \leq r$, and for all edges e of each face $F \in \mathcal{F}$

(H5$_{\Delta_A}$) $D^{\beta}_{e^{\hat{F}}} s(\varsigma^i_{e,j}) = D^{\beta}_{e^{\hat{F}}} f(\varsigma^i_{e,j})$, for all $|\beta| = \mu + 2i$, $1 \le j \le \mu + 2i$, $1 \le i \le \lfloor \frac{r}{3} \rfloor$, and for all edges e of each tetrahedron $T \in \Delta$

(H6$_{\Delta_A}$) $D^{r+i}_{e^{\hat{F}}} D^{\mu-r+i+j}_{e\perp F} s(\varsigma^{\mu+2i+j}_{e,k}) = D^{r+i}_{e^{\hat{F}}} D^{\mu-r+i+j}_{e\perp F} f(\varsigma^{\mu+2i+j}_{e,k})$, for all $1 \le k \le \mu + 2i + j$, $1 \le j \le r - 3i - 1$, $1 \le i \le \lfloor \frac{r}{3} \rfloor$, for all edges e and all faces F of each tetrahedron $T \in \Delta$

(H7$_{\Delta_A}$) $D^{r+\beta}_F s(\xi) = D^{r+\beta}_F f(\xi)$, for all $\xi \in B^F_\beta$, $1 \le \beta \le \lfloor \frac{r}{3} \rfloor$, and for all faces F of each tetrahedron $T \in \Delta$

(H8$_{\Delta_A}$) $D^\alpha s(v_T) = D^\alpha f(v_T)$, for all $|\alpha| \le \eta - 4(m + \lfloor \frac{r}{3} \rfloor + 1)$, and all vertices $v_T \in \mathcal{V}_I$

Following Theorem 7.12, this Hermite interpolation method is 1-local and stable.

This defines a linear projector $\mathcal{I} : C^\rho \longrightarrow \mathcal{S}^{r,\rho,\mu,\eta}_d(\Delta_A)$. The construction ensures that $\mathcal{I}s = s$ for every spline $s \in \mathcal{S}^{r,\rho,\mu,\eta}_d(\Delta_A)$ and thus, especially, $\mathcal{I}$ reproduces all polynomials of degree d. Since the Hermite interpolation method is 1-local and stable, we can establish the following error bounds.

Theorem 7.14:
For every $f \in C^{k+1}(\Omega)$ with $\rho - 1 \le k \le d$,

$$\|D^\alpha(f - \mathcal{I}f)\|_\Omega \le K|\Delta|^{k+1-|\alpha|}|f|_{k+1,\Omega}, \tag{7.3}$$

for all $0 \le |\alpha| \le k$, where K depends only on r and the smallest solid and face angle in Δ.

Proof:
Let $T \in \Delta$, $\Omega_T :=$ star(T) and $f \in C^{k+1}(\Omega)$. Fix α with $|\alpha| \le k$. By Lemma 4.3.8 of [19], see also Lemma 4.6 of [57], there exists a polynomial $p \in \mathcal{P}^3_k$ such that

$$\|D^\alpha(f - p)\|_{\Omega_T} \le K_1|\Omega_T|^{k+1-|\beta|}|f|_{k+1,\Omega_T}, \tag{7.4}$$

for all $0 \le |\beta| \le k$, where $|\Omega_T|$ is the diameter of Ω_T. Since $\mathcal{I}p = p$,

$$\|D^\alpha(f - \mathcal{I}f)\|_T \le \|D^\alpha(f - p)\|_T + \|D^\alpha\mathcal{I}(f - p)\|_T$$

holds. We can estimate the first term using (7.4) with $\beta = \alpha$. Applying the Markov inequality [102] to each of the polynomials $\mathcal{I}(f - p)|_{T_i}$, where T_i, $i = 1,\ldots,4$, are the four subtetrahedra of T_A, we get

$$\|D^\alpha\mathcal{I}(f - p)\|_{T_i} \le K_2|T|^{-\alpha}\|\mathcal{I}(f - p)\|_{T_i},$$

where K_2 is a constant depending only on r and the smallest solid and face angle in Δ.

Now, let $\{c_\xi\}$ be the B-coefficients of the polynomial $\mathcal{I}(f-p)|_{T_i}$ relative to the tetrahedron T_i. Then combining (2.12) and the fact that the interpolation method is stable and that the Bernstein basis polynomials form a partition of unity, it follows that

$$\|\mathcal{I}(f-p)\|_{T_i} \le K_3 \max_{\xi \in \mathcal{D}_{d,T_{i,j}}} |c_\xi| \le K_4 \sum_{j=0}^{\rho} |T|^j |f-p|_{j,T}.$$

Now taking the maximum over i and combining this with (7.4) and $|T| \le |\Omega_T|$, we get that

$$\|\mathcal{I}(f-p)\|_T \le K_5 |T|^{k+1} |f|_{k+1,T} \le K_6 |\Omega_T|^{k+1} |f|_{k+1,\Omega_T},$$

which gives

$$\|D^\alpha \mathcal{I}(f-p)\|_T \le K_7 |\Omega_T|^{k+1-|\alpha|} |f|_{k+1,\Omega_T}.$$

Maximizing over all tetrahedra T in Δ leads to (7.3). □

8 Macro-elements of arbitrary smoothness over the Worsey-Farin split of a tetrahedron

In this chapter we describe the C^r macro-elements based on the Worsey-Farin split of a tetrahedron (see Definition 2.7) by Matt [66]. In section 8.1, we review the minimal conditions for the degree of polynomials and the degree of supersmoothness for constructing C^r macro-elements over the Worsey-Farin split. In the following section 8.2, we investigate minimal determining sets for C^r macro-elements defined on the Worsey-Farin split of tetrahedra. In the next section, we illustrate these minimal determining sets with some examples to ease their understanding. In section 8.4, we examine nodal minimal determining sets for the C^r splines considered in this chapter. Finally, in section 8.5, a Hermite interpolant set for C^r splines based on the Worsey-Farin split is constructed. Moreover, it is shown that the interpolation yields optimal approximation order.

8.1 Minimal degrees of supersmoothness and polynomials for macro-elements over the Worsey-Farin split of a tetrahedron

The constraints on the degree of polynomials, as well as those on the degree of supersmoothness at the vertices and edges of a tetrahedron, in order to construct C^r macro-elements based on the Worsey-Farin split of a tetrahedron, are considered in this section. As already mentioned in section 7.1, it is desirable to use the lowest possible degree. So, to use splines with a polynomial degree lower than $8r + 1$, the tetrahedra of a partition are refined with a Worsey-Farin split in this chapter. Using the same approach as in section 7.1, bivariate layers in the split tetrahedra are examined in order to obtain the restrictions on the degree of polynomials and the degree of supersmoothness at the vertices and edges of tetrahedra for a spline defined on a partition refined with the Worsey-Farin split.

Now, with the results from section 6.1, it is possible to determine the minimal degree of polynomials and the minimal and maximal degree of supersmoothnesses for C^r macro-elements over the Worsey-Farin split of a tetrahedron.

Theorem 8.1:
Let $T := \langle v_1, v_2, v_3, v_4 \rangle$ be a tetrahedron and T_{WF} the Worsey-Farin split of T at the point v_T strictly in the interior of T and the points v_{F_i}, $i = 1, \ldots, 4$, strictly in the interior of the faces $F_i := \langle v_i, v_{i+1}, v_{i+2} \rangle$, $i = 1, \ldots, 4$, of T, with $v_5 := v_1$ and $v_6 := v_2$. For $r \geq 0$, a trivariate C^r macro-element based on T_{WF} can be constructed if and only if the degree of polynomials d, the degree of supersmoothness ρ at the vertices of T, and the degree of supersmoothness μ at the edges of T are at least as in Table 8.1, and the degree of supersmoothness η at v_T and at the edges $\langle v_{F_i}, v_T \rangle$, $i = 1, \ldots, 4$, is at most as in Table 8.1, for $m \geq 0$.

r	μ	ρ	η	d
4 m	6 m	9 m	15 m+1	18 m+1
4 m+1	6 m+1	9 m+1	15 m+2	18 m+3
4 m+2	6 m+3	9 m+4	15 m+7	18 m+9
4 m+3	6 m+4	9 m+6	15 m+11	18 m+13

Table 8.1: Minimal degree of polynomials d, minimal supersmoothnesses ρ and μ, and maximal supersmoothness η for C^r macro-elements over Worsey-Farin split tetrahedra.

Proof:
Let F be a plane intersecting T at a point $v_{\tilde{T}}$, which is situated on the edge $\langle v_1, v_T \rangle$, where F is perpendicular to $\langle v_1, v_T \rangle$. The proof is analog for any other vertex v_i, $i = 2, 3, 4$, of T. Since T is refined with a Worsey-Farin split, a Powell-Sabin split triangle is induced on F. This triangle is shown with light gray in Figure 8.1. In Figure 8.1 (right) only this triangle is shown, where $v_{\langle v_1, v_i \rangle}$, $i = 2, 3, 4$, are the three points where F intersects the edges $\langle v_1, v_i \rangle$ of T. Then, the edges $\langle v_{\langle v_1, v_i \rangle}, v_{\tilde{T}} \rangle$, $i = 2, 3, 4$, are the intersections of the plane F with the interior triangular faces $\langle v_1, v_i, v_T \rangle$ of T_{WF}, and the edges $\langle v_{e_j}, v_{\tilde{T}} \rangle$, $j = 1, 3, 4$, are the intersections of F with the interior faces $\langle v_1, v_{F_j}, v_T \rangle$ of T_{WF}.

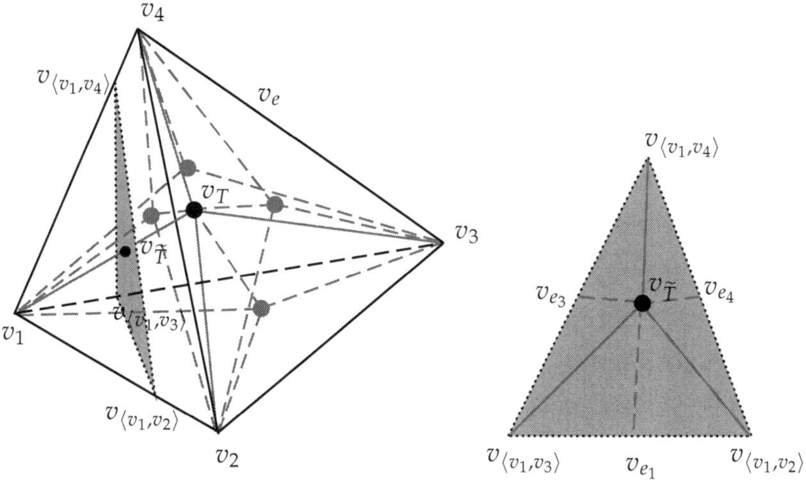

Figure 8.1: Bivariate layer with a Powell-Sabin split triangle in a tetrahedron $T := \langle v_1, v_2, v_3, v_4 \rangle$ refined with the Worsey-Farin split at the point v_T in the interior of T and the points v_{F_i}, $i = 1,\ldots,4$, in the interior of the faces F_i of T.

Due to the fact that each trivariate spline s defined on T_{WF} reduces to a bivariate spline with the same degree of polynomials and the same degree of supersmoothnesses when restricted to any triangle in T_{WF}, the spline s has to suffice the same constraints as a bivariate spline in Corollary 6.3 and 6.5. Thus, following Corollary 6.5, the degree of supersmoothness of s at the three vertices $v_{\langle v_1, v_i \rangle}$, $i = 2,3,4$, has to be at least $6m$, $6m+1$, $6m+3$, or $6m+4$ for a spline s in C^{4m}, C^{4m+1}, C^{4m+2}, or C^{4m+3}, respectively. Since the three vertices $v_{\langle v_1, v_i \rangle}$, $i = 2,3,4$, of the triangle on F lie on the edges of the tetrahedron T, the degree of supersmoothness of the spline s at the edges of the tetrahedron has to be at least $6m$, $6m+1$, $6m+3$, or $6m+4$, respectively.

Next, the triangle $\langle v_1, v_2, v_3 \rangle$, a face of T, is considered. The proof is analogue for any other face of T. The Worsey-Farin split of the tetrahedron T leads to a refinement of this triangle with a Clough-Tocher split. Therefore, following Corollary 6.3, the degree of supersmoothness of the spline

s at the vertices of the triangle $\langle v_1, v_2, v_3 \rangle$, and thus also at the vertices of the tetrahedron T, has to be at least $9m$, $9m+1$, $9m+4$, or $9m+6$, such that s has C^{6m}, C^{6m+1}, C^{6m+3}, or C^{6m+4} supersmoothness at the edges of T, respectively. The degree of supersmoothness at the vertices also leads to the minimal degree of polynomials. Again following Corollary 6.3, it can be seen that the degree of polynomials needed to construct C^{4m}, C^{4m+1}, C^{4m+2}, or C^{4m+3} macro-elements based on the Worsey-Farin split of T has to be at least equal to $18m+1$, $18m+3$, $18m+9$, or $18m+13$, respectively. Moreover, following Lemma 6.4, it can be seen that the degree of supersmoothness of s at the vertex v_T, and the edges $\langle v_{F_i}, v_T \rangle$, $i = 1, \ldots, 4$, can be at most $15m+1$, $15m+2$, $15m+7$, or $15m+11$ for s in C^{4m}, C^{4m+1}, C^{4m+2}, or C^{4m+3}, respectively. □

8.2 Minimal determining sets for C^r macro-elements over the Worsey-Farin split

In this section minimal determining sets for C^r macro-elements over the Worsey-Farin split of a tetrahedron and a tetrahedral partition are constructed. Therefore, some notations are established in order to define the considered space of supersplines. Thereafter, minimal determining sets for the C^r macro-elements based on the Worsey-Farin split of a tetrahedron are considered. In the following, some more notations are introduced and the space of supersplines used for the C^r macro-elements over the Worsey-Farin split of a tetrahedral partition is defined. Afterwards, minimal determining sets for theses macro-elements are constructed.

Notation 8.2:
Let $T := \langle v_1, v_2, v_3, v_4 \rangle$ be a tetrahedron with the four triangular faces $F_i := \langle v_i, v_{i+1}, v_{i+2} \rangle$, $i = 1, \ldots, 4$, with $v_5 := v_1$ and $v_6 := v_2$. Then let T_{WF} be the Worsey-Farin split of T at the point v_T in the interior of T and the points v_{F_i}, $i = 1, \ldots, 4$, in the interior of the face F_i, which consists of the twelve subtetrahedra $T_{i,j} := \langle u_{i,j}, u_{i,j+1}, v_{F_i}, v_T \rangle$, $i = 1, \ldots, 4$, $j = 1,2,3$, where $u_{i,j} := v_{i+j-1}$ for $i + j \leq 5$, with $u_{i,4} := u_{i,1}$, and $u_{i,j} := v_{i+j-5}$ for $i + j > 5$ and $j \neq 4$. Moreover, let $\mathcal{V}_T$ and $\mathcal{E}_T$ be the set of vertices and edges of T, $\mathcal{V}_{T,F}$ the set of four split points on the faces of T, and let $\mathcal{F}_{T,I}$ be the set of six interior faces of T_{WF} of the form $\langle u, w, v_T \rangle$, where u and w are vertices of T.

The C^r macro-elements based on the Worsey-Farin split of a tetrahedron T considered here are in the superspline space

$$\begin{aligned}\mathcal{S}_d^{r,\rho,\mu,\eta}(T_{WF}) := \{s \in C^r(\widetilde{T}) :\ & s|_{\widetilde{T}} \in \mathcal{P}_d^3 \text{ for all } \widetilde{T} \in T_{WF},\\ & s \in C^\mu(e),\ \forall e \in \mathcal{E}_T,\\ & s \in C^\rho(v),\ \forall v \in \mathcal{V}_T,\\ & s \in C^\eta(\widetilde{e}),\ \forall \widetilde{e} \in \mathcal{E}_{T,I},\\ & s \in C^\eta(v_T)\},\end{aligned}$$

where r,ρ,μ,η, and d are as in Table 8.1.

Theorem 8.3:
Let $\mathcal{M}_{T_{WF}}$ be the union of the following sets of domain points:

(M1$_{T_{WF}}$) $D_\rho^{T_{i,1}}(v_i)$, $i = 1,\dots,4$

(M2$_{T_{WF}}$) $\bigcup\limits_{e:=\langle u,v\rangle \in \mathcal{E}_T} E_\mu^{T_e}(e) \setminus (D_\rho^{T_e}(u) \cup D_\rho^{T_e}(v))$, with $e \in T_e$

(M3$_{T_{WF}}$) $\{\xi_{i,j,k,l}^{T_{\alpha,1}} \in E_{\mu-2}^{T_{\alpha,1}}(\langle v_{F_\alpha}, v_T\rangle)\}$, $\alpha = 1,\dots,4$, with $0 \le l \le d-\eta$

(M4$_{T_{WF}}$) $\{\xi_{i,j,k,l}^{\widetilde{T}}\}$ $\forall \widetilde{T} \in T_{WF}$, with $\mu - l + 1 \le k \le \mu - l + \left\lceil \frac{l-d+\eta-1}{2} \right\rceil$,
$i,j < d-l-\rho$, $l = d-\eta+1,\dots,r$, and $i+j+k+l = d$

(M5$_{T_{WF}}$) $\{\xi_{i,j,k,l}^{T_{\alpha,1}} \in E^{T_{\alpha,1}}_{\mu+2l-2d+2\eta-2-3\left\lceil \frac{l-d+\eta-1}{2}\right\rceil}(\langle v_{F_\alpha}, v_T\rangle)\}$, $\alpha = 1,\dots,4$,
with $l = d-\eta+1,\dots,r$

(M6$_{T_{WF}}$) for $r = 4m$ and $r = 4m+2$:
$$\bigcup_{F:=\langle u,w,v_T\rangle \in \mathcal{F}_{I,T}} \{\xi^{T_F}_{\rho-\mu+j-i,d-\rho-j,\mu+i,0} : i = 1,\dots,m,\ j = 1,\dots,\mu+i\},$$
with $T_F := \langle u,w,v_T,v_F\rangle$ for $v_F \in \mathcal{V}_{T,F}$

(M7$_{T_{WF}}$) $\{\xi_{i,j,k,l}^{\widetilde{T}}\}$ $\forall \widetilde{T} \in T_{WF}$, with $\mu - r + 1 \le k \le \mu - r + \left\lceil \frac{2r-(d-\eta)-l-1}{2} \right\rceil$,
$i,j < \rho+1-l$, $l = r+1,\dots,r+m$, and $i+j+k+l = d$

(M8$_{T_{WF}}$) $\{\xi_{i,j,k,l}^{T_{\alpha,1}} \in E^{T_{\alpha,1}}_{d+3r-l-3\left\lceil \frac{\rho+2r-l-1}{2}\right\rceil-3}(\langle v_{F_\alpha}, v_T\rangle)\}$, $\alpha = 1,\dots,4$
with $l = r+1,\dots,r+m$

(M9$_{T_{WF}}$) $D^{T_{1,1}}_{\eta-4(r+m-(d-\eta)+1)}(v_T)$,

where m is as in Table 8.1. Then $\mathcal{M}_{T_{WF}}$ is a stable minimal determining set for the space $\mathcal{S}_d^{r,\rho,\mu,\eta}(T_{WF})$ and the dimension of the superspline space is given as

$$\dim \mathcal{S}_d^{r,\rho,\mu,\eta}(T_{WF}) = \begin{cases} \frac{2797m^3+1161m^2+164m+8}{2}, & \text{for } r = 4m, \\ \frac{2797m^3+2292m^2+623m+56}{2}, & \text{for } r = 4m+1, \\ \frac{2797m^3+5115m^2+3116m+632}{2}, & \text{for } r = 4m+2, \\ \frac{2797m^3+6729m^2+5414m+1456}{2}, & \text{for } r = 4m+3. \end{cases} \quad (8.1)$$

Proof:
First, it is proved that $\mathcal{M}_{T_{WF}}$ is a minimal determining set for the superspline space $\mathcal{S}_d^{r,\rho,\mu,\eta}(T_{WF})$. To this end the B-coefficients c_ξ of a spline $s \in \mathcal{S}_d^{r,\rho,\mu,\eta}(T_{WF})$ are fixed to arbitrary values for each domain point ξ contained in the set $\mathcal{M}_{T_{WF}}$. Then it is shown that all remaining B-coefficients can be uniquely and stably determined.

The undetermined B-coefficients of s associated with domain points in the balls $D_\rho(v_i)$, $i = 1,\ldots,4$, can be uniquely and stably determined from those corresponding to the domain points in ($M1_{T_{WF}}$) using the C^ρ supersmoothness at the four vertices of T.

Now, the remaining undetermined B-coefficients corresponding to the domain points in the tubes $E_\mu(e)$, $e \in \mathcal{E}_T$, are investigated. These can be uniquely and stably determined from the B-coefficients associated with the domain points contained in ($M2_{T_{WF}}$) using the C^μ supersmoothness at the edges of T.

Next, the B-coefficients corresponding to the domain points on the faces of the tetrahedron T are considered. Each face F_i, $i = 1,\ldots,4$, is refined with a Clough-Tocher split. The resulting three subtriangles forming F_i are degenerated faces of the three subtetrahedra $T_{i,j}$, $j = 1,2,3$, since they lie in one common plane. Thus, as in the proof of Theorem 8.1, the B-coefficients of $s|_{F_i}$ can be regarded as those of a bivariate spline $g_{i,0}$ defined on F_i. Then the spline $g_{i,0}$ has the same degree of polynomials d, the degree of smoothness μ at the edges, ρ at the vertices, and η at the vertex v_{F_i} in the interior of F_i. Now, together with the already determined B-coefficients of $s|_{F_i}$ and those corresponding to the domain points in ($M3_{T_{WF}}$) lying on the face F_i, those with $l = 0$, all B-coefficients corresponding to a minimal determining set for a bivariate Clough-Tocher macro-element of smoothness μ, supersmoothness ρ at the vertices, supersmoothness η at the split point, and degree of polynomials d are determined. Thus, following Theorem 6.6, all B-coefficients of $g_{i,0}$ are uniquely and stably determined, and therefore, all B-coefficients of $s|_{F_i}$.

Now, let F_{i,ν_1} be the triangle formed by the domain points in the three shells $R^{T_{i,1}}_{d-\nu_1}(v_T)$, $R^{T_{i,2}}_{d-\nu_1}(v_T)$, and $R^{T_{i,3}}_{d-\nu_1}(v_T)$, for $\nu_1 = 1,\ldots,d-\eta$. Then again, the B-coefficients of $s|_{F_{i,\nu_1}}$ can be regarded as those of a bivariate spline g_{ν_1,F_i} which has $\mu-\nu_1$ smoothness at the edges, $\rho-\nu_1$ smoothness at the vertices and η smoothness at the split point. Moreover, the degree of polynomials of the bivariate spline g_{ν_1,F_i} is equal to $d-\nu_1$. Considering the already determined B-coefficients of $s|_{F_{i,\nu_1}}$ and the B-coefficients associated with the domain points in $(M3_{T_{WF}})$ lying in F_{i,ν_1}, those with $l=\nu_1$, it can be seen from Theorem 6.6, that these form a minimal determining set for a Clough-Tocher macro-element of smoothness $\mu-\nu_1$. Then, for the values of p and q in Theorem 6.6, $p=\nu_1$ and $q=\lfloor\frac{\nu_1+1}{2}\rfloor$ if μ is even, and $q=\lfloor\frac{\nu_1}{2}\rfloor$ in case μ is odd, holds. Thus, g_{ν_1,F_i} is uniquely and stably determined. So now, all B-coefficients of s restricted to the shells $R_d(v_T),\ldots,R_\eta(v_T)$ are uniquely and stably determined.

Next, the B-coefficients of s restricted to the domain points in the shells $R_{d-\nu_2}(v_T)$, $\nu_2=d-\eta+1,\ldots,r$, are examined. Therefore, let F_{i,ν_2} be the triangle formed by the domain points in the shells $R^{T_{i,1}}_{d-\nu_2}(v_T)$, $R^{T_{i,2}}_{d-\nu_2}(v_T)$, and $R^{T_{i,3}}_{d-\nu_2}(v_T)$. Again, the B-coefficients of the spline s restricted to the triangle F_{i,ν_2} can be regarded as those of a bivariate spline g_{ν_2,F_i} of degree $d-\nu_2$ with smoothness $\mu-\nu_2$ at the edges, $\rho-\nu_2$ at the vertices, and η at the split point of F_{i,ν_2}, at which F_{i,ν_2} is subjected to a Clough-Tocher split. Note that g_{ν_2,F_i} is actually a polynomial, due to the C^η supersmoothness at the split point in F_{i,ν_2}. Considering g_{ν_2,F_i}, it can be seen that the B-coefficients corresponding to the domain points within a distance of $\rho-\nu_2$ from the vertices of F_{i,ν_2} and within a distance of $\mu-\nu_2$ from the edges of F_{i,ν_2} are already uniquely determined. Together with the B-coefficients associated with the domain points in $(M4_{T_{WF}})$ lying in F_{i,ν_2}, those with $l=\nu_2$, the B-coefficients of g_{ν_2,F_i} corresponding to the domain points within a distance of $\mu-\nu_2+\lceil\frac{\nu_2-d+\eta-1}{2}\rceil$, from the edges of F_{i,ν_2} are already uniquely and stably determined. Now, the B-coefficients corresponding to the domain points in $(M5_{T_{WF}})$ lying in F_{i,ν_2} uniquely and stably determine the derivatives of g_{ν_2,F_i} at the split point in F_{i,ν_2} up to order $\mu+2(\nu_2-d+\eta-1)-3\lceil\frac{\nu_2-d+\eta-1}{2}\rceil = d-\nu_2-3(\mu-\nu_2+\lceil\frac{\nu_2-d+\eta-1}{2}\rceil+1)$. Thus, following Lemma 2.42 for $n=3$, the remaining undetermined B-coefficients of g_{ν_2,F_i} can be uniquely and stably computed from these derivatives.

At this point, all B-coefficients of s corresponding to the domain points within a distance of r from the faces of T, within a distance of μ from the edges of T, and within a distance of ρ from the vertices of T are uniquely and stably determined.

Next, the B-coefficients of s associated with the domain points in the shells $R_{d-\nu_3}(v_T)$, $\nu_3 = r+1,\dots,r+m$, are considered. Thus, the following is repeated for $\nu_3 = r+1,\dots,r+m$.

The spline s has C^r smoothness at the inner faces of T_{WF}. Therefore, Lemma 2.33 can be used in order to determine the remaining undetermined B-coefficients corresponding to the domain points within a distance of $\mu + \nu_3 - r$ from the edges of T.

For the case that r is odd, following Lemma 2.33 for the case $j = 0$, the undetermined B-coefficients of s corresponding to the domain points within a distance of $\mu + \nu_3 - r$ from the edges of T can be directly computed from those already determined by C^r smoothness conditions. Thereby, the B-coefficients are determined uniquely and stably.

If r is even, following Lemma 2.33 for the case $j = 1$, the remaining undetermined B-coefficients of s corresponding to the domain points within a distance of $\mu + \nu_3 - r$ from the edges of T can be uniquely and stably determined from those already known together with the ones associated with the domain points in ($M6_{T_{WF}}$) for $i = \nu_3 - r$. Thereby, a linear system based on the C^r smoothness conditions across the interior faces of T_{WF} has to be solved, where, by Lemma 2.33 for $j = 1$, the corresponding determinant is not equal to zero.

Now, let F_{i,ν_3} be the triangle formed by the domain points in the shells $R^{T_{i,1}}_{d-\nu_3}$, $R^{T_{i,2}}_{d-\nu_3}$, and $R^{T_{i,3}}_{d-\nu_3}$, for $i = 1,\dots,4$. Again, the spline s can be considered as a bivariate spline g_{ν_3,F_i} of degree $d - \nu_3$ defined on F_{i,ν_3}. The Worsey-Farin split of T induces a Clough-Tocher split for the triangle F_{i,ν_3}. Since g_{ν_3,F_i} has supersmoothness of degree $\eta > d - \nu_3$ at the split point in F_{i,ν_3}, the bivariate spline g_{ν_3,F_i} reduces to a bivariate polynomial. Considering the polynomial g_{ν_3,F_i}, it can be seen that the B-coefficients corresponding to the domain points within a distance of $\rho - \nu_3$ from the vertices of F_{i,ν_3}, and those corresponding to the domain points within a distance of $\mu - r$ from the edges of F_{i,ν_3} are already uniquely determined. Together with the B-coefficients associated with the domain points in ($M7_{T_{WF}}$) lying in F_{i,ν_3}, those with $l = \nu_3$, all B-coefficients of g_{ν_3,F_i} within a distance of $\mu - r + \lceil \frac{2r-(d-\eta)-\nu_3-1}{2} \rceil = \lceil \frac{\rho+2r-\nu_3-1}{2} \rceil - r$ from the edges of F_{i,ν_3} are uniquely and stably determined so far. Now, the B-coefficients corresponding to the domain points in ($M8_{T_{WF}}$) lying in F_{i,ν_3}, those with $l = \nu_3$, uniquely and stably determined the derivatives of g_{ν_3,F_i} at the split point in F_{i,ν_3} up to order $d + 3r - \nu_3 - 3\lceil \frac{\rho+2r-\nu_3-1}{2} \rceil - 3$. Thus, following Lemma 2.42 for the case $n = 3$, the remaining undetermined B-coefficients of g_{ν_3,F_i} are uniquely and stably determined from these derivatives.

So far, all B-coefficients of s corresponding to the domain points within a distance of $r + m$ from the faces of T, within a distance of $\mu + m$ from the edges of T, and within a distance of ρ from the vertices of T are uniquely and stably determined.

Now, the last undetermined B-coefficients of the spline s are determined. These B-coefficients are associated with domain points in the ball $D_\eta(v_T)$. Since the spline s has C^η supersmoothness at the split point v_T, the B-coefficients of $s|_{D_\eta(v_T)}$ can be regarded as those of a trivariate polynomial p of degree η defined on the tetrahedron $\widetilde{T}$, which is formed by the domain points in $D_\eta(v_T)$. Due to the reason that T was refined with a Worsey-Farin split, the tetrahedron $\widetilde{T}$ is also subjected to the Worsey-Farin.

Since some of the B-coefficients of s associated with domain points in the ball $D_\eta(v_T)$ are already known, these B-coefficients are also known for the polynomial p. Thus, the B-coefficients of p corresponding to the domain points within a distance of $r + m - (d - \eta)$ from the faces of $\widetilde{T}$, corresponding to the domain points within a distance of $\mu + m - (d - \eta)$ from the edges of $\widetilde{T}$, and corresponding to those domain points within a distance of $\rho - (d - \eta)$ from the vertices of $\widetilde{T}$ are already known. Now, from the B-coefficient associated with the domain points in ($M9_{T_{WF}}$) the derivatives of p at v_T up to order $\eta - 4(r + m - (d - \eta) + 1)$ can be uniquely and stably computed. Then, following Lemma 2.42 for the case $n = 4$, the remaining undetermined B-coefficients of p are uniquely and stably determined by these derivatives. Therefore, also all B-coefficients of s can be uniquely and stably determined.

Next, the dimension of the spline space $\mathcal{S}_d^{r,\rho,\mu,\eta}(T_{WF})$ is considered. Since $\mathcal{M}_{T_{WF}}$ is a minimal determining set for this space of supersplines,

$$\dim \mathcal{S}_d^{r,\rho,\mu,\eta}(T_{WF}) = \#\mathcal{M}_{T_{WF}}$$

holds. Then, by adding the number of domain points contained in the sets ($M1_{T_{WF}}$) - ($M9_{T_{WF}}$), it can be obtained that

$$\begin{aligned}
\#\mathcal{M}_{T_{WF}} &= 4\binom{\rho+3}{3} + 12\binom{\mu+2}{3} + 4(d-\eta+1)\binom{\mu}{2} \\
&+ 12\sum_{\beta=0}^{r-(d-\eta)-1}\sum_{i=1}^{\left\lceil\frac{\beta}{2}\right\rceil} i+\mu \\
&+ 4\sum_{\beta=0}^{r-(d-\eta)-1}\binom{\mu+2\beta-3\left\lceil\frac{\beta}{2}\right\rceil+2}{2} \\
&+ ((r+1) \mod 2)\cdot 6\sum_{\beta=1}^{m}(\beta+\mu) \\
&+ 12\sum_{\beta=1}^{m}\sum_{i=1}^{\left\lceil\frac{r-(d-\eta)-\beta-1}{2}\right\rceil}(i+\mu) \\
&+ 4\sum_{\beta=1}^{m}\binom{d-3\mu+2r-\beta-3\left\lceil\frac{r-(d-\eta)-\beta-1}{2}\right\rceil-1}{2} \\
&+ \binom{\eta-4(r+m-(d-\eta))-1}{3},
\end{aligned}$$

which reduces to the number in (8.1). □

The construction of the macro-elements shown above for a single tetrahedron T can be extended to a tetrahedral partition Δ, since all B-coefficients of a spline corresponding to the domain points of a tetrahedron T within a distance of r from the faces, within a distance of μ from the edges, and within a distance of ρ from the vertices of this tetrahedron can be determined from the B-coefficients associated with the domain points in ($M1_{T_{WF}}$) - ($M5_{T_{WF}}$). Thus, all B-coefficients involved in the smoothness conditions connecting two individual tetrahedra of a partition are contained in the minimal determining set, and therefore these smoothness conditions can be fulfilled.

Next, minimal determining sets for C^r macro-elements based on the Worsey-Farin split of a tetrahedral partition are considered. Therefore, some more notations are needed.

Notation 8.4:
Let Δ be an arbitrary tetrahedral partition of a polyhedral domain Ω. Then, let $\mathcal{V}$, $\mathcal{E}$, and $\mathcal{F}$ be the set of vertices, edges, and faces of Δ, respectively.

Moreover, let Δ_{WF} be the partition obtained by subjecting each tetrahedron $T \in \Delta$ to the Worsey-Farin split. Then let $\mathcal{V}_I$ be the set of split points in the interior of the tetrahedra in Δ, $\mathcal{V}_F$ the set of split points in the interior of the faces of the tetrahedra of Δ, $\mathcal{E}_I$ the set of interior edges of Δ of the form $\langle v_F, v_T \rangle$, with $v_F \in \mathcal{V}_F$ and $v_T \in \mathcal{V}_I$, and let $\mathcal{F}_I$ be the set of interior faces of Δ of the form $\langle u, v, w \rangle$, with $u, v \in \mathcal{V}$ and $w \in \mathcal{V}_I$. Moreover, let $\mathcal{E}_{I,F}$ be the subset of $\mathcal{E}_I$, which contains exactly one edge for each vertex $v_F \in \mathcal{V}_F$. Finally, let $\#\mathcal{V}$, $\#\mathcal{E}$, and $\#\mathcal{F}$ be the cardinalities of the sets $\mathcal{V}$, $\mathcal{E}$, and $\mathcal{F}$, and let $\#\mathcal{N}$ be the number of tetrahedra contained in Δ.

Note that the set $\mathcal{E}_{I,F}$ contains exactly one edge for each face in $\mathcal{F}$.

The C^r macro-elements based on tetrahedral partition Δ_{WF} are in the superspline space

$$\begin{aligned} \mathcal{S}_d^{r,\rho,\mu,\eta}(\Delta_{WF}) := \{ & s \in C^r(T) : \ s|_T \in \mathcal{P}_d^3 \text{ for all } T \in \Delta_{WF}, \\ & s \in C^{\mu}(e), \ \forall e \in \mathcal{E}, \\ & s \in C^{\rho}(v), \ \forall e \in \mathcal{V}, \\ & s \in C^{\eta}(\widetilde{e}), \ \forall \widetilde{e} \in \mathcal{E}_I, \\ & s \in C^{\eta}(\widetilde{v}), \ \forall \widetilde{v} \in \mathcal{V}_I \}, \end{aligned}$$

where r, ρ, μ, η, and d are as in Table 8.1.

Theorem 8.5:
Let $\mathcal{M}_{\Delta_{WF}}$ be the union of the following sets of domain points:

($\mathrm{M1}_{\Delta_{WF}}$) $\bigcup_{v \in \mathcal{V}} D_{\rho}^{T_v}(v)$, with $v \in T_v$

($\mathrm{M2}_{\Delta_{WF}}$) $\bigcup_{e := \langle u,v \rangle \in \mathcal{E}} E_{\mu}^{T_e}(e) \setminus (D_{\rho}^{T_e}(u) \cup D_{\rho}^{T_e}(v))$, with $e \in T_e$

($\mathrm{M3}_{\Delta_{WF}}$) $\bigcup_{\widetilde{e}_F \in \mathcal{E}_{I,F}} \{\xi_{i,j,k,l}^{T_{\widetilde{e}_F}} \in E_{\mu-2}^{T_{\widetilde{e}_F}}(\widetilde{e}_F)\}$, with $0 \le l \le d - \eta$, and $\widetilde{e}_F \in T_{\widetilde{e}_F}$

($\mathrm{M4}_{\Delta_{WF}}$) $\bigcup_{F \in \mathcal{F}} \bigcup_{\substack{T_F \in \Delta_{WF} \\ F \in T_F}} \{\xi_{i,j,k,l}^{T_F}\}$, with $\mu - l + 1 \le k \le \mu - l + \left\lceil \frac{l-d+\eta-1}{2} \right\rceil$,

$i, j < d - l - \rho$, $l = d - \eta + 1, \ldots, r$, and $i + j + k + l = d$

($\mathrm{M5}_{\Delta_{WF}}$) $\bigcup_{\widetilde{e}_F \in \mathcal{E}_{I,F}} \{\xi_{i,j,k,l}^{T_{\widetilde{e}_F}} \in E_{\mu+2l-2d+2\eta-2-3\left\lceil \frac{l-d+\eta-1}{2} \right\rceil}^{T_{\widetilde{e}_F}}(\widetilde{e}_F)\}$,

with $l = d - \eta + 1, \ldots, r$, and $\widetilde{e}_F \in T_{\widetilde{e}_F}$

($\mathrm{M6}_{\Delta_{WF}}$) for $r = 4m$ and $r = 4m + 2$:

$$\bigcup_{\widetilde{F} \in \mathcal{F}_I} \{\xi_{\rho-\mu+j-i,d-\rho-j,\mu+i,0}^{T_{\widetilde{F}}} : \ i = 1, \ldots, m, \ j = 1, \ldots, \mu + i\},$$

with $T_{\widetilde{F}} := \langle u,v,w,v_{\widetilde{F}}\rangle$ for $v_{\widetilde{F}} \in \mathcal{V}_F$

(M7$_{\Delta_{WF}}$) $\bigcup_{T\in\Delta_{WF}} \{\xi^T_{i,j,k,l}\}$, with $\mu - r + 1 \leq k \leq \mu - r + \lceil \frac{2r-(d-\eta)-l-1}{2} \rceil$, $i,j < \rho + 1 - l$, $l = r+1,\dots,r+m$, and $i+j+k+l = d$

(M8$_{\Delta_{WF}}$) $\bigcup_{\widetilde{e}\in\mathcal{E}_I} \{\xi^{T_{\widetilde{e}}}_{i,j,k,l} \in E^{T_{\widetilde{e}}}_{d+3r-l-3\lceil\frac{\rho+2r-l-1}{2}\rceil-3}(\widetilde{e})\}$, with $l = r+1,\dots,r+m$, and $\widetilde{e} \in T_{\widetilde{e}}$

(M9$_{\Delta_{WF}}$) $\bigcup_{\widetilde{v}\in\mathcal{V}_I} D^{T_{\widetilde{v}}}_{\eta-4(r+m-(d-\eta)+1)}(\widetilde{v})$, with $\widetilde{v} \in T_{\widetilde{v}}$,

where m is as in Table 8.1. Then $\mathcal{M}_{\Delta_{WF}}$ is a local and stable minimal determining set for the space $\mathcal{S}^{r,\rho,\mu,\eta}_d(\Delta_{WF})$ and the dimension of the superspline space is given as

$$\dim \mathcal{S}^{r,\rho,\mu,\eta}_d(\Delta_{WF}) = \begin{cases} \frac{243m^3+162m^2+33m+2}{2}\#\mathcal{V} + (72m^3+36m^2+4m)\#\mathcal{E} \\ +\frac{469m^3+75m^2-16m}{6}\#\mathcal{F} + \frac{1007m^3-57m^2+16m}{6}\#\mathcal{N}, \\ \text{for } r = 4m, \\ \\ \frac{243m^3+243m^2+78m+8}{2}\#\mathcal{V} + (72m^3+72m^2+22m \\ +2)\#\mathcal{E} + \frac{469m^3+333m^2+44m}{6}\#\mathcal{F} \\ +\frac{1007m^3+36m^2-35m}{6}\#\mathcal{N}, \text{ for } r = 4m+1, \\ \\ \frac{243m^3+486m^2+321m+70}{2}\#\mathcal{V} + (72m^3+144m^2 \\ +94m+20)\#\mathcal{E} + \frac{469m^3+741m^2+362m+54}{6}\#\mathcal{F} \\ +\frac{1007m^3+1365m^2+664m+120}{6}\#\mathcal{N}, \text{ for } r = 4m+2, \\ \\ \frac{243m^3+648m^2+573m+168}{2}\#\mathcal{V} + (72m^3+180m^2 \\ +148m+40)\#\mathcal{E} + \frac{469m^3+1074m^2+803m+198}{6}\#\mathcal{F} \\ +\frac{1007m^3+1635m^2+826m+120}{6}\#\mathcal{N}, \text{ for } r = 4m+3. \end{cases} \tag{8.2}$$

Proof:
At first, it is shown that $\mathcal{M}_{\Delta_{WF}}$ is a minimal determining set for the space $\mathcal{S}_d^{r,\rho,\mu,\eta}(\Delta_{WF})$ of supersplines. Therefore, the B-coefficients c_ξ of a spline $s \in \mathcal{S}_d^{r,\rho,\mu,\eta}(\Delta_{WF})$ are set to arbitrary values for each domain point ξ contained in the set $\mathcal{M}_{\Delta_{WF}}$. Then all other B-coefficients can be uniquely and stably determined using smoothness conditions.

Considering the balls $D_\rho(v)$, $v \in \mathcal{V}$, of domain points, it can be seen that these do not overlap. Thus, the B-coefficients associated with the domain points in (M1$_{\Delta_{WF}}$) can be set independently and used to uniquely and stably determined the remaining undetermined B-coefficients corresponding to domain points in $D_\rho(v)$, $v \in \mathcal{V}$. These B-coefficients are computed by using the C^ρ supersmoothness conditions of s at the vertices contained in $\mathcal{V}$. Moreover, the computations are 1-local, since the B-coefficients $\xi \in D_\rho(v)$, $v \in \mathcal{V}$, only depend on B-coefficients associated with domain points contained in $\mathcal{M}_{\Delta_{WF}} \cap \text{star}(T_v)$, for $v \in T_v$.

Next, the remaining undetermined B-coefficients of s corresponding to the domain points in the tubes $E_\mu(e)$, $e \in \mathcal{E}$, are examined. Since the set of domain points (M2$_{\Delta_{WF}}$) does not overlap any of the balls $D_\rho(v)$, $v \in \mathcal{V}$, and since the sets $E_\mu(\langle u,v\rangle) \setminus (D_\rho(u) \cup D_\rho(v))$, $u,v \in \mathcal{V}$, do not overlap each other, all B-coefficients of s corresponding to the domain points in (M2$_{\Delta_{WF}}$) can be set to arbitrary values and used to uniquely determine the remaining B-coefficients of $s|_{E_\mu(e)}$, $e \in \mathcal{E}$. The computation of the undetermined B-coefficients of $s|_{E_\mu(e)}$, $e \in \mathcal{E}$, is stable and 1-local, since only C^μ smoothness conditions and B-coefficients corresponding to domain points in $\mathcal{M}_{\Delta_{WF}} \cap \text{star}(T_e)$ are involved, for $e \in T_e$.

Now, since the sets of undetermined domain points of s within a distance of r from two distinct faces contained in $\mathcal{F}$ do not overlap, and since the sets (M3$_{\Delta_{WF}}$), (M4$_{\Delta_{WF}}$), and (M5$_{\Delta_{WF}}$) do not overlap any of the balls $D_\rho(v)$, $v \in \mathcal{V}$, or any of the tubes $E_\mu(e)$, $e \in \mathcal{E}$, the B-coefficients of s restricted to the domain points in (M3$_{\Delta_{WF}}$) - (M5$_{\Delta_{WF}}$) can be set to arbitrary values. Thus, following the same arguments as above in the proof of Theorem 8.3, the B-coefficients of s corresponding to the domain points within a distance of r from the faces in $\mathcal{F}$ can be uniquely and stably determined for one tetrahedron T_F containing F for each face F in $\mathcal{F}$. Then the remaining undetermined B-coefficients of s corresponding to domain points within a distance of r from the faces in $\mathcal{F}$ can be uniquely and stably determined by the C^r smoothness conditions connecting the polynomial pieces of s across the faces in $\mathcal{F}$. The computation of the B-coefficients of s corresponding to domain points within a distance of r from the faces F in $\mathcal{F}$, examined in this step, is 1-local, since they only depend on B-coefficients of s associated with domain points contained in $\mathcal{M}_{\Delta_{WF}} \cap \text{star}T_F$, where $F \in T_F$.

At this point, all B-coefficients of s are uniquely determined, such that the splines $s|_T$ and $s|_{\widetilde{T}}$ join with C^ρ, C^μ, and C^r smoothness at the common vertices, edges, and faces of T and $\widetilde{T}$, for each two distinct tetrahedra T and $\widetilde{T}$ in Δ.

For each tetrahedron $T \in \Delta$ the sets ($M6_{\Delta_{WF}}$), ($M7_{\Delta_{WF}}$), ($M8_{\Delta_{WF}}$), and ($M9_{\Delta_{WF}}$) consist of the sets ($M6_{T_{WF}}$), ($M7_{T_{WF}}$), ($M8_{T_{WF}}$), and ($M9_{T_{WF}}$) of domain points from Theorem 8.3. Thus, the remaining undetermined B-coefficients of s can be determined in the same way as in the proof of Theorem 8.3, separately on each tetrahedron in Δ. Therefore, the computation of these B-coefficients is 1-local, since they only depend on the B-coefficients of s associated with domain points in the set $\mathcal{M}_{\Delta_{WF}} \cap \text{star}(T)$, for each tetrahedron $T \in \Delta$.

Now, in order to compute the dimension of the space of superspline $\mathcal{S}_d^{r,\rho,\mu,\eta}(\Delta_{WF})$, it suffices to sum up the cardinalities of the sets ($M1_{\Delta_{WF}}$) - ($M9_{\Delta_{WF}}$) for each vertex, edge, face and tetrahedron in Δ. For each vertex $v \in \mathcal{V}$, the set ($M1_{\Delta_{WF}}$) contains exactly

$$\binom{\rho+3}{3}$$

domain points. For each edge $e \in \mathcal{E}$, the set ($M2_{\Delta_{WF}}$) contains

$$2\binom{\mu+2}{3}$$

domain points. For each face $F \in \mathcal{F}$, the three sets ($M3_{\Delta_{WF}}$) - ($M5_{\Delta_{WF}}$) contain a total number of

$$(d-\eta+1)\binom{\mu}{2} + 3\sum_{\beta=0}^{r-(d-\eta)-1}\sum_{i=1}^{\lceil \frac{\beta}{2} \rceil} i + \mu + \sum_{\beta=0}^{r-(d-\eta)-1}\binom{\mu+2\beta-3\lceil \frac{\beta}{2} \rceil+2}{2}$$

domain points.

For each tetrahedron $T \in \Delta$, the sets (M6$_{\Delta_{WF}}$) - (M9$_{\Delta_{WF}}$) contain exactly

$$((r+1) \mod 2) \cdot 6 \sum_{\beta=1}^{m} (i+\mu) + 12 \sum_{\beta=1}^{m} \sum_{i=1}^{\left\lceil \frac{r-(d-\eta)-\beta-1}{2} \right\rceil} (i+\mu)$$

$$+ 4 \sum_{\beta=1}^{m} \binom{d - 3\mu + 2r - \beta - 3\left\lceil \frac{r-(d-\eta)-\beta-1}{2} \right\rceil - 1}{2}$$

$$+ \binom{\eta - 4(r+m-(d-\eta)) - 1}{3},$$

domain points.

Adding these numbers multiplied with the number of vertices, edges, faces, and tetrahedra of Δ, respectively, leads to the dimension of the superspline space $\mathcal{S}_d^{r,\rho,\mu,\eta}(\Delta_{WF})$, as given in (8.2). □

Remark 8.6:
The C^r macro-elements considered in this section are consistent with those constructed in [62] for the cases $r = 1$ and $r = 2$. The dimension of the space $\mathcal{S}_3^{1,1,1,2}(\Delta_{WF})$ of supersplines used here to construct the C^1 macro-elements based on the Worsey-Farin split is equal to $4\#\mathcal{V} + 2\#\mathcal{E}$. The same space of supersplines is used in [62] in order to construct C^1 macro-elements over the Worsey-Farin split of a tetrahedral partition. The superspline space used here to construct the C^2 macro-elements based on the Worsey-Farin split is $\mathcal{S}_9^{2,4,3,7}(\Delta_{WF})$, which has a dimension of $35\#\mathcal{V} + 20\#\mathcal{E} + 9\#\mathcal{F} + 20\#\mathcal{N}$. Again, the same space was used in [62], in order to construct C^2 macro-elements. Another C^2 macro-element was constructed in [9], which was also presented in [62]. In their work, Alfeld and Schumaker [9] use six additional smoothness conditions per face in order to define their spline space. Thus, the dimension of their macro-elements is equal to $35\#\mathcal{V} + 20\#\mathcal{E} + 3\#\mathcal{F} + 20\#\mathcal{N}$, which can also be obtained by the macro-elements presented here by adding these additional smoothness conditions, though this would lead to a much more complicated spline space for higher r.

8.3 Examples for macro-elements over the Worsey-Farin split of a tetrahedron

In this section some examples for the macro-elements constructed in section 8.2 are shown, in order to facilitate their understanding. As in [66], the macro-elements are considered on a single tetrahedron $T := \langle v_1, v_2, v_3, v_4 \rangle$ for the cases $r = 1, \ldots, 4$. Moreover, also the cases $r = 5$ and $r = 6$ are investigated. For the cases $r = 1$ and $r = 2$, it is also shown that the macro-elements examined in this chapter reduce to those from Lai and Schumaker [62].

8.3.1 C^1 macro-elements over the Worsey-Farin split

The macro-elements are in the space $\mathcal{S}_d^{r,\rho,\mu,\eta}(T_{WF})$ of supersplines. For $r = 1$, following Table 8.1 (third row), m is equal to zero and $\mathcal{S}_d^{r,\rho,\mu,\eta}(T_{WF})$ reduces to $\mathcal{S}_3^{1,1,1,2}(T_{WF})$. In Theorem 8.3 is shown, that the dimension of this spline space is equal to 28 and a corresponding minimal determining set $\mathcal{M}_{T_{WF},1}$ is given by the union of the following sets of domain points:

(C^1:M1$_{T_{WF}}$) $D_1^{T_{i,1}}(v_i)$, $i = 1, \ldots, 4$

(C^1:M2$_{T_{WF}}$) $\bigcup_{e:=\langle u,v \rangle \in \mathcal{E}_T} E_1^{T_e}(e) \setminus (D_1^{T_e}(u) \cup D_1^{T_e}(v))$, with $e \in T_e$

Proof:
It is first shown that $\mathcal{M}_{T_{WF},1}$ is a stable minimal determining set for the superspline space $\mathcal{S}_3^{1,1,1,2}(T_{WF})$. Due to this reason, the B-coefficients c_ξ of a spline $s \in \mathcal{S}_3^{1,1,1,2}(T_{WF})$ are set to arbitrary values for each domain point $\xi \in \mathcal{M}_{T_{WF},1}$. Then it is shown that all other B-coefficients of s are uniquely and stably determined.

Considering the B-coefficients of s corresponding to the domain points in the balls $D_1(v_i)$, $i = 1, \ldots, 4$, it can be seen that these are uniquely determined from the B-coefficients associated with the domain points in (C^1:M1$_{T_{WF}}$). Thereby, the computation of the undetermined B-coefficients is stable, since they are directly determined from C^1 smoothness conditions at the four vertices of T.

Next, the remaining B-coefficients of s associated with domain points in the six tubes $E_1(e)$, $e \in \mathcal{E}_T$, are examined. These B-coefficients can be uniquely and stably determined from the B-coefficients of s corresponding to the domain points in (C^1:M2$_{T_{WF}}$) by the C^1 smoothness conditions at the edges of T.

Now, the remaining undetermined B-coefficients of s associated with domain points on the faces of T are considered. Each of the four faces F_i, $i = 1,\dots,4$, is refined with a Clough-Tocher split, due to the Worsey-Farin split of T. For each face F_i, the three resulting subtriangles are degenerated faces of the subtetrahedra $T_{i,j}$, $j = 1,2,3$. Therefore, the B-coefficients of $s|_{F_i}$, $i = 1,\dots,4$, can be regarded as those of a bivariate spline g_{0,F_i} defined on F_i. The spline g_{0,F_i} has a polynomial degree of three, C^1 smoothness, and C^2 supersmoothness at the vertex v_{F_i} in the interior of F_i. Some of the B-coefficients of $s|_{F_i}$, $i = 1,\dots,4$, and thus also some of g_{0,F_i}, are already determined. That are the B-coefficients corresponding to the domain points within a distance of one from the edges of F_i. Following Theorem 6.6, these domain points form a minimal determining set for the spline g_{0,F_i}. Thus, all B-coefficients of g_{0,F_i}, and consequently all B-coefficients of $s|_{F_i}$, are uniquely and stably determined (see Example 6.8).

Next, let $F_{i,1}$ be the triangle formed by the domain points in the three shells $R_2^{T_{i,1}}(v_T)$, $R_2^{T_{i,2}}(v_T)$, and $R_2^{T_{i,3}}(v_T)$. Due to the Worsey-Farin split of T, this triangle is also subjected to a Clough-Tocher split. Then the B-coefficients of $s|_{F_{i,1}}$, $i = 1,\dots,4$, can be considered as those of a bivariate spline g_{v_1,F_i}, which has a polynomial degree of two, C^0 smoothness, and C^2 supersmoothness at the split point of $F_{i,1}$. For the bivariate spline g_{v_1,F_i}, which is in fact just a polynomial, since the smoothness at the split point is equal to the degree of polynomials, all B-coefficients corresponding to the domain points on the edges of $F_{i,1}$ are already determined. Since these domain points form a minimal determining set for this spline, by Theorem 6.6, all other B-coefficients can be uniquely and stably determined in the way described in the proof of Theorem 6.6. Thus, all B-coefficients of $g_{i,1}$, and therefore all B-coefficients of $s|_{F_{i,1}}$, are uniquely and stably determined (see Example 6.7).

So far, all B-coefficients of s corresponding to the domain points in the shells $R_3(v_T)$ and $R_2(v_T)$ are uniquely and stably determined.

Now, the last undetermined B-coefficients of the spline s are considered. Those are the B-coefficients associated with domain points in the ball $D_2(v_T)$. Since s has C^2 supersmoothness at v_T, these B-coefficients can be regarded as those of a quadratic trivariate polynomial p of degree two defined on $\widetilde{T}$, where $\widetilde{T}$ is bounded by the domain points in the shell $R_2(v_T)$. The tetrahedron $\widetilde{T}$ is also refined with a Worsey-Farin split.

Since the B-coefficients of s associated with the domain points in the shell $R_2(v_T)$ are already determined, these B-coefficients are also known for the polynomial p. Thus, the B-coefficients of p corresponding to the domain points on the faces of $\widetilde{T}$ already known. Now, following the same arguments as in the proof of Lemma 2.42, it can be seen that all other B-coefficients of p are uniquely and stably determined. The dimension of a trivariate polynomial of degree two is equal to 10. Since a total number of 10 B-coefficients of $p|_{\widetilde{T}_{WF}}$ are already determined, all other B-coefficients of p can be uniquely and stably determined from these by using smoothness conditions. Therefore, also all B-coefficients of s are uniquely and stably determined.

Now, the dimension of the space $\mathcal{S}_3^{1,1,1,2}(T_{WF})$ of supersplines is equal to the cardinality of the minimal determining set $\mathcal{M}_{T_{WF},1}$. The set (C^1:M1$_{T_{WF}}$) contains exactly

$$4\binom{1+3}{3} = 16$$

domain points. The cardinality of the set (C^1:M2$_{T_{WF}}$) is equal to

$$12\binom{1+2}{3} = 12.$$

Therefore, the dimension of $\mathcal{S}_3^{1,1,1,2}(T_{WF})$ is equal to 28. □

Remark 8.7:
The macro-elements considered in this example are exactly the same as the ones constructed by Lai and Schumaker [62] in section 18.4.

8.3.2 C^2 macro-elements over the Worsey-Farin split

The C^2 macro-elements over the Worsey-Farin split of a tetrahedron T are in the superspline space $\mathcal{S}_9^{2,4,3,7}(T_{WF})$, which is derived from the space $\mathcal{S}_d^{r,\rho,\mu,\eta}(T_{WF})$ with the values from the fourth row of Table 8.1 and $m = 0$. By Theorem 8.3, the dimension of this spline space is equal to 316 and a corresponding minimal determining set $\mathcal{M}_{T_{WF},2}$ is given by the union of the following sets of domain points:

(C^2:M1$_{T_{WF}}$) $D_4^{T_{i,1}}(v_i)$, $i = 1,\dots,4$

(C^2:M2$_{T_{WF}}$) $\bigcup_{e:=\langle u,v\rangle \in \mathcal{E}_T} E_3^{T_e}(e) \setminus (D_4^{T_e}(u) \cup D_4^{T_e}(v))$, with $e \in T_e$

(C^2:M3$_{T_{WF}}$) $\{\xi^{T_{\alpha,1}}_{i,j,k,l} \in E^{T_{\alpha,1}}_1(\langle v_{F_\alpha}, v_T\rangle)\}$, $\alpha = 1,\dots,4$, with $0 \leq l \leq 2$

(C^2:M4$_{T_{WF}}$) $D^{T_{1,1}}_3(v_T)$,

Proof:
First, it is shown that $\mathcal{M}_{T_{WF},2}$ is a stable minimal determining set for the space $\mathcal{S}^{2,4,3,7}_9(T_{WF})$ of supersplines. Therefore, the B-coefficients c_ξ of a spline $s \in \mathcal{S}^{2,4,3,7}_9(T_{WF})$ are set to arbitrary values for each domain point $\xi \in \mathcal{M}_{T_{WF},2}$. Then it is shown that all other B-coefficients of s are uniquely and stably determined.

The undetermined B-coefficients of s associated with the domain points in the four balls $D_4(v_i)$, $i = 1,\dots,4$, can be uniquely and stably computed from the B-coefficients corresponding to the domain points in the set (C^2:M1$_{T_{WF}}$) by the C^4 smoothness conditions at the vertices of T.

Now, the undetermined B-coefficients of s corresponding to domain points in the tubes $E_3(e)$, $e \in \mathcal{E}_T$, are considered. Together with the already determined B-coefficients and those associated with the domain points in the set (C^2:M2$_{T_{WF}}$), all B-coefficients associated with the domain points in $E_3(e)$, $e \in \mathcal{E}_T$, can be uniquely and stably determined by the C^3 smoothness conditions at the six edges of the tetrahedron T.

Next, the B-coefficients of s restricted to the faces of T are considered. Due to the Worsey-Farin split of T, each face F_i, $i = 1,\dots,4$, is refined with a Clough-Tocher split and the three subtriangles forming F_i are degenerated faces of the subtetrahedra $T_{i,j}$, $j = 1,2,3$, since they lie in one common plane. Therefore, the B-coefficients of $s|_{F_i}$ can be regarded as those of a bivariate spline $g_{i,0}$ defined on the face F_i. The spline $g_{i,0}$ has polynomial degree nine, C^3 smoothness at the edges, C^4 smoothness at the vertices, and C^7 smoothness at the split point v_{F_i} in the interior of F_i. Note that the B-coefficients of $g_{i,0}$ corresponding to the domain points within a distance of four from the vertices, and three from the edges of F_i are already uniquely determined. Together with the B-coefficients associated with the domain points in (C^2:M3$_{T_{WF}}$) lying on the face F_i, those with $l = 0$, all B-coefficients corresponding to a minimal determining set for a bivariate Clough-Tocher macro-element of degree nine, smoothness three, supersmoothness four at the vertices, and supersmoothness seven at the split point are determined. Therefore, following Theorem 6.6, all B-coefficients of $g_{i,0}$ are uniquely and stably determined, and therefore, all B-coefficients of $s|_{F_i}$ (see Example 6.12).

Next, the B-coefficients of s corresponding to the domain points in the shells $R_8(v_T)$ and $R_7(v_T)$ are considered. For each face F_i, $i = 1,\ldots,4$, let F_{i,ν_1} be the triangle formed by the domain points in the shells $R^{T_{i,1}}_{9-\nu_1}(v_T)$, $R^{T_{i,2}}_{9-\nu_1}(v_T)$, and $R^{T_{i,3}}_{9-\nu_1}(v_T)$, for $\nu_1 = 1,2$. This triangle is also split with a Clough-Tocher split due to the Worsey-Farin split of T. Again, the B-coefficients of $s|_{F_{i,\nu_1}}$ can be regarded as those of a bivariate spline g_{ν_1,F_i} of degree $9-\nu_1$ which has $C^{3-\nu_1}$ smoothness at the edges, $C^{4-\nu_1}$ smoothness at the vertices and C^7 smoothness at the split point. The B-coefficients of $s|_{F_{i,\nu_1}}$ corresponding to the domain points within a distance of $4-\nu_1$ from the vertices and $3-\nu_1$ from the edges of F_{i,ν_1} are already uniquely and stably determined. Thus, these B-coefficients are also known for the bivariate spline g_{ν_1,F_i}. Together with the B-coefficients associated with the domain points in (C^2:M3$_{T_{WF}}$) lying in F_{i,ν_1}, those with $l = \nu_1$, all B-coefficients corresponding to a minimal determining set, as constructed in Theorem 6.6, for the spline g_{ν_1,F_i} are known. Therefore, all other B-coefficients of g_{ν_1,F_i} can be uniquely and stably determined in the way described in the proof of Theorem 6.6. For the values p and q in Theorem 6.6, $p = \nu_1$ and $q = \lfloor \frac{\nu_1}{2} \rfloor$ holds (see Examples 6.9 and 6.10). So far, all B-coefficients of s associated with the domain points in the shells $R_9(v_T)$, $R_8(v_T)$, and $R_7(v_T)$ are uniquely and stably determined.

Now, the remaining undetermined B-coefficients of s are examined. These B-coefficients are associated with domain points in the ball $D_7(v_T)$. Due to the fact that s has C^7 supersmoothness at v_T, the B-coefficients associated with the domain points in $D_7(v_T)$ can be considered as those of a trivariate polynomial p of degree seven defined on the tetrahedron $\widetilde{T}$, bounded by the domain points in $R_7(v_T)$. Note that the Worsey-Farin split of T induces a Worsey-Farin split of $\widetilde{T}$. Since the B-coefficients of s corresponding to the domain points within a distance of two from the faces, within a distance of three from the edges, and within a distance of four from the vertices of T are already determined, the B-coefficients of p corresponding to the domain points on the faces, within a distance of one from the edges, and within a distance of two from the vertices of $\widetilde{T}$ are also determined. Now, the B-coefficient associated with the domain points in (C^2:M4$_{T_{WF}}$) uniquely and stably determined the derivatives of p at v_T up to order three. Thus, following Lemma 2.42 for $n = 4$, all remaining undetermined B-coefficients of p can be uniquely and stably determined by these derivatives. Consequently, all B-coefficients of s are uniquely and stably determined.

In order to compute the dimension of the spline space $\mathcal{S}_9^{2,4,3,7}(T_{WF})$, it suffices consider the cardinality of the minimal determining set $\mathcal{M}_{T_{WF},2}$. The set (C^2:M1$_{T_{WF}}$) contains a total number of

$$4\binom{4+3}{3} = 140$$

domain points. The set (C^2:M2$_{T_{WF}}$) consists of

$$12\binom{3+2}{3} = 120$$

domain points. The cardinality of the set (C^2:M3$_{T_{WF}}$) is equal to

$$12\binom{3}{2} = 36.$$

Finally, the set (C^2:M4$_{T_{WF}}$) contains

$$\binom{3+3}{3} = 20$$

domain points. Thus, the dimension of $\mathcal{S}_9^{2,4,3,7}(T_{WF})$ is equal to 316. □

Remark 8.8:
The macro-elements examined in this example are equal to those considered by Lai and Schumaker [62] in section 18.8. They differ from those constructed by Alfeld and Schumaker [9], which were also presented by Lai and Schumaker [62] in chapter 18.9, since additional smoothness conditions in the shells $R_9(v_T)$ and $R_8(v_T)$ are used there. This leads to a smaller dimension but also to a more complex superspline space.

8.3.3 C^3 macro-elements over the Worsey-Farin split

For the C^3 macro-elements the superspline space $\mathcal{S}_d^{r,\rho,\mu,\eta}(T_{WF})$ reduces to the space $\mathcal{S}_{13}^{3,6,4,11}(T_{WF})$, by the fifth row of Table 8.1 and $m = 0$. As shown in Theorem 8.3, the dimension of the space is equal to 728 and the corresponding minimal determining set $\mathcal{M}_{T_{WF},3}$ is the union of the following sets of domain points:

(C^3:M1$_{T_{WF}}$) $D_6^{T_{i,1}}(v_i)$, $i = 1,\ldots,4$

(C^3:M2$_{T_{WF}}$) $\bigcup_{e:=\langle u,v\rangle\in\mathcal{E}_T} E_4^{T_e}(e) \setminus (D_6^{T_e}(u) \cup D_6^{T_e}(v))$, with $e \in T_e$

(C^3:M3$_{T_{WF}}$) $\{\xi_{i,j,k,l}^{T_{\alpha,1}} \in E_2^{T_{\alpha,1}}(\langle v_{F_\alpha}, v_T\rangle)\}$, $\alpha = 1,\dots,4$, with $0 \le l \le 2$

(C^3:M4$_{T_{WF}}$) $\{\xi_{i,j,k,3}^{T_{\alpha,1}} \in E_4^{T_{\alpha,1}}(\langle v_{F_\alpha}, v_T\rangle)\}$, $\alpha = 1,\dots,4$

(C^3:M5$_{T_{WF}}$) $D_3^{T_{1,1}}(v_T)$,

Proof:
First, it is shown that the set $\mathcal{M}_{T_{WF},3}$ is a minimal determining set for the superspline space $\mathcal{S}_{13}^{3,6,4,11}(T_{WF})$. Therefore, all B-coefficients of a spline $s \in \mathcal{S}_{13}^{3,6,4,11}(T_{WF})$ corresponding to the domain points in $\mathcal{M}_{T_{WF},3}$ are set to arbitrary values. Then it is shown that all other B-coefficients are uniquely and stably determined.

The undetermined B-coefficients of s associated with domain points in the balls $D_6(v_i)$, $i = 1,\dots,4$, can be uniquely determined by those B-coefficients of s corresponding to the domain points in (C^3:M1$_{T_{WF}}$) using the C^6 supersmoothness conditions at the vertices of T. The computation of these B-coefficients is stable, since they are directly determined from smoothness conditions.

Next, the remaining undetermined B-coefficients of s corresponding to domain points in the tubes $E_4(e)$, $e \in \mathcal{E}_T$, are examined. Using the C^4 supersmoothness conditions at the edges of T, these B-coefficients can be uniquely and stably determined from the already determined B-coefficients and those associated with the domain points in (C^3:M2$_{T_{WF}}$).

Now, the B-coefficients of the spline s restricted to the four faces of T are considered. Due to the Worsey-Farin split of T, each face F_i, $i = 1,\dots,4$, is refined with a Clough-Tocher split. Thereby, the three subtriangles forming a face F_i are degenerated faces of the three subtetrahedra $T_{i,j}$, $j = 1,2,3$, which lie in a common plane. Then the spline $s|_{F_i}$ can be considered as a bivariate spline $g_{i,0}$ defined on F_i with C^4 smoothness at the edges, C^6 smoothness at the vertices and C^{11} smoothness at the split point of F_i. Moreover, $g_{i,0}$ has degree thirteen. Note that the B-coefficients of $g_{i,0}$ corresponding to the domain points within a distance of four from the edges of F_i and within a distance of six from the vertices of F_i are already uniquely and stably determined. Thus, together with the B-coefficients associated with the domain points in (C^3:M3$_{T_{WF}}$) with $l = 0$, those on the faces F_i, $i = 1,\dots,4$, all B-coefficients of a minimal determining set, as constructed in Theorem 6.6, for the bivariate spline $g_{i,0}$ are uniquely and stably determined (see Example 6.16). Therefore, all B-coefficients of s associated with the domain points on the faces of T can be uniquely and stably computed.

Next, the remaining undetermined B-coefficients of s restricted to the shells $R_{12}(v_T)$ and $R_{11}(v_T)$ are considered. Therefore, for $i = 1,\ldots,4$, let $F_{i,1}$ be the triangle formed by the domain points $R_{12}^{T_{i,1}}(v_T) \cup R_{12}^{T_{i,2}}(v_T) \cup R_{12}^{T_{i,3}}(v_T)$, and $F_{i,2}$ the triangle formed by the domain points $R_{11}^{T_{i,1}}(v_T) \cup R_{11}^{T_{i,2}}(v_T) \cup R_{11}^{T_{i,3}}(v_T)$. Then the B-coefficients of $s|_{F_{i,\nu_1}}$, $\nu_1 = 1,2$, can be regarded as those of a bivariate spline g_{ν_1,F_i} of degree $13 - \nu_1$ which has $C^{4-\nu_1}$ smoothness at the edges, $C^{6-\nu_1}$ smoothness at the vertices and C^{11} smoothness at the split point of F_{i,ν_1}, which is subjected to a Clough-Tocher split due to the Worsey-Farin split of T. For the bivariate spline g_{ν_1,F_i}, $\nu_1 = 1,2$, the B-coefficients corresponding to the domain points within a distance of $4 - \nu_1$ from the edges of F_{i,ν_1}, and within a distance of $6 - \nu_1$ from the vertices of F_{i,ν_1} are already uniquely and stably determined. Consequently, together with the B-coefficients associated with the domain points in (C^3:M3$_{TWF}$) with a distance of ν_1 from the face F_i, those with $l = \nu_1$, all B-coefficients associated with the domain points of a minimal determining set for the bivariate spline g_{ν_1,F_i} are known (see Examples 6.11 and 6.13). Thus, the remaining undetermined B-coefficients of s corresponding to domain points in the shells $R_{12}(v_T)$ and $R_{11}(v_T)$ can be uniquely and stably determined in the way described in the proof of Theorem 6.6.

Now, the B-coefficients of s corresponding to the domain points in the shell $R_{10}(v_T)$ are examined more closely. Thus, for $i = 1,\ldots,4$, let $F_{i,3}$ be the triangle formed by the domain points in the shells $R_{10}^{T_{i,1}}(v_T)$, $R_{10}^{T_{i,2}}(v_T)$, and $R_{10}^{T_{i,3}}(v_T)$. Then, the trivariate spline s can be regarded as a bivariate spline g_{3,F_i} of degree ten, which has C^1 smoothness at the edges, C^3 smoothness at the vertices, and C^{10} smoothness at the split point of the triangle $F_{i,3}$, which is subjected to a Clough-Tocher split due to the Worsey-Farin split of the tetrahedron T. For the spline g_{3,F_i}, the B-coefficients corresponding to the domain points within a distance of one from the edges of $F_{i,3}$ and within a distance of three from the vertices of $F_{i,3}$ are already uniquely determined. Moreover, from the B-coefficients corresponding to the domain points in (C^3:M4$_{TWF}$), the derivatives of g_{3,F_i} at the split point in $F_{i,3}$ up to order four can be uniquely and stably computed. Therefore, following Lemma 2.42 for the case $n = 3$, all remaining undetermined B-coefficients of g_{3,F_i} can be uniquely and stably determined from these derivatives. Thus, all B-coefficients of s associated with the domain points in the shells $R_{13}(v_T)$,

$R_{12}(v_T)$, $R_{11}(v_T)$, and $R_{10}(v_T)$ are uniquely and stably determined so far.

Finally, the last undetermined B-coefficients of s are considered. These are associated with domain points in the ball $D_{11}(v_T)$. Due to the fact that s has C^{11} supersmoothness at the split point v_T of T, the B-coefficients of s restricted to the domain points in the ball $D_{11}(v_T)$ can be regarded as those of a trivariate polynomial p of degree eleven defined on the tetrahedron $\widetilde{T}$, which is bounded by the domain points in $R_{11}(v_T)$. Since T is subjected to a Worsey-Farin split, the tetrahedron $\widetilde{T}$ is also refined with the same split. Some of the B-coefficients of s corresponding to domain points in the ball $D_{11}(v_T)$ are already uniquely determined. Thus, these B-coefficients are also already known for the polynomial p. In particular, these are the B-coefficients corresponding to the domain points within a distance of one from the faces of $\widetilde{T}$, within a distance of two from the edges of $\widetilde{T}$, and within a distance of four from the vertices of $\widetilde{T}$. Now, the derivatives of the polynomial p at the split point v_T of $\widetilde{T}$ up to order three can be uniquely and stably computed from the B-coefficients associated with the domain points in (C^3:M5$_{T_{WF}}$). These derivatives, following Lemma 2.42 for the case $n = 4$, uniquely and stably determine the remaining undetermined B-coefficients of p. Thus, the remaining undetermined B-coefficients of the spline s can be uniquely and stably determined by the de Casteljau algorithm.

Next, the dimension of the superspline space $\mathcal{S}_{13}^{3,6,4,11}(T_{WF})$ is examined. For this purpose, the cardinality of the minimal determining set $\mathcal{M}_{T_{WF},3}$ is considered. The set (C^3:M1$_{T_{WF}}$) consists of

$$4\binom{6+3}{3} = 336$$

domain points. The set (C^3:M2$_{T_{WF}}$) contains a total of

$$12\binom{4+2}{3} = 240$$

domain points. The cardinality of the set (C^3:M3$_{T_{WF}}$) is equal to

$$12\binom{4}{2} = 72.$$

The set (C^3:M4$_{T_{WF}}$) consists of

$$4\binom{6}{2} = 60$$

domain points. The last set (C^3:M5$_{T_{WF}}$) contains a total of

$$\binom{3+3}{3} = 20$$

domain points. Therefore, the dimension of the spline space $\mathcal{S}_{13}^{3,6,4,11}(T_{WF})$ is equal to 728. □

8.3.4 C^4 macro-elements over the Worsey-Farin split

For the case of C^4 macro-elements, the superspline space $\mathcal{S}_d^{r,\rho,\mu,\eta}(T_{WF})$ reduces to the space $\mathcal{S}_{19}^{4,9,6,16}(T_{WF})$, by the second row of Table 8.1 and $m = 1$. From Theorem 8.3, it can be seen that the dimension of the spline space is equal to 2065 and the corresponding minimal determining set $\mathcal{M}_{T_{WF},4}$ is the union of the following sets of domain points:

(C^4:M1$_{T_{WF}}$) $D_9^{T_{i,1}}(v_i)$, $i = 1,\ldots,4$

(C^4:M2$_{T_{WF}}$) $\bigcup_{e:=\langle u,v\rangle \in \mathcal{E}_T} E_6^{T_e}(e) \setminus (D_9^{T_e}(u) \cup D_9^{T_e}(v))$, with $e \in T_e$

(C^4:M3$_{T_{WF}}$) $\{\xi_{i,j,k,l}^{T_{\alpha,1}} \in E_4^{T_{\alpha,1}}(\langle v_{F_\alpha}, v_T\rangle)\}$, $\alpha = 1,\ldots,4$, with $0 \le l \le 3$

(C^4:M4$_{T_{WF}}$) $\{\xi_{i,j,k,4}^{T_{\alpha,1}} \in E_6^{T_{\alpha,1}}(\langle v_{F_\alpha}, v_T\rangle)\}$, $\alpha = 1,\ldots,4$

(C^4:M5$_{T_{WF}}$) $\bigcup_{F:=\langle u,w,v_T\rangle \in \mathcal{F}_{I,T}} \{\xi_{2+j,3-j,7,0}^{T_F} : j = 1,\ldots,7\}$,

with $T_F := \langle u, w, v_T, v_F\rangle$ for $v_F \in \mathcal{V}_{T,F}$

(C^4:M6$_{T_{WF}}$) $\{\xi_{i,j,k,5}^{T_{\alpha,1}} \in E_5^{T_{\alpha,1}}(\langle v_{F_\alpha}, v_T\rangle)\}$, $\alpha = 1,\ldots,4$

(C^4:M7$_{T_{WF}}$) $D_4^{T_{1,1}}(v_T)$,

Proof:
First, it is shown that the set $\mathcal{M}_{T_{WF},4}$ is a stable minimal determining set for the space $\mathcal{S}_{19}^{4,9,6,16}(T_{WF})$ of supersplines. Therefore, let $s \in \mathcal{S}_{19}^{4,9,6,16}(T_{WF})$ be a spline whose B-coefficients c_ξ are set to arbitrary values for each domain point $\xi \in \mathcal{M}_{T_{WF},4}$. Then it is shown that all other B-coefficients of s are uniquely and stably determined.

The undetermined B-coefficients of s corresponding to domain points in the balls $D_9(v_i)$, $i = 1,\ldots,4$, can be uniquely determined from the B-coefficients associated with the domain points in (C^4:M1$_{T_{WF}}$) using the C^9 supersmoothness conditions at the vertices of T. The computation of these B-coefficients is stable, since they are directly computed from smoothness conditions.

Next, the remaining undetermined B-coefficients associated with domain points in the tubes $E_6(e)$, $e \in \mathcal{E}_T$, are computed. They can be uniquely and stably determined from the already known B-coefficients and those corresponding to the domain point in the set (C^4:M2$_{T_{WF}}$) by using the C^6 supersmoothness conditions at the edges of the tetrahedron T.

Now, the B-coefficients of s restricted to the faces of the tetrahedron T are considered. Since T is refined with a Worsey-Farin split, each face F_i, $i = 1,\ldots,4$, is subjected to a Clough-Tocher split. Then for the face F_i, the three subtriangles are degenerated faces of the three subtetrahedra $T_{i,j}$, $j = 1,2,3$, of T_A. Therefore, the B-coefficients of s associated with the domain points on the faces of T can be regarded as those of bivariate splines $g_{i,0}$ defined on the face F_i, $i = 1,\ldots,4$. The bivariate spline $g_{i,0}$ has C^6 smoothness at the edges of F_i, C^9 smoothness at the vertices of F_i, and C^{16} smoothness at the split point v_{F_i} in the interior of F_i. Moreover, the polynomial degree of $g_{i,0}$ is equal to 19. Since some of the B-coefficients of s restricted to the faces of T are already uniquely determined, the same B-coefficients are also already known for the bivariate spline $g_{i,0}$. Thus, the B-coefficients of $g_{i,0}$ corresponding to the domain points within a distance of nine from the vertices of F_i and corresponding to those within a distance of six from the edges of F_i are already determined. Then, together with the B-coefficients associated with the domain points in the set (C^4:M3$_{T_{WF}}$) with $l = 0$ corresponding to the face F_i, all B-coefficients corresponding to a minimal determining set for a bivariate Clough-Tocher macro-element of degree 19, with C^6 smoothness, C^9 supersmoothness at the vertices, C^{16} supersmoothness at the split point are determined (see Example 6.23). Then, following Theorem 6.6, the remaining undetermined B-coefficients of $g_{i,0}$ are uniquely and stably determined.

Next, the B-coefficients of s associated with the domain points in the shells $R_{18}(v_T)$, $R_{17}(v_T)$, and $R_{16}(v_T)$ are considered more closely. Therefore, let F_{i,ν_1} be the triangle formed by the domain points in the three shells $R^{T_{i,1}}_{19-\nu_1}(v_T)$, $R^{T_{i,2}}_{19-\nu_1}(v_T)$, and $R^{T_{i,3}}_{19-\nu_1}(v_T)$, for $\nu_1 = 1,2,3$, and $i = 1,\ldots,4$. Then the triangle F_{i,ν_1} is also subjected to a Clough-Tocher split, since the tetrahedron T is refined with a Worsey-Farin split. Now again, the B-coefficients of $s|_{F_{i,\nu_1}}$ can be considered as those of a bivariate spline g_{ν_1,F_i} of degree $19-\nu_1$, which has $C^{6-\nu_1}$ smoothness at the edges, $C^{9-\nu_1}$ smoothness at the vertices and C^{16} smoothness at the split point of F_{i,ν_1}. Note that the B-coefficients of g_{ν_1,F_i} corresponding to the domain points within a distance of $6-\nu_1$ from the edges and $9-\nu_1$ from the vertices of F_{i,ν_1} are already uniquely determined. Then, together with the B-coefficients associated with the domain points in the set (C^4:M$3_{T_{WF}}$) with $l=\nu_1$ corresponding to the face F_i, all B-coefficients associated with a minimal determining set for a bivariate Clough-Tocher macro-element of degree $19-\nu_1$, with $C^{6-\nu_1}$ smoothness, $C^{9-\nu_1}$ supersmoothness at the vertices, and C^{16} supersmoothness at the split point are determined. See Example 6.20 for $\nu_1=1$, Example 6.17 for $\nu_1=2$, and Example 6.14 for $\nu_1=3$. Thus, following Theorem 6.6 with $p=\nu_1$ and $q=\lfloor\frac{\nu_1+1}{2}\rfloor$, all B-coefficients of the bivariate spline g_{ν_1,F_i} are uniquely and stably determined. Then, all B-coefficients of s associated with the domain points in the shells $R_{19}(v_T),\ldots,R_{16}(v_T)$ are uniquely and stably determined.

In the following, the B-coefficients of the spline s corresponding to the domain points in the shell $R_{15}(v_T)$ are investigated. To this end, let $F_{i,4}$, $i=1,\ldots,4$, be the four triangles formed by the domain points in the shells $R_{15}^{T_{i,1}}(v_T)$, $R_{15}^{T_{i,2}}(v_T)$, and $R_{15}^{T_{i,3}}(v_T)$, which are subjected to a Clough-Tocher split, due to the Worsey-Farin split of T. Then the B-coefficients of s restricted to the triangle $F_{i,4}$ can be regarded as the B-coefficients of a bivariate spline g_{4,F_i} of degree 15, with C^2 smoothness at the edges, C^5 smoothness at the vertices, and C^{15} smoothness at the split point of $F_{i,4}$. Due to the C^{15} supersmoothness at the split point of $F_{i,4}$, the bivariate spline g_{4,F_i} reduces to a bivariate polynomial. Now, considering g_{4,F_i}, it can be seen that the B-coefficients corresponding to the domain points within a distance of five from the vertices of $F_{i,4}$ and to those within a distance of two from the edges of $F_{i,4}$ are already uniquely determined. Then, from the B-coefficients associated with the domain points in (C^4:M$4_{T_{WF}}$) in $F_{i,4}$, the derivatives of g_{4,F_i} at the split point in $F_{i,4}$ up to order six can be uniquely and stably computed. Following Lemma 2.42 for the case $n=3$, the remaining undetermined B-coefficients of g_{4,F_i} can be uniquely and stably computed from these derivatives. See Figure 8.2 (left) for one of the triangles $F_{i,4}$, $i=1,\ldots,4$.

So far, the B-coefficients of s corresponding to the domain points within a distance of four from the faces of T, within a distance of six from the edges of T, and within a distance of nine from the vertices of T are uniquely and stably determined.

Now, the B-coefficients of s associated with the domain points in the shell $R_{14}(v_T)$ are examined. Therefore, first the remaining undetermined B-coefficients of s corresponding to the domain points with a distance of seven from the edges of T are determined. The still undetermined B-coefficients associated with these domain points correspond to domain points in the shell $R_{14}(v_T)$ with a distance of two from the faces in $\mathcal{F}_{T,I}$, to domain points in the shell $R_{13}(v_T)$ with a distance of one from the faces in $\mathcal{F}_{T,I}$, and to domain points in the shell $R_{12}(v_T)$ lying on the faces $\mathcal{F}_{T,I}$. The corresponding domain points are marked with ✖ in Figure 8.2 (right). Now, considering the C^4 smoothness conditions at the faces in $\mathcal{F}_{T,I}$ which cover the undetermined B-coefficients, it can be seen that each quadruple of C^4, and the corresponding C^3, C^2, and C^1 conditions connect exactly five undetermined B-coefficients. Two of these B-coefficients are associated with domain points in the shell $R_{14}(v_T)$, two correspond to domain points in the shell $R_{13}(v_T)$ and one is associated with a domain point in the shell $R_{12}(v_T)$ lying on one of the faces in $\mathcal{F}_{T,I}$. In fact, there are two B-coefficients associated with each domain point on the faces in $\mathcal{F}_{T,I}$ in the shell $R_{12}(v_T)$, but due to the C^0 smoothness, these have to be equal and can be considered as one B-coefficient. Then, the set (C^4:M5$_{T_{WF}}$) contains the domain points in $R_{12}(v_T)$ on the faces in $\mathcal{F}_{T,I}$, whose B-coefficients are not already determined. Thus, there are only four undetermined B-coefficients left for each quadruple of C^4, C^3, C^2, and C^1 smoothness conditions at the faces in $\mathcal{F}_{T,I}$. Following Lemma 2.33 for the case $j = 1$, the four equations from the smoothness conditions uniquely and stably determine the four undetermined B-coefficients. In this way, the remaining undetermined B-coefficients of s corresponding to the domain points with a distance of seven from the edges of T can be uniquely and stably determined.

Next, the remaining undetermined B-coefficients of s restricted to the domain points in the shell $R_{14}(v_T)$ are computed. Therefore, let $F_{i,5}$, $i = 1,\dots,4$, be the triangle formed by the domain points in the shells $R_{14}^{T_{i,1}}$, $R_{14}^{T_{i,2}}$, and $R_{14}^{T_{i,3}}$. Again, due to the Worsey-Farin split of T, the triangles $F_{i,5}$ are subjected to a Clough-Tocher split. Then the spline s can be regarded as a bivariate spline $g_{5,F_{i,5}}$ of degree 14 defined on $F_{i,5}$. Due to the C^{15} supersmoothness of $g_{5,F_{i,5}}$ at the split point of $F_{i,5}$, which is a higher supersmoothness than the degree of the polynomials of the bivariate spline, $g_{5,F_{i,5}}$ reduces to a bivariate polynomial. Considering the bivariate polynomial $g_{5,F_{i,5}}$, it can be seen that the B-coefficients corresponding to the domain points within a distance of four from the vertices of $F_{i,5}$, and those corresponding to the domain points within a distance of two from the edges of $F_{i,5}$ are already uniquely determined. Now, from the B-coefficients associated with the domain points in (C^4:M6$_{T_{WF}}$) lying in $F_{i,5}$, the derivatives of $g_{5,F_{i,5}}$ at the split point in $F_{i,5}$ up to the fifth order can be uniquely and stably computed. Following Lemma 2.42 for the case $n = 3$, these derivative uniquely and stably determine the remaining undetermined B-coefficients of $g_{5,F_{i,5}}$. See Figure 8.2 (right) for one of the triangles $F_{i,5}$, $i = 1, \ldots, 4$.

At this point, all B-coefficients of the spline s corresponding to the domain points within a distance of five from the faces of T, within a distance of six from the edges of T, and within a distance of nine from the vertices of T are uniquely and stably determined.

Now, the last undetermined B-coefficients of s, that are associated with some of the domain points in the ball $D_{16}(v_T)$, are considered. Due to the C^{16} supersmoothness of s at the split point v_T, the B-coefficients of s restricted to the tetrahedron $\widetilde{T}$, formed by the domain points in the ball $D_{16}(v_T)$, can be regarded as those of a trivariate polynomial p of degree 16 defined on the tetrahedron $\widetilde{T}$. Note that the tetrahedron $\widetilde{T}$ is also subjected to a Worsey-Farin split, with split point v_T. Since some of the B-coefficients of $s|_{D_{16}(v_T)}$ are already uniquely determined, these B-coefficients are also known for the polynomial p. Therefore, the B-coefficients of p corresponding to the domain points within a distance of two from the faces of $\widetilde{T}$, corresponding to the domain points within a distance of four from the edges of $\widetilde{T}$, and corresponding to the domain points within a distance of six from the vertices of $\widetilde{T}$ are already uniquely determined. Then, the derivatives of the polynomial p at v_T up to order four can be uniquely and stably computed from the B-coefficient associated with the domain points in (C^4:M7$_{T_{WF}}$). Now, following Lemma 2.42 for the case $n = 4$, the remaining B-coefficients of p are uniquely and stably determined by these derivatives. Thus, all B-coefficients of s are uniquely and stably determined.

Finally, the dimension of the spline space $\mathcal{S}_{19}^{4,9,6,16}(T_{WF})$ is examined. Since the set $\mathcal{M}_{T_{WF},4}$ is a minimal determining set for this spline space, the dimension $\dim \mathcal{S}_{19}^{4,9,6,16}(T_{WF})$ is equal to the cardinality $\#\mathcal{M}_{T_{WF},4}$ of the minimal determining set. Thus, the cardinalities of the sets (C^4:M1$_{T_{WF}}$) - (C^4:M7$_{T_{WF}}$) are considered.

The set (C^4:M1$_{T_{WF}}$) contains a total number of

$$4\binom{9+3}{3} = 880$$

domain points. The set (C^4:M2$_{T_{WF}}$) consists of

$$12\binom{6+2}{3} = 672$$

domain points. The set (C^4:M3$_{T_{WF}}$) contains

$$16\binom{6}{2} = 240$$

domain points. The cardinality of the set (C^4:M4$_{T_{WF}}$) is equal to

$$4\binom{6+2}{2} = 112.$$

The set (C^4:M5$_{T_{WF}}$) consists of

$$6 \cdot 7 = 42$$

domain points. The set (C^4:M6$_{T_{WF}}$) contains a total number of

$$4\binom{7}{2} = 84$$

domain points. The cardinality of the last set (C^4:M7$_{T_{WF}}$) is equal to

$$\binom{4+3}{3} = 35.$$

Then, the dimension of the superspline space $\mathcal{S}_{19}^{4,9,6,16}(T_{WF})$ is equal to 2065. □

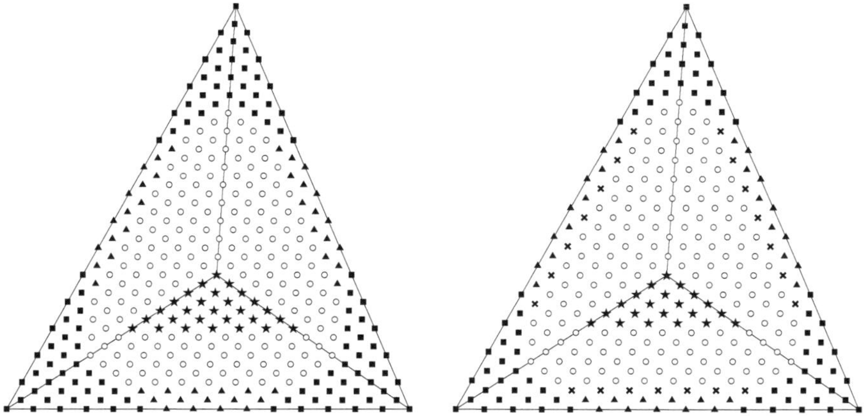

Figure 8.2: Triangles $F_{i,4}$ and $F_{i,5}$ in the shells $R_{15}(v_T)$ and $R_{14}(v_T)$, respectively. Domain points whose B-coefficients are determined from B-coefficients associated with domain points in (C^4:M1$_{T_{WF}}$) are marked with ■, those from B-coefficients associated with domain points in (C^4:M2$_{T_{WF}}$) with ▲, and those from B-coefficients associated with domain points in (C^4:M5$_{T_{WF}}$) and smoothness conditions using Lemma 2.33 for $j = 1$ with ✖. Domain points contained in (C^4:M4$_{T_{WF}}$) and (C^4:M6$_{T_{WF}}$) are marked with ★.

8.3.5 C^5 macro-elements over the Worsey-Farin split

The macro-elements are in the superspline space $\mathcal{S}_d^{r,\rho,\mu,\eta}(T_{WF})$. For the case $r = 5$, m is equal to one and following Table 8.1 (third row) the superspline space reduces to the space $\mathcal{S}_{21}^{5,10,7,17}(T_{WF})$. Following Theorem 8.3, we get that the dimension of the space is equal to 2884 and the corresponding minimal determining set $\mathcal{M}_{T_{WF},5}$ is the union of the following sets of domain points:

(C^5:M1$_{T_{WF}}$) $D_{10}^{T_{i,1}}(v_i)$, $i = 1, \dots, 4$

(C^5:M2$_{T_{WF}}$) $\bigcup\limits_{e:=\langle u,v\rangle \in \mathcal{E}_T} E_7^{T_e}(e) \setminus (D_{10}^{T_e}(u) \cup D_{10}^{T_e}(v))$, with $e \in T_e$

(C^5:M3$_{T_{WF}}$) $\{\xi_{i,j,k,l}^{T_{\alpha,1}} \in E_5^{T_{\alpha,1}}(\langle v_{F_\alpha}, v_T\rangle)\}$, $\alpha = 1, \dots, 4$, with $0 \le l \le 4$

$(C^5{:}\mathrm{M4}_{T_{WF}})$ $\{\xi^{T_{\alpha,1}}_{i,j,k,5} \in E^{T_{\alpha,1}}_7(\langle v_{F_\alpha}, v_T\rangle)\}$, $\alpha = 1,\ldots,4$

$(C^5{:}\mathrm{M5}_{T_{WF}})$ $\{\xi^{T_{\alpha,1}}_{i,j,k,6} \in E^{T_{\alpha,1}}_6(\langle v_{F_\alpha}, v_T\rangle)\}$, $\alpha = 1,\ldots,4$

$(C^5{:}\mathrm{M6}_{T_{WF}})$ $D^{T_{1,1}}_5(v_T)$,

Proof:
We first show, that the set $\mathcal{M}_{T_{WF},5}$ is a minimal determining set for the space $\mathcal{S}^{5,10,7,17}_{21}(T_{WF})$ of supersplines. Therefore, we set the B-coefficients c_ξ of a spline $s \in \mathcal{S}^{5,10,7,17}_{21}(T_{WF})$ to arbitrary values for each domain point $\xi \in \mathcal{M}_{T_{WF},5}$. Then we show that all other other B-coefficients of s can be uniquely and stably determined from those.

We can uniquely compute all undetermined B-coefficients of s associated with the domain points in the balls $D_{10}(v_i)$, $i = 1,\ldots,4$, from the B-coefficients corresponding to the domain points in $(C^5{:}\mathrm{M1}_{T_{WF}})$ using the C^{10} supersmoothness conditions at these vertices of T. Since all B-coefficients are directly determined from smoothness conditions, the computation of them is stable.

Next, we consider the remaining undetermined B-coefficients associated with domain points in the six tubes $E_7(e)$, $e \in \mathcal{E}_T$. These B-coefficients can be uniquely and stably determined from the B-coefficients corresponding to the domain points in the set $(C^5{:}\mathrm{M2}_{T_{WF}})$ using the C^7 supersmoothness of s at the edges of T.

Now, the B-coefficients of the spline s restricted to the faces of the tetrahedron T are examined. Note that each face F_i, $i = 1,\ldots,4$, of T is refined with a Clough-Tocher split, where the three subtriangles for each face F_i are just degenerated faces of the three subtetrahedra $T_{i,j}$, $j = 1,2,3$. Then, restricted to the face F_i, the B-coefficients of the trivariate spline s can be considered as those of a bivariate one, which has the same degree of polynomials and the same degrees of smoothness. Thus, let $g_{i,0}$ be the bivariate spline defined on F_i, which has a polynomial degree of 21, C^7 smoothness at the edges of F_i, C^{10} smoothness at the vertices of F_i, and C^{17} smoothness at the split point in F_i. Considering $g_{i,0}$, it can be seen that the B-coefficients corresponding to the domain points within a distance of ten from the vertices, and within a distance of seven from the edges of F_i are already uniquely and stably determined. Now, the set $(C^5{:}\mathrm{M3}_{T_{WF}})$ contains domain points lying in F_i, those with $l = 0$. Thus, considering $g_{i,0}$, the B-coefficients corresponding to the domain points in one subtriangle within a distance of five from the split point in F_i are also already known. Therefore, all B-coefficients associated with the domain points corresponding to a minimal determining set of a bivariate C^7 macro-element of degree 21 based on the Clough-Tocher split, with C^{10} supersmoothness at the vertices, and C^{17} supersmoothness at the split point, are uniquely and stably determined. Then, following Theorem 6.6, all remaining undetermined B-coefficients can be uniquely and stably computed (see Example 6.26).

In the following, the B-coefficients of s associated with the domain points in the shells $R_{21-\nu_1}(v_T)$, for $\nu_1 = 1,\ldots,4$, are considered. Therefore, let F_{i,ν_1}, $i = 1,\ldots,4$, be the triangles formed by the domain points in the shells $R^{T_{i,1}}_{21-\nu_1}(v_T)$, $R^{T_{i,2}}_{21-\nu_1}(v_T)$, and $R^{T_{i,3}}_{21-\nu_1}(v_T)$, for $\nu_1 = 1,\ldots,4$, in other words, the domain points within a distance of ν_1 from the face F_i. Note that, due to the Worsey-Farin split of T, the triangles F_{i,ν_1}, $i = 1,\ldots,4$, $\nu_1 = 1,\ldots,4$, have been subjected to a Clough-Tocher split. Again, the B-coefficients of the spline s restricted to the triangle F_{i,ν_1} can be regarded as those of a bivariate spline g_{ν_1,F_i} of degree $21-\nu_1$ that has $C^{7-\nu_1}$ smoothness at the edges, $C^{10-\nu_1}$ smoothness at the vertices, and C^{17} smoothness at the split point in the interior of F_{i,ν_1}. Since some of the B-coefficients of s restricted to F_{i,ν_1} are already known, the B-coefficients of g_{ν_1,F_i} corresponding to the domain points within a distance of $7-\nu_1$ from the edges of F_{i,ν_1}, and those corresponding to the domain points within a distance of $10-\nu_1$ from the vertices of F_{i,ν_1} are also already uniquely and stably determined. Then, the set $(C^5\text{:M3}_{T_{WF}})$ contains the domain points in one subtriangle of F_{i,ν_1} within a distance of five from the split point in F_{i,ν_1}. For these domain points in $(C^5\text{:M3}_{T_{WF}})$, we have $l = \nu_1$. Then again, all B-coefficients of g_{ν_1,F_i} associated with the domain points in the minimal determining set for a bivariate Clough-Tocher macro-element of degree $21-\nu_1$, with $C^{7-\nu_1}$ smoothness at the edges, $C^{10-\nu_1}$ supersmoothness at the vertices, and C^{17} supersmoothness at the split point of F_{i,ν_1}, are determined. Thus, following Theorem 6.6, all remaining B-coefficients of g_{ν_1,F_i} are uniquely and stably determined (see Examples 6.15, 6.18, 6.21, and 6.24).

So far, all B-coefficients of s associated with the domain points in the shells $R_{21}(v_T),\ldots,R_{17}(v_T)$, and those corresponding to the domain points within a distance of seven from the edges, and within a distance of ten from the vertices of T are determined.

Now, the B-coefficients of s associated with the domain points in the shell $R_{16}(v_T)$ are examined. To this end, let $F_{i,5}$ be the triangles formed by the domain points in the shells $R^{T_{i,1}}_{16}(v_T)$, $R^{T_{i,2}}_{16}(v_T)$, and $R^{T_{i,3}}_{16}(v_T)$, for $i = 1,\ldots,4$. Then, $F_{i,5}$ contains the domain points with a distance of five from the face F_i in the three subtetrahedra $T_{i,1}, T_{i,2}$, and $T_{i,3}$. Now, the B-coefficients of the trivariate spline s associated with the domain points in the triangle $F_{i,5}$ can be regarded as the B-coefficients of a bivariate spline g_{5,F_i} of degree 16, with C^5 smoothness at the edges, C^5 smoothness at the vertices, and C^{17} smoothness at the split point of $F_{i,5}$, at which $F_{i,5}$ is refined with a Clough-Tocher split. Note that the bivariate spline g_{5,F_i} is in fact just a bivariate polynomial, since the degree of supersmoothness at the split point of $F_{i,5}$ is equal to 17 and the degree of polynomials is just 16. Now, considering the B-coefficients of g_{5,F_i}, it can be seen that those corresponding to the domain points within a distance of five from the vertices of $F_{i,5}$ and within a distance of two from the edges of $F_{i,5}$ are already uniquely and stably determined. Then, from the B-coefficients associated with the domain points in (C^5:M4$_{T_{WF}}$), the derivatives of g_{5,F_i} at the split point in $F_{i,5}$ up to order seven can be uniquely and stably determined. Thus, following Lemma 2.42 for the case $n = 3$, the remaining B-coefficients of g_{5,F_i} can be uniquely and stably computed from these derivatives. See Figure 8.3 (left) for one of the triangles $F_{i,5}$, $i = 1,\dots,4$.

So far, the B-coefficients of s corresponding to the domain points within a distance of five from the faces of T, within a distance of seven from the edges of T, and within a distance of ten from the vertices of T are uniquely and stably determined.

Next, we consider the B-coefficients corresponding to the domain points with a distance of eight from the edges of T. These B-coefficients correspond to domain points in the shell $R_{15}(v_T)$ with a distance of two from the faces in $\mathcal{F}_{T,I}$, to domain points in the shell $R_{14}(v_T)$ with a distance of one from the faces in $\mathcal{F}_{T,I}$, and to domain points in the shell $R_{13}(v_T)$ lying on the faces in $\mathcal{F}_{T,I}$. The corresponding domain points in the shell $R_{15}(v_T)$ are marked with ✖ in Figure 8.3 (right). Then, considering exactly one of these undetermined B-coefficients associated with a domain point lying on a face in $\mathcal{F}_{T,I}$ in $R_{13}(v_T)$, it can be seen that the C^5 smoothness condition with its tip at this B-coefficient, and the corresponding $C^4,\dots,C^1$ smoothness conditions connect this B-coefficient with exactly four other undetermined B-coefficients corresponding to two domain points in the shells $R_{14}(v_T)$, and $R_{15}(v_T)$, respectively. Thus, we have a system of five equations and five undetermined B-coefficients. Following Lemma 2.33 for the case $j = 0$, the corresponding matrix is nonsingular and we can uniquely and stably determined the B-coefficients of s corresponding to the domain points with a distance of eight from the edges of T.

Now, we are ready to consider the B-coefficients of s associated with the domain points in the shell $R_{15}(v_T)$. Therefore, let $F_{i,6}$ be the triangle formed by the domain points in the shells $R_{15}^{T_{i,1}}$, $R_{15}^{T_{i,2}}$, and $R_{15}^{T_{i,3}}$, for $i = 1,\dots,4$. Here again, we can consider the B-coefficients of s restricted to $F_{i,6}$ as those of a bivariate spline g_{6,F_i} of degree 15 defined on $F_{i,6}$, which has C^5 smoothness at the edges and vertices of $F_{i,6}$, and C^{17} supersmoothness at the split point in $F_{i,6}$. Note that, due to the Worsey-Farin split of T, the triangles $F_{i,6}$, $i = 1,\dots,4$, are refined with a Clough-Tocher split. Now, since the degree of supersmoothness at the split point in $F_{i,6}$ is higher than the degree of polynomials of the spline g_{6,F_i}, the bivariate spline g_{6,F_i} reduces to a bivariate polynomial defined on $F_{i,6}$. Considering this polynomial, we see that the B-coefficients corresponding to the domain points within a distance of four from the vertices of $F_{i,6}$, and those corresponding to the domain points within a distance of two from the edges of $F_{i,6}$ are already uniquely and stably determined. Then, the derivatives of g_{6,F_i} at the split point in $F_{i,6}$ up to order six can be uniquely and stably computed from the B-coefficients associated with the domain points in (C^5:M5$_{T_{WF}}$) lying in $F_{i,6}$. Thus, following Lemma 2.42 for the case $n = 3$, we can uniquely and stably determine the remaining undetermined B-coefficients of g_{6,F_i} from these derivatives. See Figure 8.3 (right) for one of the triangles $F_{i,6}$, $i = 1,\dots,4$.

At this point, we have determined all B-coefficients of s corresponding to the domain points within a distance of six from the faces of T, within a distance of seven from the edges of T, and within a distance of ten from the vertices of T.

Now, we consider the last undetermined B-coefficients of s. These are associated with domain points in the ball $D_{17}(v_T)$. Due to the C^{17} supersmoothness of s at the split point v_T of T, the B-coefficients of the spline s corresponding to the domain points in $D_{17}(v_T)$ can be regarded as those of a trivariate polynomial p of degree 17 defined on the tetrahedron $\widetilde{T}$, that is bounded by the domain points in the shell $R_{17}(v_T)$. Since the tetrahedron T is refined with a Worsey-Farin split, $\widetilde{T}$ is also subjected to a Worsey-Farin split with split point v_T. Now, considering p, we see that some of the B-coefficients are already determined. Thus, the B-coefficients of p corresponding to the domain points within a distance of two from the faces of $\widetilde{T}$, the domain points within a distance of four from the edges of $\widetilde{T}$, and the domain points within a distance of six from the vertices of $\widetilde{T}$ are already uniquely and stably determined. Then, the derivatives of p at v_T up to order five can be uniquely and stably computed from the B-coefficients corresponding to the domain points in (C^5:M$6_{T_{WF}}$). Following Lemma 2.42 for the case $n = 4$, we can uniquely and stably determined the remaining undetermined B-coefficients of p from these derivatives. Therefore, all B-coefficients of s are uniquely and stably determined.

Finally, we examine the dimension of the spline space $\mathcal{S}_{21}^{5,10,7,17}(T_{WF})$. Since the set $\mathcal{M}_{T_{WF},5}$ is a minimal determining set for this space of supersplines, the dimension $\dim \mathcal{S}_{21}^{5,10,7,17}(T_{WF})$ is equal to the cardinality $\#\mathcal{M}_{T_{WF},5}$ of the minimal determining set. Thus, we need to consider the cardinalities of the sets (C^5:M$1_{T_{WF}}$) - (C^5:M$6_{T_{WF}}$). The set (C^5:M$1_{T_{WF}}$) consists of

$$4\binom{10+3}{3} = 1144$$

domain points. The set (C^5:M$2_{T_{WF}}$) contains

$$12\binom{7+2}{3} = 1008$$

domain points. The cardinality of the set (C^5:M$3_{T_{WF}}$) is equal to

$$20\binom{5+2}{2} = 420.$$

The set (C^5:M$4_{T_{WF}}$) consists of a total number of

$$4\binom{7+2}{2} = 144$$

domain points. The set (C^5:M$5_{T_{WF}}$) contains

$$4\binom{6+2}{2} = 112$$

domain points. The cardinality of the last set (C^5:M$6_{T_{WF}}$) is equal to

$$\binom{5+3}{3} = 56.$$

Thus, the dimension of the superspline space $\mathcal{S}_{21}^{5,10,7,17}(T_{WF})$ is equal to 2884. □

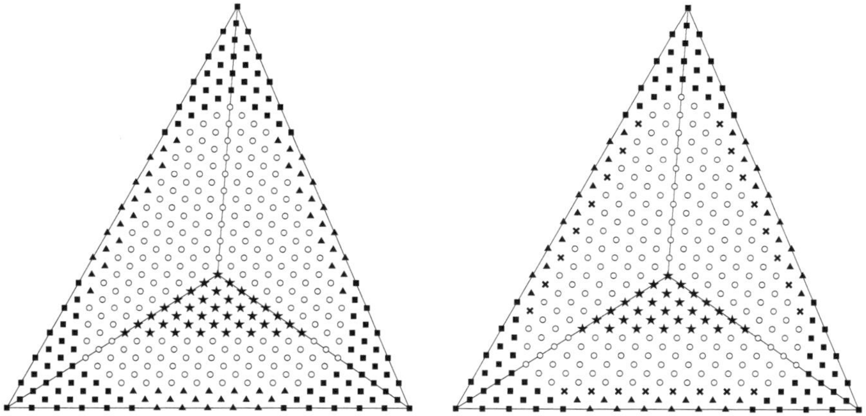

Figure 8.3: Triangles $F_{i,5}$ and $F_{i,6}$ in the shells $R_{16}(v_T)$ and $R_{15}(v_T)$, respectively. Domain points whose B-coefficients are determined from B-coefficients associated with domain points in (C^5:M1$_{T_{WF}}$) are marked with ■, those from B-coefficients associated with domain points in (C^5:M2$_{T_{WF}}$) with ▲, and those from smoothness conditions using Lemma 2.33 for $j = 0$ with ✖. Domain points contained in (C^5:M4$_{T_{WF}}$) and (C^5:M5$_{T_{WF}}$) are marked with ★.

8.3.6 C^6 macro-elements over the Worsey-Farin split

The C^6 macro-elements based on the Worsey-Farin split of a tetrahedron T are in the superspline space $\mathcal{S}_{27}^{6,13,9,22}(T_{WF})$, which we can obtain from the space $\mathcal{S}_d^{r,\rho,\mu,\eta}(T_{WF})$ with the values from the fourth row of Table 8.1 and $m = 1$. Following Theorem 8.3, the dimension of the superspline space is equal to 5830 and the corresponding minimal determining set $\mathcal{M}_{T_{WF},6}$ is the union of the following sets of domain points:

(C^6:M1$_{T_{WF}}$) $D_{13}^{T_{i,1}}(v_i)$, $i = 1,\dots,4$

(C^6:M2$_{T_{WF}}$) $\bigcup_{e:=\langle u,v\rangle \in \mathcal{E}_T} E_9^{T_e}(e) \setminus (D_{13}^{T_e}(u) \cup D_{13}^{T_e}(v))$, with $e \in T_e$

(C^6:M3$_{T_{WF}}$) $\{\xi_{i,j,k,l}^{T_{\alpha,1}} \in E_7^{T_{\alpha,1}}(\langle v_{F_\alpha}, v_T\rangle)\}$, $\alpha = 1,\dots,4$, with $0 \le l \le 5$

(C^6:M4$_{T_{WF}}$) $\{\xi_{i,j,k,6}^{T_{\alpha,1}} \in E_9^{T_{\alpha,1}}(\langle v_{F_\alpha}, v_T\rangle)\}$, $\alpha = 1,\dots,4$

(C^6:M5$_{T_{WF}}$) $\bigcup_{F:=\langle u,w,v_T\rangle\in\mathcal{F}_{I,T}} \{\xi^{T_F}_{3+j,14-j,10,0} : j=1,\dots,10\}$,
with $T_F := \langle u,w,v_T,v_F\rangle$ for $v_F \in \mathcal{V}_{T,F}$
(C^6:M6$_{T_{WF}}$) $\{\xi^{T_{\alpha,1}}_{i,j,k,7} \in E^{T_{\alpha,1}}_8(\langle v_{F_\alpha},v_T\rangle)\}$, $\alpha=1,\dots,4$
(C^6:M7$_{T_{WF}}$) $D^{T_{1,1}}_{10}(v_T)$,

Proof:
First, we show that $\mathcal{M}_{T_{WF},6}$ is a minimal determining set for the superspline space $\mathcal{S}^{6,13,9,22}_{27}(T_{WF})$. Therefore, we set all B-coefficients of a spline $s \in \mathcal{S}^{6,13,9,22}_{27}(T_{WF})$ associated with the domain points in $\mathcal{M}_{T_{WF},6}$ to arbitrary values and show that all remaining B-coefficients are uniquely and stably determined.

We can uniquely and stably determine the B-coefficients of s associated with domain points in the balls $D_{13}(v_i)$, $i=1,\dots,4$, from the B-coefficients corresponding to the domain points in (C^6:M1$_{T_{WF}}$) and the C^{13} supersmoothness conditions at the four vertices of T.

Now, we examine the B-coefficients associated with the domain points in the tubes $E_9(e)$, $e \in \mathcal{E}_T$. The remaining undetermined B-coefficients of s corresponding to domain points in these tubes can be uniquely and stably computed from those corresponding to the domain points in the set (C^6:M2$_{T_{WF}}$) and the C^9 supersmoothness conditions at the edges of T.

Next, we consider the B-coefficients of s associated with the domain points on the faces of T. Note that each face F_i, $i=1,\dots,4$, of T is refined with a Clough-Tocher split, due to the Worsey-Farin split of T. Since the three subtriangles forming each face F_i are degenerated faces of the three subtetrahedra $T_{i,j}$, $j=1,2,3$, we can regard the B-coefficients of $s|_{F_i}$ as those of a bivariate spline $g_{i,0}$ of degree 27 defined on F_i. Then, the spline $g_{i,0}$ has C^9 smoothness at the edges of F_i, C^{13} supersmoothness at the vertices of F_i, and C^{22} supersmoothness at the split point in the interior of F_i. Moreover, we have already determined the B-coefficients of $g_{i,0}$ corresponding to the domain points within a distance of nine from the edges of F_i and within a distance of thirteen from the vertices of F_i. Now, together with the B-coefficients associated with the domain points in (C^6:M3$_{T_{WF}}$) with $l=0$, those on the faces of T, all B-coefficients corresponding to a minimal determining set for a bivariate C^9 Clough-Tocher macro-element of degree 27 with C^{13} supersmoothness at the vertices, and C^{22} supersmoothness at the split point are determined. Then, following Theorem 6.6, we can uniquely and stably determined the remaining undetermined B-coefficients of $g_{i,0}$ (see Example 6.29).

Now, we consider the remaining undetermined B-coefficients of s corresponding to domain points in the shells $R_{26}(v_T),\ldots,R_{22}(v_T)$. Therefore, let F_{i,ν_1}, $i=1,\ldots,4$, be the four triangles formed by the domain points in the shells $R^{T_{i,1}}_{27-\nu_1}(v_T)$, $R^{T_{i,2}}_{27-\nu_1}(v_T)$, and $R^{T_{i,3}}_{27-\nu_1}(v_T)$, with $\nu_1=1,\ldots,5$. Again, we can consider the B-coefficients of $s|_{F_{i,\nu_1}}$ as the B-coefficients of a bivariate spline g_{ν_1,F_i} of degree $27-\nu_1$, that has $C^{9-\nu_1}$ smoothness at the edges, $C^{13-\nu_1}$ supersmoothness at the vertices, and C^{22} supersmoothness at the split point in the interior of F_{i,ν_1}. By examining g_{ν_1,F_i}, we can see that the B-coefficients corresponding to the domain points within a distance of $9-\nu_1$ from the edges and those within a distance of $13-\nu_1$ from the vertices of F_{i,ν_1} are already uniquely and stably determined. Moreover, together with the B-coefficients associated with the domain points in (C^6:M3$_{T_{WF}}$) lying in F_{i,ν_1}, those with $l=\nu_1$, following Theorem 6.6, we have determined all B-coefficients corresponding to the domain points contained in a minimal determining set of a bivariate $C^{9-\nu_1}$ Clough-Tocher macro-element of degree $27-\nu_1$, with $C^{13-\nu_1}$ supersmoothness at the vertices, and C^{22} supersmoothness at the split point of F_{i,ν_1} (see Examples 6.19, 6.22, 6.25, 6.27, and 6.28). Note that, for the values p and q in Theorem 6.6 we have $p=\nu_1$ and $q=\lfloor\frac{\nu_1}{2}\rfloor$. Thus, all B-coefficients of g_{ν_1,F_i} are uniquely and stably determined.

So far, all B-coefficients of s restricted to the shells $R_{27}(v_T),\ldots,R_{22}(v_T)$ are uniquely and stably determined.

In the following, we consider the B-coefficients of s associated with the domain points in the shell $R_{21}(v_T)$. To this end, let $F_{i,6}$, $i=1,\ldots,4$, be the four triangles formed by the domain points in the shells $R^{T_{i,1}}_{21}(v_T)$, $R^{T_{i,2}}_{21}(v_T)$, and $R^{T_{i,3}}_{21}(v_T)$. Note that, due to the Worsey-Farin split of T, the triangles $F_{i,6}$, $i=1,\ldots,4$, are refined with a Clough-Tocher split. Then, we can consider the B-coefficients of $s|_{F_{i,6}}$ as those of a bivariate spline g_{6,F_i} of degree 21, with C^3 smoothness at the edges, C^7 supersmoothness at the vertices, and C^{22} supersmoothness at the split point of $F_{i,6}$. Since the supersmoothness at the split point is equal to 22, which is greater than the degree of the bivariate spline, g_{6,F_i} can be considered as a bivariate polynomial defined on $F_{i,6}$. Now, examining g_{6,F_i}, we can see that the B-coefficients corresponding to the domain points within a distance of three from the edges of $F_{i,6}$, and within a distance of seven from the vertices of $F_{i,6}$ are already uniquely determined. Then, the B-coefficients associated with the domain points in (M5$_{T_{WF}}$) can be used to uniquely and stably determine the derivatives of

g_{6,F_i} at the split point in $F_{i,6}$ up to order nine. Thus, from Lemma 2.42 for the case $n = 3$, we see that the remaining undetermined B-coefficients of g_{6,F_i} can be uniquely and stably determined from these derivatives. See Figure 8.4 (left) for one of the triangles $F_{i,6}$, $i = 1,\dots,4$.

At this point, we have determined all B-coefficients of the spline s restricted to the shells $R_{27}(v_T),\dots,R_{21}(v_T)$, and those corresponding to the domain points within a distance of nine from the edges, and thirteen from the vertices of T.

Next, we consider the B-coefficients of s corresponding to the domain points with a distance of ten from the edges of T more closely. Here, the only undetermined B-coefficients correspond to domain points in the shell $R_{20}(v_T)$ with a distance of three from the faces in $\mathcal{F}_{T,I}$, to domain points in the shell $R_{19}(v_T)$ with a distance of two, to domain points in the shell $R_{18}(v_T)$ with a distance of one, and to the domain points in the shell $R_{17}(v_T)$ lying on the faces in $\mathcal{F}_{T,I}$. In Figure 8.4 (right), the corresponding domain points in the shell $R_{20}(v_T)$ are marked with ✖. Since the considered domain points in the shell $R_{17}(v_T)$ are all contained in the set (C^6:M5$_{T_{WF}}$), the corresponding B-coefficients are already known. Then, considering exactly one of the B-coefficients associated with a domain point in (C^6:M5$_{T_{WF}}$), we can see that the C^6 smoothness condition, and the corresponding $C^5,\dots,C^1$ smoothness conditions, with their tip at this B-coefficient, form a system of equations, where exactly six of the examined undetermined B-coefficients are involved. These are two B-coefficients associated with domain points in the shell $R_{18}(v_T)$, two associated with domain points in the shell $R_{19}(v_T)$, and two associated with domain points in the shell $R_{20}(v_T)$. Therefore, we have a system of six equations covering exactly six undetermined B-coefficients. Then, following Lemma 2.33 for the case $j = 1$, we obtain that the matrix of the system is nonsingular and we can uniquely and stably determined the B-coefficients of s corresponding to the domain points with a distance of ten from the edges of T.

Subsequently, we examine the B-coefficients of s associated with the domain points in the shell $R_{20}(v_T)$. Therefore, let $F_{i,7}$, $i = 1,\dots,4$, be the four triangles formed by the domain points in the shells $R_{20}^{T_{i,1}}$, $R_{20}^{T_{i,2}}$, and $R_{20}^{T_{i,3}}$. Due to the Worsey-Farin split of T, these triangles are refined with a Clough-Tocher split. Now, the trivariate spline s can be considered as a bivariate spline g_{7,F_i} of degree 20 defined on the triangle $F_{i,7}$, that has C^{22} supersmoothness at the split point in the interior of $F_{i,7}$. Since the supersmoothness at the split point is higher than the degree of polynomials,

g_{7,F_i} reduces to a bivariate polynomial. For this polynomial, we have already uniquely and stably determined the B-coefficients corresponding to the domain points within a distance of six from the vertices, and within a distance of three from the edges of $F_{i,7}$. Then, from the B-coefficients corresponding to the domain points in (C^6:M6$_{T_{WF}}$), we can uniquely and stably compute the derivatives of g_{7,F_i} at the split point in $F_{i,7}$ up to order eight. Subsequently, following Lemma 2.42 for the case $n = 3$, we obtain that the remaining undetermined B-coefficients of g_{7,F_i} are uniquely and stably determined from these derivatives. See Figure 8.4 (right) for one of the triangles $F_{i,7}$, $i = 1,\ldots,4$.

Until now, we have uniquely and stably determined all B-coefficients of s corresponding to the domain points within a distance of seven from the faces of T, within a distance of ten from the edges of T, and within a distance of thirteen from the vertices of T.

Finally, we determine the remaining B-coefficients of s. These undetermined B-coefficients correspond to domain points in the ball $D_{22}(v_T)$. Due to the C^{22} supersmoothness of s at the split point v_T, we can consider the B-coefficients of $s|_{D_{22}(v_T)}$ as those of a trivariate polynomial p of degree 22 defined on the tetrahedron $\widetilde{T}$, which is bounded by the domain points in the shell $R_{22}(v_T)$. Note that $\widetilde{T}$ is also refined with a Worsey-Farin split. Due to the fact that we have already determined some of the B-coefficients of s associated with the domain points in $D_{22}(v_T)$, we already know some of the B-coefficients of the polynomial p. Thus, the B-coefficients of p corresponding to the domain points within a distance of two from the faces of $\widetilde{T}$, corresponding to the domain points within a distance of five from the edges of $\widetilde{T}$, and corresponding to the domain points within a distance of eight from the vertices of $\widetilde{T}$ are already determined. Next, we consider the B-coefficients associated with the domain points in the set (C^6:M7$_{T_{WF}}$). From these B-coefficients, we can uniquely and stably compute the derivatives of p at the split point v_T up to order ten. Thus, following Lemma 2.42 for the case $n = 4$, we can uniquely and stably determine the remaining B-coefficients of p from these derivatives. Therefore, we have also obtained all remaining B-coefficients of s.

In the end, we consider the dimension of the space $\mathcal{S}_{27}^{6,13,9,22}(T_{WF})$ of supersplines. Since we have just shown that $\mathcal{M}_{T_{WF},6}$ is a minimal determining set for this superspline space, $\dim \mathcal{S}_{27}^{6,13,9,22}(T_{WF}) = \#\mathcal{M}_{T_{WF},6}$ holds. Therefore, we have to sum up the cardinalities of the sets (C^6:M1$_{T_{WF}}$) - (C^6:M7$_{T_{WF}}$) in order to obtain the dimension of $\mathcal{S}_{27}^{6,13,9,22}(T_{WF})$.

The set (C^6:M1$_{T_{WF}}$) contains exactly

$$4\binom{13+3}{3} = 2240$$

domain points. The set (C^6:M2$_{T_{WF}}$) consists of

$$12\binom{9+2}{3} = 1980$$

domain points. The cardinality of the set (C^6:M3$_{T_{WF}}$) is equal to

$$24\binom{7+2}{2} = 864.$$

The set (C^6:M4$_{T_{WF}}$) contains a total number of

$$4\binom{9+2}{2} = 220$$

domain points. The set (C^6:M5$_{T_{WF}}$) consists of

$$6 \cdot 10 = 60$$

domain points. The set (C^6:M6$_{T_{WF}}$) contains

$$4\binom{8+2}{2} = 180$$

domain points. The last set (C^6:M7$_{T_{WF}}$) consists of

$$\binom{10+3}{3} = 286$$

domain points. Therefore, the dimension of the space $\mathcal{S}_{27}^{6,13,9,22}(T_{WF})$ is equal to 5830. □

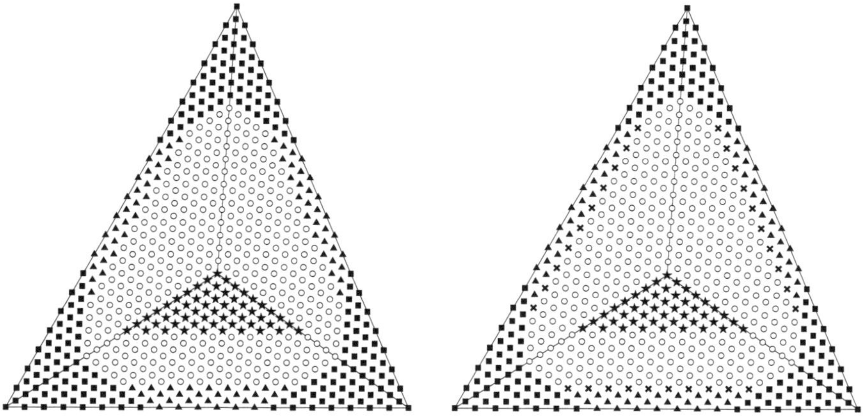

Figure 8.4: Triangles $F_{i,6}$ and $F_{i,7}$ in the shells $R_{21}(v_T)$ and $R_{20}(v_T)$, respectively. Domain points whose B-coefficients are determined from B-coefficients associated with domain points in (C^6:M1$_{T_{WF}}$) are marked with ■, those from B-coefficients associated with domain points in (C^6:M2$_{T_{WF}}$) with ▲, and those from B-coefficients associated with domain points in (C^6:M5$_{T_{WF}}$) and smoothness conditions using Lemma 2.33 for $j = 0$ with ✖. Domain points contained in (C^6:M4$_{T_{WF}}$) and (C^6:M6$_{T_{WF}}$) are marked with ★.

8.4 Nodal minimal determining sets for C^r macro-elements over the Worsey-Farin split

In this section nodal minimal determining sets for the C^r macro-elements constructed in section 8.2 are examined. Therefore, the same notations as in section 8.2 for the distinct sets of vertices, edges and faces are used, as well as some notations for certain derivatives and points, which were previously used in section 7.4. Furthermore, some more derivatives are needed in this section. Thus, in order to ease the understanding of the nodal minimal determining sets constructed here, all needed notations for derivatives are given at the beginning. In the following, we examine and prove nodal minimal determining sets for C^r macro-elements based on the Worsey-Farin split of a single tetrahedron. Then, the section ends with the investigation and the proof of nodal minimal determining sets for C^r macro-elements over a tetrahedral partition where each tetrahedron has been subjected to a Worsey-Farin split.

In the following, some derivatives are denoted, which are necessary to describe the nodal minimal determining sets.

Notation 8.9:
For any multi-index $\alpha := (\alpha_1, \alpha_2, \alpha_3)$ let D^α be the mixed partial derivative $D_x^{\alpha_1} D_y^{\alpha_2} D_z^{\alpha_3}$. For each edge $e := \langle u, v \rangle$ of a tetrahedron T let X_e be the plane perpendicular to e at the point u, which is endowed with Cartesian coordinate axes, where the origin lies in u. Then, for any multi-index $\beta := (\beta_1, \beta_2)$, let $D_{e\perp}^\beta$ be the derivative of order $|\beta| := \beta_1 + \beta_2$ in direction of the two coordinate axes in X_e. Moreover, for every face F of T, with $e \in F$, let $D_{e,F}$ be the derivative normal to e in F. Then, for an oriented triangular face $F := \langle u, v, w \rangle$ let D_F be the unit normal derivative corresponding to F. Also, let $D_{F,u}$ be the directional derivatives associated with the vector $\langle u, v_F \rangle$, where v_F is the split point in the face F. Furthermore, for each edge $e \in \mathcal{E}_T$, let $D_{e\perp F}$ be the directional derivative associated with a unit vector perpendicular to e that lies in the interior face F of T_{WF} containing e.

In order to define a nodal minimal determining set for the macro-elements considered in this chapter, the following points are also needed:

Definition 8.10:
Let

$$\varsigma_{e,j}^i := \frac{(i-j+1)u + jv}{i+1}, \qquad j = 1, \ldots, i,$$

be equally spaced points in the interior of the edge e.

Moreover, for each point $u \in \mathbb{R}^3$ let ϵ_u be the point evaluation functional defined by

$$\epsilon_u f := f(u).$$

Now, we can construct nodal minimal determining sets for the C^r macro-elements based on the Worsey-Farin split of a tetrahedron constructed in section 8.2.

Theorem 8.11:
Let $\mathcal{N}_{T_{WF}}$ be the union of the following sets of nodal data:

(N1$_{T_{WF}}$) $\bigcup\limits_{v \in \mathcal{V}_T} \{\epsilon_v D^\alpha\}_{|\alpha| \le \rho}$

(N2$_{T_{WF}}$) $\bigcup\limits_{e \in \mathcal{E}_T} \bigcup\limits_{i=1}^{\mu} \bigcup\limits_{j=1}^{i} \{\epsilon_{\varsigma^i_{e,j}} D^\beta_{e^\perp}\}_{|\beta|=i}$

(N3$_{T_{WF}}$) $\bigcup\limits_{F:=\langle u,v,w\rangle \in \mathcal{F}_T} \{\epsilon_{v_F} D^l_F D^j_{F,u} D^k_{F,v}\}$, $0 \le j+k \le \mu - 2$,
$l = 0,\dots,d-\eta$

(N4$_{T_{WF}}$) $\bigcup\limits_{F \in \mathcal{F}_T} \bigcup\limits_{e \in F} \bigcup\limits_{i=\mu-l+1}^{\mu-l+\lceil \frac{l-d+\eta-1}{2} \rceil} \bigcup\limits_{j=1}^{i} \{\epsilon_{\varsigma^i_{e,j}} D^l_{e \perp \widetilde{F}_e} D^i_{e,F}\}$, $l = d-\eta+1,\dots,r$,
where $\widetilde{F}_e$ is the interior face of T_{WF} containing e

(N5$_{T_{WF}}$) $\bigcup\limits_{F:=\langle u,v,w\rangle \in \mathcal{F}_T} \{\epsilon_{v_F} D^l_F D^j_{F,u} D^k_{F,v}\}$, $0 \le j+k \le \mu + 2l - 2d + 2\eta - 2 -$
$3\lceil \frac{l-d+\eta-1}{2} \rceil$,
$l = d-\eta+1,\dots,r$, with $v_F \in \mathcal{V}_F$ and $v_F \in F$

(N6$_{T_{WF}}$) for $r = 4m$ and $r = 4m+2$:
$\bigcup\limits_{e \in \mathcal{E}_T} \bigcup\limits_{i=1}^{m} \bigcup\limits_{j=1}^{\mu+i} \{\epsilon_{\varsigma^{\mu+i}_{e,j}} D^{\mu+i}_{e \perp \widetilde{F}_e}\}$, where $\widetilde{F}_e$ is the interior face of T_{WF} containing e

(N7$_{T_{WF}}$) $\bigcup\limits_{F \in \mathcal{F}_T} \bigcup\limits_{e \in F} \bigcup\limits_{i=\mu-r+1}^{\mu-r+\lceil \frac{2r-(d-\eta)-l-1}{2} \rceil} \bigcup\limits_{j=1}^{i} \{\epsilon_{\varsigma^i_{e,j}} D^l_{e \perp \widetilde{F}_e} D^i_{e,F}\}$, $l = r+1,\dots,r+m$,
where $\widetilde{F}_e$ is the interior face of T_{WF} containing e

(N8$_{T_{WF}}$) $\bigcup\limits_{F:=\langle u,v,w\rangle \in \mathcal{F}_T} \{\epsilon_{v_F} D^l_F D^j_{F,u} D^k_{F,v}\}$, $0 \le j+k \le d - 3\mu + 3r - l$
$-3\lceil \frac{2r-(d-\eta)-l-1}{2} \rceil - 3$, $l = r+1,\dots,r+m$, with $v_F \in F$

(N9$_{T_{WF}}$) $\{\epsilon_{v_T} D^\alpha\}_{|\alpha| \le \eta - 4(r+m-(d-\eta)+1)}$

Then $\mathcal{N}_{T_{WF}}$ is a stable nodal minimal determining set for the superspline space $\mathcal{S}^{r,\rho,\mu,\eta}_d(T_{WF})$.

Proof:
It can easily be seen that the cardinality of the set $\mathcal{N}_{T_{WF}}$ matches the dimension of the superspline space $\mathcal{S}^{r,\rho,\mu,\eta}_d(T_{WF})$ as given in Theorem 8.3. Thus, it suffices to show that setting the values $\{\lambda s\}_{\lambda \in \mathcal{N}_{T_{WF}}}$ for a spline $s \in \mathcal{S}^{r,\rho,\mu,\eta}_d(T_{WF})$ determines all B-coefficients of s.

First, we consider the B-coefficients of s corresponding to the domain points within a distance of ρ from the vertices of T. These can be uniquely and stably determined from the values of the derivatives $D^{\alpha}s(v)$, $v \in \mathcal{V}_T$, corresponding to the nodal data in ($N1_{T_{WF}}$).

Next, we examine the B-coefficients of s corresponding to the domain points within a distance of μ from the edges in $\mathcal{E}_T$. We can uniquely and stably determined these B-coefficients from the values of the derivatives of s associated with the nodal data in ($N2_{T_{WF}}$).

Now, we compute the B-coefficients of s associated with the domain points in the shells $R_d(v_T),\ldots,R_{\eta}(v_T)$. Therefore, let F_{i,ν_1}, $i=1,\ldots,4$, be the four triangles formed by the domain points in the three shells $R^{T_{i,1}}_{d-\nu_1}(v_T)$, $R^{T_{i,2}}_{d-\nu_1}(v_T)$, and $R^{T_{i,3}}_{d-\nu_1}(v_T)$, for $\nu_1 = 0,\ldots,d-\eta$. Note that the triangle F_{i,ν_1} is refined with a Clough-Tocher split due to the Worsey-Farin split of T. Then, we can regard the B-coefficients of $s|_{F_{i,\nu_1}}$ as those of a bivariate spline g_{ν_1,F_i} of degree $d-\nu_1$ defined on F_{i,ν_1}, which has $\mu-\nu_1$ smoothness at the edges, $\rho-\nu_1$ smoothness at the vertices and η smoothness at the split point. Then, from the nodal data in ($N3_{T_{WF}}$) corresponding to the triangle F_{i,ν_1}, those with $l=\nu_1$, we can uniquely and stably determined the B-coefficients of g_{ν_1,F_i} corresponding to the domain points within a distance of $\mu-2$ from the split point in F_{i,ν_1}. Now, as in the proof of Theorem 8.3, we have determined all B-coefficients associated with the domain points corresponding to a minimal determining set for a bivariate Clough-Tocher macro-element of degree $d-\nu_1$ with $\mu-\nu_1$ smoothness at the edges, $\rho-\nu_1$ smoothness at the vertices and η smoothness at the split point, as given in Theorem 6.6. This way, we can uniquely and stably determined the B-coefficients of s associated with the domain points in the shells $R_d(v_T),\ldots,R_{\eta}(v_T)$.

Next, we consider the B-coefficients of s associated with the domain points in the shells $R_{\eta+1}(v_T),\ldots,R_{d-r}(v_T)$. To this end, let F_{i,ν_2}, $i=1,\ldots,4$, be the triangle formed by the domain points in the shells $R^{T_{i,1}}_{d-\nu_2}(v_T)$, $R^{T_{i,2}}_{d-\nu_2}(v_T)$, and $R^{T_{i,3}}_{d-\nu_2}(v_T)$, with $\nu_2 = d-\eta+1,\ldots,r$. Note that the triangle F_{i,ν_2} is refined with a Clough-Tocher split. Now, we can regard the B-coefficients of s restricted to F_{i,ν_2} as the B-coefficients of a bivariate spline g_{ν_2,F_i} defined on the triangle F_{i,ν_2}. The spline g_{ν_2,F_i} has degree $d-\nu_2$, $C^{\mu-\nu_2}$ smoothness at the edges, $C^{\rho-\nu_2}$ smoothness at the vertices, and C^{η} smoothness at the split point of F_{i,ν_2}. Note due to the C^{η} smoothness at the split point in F_{i,ν_2}, the spline g_{ν_2,F_i} reduces to a bivariate polynomial, since the degree of g_{ν_2,F_i} is less than η. Investigating the polynomial g_{ν_2,F_i}, it can be seen that we have already determined the B-coefficients corresponding to the domain points within a distance of $\rho - \nu_2$ from the vertices of F_{i,ν_2} and those corresponding to the domain points within a distance of $\mu - \nu_2$ from the edges of F_{i,ν_2}. Then, from the nodal date in ($N4_{T_{WF}}$) corresponding to the triangle F_{i,ν_2}, those with $l = \nu_2$, we can uniquely and stably determine the remaining B-coefficients of g_{ν_2,F_i} within a distance of $\mu - \nu_2 + \lceil \frac{\nu_2-d+\eta-1}{2} \rceil$ from the edges of F_{i,ν_2}. Then, following Lemma 2.42 for the case $n = 3$, we can uniquely and stably compute the remaining undetermined B-coefficients of g_{ν_2,F_i} from the nodal date in ($N4_{T_{WF}}$) corresponding to F_{i,ν_2}, those with $l = \nu_2$.

Now, we consider the B-coefficients of s associated with the domain points in the shells $R_{d-\nu_3}(v_T)$, $\nu_3 = r+1,\dots,r+m$. To this end, we first have to compute the remaining B-coefficients of s corresponding to the domain points within a distance of $\mu - r + \nu_3$ from the edges of T.

For the case that r is odd, following Lemma 2.33 for the case $j = 0$, we can directly compute these B-coefficients from those already determined by using the C^r smoothness conditions of s at the interior faces of T_{WF}. The computation of these B-coefficients works in the same way as described in the proof of Theorem 8.3. Moreover, note that the computation is stable.

In case r is even, we can uniquely and stably determine the B-coefficients associated with the domain points with $i = \nu_3 - r$ in the subset ($M6_{T_{WF}}$) of the minimal determining set in Theorem 8.3 from the nodal data in ($N6_{T_{WF}}$). Therefore, in the same way as in the proof of Theorem 8.3, following Lemma 2.33 for the case $j = 1$, we can uniquely and stably compute the remaining undetermined B-coefficients of s corresponding to the domain points within a distance of $\mu - r + \nu_3$ from the edges of T using the C^r smoothness conditions of s.

Next, let F_{i,ν_3}, $i = 1,\dots,4$, be the triangles formed by the domain points in the shells $R^{T_{i,1}}_{d-\nu_3}$, $R^{T_{i,2}}_{d-\nu_3}$, and $R^{T_{i,3}}_{d-\nu_3}$. Due to the Worsey-Farin split of T, a Clough-Tocher split is induced for the triangle F_{i,ν_3}. Examining s restricted to the triangle F_{i,ν_3}, it can be seen that we have already determined the B-coefficients corresponding to the domain points within a distance of $\rho - \nu_3$ from the vertices of F_{i,ν_3}, and those corresponding to the domain points within a distance of $\mu - r$ from the edges of F_{i,ν_3}. Now, from the nodal data in ($N7_{T_{WF}}$) with $l = \nu_3$, we can uniquely and stably determine the remaining B-coefficients of s corresponding to the domain points within a distance of $\mu - r + \lceil \frac{2r-(d-\eta)-\nu_3-1}{2} \rceil$ from the edges of F_{i,ν_3}. So, now we consider the B-coefficients of the spline $s|_{F_{i,\nu_3}}$ as a bivariate spline g_{ν_3,F_i} of degree $d - \nu_3$ defined on F_{i,ν_3}. Note that g_{ν_3,F_i} has C^η supersmoothness at the split point in g_{ν_3,F_i}. Since the degree of this supersmoothness is higher than the degree of polynomials of g_{ν_3,F_i}, the bivariate spline g_{ν_3,F_i} reduces to a bivariate polynomial. Then, from the nodal data in ($N8_{T_{WF}}$) with $l = \nu_3$, we obtain the derivatives of g_{ν_3,F_i} at the split point in F_{i,ν_3} up to order $d - 3\mu + 3r - l - 3\lceil \frac{2r-(d-\eta)-l-1}{2} \rceil - 3$. Thus, following Lemma 2.42 for the case $n = 3$, we can uniquely and stably determine the remaining B-coefficients of g_{ν_3,F_i} from these derivatives.

Next, we compute the last undetermined B-coefficients of s, which correspond to domain points in the ball $D_\eta(v_T)$. Since s has C^η supersmoothness at the split point v_T in T, we can regard the B-coefficients of s corresponding to the domain points in $D_\eta(v_T)$ as those of a trivariate polynomial p of degree η defined on the tetrahedron $\widetilde{T}$, which is formed by the domain points in $D_\eta(v_T)$. Thus, we have already determined the B-coefficients of p associated with the domain points within a distance of $r + m - (d - \eta)$ from the faces of $\widetilde{T}$. Now, following Lemma 2.42 for the case $n = 4$, we can uniquely and stably compute the remaining undetermined B-coefficients of p from the nodal data in ($N9_{T_{WF}}$). □

Next, let Δ be a tetrahedral partition and Δ_{WF} the refined partition obtained by subjecting each tetrahedron in Δ to the Worsey-Farin split. Then, nodal minimal determining sets for the C^r macro-elements constructed in section 8.2, based on the partition Δ_{WF} are examined.

Theorem 8.12:
Let $\mathcal{N}_{\Delta_{WF}}$ be the union of the following sets of nodal data:

($N1_{\Delta_{WF}}$) $\bigcup\limits_{v \in \mathcal{V}} \{\epsilon_v D^\alpha\}_{|\alpha| \le \rho}$

($N2_{\Delta_{WF}}$) $\bigcup\limits_{e \in \mathcal{E}} \bigcup\limits_{i=1}^{\mu} \bigcup\limits_{j=1}^{i} \{\epsilon_{\varsigma^i_{e,j}} D^\beta_{e^\perp}\}_{|\beta|=i}$

($N3_{\Delta_{WF}}$) $\bigcup\limits_{F:=\langle u,v,w\rangle \in \mathcal{F}} \{\epsilon_{v_F} D^l_F D^j_{F,u} D^k_{F,v}\}$, $0 \le j + k \le \mu - 2$, $l = 0, \ldots, d - \eta$, with $v_F \in \mathcal{V}_F$ and $v_F \in F$

($N4_{\Delta_{WF}}$) $\bigcup\limits_{F \in \mathcal{F}} \bigcup\limits_{e \in F} \bigcup\limits_{i=\mu-l+1}^{\mu-l+\left\lceil \frac{l-d+\eta-1}{2} \right\rceil} \bigcup\limits_{j=1}^{i} \{\epsilon_{\varsigma^i_{e,j}} D^l_{e\perp \widetilde{F}_e} D^i_{e,F}\}$, $l = d - \eta + 1, \ldots, r$,
where $\widetilde{F}_e$ is an interior face of Δ_{WF} containing e

($N5_{\Delta_{WF}}$) $\bigcup\limits_{F:=\langle u,v,w\rangle \in \mathcal{F}} \{\epsilon_{v_F} D^l_F D^j_{F,u} D^k_{F,v}\}$, $0 \le j + k \le \mu + 2(l - d + \eta - 1) - 3\lceil \frac{l-d+\eta-1}{2} \rceil$, $l = d - \eta + 1, \ldots, r$, with $v_F \in \mathcal{V}_F$ and $v_F \in F$

($\mathrm{N6}_{\Delta_{WF}}$) for $r = 4m$ and $r = 4m+2$:
$\bigcup_{T\in\Delta}\bigcup_{e\in T}\bigcup_{i=1}^{m}\bigcup_{j=1}^{\mu+i}\{\epsilon_{\zeta_{e,j}^{\mu+i}} D_{e\perp\widetilde{F}_e}^{\mu+i}\}$, where $\widetilde{F}_e$ is the interior face of T_{WF} containing e

($\mathrm{N7}_{\Delta_{WF}}$) $\bigcup_{T\in\Delta}\bigcup_{F\in T}\bigcup_{e\in F}\bigcup_{i=\mu-r+1}^{\mu-r+\left\lceil\frac{2r-(d-\eta)-l-1}{2}\right\rceil}\bigcup_{j=1}^{i}\{\epsilon_{\zeta_{e,j}^{i}} D_{e\perp\widetilde{F}_e}^{l} D_{e,F}^{i}\}$, $l = r+1,\ldots,r+m$, where $\widetilde{F}_e$ is the interior face of T_{WF} containing e

($\mathrm{N8}_{\Delta_{WF}}$) $\bigcup_{T\in\Delta}\bigcup_{F:=\langle u,v,w\rangle\in T}\{\epsilon_{v_F} D_F^l D_{F,u}^j D_{F,v}^k\}$, $0 \le j+k \le d-3\mu+3r-l$
$-3\left\lceil\frac{2r-(d-\eta)-l-1}{2}\right\rceil - 3$, $l = r+1,\ldots,r+m$, with $v_F \in \mathcal{V}_F$ and $v_F \in F$

($\mathrm{N9}_{\Delta_{WF}}$) $\bigcup_{v_T\in\mathcal{V}_I}\{\epsilon_{v_T} D^{\alpha}\}_{|\alpha|\le\eta-4(r+m-(d-\eta)+1)}$

Then $\mathcal{N}_{\Delta_{WF}}$ is a 1-local and stable nodal minimal determining set for the superspline space $\mathcal{S}_d^{r,\rho,\mu,\eta}(\Delta_{WF})$.

Proof:
It can easily be seen that the dimension of the spline space $\mathcal{S}_d^{r,\rho,\mu,\eta}(\Delta_{WF})$, as given in (8.2) in Theorem 8.5, is equal to the cardinality of the set $\mathcal{N}_{\Delta_{WF}}$. Therefore, it suffices to show that setting $\{\lambda s\}_{\lambda\in\mathcal{N}_{\Delta_{WF}}}$ to arbitrary values stably and locally determines all B-coefficients of a spline $s \in \mathcal{S}_d^{r,\rho,\mu,\eta}(\Delta_{WF})$.

For each vertex $v \in \mathcal{V}$ and a tetrahedron T_v containing v, the B-coefficients of s associated with the domain points in the ball $D_\rho^{T_v}(v)$ can be uniquely and stably determined from the derivatives in the set ($\mathrm{N1}_{\Delta_{WF}}$). Now, the undetermined B-coefficients of s corresponding to domain points in $D_\rho(v) \setminus D_\rho^{T_v}(v)$ can be uniquely and stably computed from those already determined by using the C^ρ supersmoothness conditions at the vertices in $\mathcal{V}$. Note that these computations are 1-local, since each B-coefficient $c_{i,j,k,l}^T$ of s determined so far only depends on the nodal data from the set ($\mathrm{N1}_{\Delta_{WF}}$) contained in $\mathrm{star}(T)$.

Next, the remaining undetermined B-coefficients corresponding to domain points in the tubes $E_\mu(e), e \in \mathcal{E}$, are determined. For each edge $e \in \mathcal{E}$ and a tetrahedron T_e containing e the undetermined B-coefficients of s associated with the domain points in the tube $E_\mu^{T_e}(e)$ can be uniquely and stably determined from the nodal data in ($\mathrm{N2}_{\Delta_{WF}}$). Then, the remaining B-coefficients of s associated with domain points in the tube $E_\mu(e)$ can be uniquely and stably computed from those already determined by using the C^μ supersmoothness conditions at the edges in $\mathcal{E}$. These computations are also 1-local, since each B-coefficient $c^T_{i,j,k,l}$ of s associated with a domain point $\xi^T_{i,j,k,l} \in E_\mu(e)$ only depends on the nodal data from the sets (N1$_{\Delta_{WF}}$) and (N2$_{\Delta_{WF}}$) contained in star(T).

Now, the B-coefficients of s associated with the domain points in the shells $R_d(v_T),\ldots,R_{d-r}(v_T)$, $v_T \in \mathcal{V}_I$, are computed. For each face $F \in \mathcal{F}$, the sets (N3$_{\Delta_{WF}}$), (N4$_{\Delta_{WF}}$), and (N5$_{\Delta_{WF}}$) contain the same nodal data for exactly one tetrahedron T_F containing F, as the sets (N3$_{T_{WF}}$), (N4$_{T_{WF}}$), and (N5$_{T_{WF}}$) in Theorem 8.11. Thus, the B-coefficients of $s|_{T_F}$ corresponding to the domain points within a distance of r from F can be uniquely and stably determined in the same way as in the proof of Theorem 8.11. Then, in case F is not a boundary face of Δ, the B-coefficients of s corresponding to the domain points within a distance of r from F in the tetrahedron $\widetilde{T}_F$, sharing the face F with T_F, can be uniquely and stably computed from those already determined by using the C^r smoothness conditions of s at F. Note that the computation of the B-coefficients determined so far is 1-local, since each B-coefficient $c^T_{i,j,k,l}$ of s already determined determined only depends on the nodal data from the set (N1$_{\Delta_{WF}}$), (N2$_{\Delta_{WF}}$), (N3$_{\Delta_{WF}}$), (N4$_{\Delta_{WF}}$), and (N5$_{\Delta_{WF}}$) contained in star(T).

At this point, all B-coefficients of s are determined, such that the smoothness conditions at the vertices, edges and faces of Δ are satisfied.

Then, since the sets (N6$_{\Delta_{WF}}$) - (N9$_{\Delta_{WF}}$) contain the same nodal data as the sets (N6$_{T_{WF}}$) - (N9$_{T_{WF}}$) from Theorem 8.11 for each tetrahedron $T \in \Delta$, the remaining undetermined B-coefficients of s can be uniquely and stably determined in the same way as in the proof of Theorem 8.11. Thus, all B-coefficients of s are uniquely and stably determined. Moreover, since the B-coefficients determined in this last step in each tetrahedron $T \in \Delta$ are computed from those already determined so far and the nodal data in the sets (N6$_{\Delta_{WF}}$) - (N9$_{\Delta_{WF}}$) contained T, the computation of all B-coefficients $c^T_{i,j,k,l}$ of s is 1-local, since each of these B-coefficients only depends on nodal data contained in $\mathcal{N}_{\Delta_{WF}} \cap \text{star}(T)$. □

Remark 8.13:

For the cases $r = 1$ and $r = 2$, the nodal minimal determining sets presented in this chapter are consistent with those in [62]. Considering the nodal data corresponding to a C^1 macro-element over the Worsey-Farin split of a tetrahedron, it can be seen that the same nodal data as investigated here, are used in [103] in order to construct C^1 macro-elements in $\mathbb{R}^3$. The nodal minimal determining set for the case $r = 2$ differs from the one used in [9], since additional smoothness conditions are used there (cf. Remark 8.6). Moreover, the nodal data used in [9] also differ from those used here, since the B-coefficients of a spline corresponding to the domain points with a distance of two from the faces of a tetrahedron are computed in another way. This approach can be used in [9], since Conjecture 2.43 has been proved for the case arising there for these B-coefficients. Since the conjecture has not been proved yet, it is not possible to extend the method examined in [9] to higher order macro-elements.

8.5 Hermite interpolation with C^r macro-elements based on the Worsey-Farin split

In this section we consider a Hermite interpolation method corresponding to the C^r macro-elements on the Worsey-Farin split examined in this chapter. We also examine the error bounds for the Hermite interpolation method.

Theorem 8.12 in section 8.4 shows that for any function $f \in C^\rho(\Omega)$, there is a unique spline $s \in \mathcal{S}_d^{r,\rho,\mu,\eta}(\Delta_{WF})$ solving the Hermite interpolation problem

$$\lambda s = \lambda f, \quad \text{for all } \lambda \in \mathcal{N}_{\Delta_{WF}},$$

or equivalently,

(H1$_{\Delta_{WF}}$) $D^\alpha s(v) = D^\alpha f(v)$, for all $|\alpha| \leq \rho$, and all vertices $v \in \mathcal{V}$

(H2$_{\Delta_{WF}}$) $D^\beta_{e^\perp} s(\varsigma^i_{e,j}) = D^\beta_{e^\perp} f(\varsigma^i_{e,j})$, for all $|\beta| = i$, $1 \leq j \leq i$, $1 \leq i \leq \mu$, and all edges $e \in \mathcal{E}$

(H3$_{\Delta_{WF}}$) $D^l_F D^j_{F,u} D^k_{F,v} s(v_F) = D^l_F D^j_{F,u} D^k_{F,v} f(v_F)$, for all $0 \leq j+k \leq \mu-2$, and $l = 0,\dots,d-\eta$, for all faces $F := \langle u,v,w \rangle \in \mathcal{F}$, with $v_F \in \mathcal{V}_F$ and $v_F \in F$

(H4$_{\Delta_{WF}}$) $D^l_{e\perp\widetilde{F}_e} D^i_{e,F} s(\varsigma^i_{e,j}) = D^l_{e\perp\widetilde{F}_e} D^i_{e,F} f(\varsigma^i_{e,j})$, for all $j = 1,\dots,i$, $i = \mu - l + 1, \dots, \mu - l + \left\lceil \frac{l-d+\eta-1}{2} \right\rceil$, $l = d-\eta+1,\dots,r$, and all edges e in all faces $F \in \mathcal{F}$, where $\widetilde{F}_e \ni e$ is an interior face of Δ_{WF}

(H5$_{\Delta_{WF}}$) $D^l_F D^j_{F,u} D^k_{F,v} s(v_F) = D^l_F D^j_{F,u} D^k_{F,v} f(v_F)$, for all $0 \leq j+k \leq \mu + 2l - 2d + 2\eta - 2 - 3\lceil \frac{l-d+\eta-1}{2} \rceil$, and $l = d-\eta+1,\dots,r$, for all faces $F := \langle u,v,w \rangle \in \mathcal{F}$, with $v_F \in \mathcal{V}_F$ and $v_F \in F$

($\mathrm{H6}_{\Delta_{WF}}$) for $r = 4m$ and $r = 4m+2$:
$D^{\mu+i}_{e\perp\widetilde{F}_e} s(\varsigma^{\mu+i}_{e,j}) = D^{\mu+i}_{e\perp\widetilde{F}_e} f(\varsigma^{\mu+i}_{e,j})$, for all $j = 1,\dots,\mu+i$, and $i = 1,\dots,m$, for all edges e in all tetrahedra $T \in \Delta$, where $\widetilde{F}_e \ni e$ is an interior face of Δ_{WF}

($\mathrm{H7}_{\Delta_{WF}}$) $D^{l}_{e\perp\widetilde{F}_e} D^{i}_{e,F} s(\varsigma^{i}_{e,j}) = D^{l}_{e\perp\widetilde{F}_e} D^{i}_{e,F} f(\varsigma^{i}_{e,j})$,
for all $j = 1,\dots,i$, $i = \mu - r + 1,\dots,\mu - r + \left\lceil \frac{2r-(d-\eta)-l-1}{2} \right\rceil$, and $l = r+1,\dots,r+m$, for all edges e in all faces F in all tetrahedra $T \in \Delta$, where $\widetilde{F}_e \ni e$ is an interior face of Δ_{WF}

($\mathrm{H8}_{\Delta_{WF}}$) $D^{l}_{F} D^{j}_{F,u} D^{k}_{F,v} s(v_F) = D^{l}_{F} D^{j}_{F,u} D^{k}_{F,v} f(v_F)$, for all
$0 \le j+k \le d - 3\mu + 3r - l - 3\lceil \frac{2r-(d-\eta)-l-1}{2} \rceil - 3$, and $l = r+1,\dots,r+m$, for all faces $F := \langle u,v,w \rangle$ in all tetrahedra $T \in \Delta$, with $v_F \in \mathcal{V}_F$ and $v_F \in F$

($\mathrm{H9}_{\Delta_{WF}}$) $D^{\alpha} s(v_T) = D^{\alpha} f(v_T)$, for all $|\alpha| \le \eta - 4(r+m-(d-\eta)+1)$, and all vertices $v_T \in \mathcal{V}_I$.

Moreover, following Theorem 8.12 this Hermite interpolation method is 1-local and stable.

Thus, a linear projector $\mathcal{I} : C^{\rho} \longrightarrow \mathcal{S}^{r,\rho,\mu,\eta}_{d}(\Delta_{WF})$ is defined. Due to the construction of $\mathcal{I}$, $\mathcal{I}s = s$ holds for every spline $s \in \mathcal{S}^{r,\rho,\mu,\eta}_{d}(\Delta_{WF})$. Therefore, especially, $\mathcal{I}$ reproduces all polynomials of degree d. Next, we can establish the following error bounds for the Hermite interpolation method.

Theorem 8.14:
For every $f \in C^{k+1}(\Omega)$ with $\rho - 1 \le k \le d$,

$$\|D^{\alpha}(f - \mathcal{I}f)\|_{\Omega} \le K|\Delta|^{k+1-|\alpha|}|f|_{k+1,\Omega}, \tag{8.3}$$

for all $0 \le |\alpha| \le k$, where K depends only on r and the smallest solid and face angle in Δ.

Proof:
Let $T \in \Delta$, $\Omega_T := \mathrm{star}(T)$ and $f \in C^{k+1}(\Omega)$. Fix α with $|\alpha| \le k$. By Lemma 4.3.8 of [19], see also Lemma 4.6 of [57], there exists a polynomial $p \in \mathcal{P}^3_k$ such that

$$\|D^{\alpha}(f - p)\|_{\Omega_T} \le K_1|\Omega_T|^{k+1-|\beta|}|f|_{k+1,\Omega_T}, \tag{8.4}$$

for all $0 \le |\beta| \le k$, where $|\Omega_T|$ is the diameter of Ω_T. Since $\mathcal{I}p = p$,

$$\|D^{\alpha}(f - \mathcal{I}f)\|_T \le \|D^{\alpha}(f - p)\|_T + \|D^{\alpha}\mathcal{I}(f - p)\|_T$$

holds.

We can estimate the first term using (8.4) with $\beta = \alpha$. Applying the Markov inequality [102] to each of the polynomials $\mathcal{I}(f-p)|_{T_{i,j}}$, where $T_{i,j}$, $i = 1,\ldots,4$, $j = 1,2,3$, are the twelve subtetrahedra of T_{WF}, we get

$$\|D^\alpha \mathcal{I}(f-p)\|_{T_{i,j}} \le K_2 |T|^{-\alpha} \|\mathcal{I}(f-p)\|_{T_{i,j}},$$

where K_2 is a constant depending only on r and the smallest solid and face angle in Δ. Next, let $\{c_\xi\}$ be the set of B-coefficients of the polynomial $\mathcal{I}(f-p)|_{T_{i,j}}$ relative to the tetrahedron $T_{i,j}$. Then combining (2.12) and the fact that the interpolation method is stable and that the Bernstein basis polynomials form a partition of unity, it follows that

$$\|\mathcal{I}(f-p)\|_{T_{i,j}} \le K_3 \max_{\xi \in \mathcal{D}_{d,T_{i,j}}} |c_\xi| \le K_4 \sum_{l=0}^{\rho} |T|^l |f-p|_{l,T}.$$

Now taking the maximum over i and j, and combining this with (8.4) and $|T| \le |\Omega_T|$, we get that

$$\|\mathcal{I}(f-p)\|_T \le K_5 |T|^{k+1} |f|_{k+1,T} \le K_6 |\Omega_T|^{k+1} |f|_{k+1,\Omega_T},$$

which gives

$$\|D^\alpha \mathcal{I}(f-p)\|_T \le K_7 |\Omega_T|^{k+1-|\alpha|} |f|_{k+1,\Omega_T}.$$

Now, maximizing over all tetrahedra T in Δ leads to (8.3). □

References

[1] P. Alfeld. A trivariate Clough-Tocher scheme for tetrahedral data. *Comp. Aided Geom. Design*, 1(2):169–181, 1984.

[2] P. Alfeld. A bivariate C^2 Clough-Tocher scheme. *Comp. Aided Geom. Design*, 1(3):257–267, 1984.

[3] P. Alfeld. Bivariate spline spaces and minimal determining sets. *J. Comput. Appl. Math.*, 119(1-2):13–27, 2000.

[4] P. Alfeld and L.L. Schumaker. The dimension of bivariate spline spaces of smoothness r for degree $d \geq 4r + 1$. *Constr. Approx.*, 3(1): 189–197, 1987.

[5] P. Alfeld and L.L. Schumaker. On the dimension of bivariate spline spaces of smoothness r and degree $d = 3r + 1$. *Numer. Math.*, 57(1): 651–661, 1990.

[6] P. Alfeld and L.L. Schumaker. Smooth macro-elements based on Clough-Tocher triangle splits. *Numer. Math.*, 90(4):597–616, 2002.

[7] P. Alfeld and L.L. Schumaker. Smooth macro-elements based on Powell-Sabin triangle splits. *Adv. Comput. Math.*, 16(1):29–46, 2002.

[8] P. Alfeld and L.L. Schumaker. Upper and lower bounds on the dimension of superspline spaces. *Constr. Approx.*, 19(1):145–161, 2003.

[9] P. Alfeld and L.L. Schumaker. A C^2 trivariate macro-element based on the Worsey-Farin split of a tetrahedron. *SIAM J. Numer. Anal.*, 43 (4):1750–1765, 2005.

[10] P. Alfeld and L.L. Schumaker. A C^2 trivariate macro-element based on the Clough-Tocher split of a tetrahedron. *Comp. Aided Geom. Design*, 22(7):710–721, 2005.

[11] P. Alfeld and L.L. Schumaker. A C^2 trivariate double-Clough-Tocher macro-element. In C. Chui, M. Neamtu, and L.L. Schumaker, editors, *Approximation Theory XI: Gatlinburg 2004*, pages 1–14, Brentwood, 2005. Nashboro Press.

[12] P. Alfeld and L.L. Schumaker. Bounds on the dimensions of trivariate spline spaces. *Adv. Comput. Math.*, 29(4):315–335, 2008.

[13] P. Alfeld and T. Sorokina. Two tetrahedral C^1 cubic macro elements. *J. Approx. Theory*, 157(1):53–69, 2009.

[14] P. Alfeld, B. Piper, and L.L. Schumaker. Minimally supported bases for spaces of bivariate piecewise polynomials of smoothness r and degree $d \geq 4r+1$. *Comp. Aided Geom. Design*, 4(1-2):105–123, 1987.

[15] P. Alfeld, B. Piper, and L.L. Schumaker. An explicit basis for C^1 quartic bivariate splines. *SIAM J. Numer. Anal.*, 24(4):891–911, 1987.

[16] P. Alfeld, L.L. Schumaker, and M. Sirvent. On dimension and existence of local bases for multivariate spline spaces. *J. Approx. Theory*, 70(2):243–264, 1992.

[17] P. Alfeld, L.L. Schumaker, and W. Whitley. The generic dimension of the space of C^1 splines of degree $d \geq 8$ on tetrahedral decompositions. *SIAM J. Numer. Anal.*, 30(3):889–920, 1993.

[18] G. Awanou and M.-J. Lai. C^1 quintic spline interpolation over tetrahedral partitions. In C.K. Chui, L.L. Schumaker, and J. Stöckler, editors, *Approximation Theory X: Wavelets, Splines and Applications*, pages 1–16. Vanderbilt University Press, 2002.

[19] S.C. Brenner and L.R. Scott. *The Mathematical Theory of Finite Element Methods*, volume 15 of *Texts in Applied Mathematics*. Springer-Verlag, New York, 1994.

[20] C.K. Chui and M.-J. Lai. On bivariate super vertex splines. *Constr. Approx.*, 6(4):399–419, 1990.

[21] C.K. Chui and M.-J. Lai. Multivariate vertex splines and finite elements. *J. Approx. Theory*, 60(3):245–343, 1990.

[22] C.K. Chui, G. Hecklin, G. Nürnberger, and F. Zeilfelder. Optimal Lagrange interpolation by quartic C^1 splines on triangulations. *J. Comput. Appl. Math.*, 216(2):344–363, 2008.

[23] P. G. Ciarlet and P.A. Raviart. General Lagrange and Hermite interpolation in $\mathbb{R}^n$ with applications to finite element methods. *Arch. Rational Mech. Anal.*, 46(3):177–199, 1972.

[24] R. W. Clough and J. L. Tocher. Finite element stiffness matrices for analysis of plates in bending. In *Proceedings of Conference on Matrix Methods in Structural Mechanics*, pages 515–545, Wright Patterson Air Force Base, Dayton, Ohio, 1965.

[25] O. Davydov and G. Nürnberger. Interpolation by C^1 splines of degree $q \geq 4$ on triangulations. *J. Comput. Appl. Math.*, 126(1-2):159–183, 2000.

[26] O. Davydov and L.L. Schumaker. Locally linearly independent bases for bivariate polynomial spline spaces. *Adv. Comput. Math.*, 15(4): 355–373, 2000.

[27] O. Davydov and L.L. Schumaker. Stable local nodal bases for C^1 bivariate polynomial splines. In A. Cohen, C. Rabut, and L.L. Schumaker, editors, *Curve and Surface Fitting: Saint-Malo 99*, 2000.

[28] O. Davydov and L.L. Schumaker. On stable local bases for bivariate polynomial spline spaces. *Constr. Approx.*, 18(1):87–116, 2002.

[29] O. Davydov and L.L. Schumaker. Stable approximation and interpolation with C^1 quartic bivariate splines. *SIAM J. Numer. Anal.*, 39(5): 1732–1748, 2002.

[30] O. Davydov, G. Nürnberger, and F. Zeilfelder. Approximation order of bivarate spline interpolation for arbitrary smoothness. *J. Comput. Appl. Math.*, 90(2):117–134, 1998.

[31] O. Davydov, G. Nürnberger, and F. Zeilfelder. Bivariate spline interpolation with optimal approximation order. *Constr. Approx.*, 17(2): 181–208, 2001.

[32] C. de Boor. *A Practical Guide to Splines*. Number 27 in Applied Mathematical Sciences. Springer-Verlag, 1978.

[33] C. de Boor. B-form basics. In G.E. Farin, editor, *Geometric Modeling: Algorithms and New Trends*, pages 131–148. SIAM, Philadelphia, 1987.

[34] C. de Boor and K. Höllig. Approximation order from bivariate C^1 cubics: a counterexample. *Proceedings of the American Mathematical Society*, 87:649–655, 1983.

[35] C. de Boor and K. Höllig. Approximation power of smooth bivariate pp functions. *Math. Z.*, 197(3):343–363, September 1988.

[36] C. de Boor and R.Q. Jia. A sharp upper bound on the approximation order of smooth bivariate pp functions. *J. Approx. Theory*, 72(1):24–33, 1993.

[37] P. de Casteljau. Courbes et surfaces à pôles. Technical report, André Citroën Automobiles SA, Paris, 1963.

[38] G. Farin. A modified Clough-Tocher interpolant. *Comp. Aided Geom. Design*, 2(1-3):19–27, 1985.

[39] G. Farin. Triangular Bernstein-Bézier patches. *Comp. Aided Geom. Design*, 3(2):83–127, 1986.

[40] G. Hecklin, G. Nürnberger, and F. Zeilfelder. The structure of C^1 spline spaces on Freudenthal partitions. *SIAM J. Math. Anal.*, 38(2): 347–367, 2006.

[41] G. Hecklin, G. Nürnberger, and F. Zeilfelder. Local data interpolation by quintic C^1-splines on tetrahedral partitions. In P. Chenin, T. Lyche, and L.L. Schumaker, editors, *Curve and Surface Design: Avignon 2006*, pages 163–172, 2007.

[42] G. Hecklin, G. Nürnberger, L.L. Schumaker, and F. Zeilfelder. A local Lagrange interpolation method based on C^1 cubic splines on Freudenthal partitions. *Math. Comp.*, 77(262):1017–1036, 2008.

[43] G. Hecklin, G. Nürnberger, L.L. Schumaker, and F. Zeilfelder. Local Lagrange interpolation with cubic C^1 splines on tetrahedral partitions. *J. Approx. Theory*, 160(1-2):89–102, 2009.

[44] X.-L. Hu, D.-F. Han, and M.-J. Lai. Bivariate splines of various degrees for numerical solution of partial differential equations. *SIAM J. Sci. Comput.*, 29(3):1338–1354, 2007.

[45] A.Kh. Ibrahim and L.L. Schumaker. Super spline spaces of smoothness r and degree $d \geq 3r+2$. *Constr. Approx.*, 7(1):401–423, 1991.

[46] M. Laghchim-Lahlou. The C^r-fundamental splines of Clough-Tocher and Powell-Sabin types for Lagrange interpolation on a three direction mesh. *Adv. Comput. Math.*, 8(4):353–366, 1998.

[47] M. Laghchim-Lahlou and P. Sablonnière. Triangular finite elements of HCT type and class C^ρ. *Adv. Comput. Math.*, 2(1):101–122, 1994.

[48] M. Laghchim-Lahlou and P. Sablonnière. Quadrilateral finite elements of FVS type and class C^ρ. *Numer. Math.*, 70(2):229–243, 1995.

[49] M. Laghchim-Lahlou and P. Sablonnière. C^r-finite elements of Powell-Sabin type on the three direction mesh. *Adv. Comput. Math.*, 6(1):191–206, 1996.

[50] M.-J. Lai. A serendipity family of locally supported splines in $S_8^2(\delta)$. *Approx. Theory Appl.*, 10(2):43–53, 1994.

[51] M.-J. Lai. Approximation order from bivariate C^1-cubics on a four-directional mesh is full. *Comp. Aided Geom. Design*, 11(2):215–223, 1994.

[52] M.-J. Lai. Scattered data interpolation and approximation using bivariate C^1 piecewise cubic polynomials. *Comp. Aided Geom. Design*, 13(1):81–88, 1996.

[53] M.-J. Lai. On C^2 quintic spline functions over triangulations of Powell-Sabin's type. *J. Comput. Appl. Math.*, 73(1-2):135–155, 1996.

[54] M.-J. Lai and M.A. Matt. A C^r trivariate macro-element based on the Alfeld split of a tetrahedron. 2012. submitted.

[55] M.-J. Lai and A. Le Méhauté. A new kind of trivariate C^1 macro-element. *Adv. Comput. Math.*, 21(3-4):273–292, 2004.

[56] M.-J. Lai and L.L. Schumaker. Scattered data interpolation using C^2 supersplines of degree six. *SIAM J. Numer. Anal.*, 34(3):905–921, 1997.

[57] M.-J. Lai and L.L. Schumaker. On the approximation power of bivariate splines. *Adv. Comput. Math.*, 9(3-4):251–279, 1998.

[58] M.-J. Lai and L.L. Schumaker. On the approximation power of splines on triangulated quadrangulations. *SIAM J. Numer. Anal.*, 36 (1):143–159, 1998.

[59] M.-J. Lai and L.L. Schumaker. Macro-elements and stable local bases for splines on Clough-Tocher triangulations. *Numer. Math.*, 88(1): 105–119, 2001.

[60] M.-J. Lai and L.L. Schumaker. Quadrilateral macroelements. *SIAM J. Math. Anal.*, 33(5):1107–1116, 2001.

[61] M.-J. Lai and L.L. Schumaker. Macro-elements and stable local bases for splines on Powell-Sabin triangulations. *Math. Computation*, 72 (241):335–354, 2003.

[62] M.-J. Lai and L.L. Schumaker. *Spline Functions on Triangulations*. Encyclopedia of Mathematics. Cambridge University Press, 2007.

[63] M.-J. Lai and L.L. Schumaker. Trivariate C^r polynomial macro-elements. *Constr. Approx.*, 26(1):11–28, 2007.

[64] M.-J. Lai, A. Le Méhauté, and T. Sorokina. An octahedral C^2 macro-element. *Comp. Aided Geom. Design*, 23(8):640–654, 2006.

[65] S.R. Marschner and R.J. Lobb. An evaluation of reconstruction filters for volume rendering. In R.D. Bergeron and A.E. Kaufman, editors, *Proceedings of the Conference on Visualization*, pages 100–107. IEEE Computer Society, October 1994.

[66] M.A. Matt. A C^r trivariate macro-element based on the Worsey-Farin split of a tetrahedron. *Numer. Math.*, 2011. submitted.

[67] M.A. Matt and G. Nürnberger. Local Lagrange interpolation using cubic C^1 splines on type-4 cube partitions. *J. Approx. Theory*, 162(3): 494–511, 2010.

[68] J. Morgan and R. Scott. A nodal basis for C^1 piecewise polynomials of degree $n \geq 5$. *Math. Comput.*, 29(131):736–740, 1975.

[69] G. Nürnberger. *Approximation by Spline Functions*. Springer Verlag, Berlin, 1989.

[70] G. Nürnberger and T. Rießinger. Lagrange and Hermite interpolation by bivariate splines. *Num. Funct. Anal. and Optimiz.*, 13(1-2): 75–96, 1992.

[71] G. Nürnberger and T. Rießinger. Bivariate spline interpolation at grid points. *Numer. Math.*, 71(1):91–119, 1995.

[72] G. Nürnberger and G. Schneider. A Lagrange interpolation method by trivariate cubic C^1 splines of low locality. In M. Neamtu and L.L. Schumaker, editors, *Approximation Theory XIII: San Antonio 2010*, volume 13 of *Springer Proceedings in Mathematics*, pages 231–248. Springer New York, 2012.

[73] G. Nürnberger and F. Zeilfelder. Interpolation by spline spaces on classes of triangulations. *J. Comput. Appl. Math.*, 119(1-2):347–376, 2000.

[74] G. Nürnberger and F. Zeilfelder. Developments in bivariate spline interpolation. *J. Comput. Appl. Math.*, 121(1-2):125–152, 2000.

[75] G. Nürnberger and F. Zeilfelder. Local Lagrange interpolation by cubic splines on a class of triangulations. In K. Kopotun, T. Lyche, and M. Neamtu, editors, *Trends in Approximation Theory*, pages 333–342. Vanderbilt University Press, Nashville, TN, 2001.

[76] G. Nürnberger and F. Zeilfelder. Lagrange interpolation by bivariate C^1-splines with optimal approximation order. *Adv. Comput. Math.*, 21(3-4):381–419, 2004.

[77] G. Nürnberger, L.L. Schumaker, and F. Zeilfelder. Local Lagrange interpolation by bivariate C^1 cubic splines. In T. Lyche and L.L. Schumaker, editors, *Mathematical Methods in CAGD: Oslo 2000*, 2001.

[78] G. Nürnberger, L.L. Schumaker, and F. Zeilfelder. Lagrange interpolation by C^1 cubic splines on triangulations of separable quadrangulations. In C.K. Chui, L.L. Schumaker, and J. Stöckler, editors, *Approximation Theory X: Wavelets, Splines and Applications*, pages 405–424. Vanderbilt University Press, 2002.

[79] G. Nürnberger, V. Rayevskaya, L.L. Schumaker, and F. Zeilfelder. Local Lagrange interpolation with C^2 splines of degree seven on triangulations. In M. Neamtu and E. Staff, editors, *Advances in Constructive Approximation, Vanderbilt 2003*, pages 345–370. Nashboro Press, Brentwood, TN, 2004.

[80] G. Nürnberger, L.L. Schumaker, and F. Zeilfelder. Lagrange interpolation by C^1 cubic splines on triangulated quadrangulations. *Adv. Comput. Math.*, 21(3-4):357–380, 2004.

[81] G. Nürnberger, L.L. Schumaker, and F. Zeilfelder. Two Lagrange interpolation methods based on C^1 splines on tetrahedral partitions. In C.K. Chui, M. Neamtu, and L.L. Schumaker, editors, *Approximation Theory XI: Gatlinburg 2004*, pages 327–344, 2005.

[82] G. Nürnberger, V. Rayevskaya, L.L. Schumaker, and F. Zeilfelder. Local Lagrange interpolation with bivariate splines of arbitrary smoothness. *Constr. Approx.*, 23(1):33–59, 2006.

[83] G. Nürnberger, M. Rhein, and G. Schneider. Local Lagrange interpolation by quintic C^1 splines on type-6 tetrahedral partitions. *J. Comput. Appl. Math.*, 236(4):529–542, 2011.

[84] M. J. D. Powell. *Approximation Theory and Methods*. Cambridge University Press, 1981.

[85] M.J.D. Powell and M.A. Sabin. Piecewise quadratic approximations on triangles. *ACM Trans. Math. Software*, 3(4):316–325, 1977.

[86] W. Quade and L. Collatz. Zur Interpolationstheorie der reellen Funktionen. *Sitzungsber. der Preussischen Akad. der Wiss., Phys.-Math. Kl.*, pages 383–429, 1938.

[87] V. Rayevskaya and L.L. Schumaker. Multi-sided macro-element spaces based on Clough-Tocher triangle splits with application to hole filling. *Comp. Aided Geom. Design*, 22(1):57–79, 2005.

[88] C. Runge. Über empirische Funktionen und die Interpolation zwischen äquidistanten Ordinaten. *Zeitschrift für Mathematik und Physik*, 46:224–243, 1901.

[89] P. Sablonnière. Composite finite elements of class C^k. *J. Comput. Appl. Math.*, 12-13:541–550, 1985.

[90] P. Sablonnière. Composite finite elements of class C^2. In C.K. Chui, L.L. Schumaker, and F.I. Utreras, editors, *Topics in multivariate approximation*, pages 207–217. Academic Press, Boston, 1987.

[91] I.J. Schoenberg. Contributions to the problem of approximation of equidistant data by analytic functions, part a: On the problem of smoothing of graduation, a first class of analytic approximation formulae. *Quart. Appl. Math.*, 4:45–99, 1946.

[92] I.J. Schoenberg. Contributions to the problem of approximation of equidistant data by analytic functions, part b: On the problem of osculatory interpolation, a second class of analytic approximation formulae. *Quart. Appl. Math.*, 4:112–141, 1946.

[93] L.L. Schumaker. *Spline Functions: Basic Theory*. Wiley, 1981.

[94] L.L. Schumaker. Dual bases for spline spaces on cells. *Comp. Aided Geom. Design*, 5(4):277–284, 1988.

[95] L.L. Schumaker and T. Sorokina. C^1 quintic splines on type-4 tetrahedral partitions. *Adv. Comput. Math.*, 21(3-4):421–444, 2004.

[96] L.L. Schumaker and T. Sorokina. A trivariate box macroelement. *Constr. Approx.*, 21(3):413–431, 2005.

[97] L.L. Schumaker, T. Sorokina, and A.J. Worsey. A C^1 quadratic trivariate macro-element space defined over arbitrary tetrahedral partitions. *J. Approx. Theory*, 158(1):126–142, 2009.

[98] T. Sorokina. A C^1 multivariate Clough-Tocher interpolant. *Constr. Approx.*, 29(1):41–59, 2009.

[99] T. Sorokina and A.J. Worsey. A multivariate Powell-Sabin interpolant. *Adv. Comput. Math.*, 29(1):71–89, 2008.

[100] M. von Golitschek, M.-J. Lai, and L.L. Schumaker. Error bounds for minimal energy bivariate polynomial splines. *Numer. Math.*, 93(2): 315–331, 2002.

[101] T. Wang. A C^2 quintic spline interpolation scheme on triangulation. *Comp. Aided Geom. Design*, 9(5):379–386, 1992.

[102] D.R. Wilhelmsen. A Markov inequality in several dimensions. *J. Approx. Theory*, 11(3):216–220, 1974.

[103] A.J. Worsey and G. Farin. An n-dimensional Clough-Tocher interpolant. *Constr. Approx.*, 3(1):99–110, 1987.

[104] A.J. Worsey and B. Piper. A trivariate Powell-Sabin interpolant. *Comp. Aided Geom. Design*, 5(3):177–186, 1988.

[105] A. Ženíšek. Polynomial approximation on tetrahedrons in the finite element method. *J. Approx. Theory*, 7(4):334–351, 1973.

[106] A. Ženíšek. A general theorem on triangular finite $C^{(m)}$-elements. *Rev. Française Automat. Informat. Recherche Opérationnelle Sér. Rouge*, 8 (2):119–127, 1974.

A Bivariate lemmata

In this chapter, we examine some bivariate lemmata, from which three are taken from Alfeld and Schumaker [9], which are needed for the proofs of the minimal determining sets of the C^2 splines on the partial Worsey-Farin splits in section 3.2. Therefore, let $F := \langle v_1, v_2, v_3 \rangle$ be a triangle in $\mathbb{R}^2$ and let F_{CT} be the Clough-Tocher split of F at the split point v_F in the interior of F. Then we define $F_i := \langle v_i, v_{i+1}, v_F \rangle$, $i = 1,2,3$, with $v_4 := v_1$, to be the three subtriangles of F_{CT}.

Lemma A.1:
The set

$$\bigcup_{i=1}^{3} D_1^{F_i}(v_i) \cup \bigcup_{i=1}^{3} \{\xi_{2,2,0}^{F_i}\} \cup \bigcup_{i=1}^{3} \{\xi_{2,0,2}^{F_i}\} \tag{A.1}$$

is a stable minimal determining set for the space of polynomials $\mathcal{P}_4^2$ defined on F_{CT}.

Proof:
It is easy to see that the cardinality of the set of domain points in (A.1) is equal to the dimension of the space of bivariate polynomials of degree four, which is equal to 15. Thus, we only have to show that setting the B-coefficients associated with the domain points in (A.1) of a polynomial $p \in \mathcal{P}_4^2$ to arbitrary values, all other B-coefficients of p are stably determined.

First, the remaining undetermined B-coefficients of p corresponding to domain points in the disks $D_1(v_i)$, $i = 1,2,3$, are stably computed from the B-coefficients associated with the domain points in the first subset of A.1 using the C^1 smoothness conditions at the vertices of F. The domain points corresponding to the B-coefficients determined in this step are marked with □ in Figure A.1. Now, together with the B-coefficients associated with the domain points $\xi_{2,2,0}^{F_1}$, $\xi_{2,2,0}^{F_2}$, and $\xi_{2,2,0}^{F_3}$, which are contained in (A.1), all B-coefficients of p corresponding to the domain points on the edges of F are uniquely and stably determined.

Next, following Lemma 2.28, we can stably compute the remaining undetermined B-coefficients of p corresponding to domain points in the disks $D_2(v_i)$, $i = 1,2,3$, from those already determined and the B-coefficients associated with the domain points $\xi_{2,0,2}^{F_1}$, $\xi_{2,0,2}^{F_2}$, and $\xi_{2,0,2}^{F_3}$ using the C^1 and C^2 smoothness conditions at the vertices of F (see Example 2.31). The domain points in the disks $D_2(v_i)$, $i = 1,2,3$, corresponding to the B-coefficients determined in this step are marked with ✖ in Figure A.1.

In the same way we can stably compute the remaining undetermined B-coefficients of p corresponding to domain points in the disks $D_3(v_1)$ and $D_4(v_1)$ from the smoothness conditions at the edge $\langle v_1, v_F \rangle$ and the already determined B-coefficients, following Lemma 2.28 (see Example 2.32). The domain points corresponding to these B-coefficients are marked with ◆ in Figure A.1.

Then the last undetermined B-coefficient can be directly and stably computed from smoothness conditions across the interior edges of F_{CT}. □

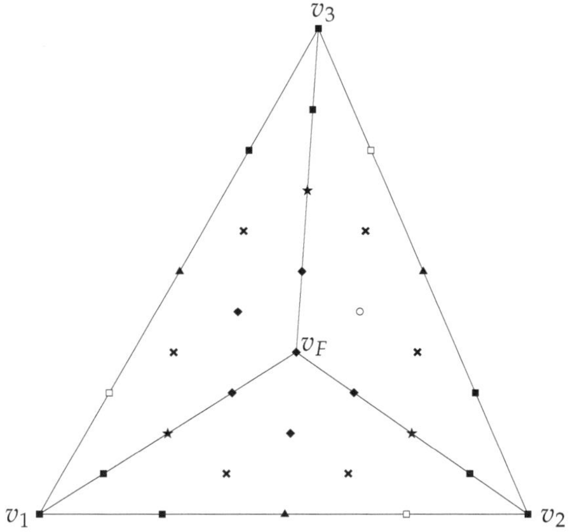

Figure A.1: Minimal determining set of a polynomial $p \in \mathcal{P}_4^2$ defined on F_{CT}. Domain points contained in the minimal determining set are marked with ■ (those in $D_1^{F_i}(v_i)$), with ▲ (those in the second subset of (A.1)), and with ★ (those in the third subset of (A.1)).

Lemma A.2:
The set

$$\bigcup_{i=1}^{3} D_1^{F_i}(v_i) \cup \bigcup_{i=1}^{3} \{\xi_{3,2,0}^{F_i}, \xi_{2,3,0}^{F_i}\} \cup \bigcup_{i=1}^{3} \{\xi_{3,0,2}^{F_i}\} \cup \bigcup_{i=1}^{3} \{\xi_{2,2,1}^{F_i}\} \tag{A.2}$$

is a stable minimal determining set for the space of polynomials $\mathcal{P}_5^2$ defined on F_{CT}.

Proof:
It can easily be seen that the cardinality of the set of domain points in (A.2) is equal to 21, which is equal to the dimension of the space of bivariate polynomials of degree five. Hence, it suffices to show that setting the B-coefficients associated with the domain points in (A.2) of a polynomial $p \in \mathcal{P}_5^2$ to arbitrary values, all other B-coefficients of p are stably determined.

First, using the C^1 smoothness conditions at the vertices of F, we stably compute the remaining undetermined B-coefficients of p corresponding to the domain points in the disks $D_1(v_i)$, $i = 1,2,3$. The corresponding domain points are marked with □ in Figure A.2. At this point, together with the B-coefficients of p associated with the domain points in the second subset of (A.2), we have stably determined all B-coefficients of p corresponding to the domain points on the edges of F.

Now, following Lemma 2.28, we can stably compute the remaining undetermined B-coefficients of p associated with the domain points in the disks $D_2(v_i)$, $i = 1,2,3$, from the already determined B-coefficients and those corresponding to the domain points $\xi^{F_1}_{3,0,2}$, $\xi^{F_2}_{3,0,2}$, and $\xi^{F_3}_{3,0,2}$ using the C^1 and C^2 smoothness conditions at the vertices of F. The corresponding domain points in the disks $D_2(v_i)$, $i = 1,2,3$, are marked with ✖ in Figure A.2.

Next, following Lemma 2.28, we can use the C^1, C^2, and C^3 smoothness conditions at the vertices of F in order to stably compute the remaining undetermined B-coefficients of p corresponding to domain points in the disks $D_3(v_i)$, $i = 1,2,3$, from the already determined B-coefficients and those associated with the domain points $\xi^{F_1}_{2,2,1}$, $\xi^{F_2}_{2,2,1}$, and $\xi^{F_3}_{2,2,1}$. The domain points of the B-coefficients determined in this step are marked with ✚ in Figure A.2.

In the same way the remaining undetermined B-coefficients of p corresponding to the domain points in the disks $D_4(v_1)$ and $D_5(v_1)$ can be stably computed from the smoothness conditions at the edge $\langle v_1, v_F \rangle$ and the already determined B-coefficients, following Lemma 2.28. The corresponding domain points are marked with ◆ in Figure A.2.

Now, the last undetermined B-coefficient can be directly and stably computed from smoothness conditions across the interior edges of F_{CT}. □

Lemma A.3:
The set

$$\bigcup_{i=1}^{3} D_1^{F_i}(v_i) \cup \bigcup_{i=1}^{3} \{\xi^{F_i}_{3,2,0}, \xi^{F_i}_{2,3,0}\} \cup \bigcup_{i=0}^{2} \{\xi^{F_1}_{3-i,1+i,1}\} \cup \{\xi^{F_1}_{2,1,2}, \xi^{F_1}_{1,2,2}, \xi^{F_1}_{1,1,3}\} \tag{A.3}$$

is a stable minimal determining set for the space of polynomials $\mathcal{P}_5^2$ defined on F_{CT}.

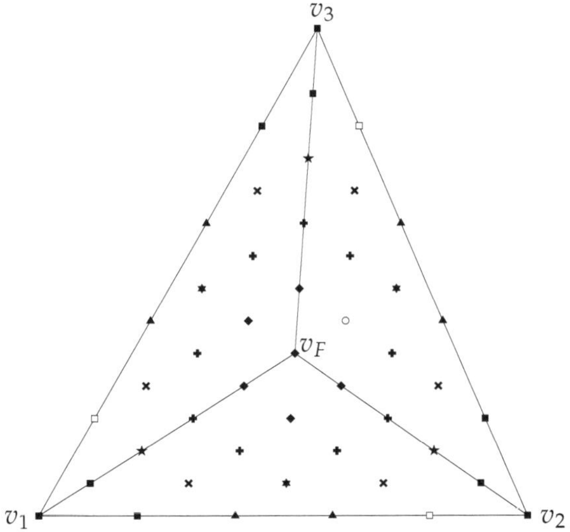

Figure A.2: Minimal determining set of a polynomial $p \in \mathcal{P}_5^2$ defined on F_{CT}. Domain points contained in the minimal determining set are marked with ■ (those in $D_1^{F_i}(v_i)$), with ▲ (those in the second subset of (A.2)), with ★ (those in the third subset of (A.2)), and with ✸(those in the fourth subset of (A.2)).

Proof:
It is easy to see that the cardinality of the set of domain points in (A.3) is equal to 21, which is equal to the dimension of the space of bivariate polynomials of degree five. Thus, it suffices to show that setting the B-coefficients associated with the domain points in (A.3) of a polynomial $p \in \mathcal{P}_5^2$ to arbitrary values, all other B-coefficients of p are stably determined.

From the C^1 smoothness conditions at the vertices of F, the remaining undetermined B-coefficients of p corresponding to the domain points in the disks $D_1(v_i)$, $i = 1,2,3$, can be stably determined. The corresponding domain points are marked with □ in Figure A.3. Together with the B-coefficients of p associated with the domain points in the second and third subset of (A.3) the remaining B-coefficients of p corresponding to the domain points on the edges of F and with a distance of one from the edge $\langle v_1, v_2 \rangle$ are stably determined.

Then, following Lemma 2.28, we can use the C^1 and C^2 smoothness conditions at the vertices v_1 and v_2 of F in order to stably compute the remaining undetermined B-coefficients corresponding to the domain points in the disks $D_2(v_i)$, $i = 1,2,3$, from those already determined. The corresponding domain points are marked with ◆ in Figure A.3.

In the same way, following Lemma 2.28 and using the C^1, C^2, C^3, and C^4 smoothness conditions at the interior edges $\langle v_1, v_F \rangle$ and $\langle v_2, v_F \rangle$, the remaining undetermined B-coefficients of p corresponding to the domain points in the disks $D_j(v_i)$, $i = 1,2$, $j = 3,4$, can be stably computed from those already determined and the B-coefficients corresponding to the domain points in the fourth subset of (A.3) (see Example 2.30). The domain points corresponding to the B-coefficients determined in this step are marked with ✖ in Figure A.3.

Finally, we can stably determined the remaining B-coefficients of p from the C^1 smoothness conditions at the edge $\langle v_3, v_F \rangle$. The corresponding domain points are marked with ✚ in Figure A.3. □

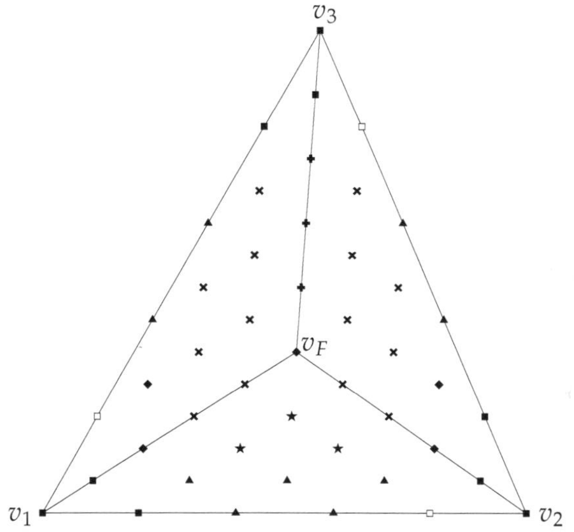

Figure A.3: Minimal determining set of a polynomial $p \in \mathcal{P}_5^2$ defined on F_{CT}. Domain points contained in the minimal determining set are marked with ■ (those in $D_1^{F_i}(v_i)$), with ▲ (those in the second and third subset of (A.3)), and with ★ (those in the fourth subset of (A.3)).

Lemma A.4:
The set

$$\bigcup_{i=1}^{3} D_1^{F_i}(v_i) \cup \bigcup_{i=1}^{3} \{\xi_{4-j,2+j,0}^{F_i}\}_{j=0}^{2} \cup \bigcup_{i=0}^{3} \{\xi_{4-i,1+1,1}^{F_1}\} \cup \bigcup_{i=0}^{2} \{\xi_{3,0,3}^{F_1}, \xi_{0,3,3}^{F_1}, \xi_{2,2,2}^{F_1}, \xi_{2-i,i,4}^{F_1}\} \tag{A.4}$$

is a stable minimal determining set for the space of polynomials $\mathcal{P}_6^2$ defined on F_{CT}.

Proof:
It is easy to see that the cardinality of the set of domain points in (A.4) is equal to 28, which is equal to the dimension of the space of bivariate polynomials of degree six. Therefore, it suffices to show that setting the B-coefficients associated with the domain points in (A.4) of a polynomial $p \in \mathcal{P}_6^2$ to arbitrary values, all other B-coefficients of p are stably determined.

Using the C^1 smoothness conditions at the vertices of F, the remaining undetermined B-coefficients of p corresponding to the domain points in the disks $D_1(v_i)$, $i = 1,2,3$, can be stably determined. The corresponding domain points are marked with □ in Figure A.4. Together with the B-coefficients of p associated with the domain points in the second and third subset of (A.4) the remaining B-coefficients of p corresponding to the domain points on the edges of F and with a distance of one from the edge $\langle v_1, v_2 \rangle$ are stably determined.

Now, following Lemma 2.28, we can use the C^1 and C^2 smoothness conditions at the vertices v_1 and v_2 of F in order to stably compute the remaining undetermined B-coefficients associated with the domain points in the disks $D_2(v_i)$, $i = 1,2,3$, from those already determined. The domain points corresponding to the B-coefficients determined in this step are marked with ◆ in Figure A.4.

In the same way, following Lemma 2.28 and using the C^1, C^2, C^3, C^4, and C^5 smoothness conditions at the interior edges $\langle v_1, v_F \rangle$ and $\langle v_2, v_F \rangle$, we can step by step stably compute the remaining undetermined B-coefficients of p corresponding to the domain points in the disks $D_j(v_i)$, $i = 1,2$, $j = 3,4,5$, from those already determined and the B-coefficients corresponding to the domain points in the fourth subset of (A.4). The corresponding domain points are marked with ✖ in Figure A.4.

Then the remaining B-coefficients of p can be stably determined from the C^1 smoothness conditions at the edge $\langle v_3, v_F \rangle$. The corresponding domain points are marked with ✚ in Figure A.4. □

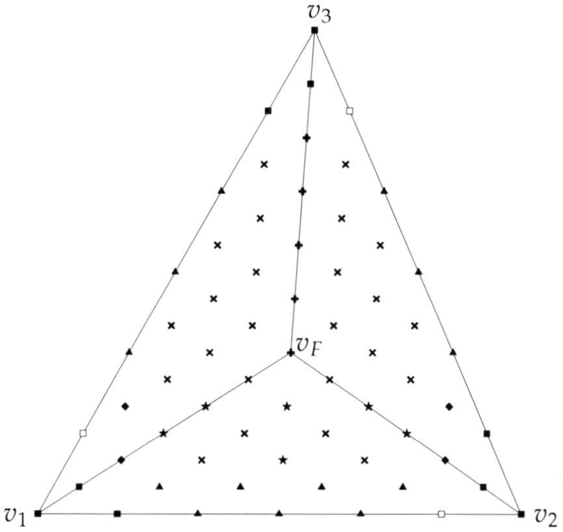

Figure A.4: Minimal determining set of a polynomial $p \in \mathcal{P}_6^2$ defined on F_{CT}. Domain points contained in the minimal determining set are marked with ■ (those in $D_1^{F_i}(v_i)$), with ▲ (those in the second and third subset of (A.4)), and with ★ (those in the fourth subset of (A.4)).

Lemma A.5:
The set

$$\bigcup_{i=1}^{3} D_1^{F_i}(v_i) \cup \bigcup_{i=1}^{3} \{\xi_{4,2,0}^{F_i}, \xi_{3,3,0}^{F_i}, \xi_{2,4,0}^{F_i}\} \cup \bigcup_{i=1}^{3} \{\xi_{4,0,2}^{F_i}, \xi_{3,0,3}^{F_i}, \xi_{3,1,2}^{F_i}\} \cup \{\xi_{2,0,4}^{F_1}\} \tag{A.5}$$

is a stable minimal determining set for the space of polynomials $\mathcal{P}_6^2$ defined on F_{CT}.

Proof:
It is easy to see that the cardinality of the set of domain points in (A.5) is equal to the dimension of the space of bivariate polynomials of degree six, that is equal to 28. Thus, it suffices to show that setting the B-coefficients associated with the domain points in (A.5) of a polynomial $p \in \mathcal{P}_6^2$ to arbitrary values, all other B-coefficients of p are stably determined.

Using the C^1 smoothness conditions at the vertices of F, the remaining undetermined B-coefficients of p corresponding to the domain points in the disks $D_1(v_i)$, $i = 1,2,3$, can be stably determined. The corresponding domain points are marked with □ in Figure A.5. Then, together with the B-coefficients of p associated with the domain points in the second subset of (A.5), the remaining B-coefficients of p corresponding to the domain points on the edges of F are stably determined.

Now, following Lemma 2.28, we can use the C^1 and C^2 smoothness conditions at the vertices of F in order to stably compute the remaining undetermined B-coefficients associated with the domain points in the disks $D_2(v_i)$, $i = 1,2,3$, from those already determined and the B-coefficients corresponding to the domain points $\xi^{F_1}_{4,0,2}$, $\xi^{F_2}_{4,0,2}$, and $\xi^{F_3}_{4,0,2}$. The domain points to these B-coefficients are marked with ✖ in Figure A.5. In the same way, following Lemma 2.28 and using the C^1, C^2, and C^3 smoothness conditions, we can stably determine the remaining undetermined B-coefficients of p corresponding to the domain points in the disks $D_3(v_i)$, $i = 1,2,3$, from those already determined and the remaining B-coefficients corresponding to the domain points in the third subset of (A.5) (see Example 2.29). The corresponding domain points are marked with ✚ in Figure A.5.

Next, we can use the C^1, C^2, C^3, and C^4 smoothness conditions at the vertex v_1 in order to stably compute the remaining undetermined B-coefficients of p corresponding to the domain points in the disk $D_4(v_1)$ from the already determined B-coefficients and the one associated with the domain point $\xi^{F_1}_{2,0,4}$. The domain points corresponding to these B-coefficients are marked with ▼ in Figure A.5.

In the same way, following Lemma 2.28, the remaining undetermined B-coefficients of p corresponding to the domain points in the disks $D_5(v_1)$ and $D_6(v_1)$ can be stably determined from the smoothness conditions at the edge $\langle v_1, v_F \rangle$. The corresponding domain points are marked with ◆ in Figure A.5.

Then, the last undetermined B-coefficients can be directly and stably computed from smoothness conditions across the interior edges of F_{CT}. □

Lemma A.6 (Lemma 4.3 [9]):
Let

$$\mathcal{P}_7^2 = \widetilde{\mathcal{S}}_7^2(F_{CT}) := \{s \in C^2(F_{CT}) : s|_{\widetilde{F}} \in \mathcal{P}_7^2, \forall \widetilde{F} \in F_{CT}, \\ s \in C^7(v_F)\}. \tag{A.6}$$

Then $\dim \widetilde{\mathcal{S}}_7^2(F_{CT}) = 36$, and the set

$$\bigcup_{i=1}^{3} D_2^{F_i}(v_i) \cup \bigcup_{i=1}^{3} \{\xi^{F_i}_{4,3,0}, \xi^{F_i}_{4,2,1}, \xi^{F_i}_{3,4,0}, \xi^{F_i}_{3,3,1}, \xi^{F_i}_{2,4,1}, \xi^{F_i}_{3,0,4}\} \tag{A.7}$$

is a minimal determining set for $\widetilde{\mathcal{S}}_7^2(F_{CT})$ (see Figure A.6).

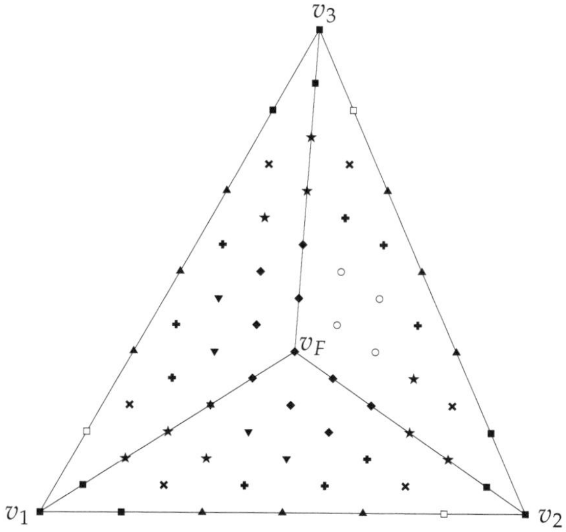

Figure A.5: Minimal determining set of a polynomial $p \in \mathcal{P}_6^2$ defined on F_{CT}. Domain points contained in the minimal determining set are marked with ■ (those in $D_1^{F_i}(v_i)$), with ▲ (those in the second subset of (A.5)), with ★ (those in the third subset of (A.5)), and with ✸($\xi_{2,0,4}^{F_1}$).

Lemma A.7:
Let

$$\mathcal{P}_7^2 = \hat{\mathcal{S}}_7^2(F_{CT}) := \{s \in C^2(F_{CT}) : s|_{\widetilde{F}} \in \mathcal{P}_7^2,\ \forall \widetilde{F} \in F_{CT},\ s \in C^7(v_F)\}. \tag{A.8}$$

Then $\dim \hat{\mathcal{S}}_7^2(F_{CT}) = 36$, and the set

$$\begin{aligned} &\bigcup_{i=1}^{3} D_2^{F_i}(v_i) \cup \bigcup_{i=1}^{2} \{\xi_{4,3,0}^{F_i}, \xi_{3,4,0}^{F_i}, \xi_{4-j,2+j,1}^{F_i}\}_{j=0}^{2} \\ &\cup \bigcup_{i=0}^{1} \{\xi_{4,0,3}^{F_1}, \xi_{4-i,1+i,2}^{F_1}, \xi_{0,1,6}^{F_1}, \xi_{1+i,4-i,2}^{F_2}, \xi_{0,3+i,4-i}^{F_{1+i}}\} \end{aligned} \tag{A.9}$$

is a stable minimal determining set for $\hat{\mathcal{S}}_7^2(F_{CT})$.

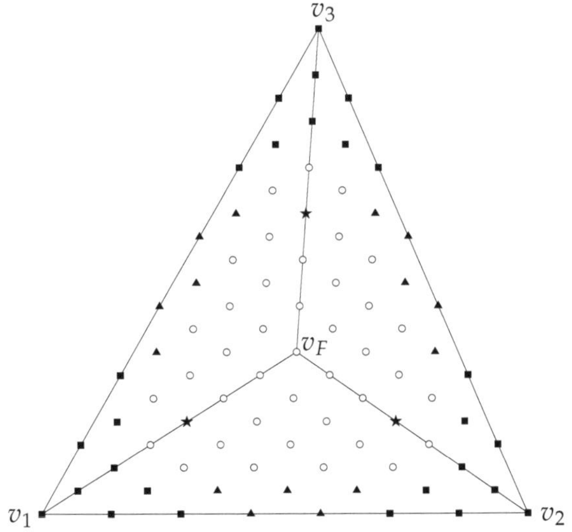

Figure A.6: Minimal determining set of a spline $s \in \widetilde{\mathcal{S}}_7^2(F_{CT})$. Domain points contained in the minimal determining set are marked with ■ (those in $D_2^{F_i}(v_i)$), with ★ ($\xi_{3,0,4}^{F_i}$), and with ▲ (the remaining domain points).

Proof:
Since the space $\hat{\mathcal{S}}_7^2(F_{CT})$ is just the space of polynomials of degree seven defined on F_{CT}, it follows that the dimension of $\hat{\mathcal{S}}_7^2(F_{CT})$ is equal to 36. Now, due to the fact that the cardinality of the set of domain points in (A.9) is equal to 36, it suffices to show that this set is a stable determining set. To this end, we set the B-coefficients of a spline $s \in \hat{\mathcal{S}}_7^2(F_{CT})$ associated with the domain points in (A.9) to arbitrary values and show that the remaining B-coefficients are stably determined.

From the B-coefficients corresponding to the domain points in the first subset of (A.9), we can stably determine the remaining B-coefficients of s associated with the domain points in the disks $D_2(v_i)$, $i = 1,2,3$. The domain points associated with these B-coefficients are marked with □ in Figure A.7. Together with the B-coefficients of s associated with the domain points in the second subset of (A.9), we have stably determined the B-coefficients of s corresponding to the domain points within a distance of one from the edges $\langle v_1, v_2 \rangle$ and $\langle v_2, v_3 \rangle$ so far.

Next, following Lemma 2.28, we can stably compute the remaining undetermined B-coefficients of s associated with the domain points in the disk $D_3(v_2)$, from the already determined B-coefficients using the supersmoothness conditions of s at the split point v_F. The domain points associated with the B-coefficients determined in this step are marked with ◆ in Figure A.7.

Now, again following Lemma 2.28, we can stably compute the remaining undetermined B-coefficients of s associated with the domain points in the disks $D_{3+i}(v_2)$, $i = 1,\ldots,4$, step by step, from the already determined B-coefficients and those associated with the domain points in the last subset of (A.9) using the supersmoothness conditions of s at the split point v_F. The domain points corresponding to these B-coefficients are marked with ✖ in Figure A.7.

Then, the last undetermined B-coefficients of s can be stably computed using the supersmoothness conditions at v_F. □

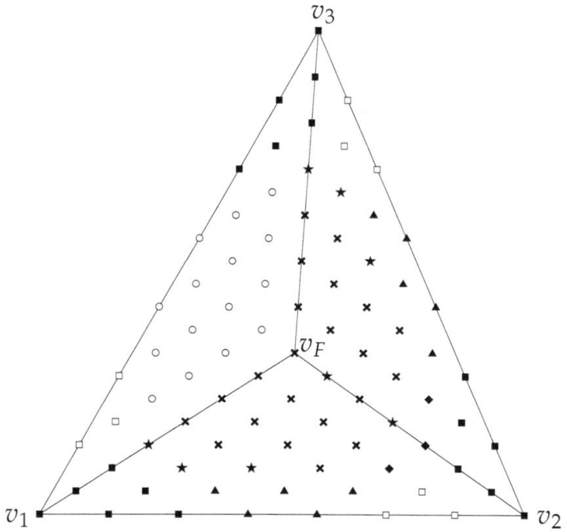

Figure A.7: Minimal determining set of a spline $s \in \hat{\mathcal{S}}_7^2(F_{CT})$. Domain points contained in the minimal determining set are marked with ■ (those in $D_2^{F_i}(v_i)$), with ▲ (those in the second subset of (A.9)), and with ★ (those in the third subset of (A.9)).

Lemma A.8 (Lemma 4.2 [9]):
Let

$$\begin{aligned}\widetilde{\mathcal{S}}_8^2(F_{CT}) := \{s \in C^2(F_{CT}) : s|_{\widetilde{F}} \in \mathcal{P}_8^2,\ \forall \widetilde{F} \in F_{CT},\\ s \in C^3(v_i),\ i = 1,2,3,\\ s \in C^7(v_F),\\ \tau_{3,5,0}^{F_i,F_{i+1}} s = 0,\ i = 1,\dots,3, \text{ with } F_4 := F_1\}.\end{aligned} \tag{A.10}$$

Then $\dim \widetilde{\mathcal{S}}_8^2(F_{CT}) = 48$, and the set

$$\bigcup_{i=1}^{3} D_3^{F_i}(v_i) \cup \bigcup_{i=1}^{3} \{\xi_{4,4,0}^{F_i}, \xi_{4,3,1}^{F_i}, \xi_{3,4,1}^{F_i}, \xi_{4,2,2}^{F_i}, \xi_{3,3,2}^{F_i}, \xi_{2,4,2}^{F_i}\} \tag{A.11}$$

is a minimal determining set for $\widetilde{\mathcal{S}}_8^2(F_{CT})$ (see Figure A.8).

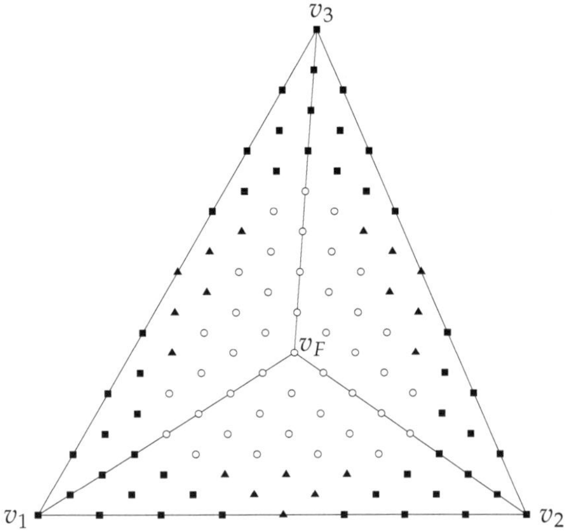

Figure A.8: Minimal determining set of a spline $s \in \widetilde{\mathcal{S}}_8^2(F_{CT})$. Domain points contained in the minimal determining set are marked with ■ (those in $D_3^{F_i}(v_i)$) and with ▲ (the remaining domain points).

Lemma A.9:
Let

$$\hat{\mathcal{S}}_8^2(F_{CT}) := \{s \in C^2(F_{CT}) : s|_{\widetilde{F}} \in \mathcal{P}_8^2,\ \forall \widetilde{F} \in F_{CT},\quad s \in C^3(v_i),\ i = 1,2,3,\quad s \in C^7(v_F),\quad \tau_{3-i,5+i,0}^{F_1,F_2} s = 0,\ i = 0,\dots,3,\quad \tau_{4,4,0}^{F_2,F_3} s = 0\}. \tag{A.12}$$

Then $\dim \hat{\mathcal{S}}_8^2(F_{CT}) = 46$, and the set

$$\bigcup_{i=1}^{3} D_3^{F_i}(v_i) \cup \bigcup_{i=1}^{2} \{\xi_{4,4,0}^{F_i}, \xi_{4,3,1}^{F_i}, \xi_{3,4,1}^{F_i}, \xi_{4,2,2}^{F_i}, \xi_{3,3,2}^{F_i}, \xi_{2,4,2}^{F_i}\} \cup \{\xi_{4,1,3}^{F_1}, \xi_{0,2,6}^{F_1}, \xi_{0,0,8}^{F_1}, \xi_{1,4,3}^{F_2}\} \tag{A.13}$$

is a stable minimal determining set for $\hat{\mathcal{S}}_8^2(F_{CT})$.

Proof:
First, we show that the set of domain points in (A.13) is a minimal determining set for the superspline space $\hat{\mathcal{S}}_8^2(F_{CT})$. To this end, we set the B-coefficients of a spline $s \in \hat{\mathcal{S}}_8^2(F_{CT})$ corresponding to the domain points in (A.13) to arbitrary values and show that the remaining B-coefficients are uniquely and stably determined.

Using the C^3 supersmoothness conditions at the three vertices of F, the remaining B-coefficients of s corresponding to the domain points in the disks $D_3(v_i)$, $i = 1,2,3$, can be uniquely and stably determined. The corresponding domain points are marked with □ in Figure A.9. Then, together with the B-coefficients of s corresponding to the domain points in the second subset of (A.13), all B-coefficients of s corresponding to the domain points within a distance of two from the edges $\langle v_1, v_2 \rangle$ and $\langle v_2, v_3 \rangle$ are uniquely stably determined.

Next, following Lemma 2.28, we can uniquely and stably determine the remaining B-coefficients of s associated with the domain points in the disks $D_{3+i}(v_2)$, $i = 1,2$, from the already determined B-coefficients using the additional smoothness condition $\tau_{3,5,0}^{F_1,F_2} s = 0$, and the supersmoothness conditions of s at the split point v_F. The domain points associated with the B-coefficients determined in this step are marked with ◆ in Figure A.9.

We can also apply Lemma 2.28, to uniquely and stably compute the remaining undetermined B-coefficients of s associated with the domain points in the disks $D_{5+i}(v_2)$, $i = 1,2,3$, from the already determined B-coefficients and those associated with the domain points $\xi_{4,1,3}^{F_1}$, $\xi_{0,2,6}^{F_1}$, $\xi_{0,0,8}^{F_1}$, and $\xi_{1,4,3}^{F_2}$ by using the additional smoothness conditions $\tau_{2,6,0}^{F_1,F_2} s = 0$, $\tau_{1,7,0}^{F_1,F_2} s = 0$, $\tau_{0,8,0}^{F_1,F_2} s = 0$, and the supersmoothness conditions of s at the vertex v_F. The domain points corresponding to these B-coefficients are marked with ✖ in Figure A.9.

Then, the last undetermined B-coefficients of s can be uniquely and stably computed from the supersmoothness conditions at the split point v_F of F_{CT} and the additional smoothness condition $\tau_{4,4,0}^{F_2,F_3} s = 0$.

Considering the dimension of $\hat{\mathcal{S}}_8^2(F_{CT})$, we see that $\dim \hat{\mathcal{S}}_8^2(F_{CT}) = 46$, since the set of domain points in (A.13) is a minimal determining set for this superspline space and the cardinality of this set is equal to 46. □

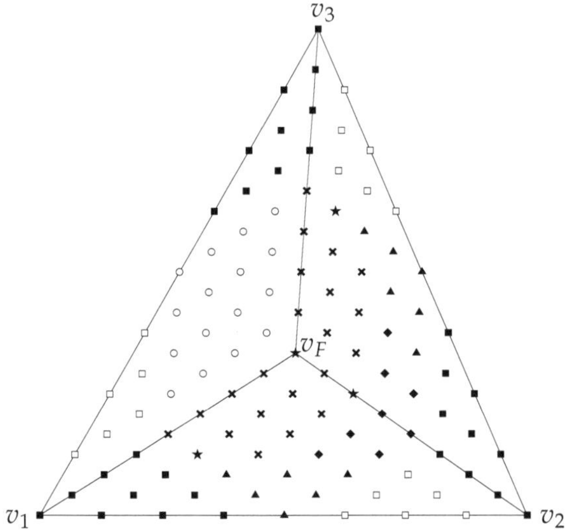

Figure A.9: Minimal determining set of a spline $s \in \hat{\mathcal{S}}_8^2(F_{CT})$. Domain points contained in the minimal determining set are marked with ■ (those in $D_3^{F_i}(v_i)$), with ▲ (those in the second subset of (A.13)), and with ★ (those in the third subset of (A.13)).

Lemma A.10:
Let

$$\bar{\mathcal{S}}_8^2(F_{CT}) := \{s \in C^2(F_{CT}) : s|_{\widetilde{F}} \in \mathcal{P}_8^2,\ \forall \widetilde{F} \in F_{CT}, \quad s \in C^3(v_i),\ i = 1,2,3, \quad s \in C^7(v_F), \quad \tau_{3-i,5+i,0}^{F_1,F_2} s = 0,\ i = 0,\dots,3\}. \tag{A.14}$$

Then $\dim \bar{\mathcal{S}}_8^2(F_{CT}) = 47$, and the set

$$\bigcup_{i=1}^{3} D_3^{F_i}(v_i) \cup \bigcup_{i=1}^{2} \{\xi_{4,4,0}^{F_i}, \xi_{4-j,3+j,1}^{F_i}, \xi_{4-j,2+j,2}^{F_i}, \xi_{2,4,2}^{F_i}\}_{j=0}^{1} \cup \{\xi_{4,1,3}^{F_1}, \xi_{0,2,6}^{F_1}, \xi_{0,0,8}^{F_1}, \xi_{1,4,3}^{F_2}\} \cup \{\xi_{4,4,0}^{F_3}\} \tag{A.15}$$

is a stable minimal determining set for $\bar{\mathcal{S}}_8^2(F_{CT})$.

Proof:
The proof is analogue to the proof of Lemma A.9. The only difference is that the B-coefficient associated with the domain point $\xi_{4,4,0}^{F_3}$ is not determined from the additional smoothness condition $\tau_{4,4,0}^{F_2,F_3} s = 0$, since it is contained in the minimal determining set. Thus, the dimension of $\bar{\mathcal{S}}_8^2(F_{CT})$ is equal to 47, which is exactly one higher than the dimension of $\hat{\mathcal{S}}_8^2(F_{CT})$. □

Lemma A.11:
Let

$$\check{\mathcal{S}}_8^2(F_{CT}) := \{s \in C^2(F_{CT}) : s|_{\widetilde{F}} \in \mathcal{P}_8^2,\ \forall \widetilde{F} \in F_{CT}, \quad s \in C^3(v_i),\ i = 1,2,3, \quad s \in C^7(v_F), \quad \tau_{3-i,5+i,0}^{F_1,F_2} s = 0,\ i = 0,\dots,2\}. \tag{A.16}$$

Then $\dim \check{\mathcal{S}}_8^2(F_{CT}) = 47$, and the set

$$\bigcup_{i=1}^{3} D_3^{F_i}(v_i) \cup \bigcup_{i=1}^{2} \{\xi_{4,4-j,j}^{F_i}, \xi_{3,4-j,1+j}^{F_i}, \xi_{4,2,2}^{F_i}, \xi_{2,4,2}^{F_i}\}_{j=0}^{1} \cup \{\xi_{4,1,3}^{F_1}, \xi_{0,2,6}^{F_1}, \xi_{1,4,3}^{F_2}\} \cup \{\xi_{4,4,0}^{F_3}, \xi_{4,3,1}^{F_3}, \xi_{3,4,1}^{F_3}\} \tag{A.17}$$

is a stable minimal determining set for $\bar{\mathcal{S}}_8^2(F_{CT})$.

Proof:
First, we show that the set of domain points in (A.17) is a minimal determining set for the superspline space $\check{\mathcal{S}}_8^2(F_{CT})$. Therefore, we set the B-coefficients of a spline $s \in \check{\mathcal{S}}_8^2(F_{CT})$ associated with the domain points in (A.17) to arbitrary values and show that the remaining B-coefficients are uniquely and stably determined.

The B-coefficients corresponding to the domain points in the disks $D_3(v_1)$, $D_3(v_3)$, and $D_7(v_2)$ and those corresponding to the domain points within a distance of two from the edges $\langle v_1, v_2 \rangle$ and $\langle v_2, v_3 \rangle$ can be uniquely and stably determined in the same way as in the proof of Lemma A.9. Moreover, from the B-coefficients in the fourth subset of (A.17), the B-coefficients corresponding to the domain points within a distance of one from the edge $\langle v_3, v_1 \rangle$ are uniquely and stably determined.

Now, following Lemma 2.28, we can uniquely and stably compute the remaining undetermined B-coefficients of s corresponding to the domain points in the disks $D_4(v_1)$ and $D_4(v_3)$ using the C^3 smoothness conditions at the corresponding interior edges of F_{CT}.

Then, by Lemma 2.28, we can uniquely and stably determine the remaining B-coefficients of s corresponding to the domain points in the ring $R_8(v_2)$ from those already determined by using the C^7 supersmoothness conditions at v_F.

The remaining undetermined B-coefficients of s can be computed from the C^7 supersmoothness conditions at v_T.

Since set of domain points in (A.17) is a minimal determining set for $\check{\mathcal{S}}_9^2(F_{CT})$, we see that $\dim \check{\mathcal{S}}_9^2(F_{CT}) = 48$, which is equal to the cardinality of the set of domain points in (A.17). □

Lemma A.12:
Let

$$\dot{\mathcal{S}}_8^2(F_{CT}) := \{s \in C^2(F_{CT}) : s|_{\widetilde{F}} \in \mathcal{P}_8^2,\ \forall \widetilde{F} \in F_{CT}, \quad s \in C^3(v_i),\ i = 1,2,3, \quad s \in C^7(v_F)\}. \tag{A.18}$$

Then $\dim \dot{\mathcal{S}}_8^2(F_{CT}) = 51$, and the set

$$\bigcup_{i=1}^{3} D_3^{F_i}(v_i) \cup \bigcup_{i=1}^{3} \{\xi_{4,4,0}^{F_i}, \xi_{4,3,1}^{F_i}, \xi_{3,4,1}^{F_i}, \xi_{4,2,2}^{F_i}, \xi_{3,3,2}^{F_i}, \xi_{2,4,2}^{F_i}\} \cup \{\xi_{1,0,7}^{F_1}, \xi_{0,1,7}^{F_1}, \xi_{0,0,8}^{F_1}\} \tag{A.19}$$

is a stable minimal determining set for $\dot{\mathcal{S}}_8^2(F_{CT})$.

Proof:
We first show that the domain points in (A.19) form a minimal determining set for the space $\dot{\mathcal{S}}_8^2(F_{CT})$. Therefore, we set the B-coefficients of a spline $s \in \dot{\mathcal{S}}_8^2(F_{CT})$

corresponding to the domain points in (A.19) to arbitrary values and show that the remaining B-coefficients are uniquely and stably determined.

From the B-coefficients associated with the domain points in the first subset of (A.19), we can uniquely and stably determine the remaining B-coefficients of s corresponding to domain points in the disks $D_3(v_i)$, $i = 1,2,3$, by using the C^3 supersmoothness conditions at the vertices of F. The domain points corresponding to the B-coefficients determined in this step are marked with □ in Figure A.10. Together with the B-coefficients of s corresponding to the domain points in the second subset of (A.19), all B-coefficients of s corresponding to the domain points within a distance of two from the edges of F are uniquely stably determined.

Since the spline s has C^7 supersmoothness at the split point v_F, we can regard the B-coefficients corresponding to the domain points in the disk $D_7(v_F)$ as those of a bivariate polynomial g of degree seven defined on the triangle $\widetilde{F}$, which is bounded by the domain points in the ring $R_7(v_T)$. Moreover, $\widetilde{F}$ is also refined with a Clough-Tocher split at v_F. Now, examining g it can be seen that the B-coefficients corresponding to the domain points within a distance of one from the edges of $\widetilde{F}$ are already uniquely and stably determined. Then, from the B-coefficients in the third subset of A.19 we can uniquely and stably compute the partial derivatives of g at the split point v_F up to order one. Thus, following Lemma 2.42, the remaining undetermined B-coefficients of g can be uniquely and stably determined by smoothness conditions.

Considering the dimension of $\dot{\mathcal{S}}_8^2(F_{CT})$, it can be seen that $\dim \dot{\mathcal{S}}_8^2(F_{CT}) = 51$, since the set (A.19) is a minimal determining set for $\hat{\mathcal{S}}_8^2(F_{CT})$ and the cardinality of this set is equal to 51. □

Lemma A.13 (Lemma 4.1 [9]):
Let

$$\widetilde{\mathcal{S}}_9^2(F_{CT}) := \{s \in C^2(F_{CT}) : s|_{\widetilde{F}} \in \mathcal{P}_9^2, \forall \widetilde{F} \in F_{CT},\quad \text{(A.20)}$$
$$s \in C^4(v_i),\ i = 1,2,3,$$
$$s \in C^7(v_F),$$
$$\tau_{3,5,1}^{F_i,F_{i+1}} s = 0,\ i = 1,\ldots,3, \text{ with } F_4 := F_1\}.$$

Then $\dim \widetilde{\mathcal{S}}_9^2(F_{CT}) = 63$, and the set

$$\bigcup_{i=1}^{3} D_4^{F_i}(v_i) \cup \bigcup_{i=1}^{3} \{\xi_{4,4,1}^{F_i}, \xi_{4,3,2}^{F_i}, \xi_{3,4,2}^{F_i}, \xi_{4,2,3}^{F_i}, \xi_{3,3,3}^{F_i}, \xi_{2,4,3}^{F_i}\} \quad \text{(A.21)}$$

is a minimal determining set for $\widetilde{\mathcal{S}}_9^2(F_{CT})$ (see Figure A.11).

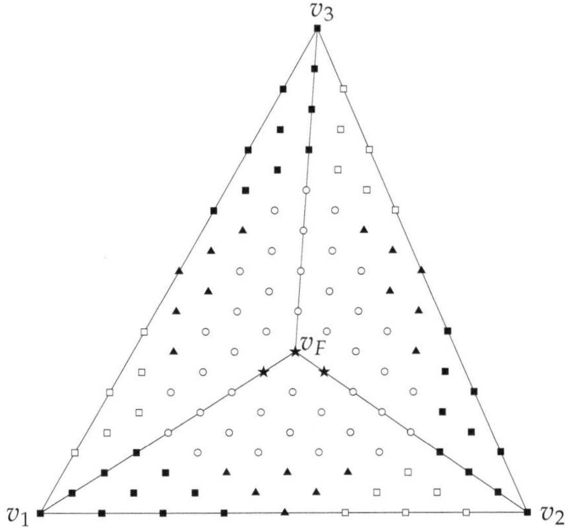

Figure A.10: Minimal determining set of a spline $s \in \dot{\mathcal{S}}_8^2(F_{CT})$. Domain points contained in the minimal determining set are marked with ■ (those in $D_3^{F_i}(v_i)$), with ▲ (those in the second subset of (A.19)), and with ★ (those in the second subset of (A.19)).

Lemma A.14:
Let

$$
\begin{aligned}
\hat{\mathcal{S}}_9^2(F_{CT}) := \{s \in C^2(F_{CT}) : \; & s|_{\widetilde{F}} \in \mathcal{P}_9^2, \; \forall \widetilde{F} \in F_{CT}, \\
& s \in C^4(v_i), \; i = 1,2,3, \\
& s \in C^7(v_F), \\
& \tau_{3-i,5+i,1}^{F_1,F_2} s = 0, \; i = 0,\dots,3, \\
& \tau_{4,4,1}^{F_2,F_3} s = 0\}.
\end{aligned} \tag{A.22}
$$

Then $\dim \hat{\mathcal{S}}_9^2(F_{CT}) = 61$, and the set

$$
\bigcup_{i=1}^{3} D_4^{F_i}(v_i) \cup \bigcup_{i=1}^{2} \{\xi_{4,4,1}^{F_i}, \xi_{4,3,2}^{F_i}, \xi_{3,4,2}^{F_i}, \xi_{4,2,3}^{F_i}, \xi_{3,3,3}^{F_i}, \xi_{2,4,3}^{F_i}\} \cup \{\xi_{4,1,4}^{F_1}, \xi_{0,2,7}^{F_1}, \xi_{0,0,9}^{F_1}, \xi_{1,4,4}^{F_2}\} \tag{A.23}
$$

is a stable minimal determining set for $\hat{\mathcal{S}}_9^2(F_{CT})$.

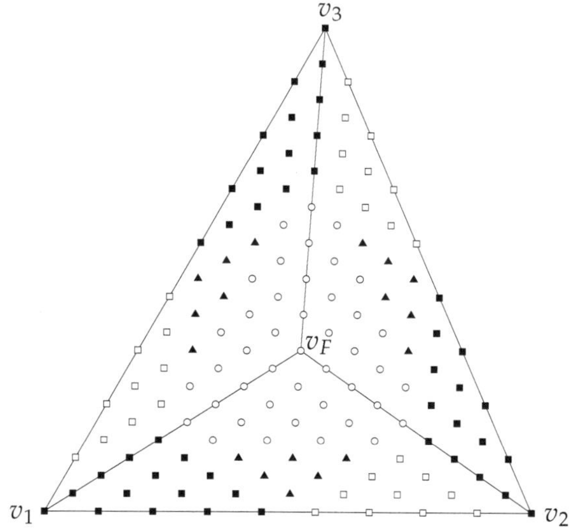

Figure A.11: Minimal determining set of a spline $s \in \widetilde{\mathcal{S}}_9^2(F_{CT})$. Domain points contained in the minimal determining set are marked with ■ (those in $D_4^{F_i}(v_i)$) and with ▲ (the remaining domain points).

Proof:
First, we show that the domain points in (A.23) form a minimal determining set for $\hat{\mathcal{S}}_9^2(F_{CT})$. Therefore, we set the B-coefficients of a spline $s \in \hat{\mathcal{S}}_9^2(F_{CT})$ associated with the domain points in (A.23) to arbitrary values and show that the remaining B-coefficients are uniquely and stably determined.

Using the C^4 supersmoothness conditions at the vertices of F, we can uniquely and stably compute the remaining undetermined B-coefficients of s corresponding to the domain points in the disks $D_4(v_i)$, $i = 1, 2, 3$. The domain points associated with the B-coefficients determined in this step are marked with □ in Figure A.12. Together with the B-coefficients of s associated with the domain points in the second subset of (A.23), the remaining B-coefficients of s corresponding to the domain points within a distance of three from the edges $\langle v_1, v_2 \rangle$ and $\langle v_2, v_3 \rangle$ are uniquely stably determined.

Now, following Lemma 2.28, we can uniquely and stably compute the remaining undetermined B-coefficients of s associated with the domain points in the disks $D_{4+i}(v_2)$, $i = 1, 2$, step by step, from the already determined B-coefficients using the additional smoothness condition $\tau_{3,5,1}^{F_1,F_2} s = 0$, and the supersmoothness conditions of s at the vertex v_F. The domain points corresponding to these B-coefficients are marked with ◆ in Figure A.12.

In the same way, following Lemma 2.28, we can uniquely and stably compute the undetermined B-coefficients of s corresponding to the domain points in the disks $D_7(v_2)$, $D_8(v_2)$ and $D_9(v_2)$, from the already determined B-coefficients and those associated with the domain points $\xi^{F_1}_{4,1,4}$, $\xi^{F_1}_{0,2,7}$, $\xi^{F_1}_{0,0,9}$, and $\xi^{F_2}_{1,4,4}$ by using the additional smoothness conditions $\tau^{F_1,F_2}_{2,6,1}s = 0$, $\tau^{F_1,F_2}_{1,7,1}s = 0$, $\tau^{F_1,F_2}_{0,8,1}s = 0$ and the supersmoothness conditions of s at the vertex v_F. The corresponding domain points are marked with ✖ in Figure A.12.

Now, the last undetermined B-coefficients can be uniquely and stably computed from supersmoothness conditions at the split point v_F of F_{CT} and the additional smoothness condition $\tau^{F_2,F_3}_{4,4,1}s = 0$.

Since the set of domain points in (A.23) is a minimal determining set for the superspline space $\hat{\mathcal{S}}^2_9(F_{CT})$, it can easily be seen that $\dim \hat{\mathcal{S}}^2_9(F_{CT}) = 61$. □

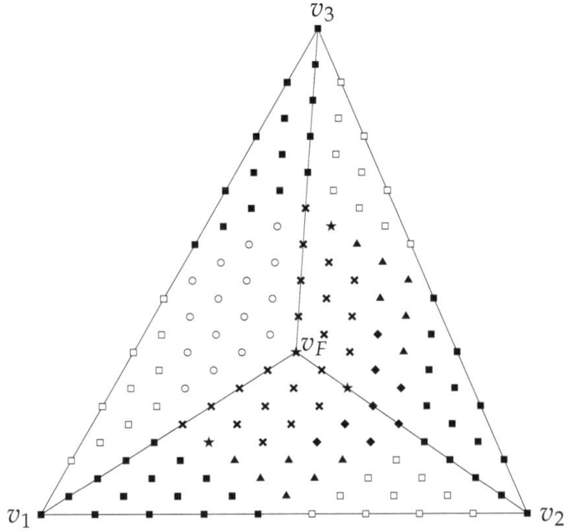

Figure A.12: Minimal determining set of a spline $s \in \hat{\mathcal{S}}^2_9(F_{CT})$. Domain points contained in the minimal determining set are marked with ■ (those in $D^{F_i}_4(v_i)$), with ▲ (those in the second subset of (A.23)), and with ★ (those in the second subset of (A.23)).

Lemma A.15:
Let

$$\check{\mathcal{S}}_9^2(F_{CT}) := \{s \in C^2(F_{CT}) : s|_{\widetilde{F}} \in \mathcal{P}_9^2,\ \forall \widetilde{F} \in F_{CT},\quad s \in C^4(v_i),\ i=1,2,3,\quad s \in C^7(v_F)\}. \tag{A.24}$$

Then $\dim \check{\mathcal{S}}_9^2(F_{CT}) = 66$, and the set

$$\bigcup_{i=1}^{3} D_4^{F_i}(v_i) \cup \bigcup_{i=1}^{3} \{\xi_{4,4,1}^{F_i}, \xi_{4,3,2}^{F_i}, \xi_{3,4,2}^{F_i}, \xi_{4,2,3}^{F_i}, \xi_{3,3,3}^{F_i}, \xi_{2,4,3}^{F_i}\} \cup \{\xi_{1,0,8}^{F_1}, \xi_{0,1,8}^{F_1}, \xi_{0,0,9}^{F_1}\} \tag{A.25}$$

is a stable minimal determining set for $\check{\mathcal{S}}_9^2(F_{CT})$.

Proof:
First, we show that the domain points in (A.25) form a minimal determining set for the superspline space $\check{\mathcal{S}}_9^2(F_{CT})$. To this end we set the B-coefficients of a spline $s \in \check{\mathcal{S}}_9^2(F_{CT})$ associated with the domain points in (A.25) to arbitrary values and show that the remaining B-coefficients are uniquely and stably determined.

We can uniquely and stably determine the B-coefficients of s associated with the domain points in the disks $D_4(v_i)$, $i=1,2,3$, from those in the first subset of A.25 by using the C^4 supersmoothness conditions at the vertices of F. The corresponding domain points are marked with □ in Figure A.13. Together with the B-coefficients of s associated with the domain points in the second subset of (A.25) all B-coefficients of s corresponding to the domain points within a distance of three from the edges of F are uniquely stably determined.

Now, since s has C^7 supersmoothness at the split point v_F, we can consider the B-coefficients corresponding to the domain points in the disk $D_7(v_F)$ as those of a bivariate polynomial g of degree seven defined on the Clough-Tocher split triangle $\widetilde{F}_{CT}$, which is bounded by the domain points in the ring $R_7(v_T)$. Considering g it can be seen that the B-coefficients corresponding to the domain points within a distance of one from the edges of $\widetilde{F}$ are already uniquely and stably determined. Moreover, from the B-coefficients in the third subset of A.25 we can uniquely and stably compute the partial derivatives of g at the split point v_F up to order one. Thus, following Lemma 2.42, the remaining undetermined B-coefficients of g can be uniquely and stably determined by smoothness conditions.

Since the set (A.25) is a minimal determining set for the superspline space $\check{\mathcal{S}}_9^2(F_{CT})$, it can easily be seen that $\dim \check{\mathcal{S}}_9^2(F_{CT}) = 66$. □

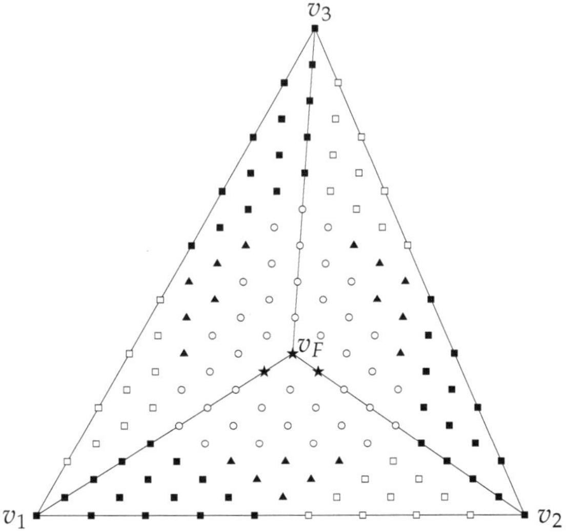

Figure A.13: Minimal determining set of a spline $s \in \check{\mathcal{S}}_9^2(F_{CT})$. Domain points contained in the minimal determining set are marked with ■ (those in $D_4^{F_i}(v_i)$), with ▲ (those in the second subset of (A.25)), and with ★ (those in the third subset of (A.25)).

Remark A.16:
It can be seen from the proofs of Lemma 4.1, 4.2 and 4.3 in [9] that the computation of the B-coefficients of a spline s in $\widetilde{\mathcal{S}}_9^2(F_{CT})$, $\widetilde{\mathcal{S}}_8^2(F_{CT})$, or $\widetilde{\mathcal{S}}_7^2(F_{CT})$, respectively, is a stable process.

Printed by Publishers' Graphics LLC